Recursive Estimation and Time-Series Analysis

Peter C. Young

Recursive Estimation and Time-Series Analysis

An Introduction for the Student and Practitioner

Second edition

Prof. Peter C. Young
Green Meadows, Stanmore Drive
LA1 5BL Haverbreaks, Lancaster
United Kingdom
p.young@lancaster.ac.uk

ISBN 978-3-642-44129-5 ISBN 978-3-642-21981-8 (eBook)
DOI 10.1007/978-3-642-21981-8
Springer Heidelberg Dordrecht London New York

Cover design: SPi Publisher Services

Printed on acid-free paper

Springer is part of Springer Science+Business Media (www.springer.com)

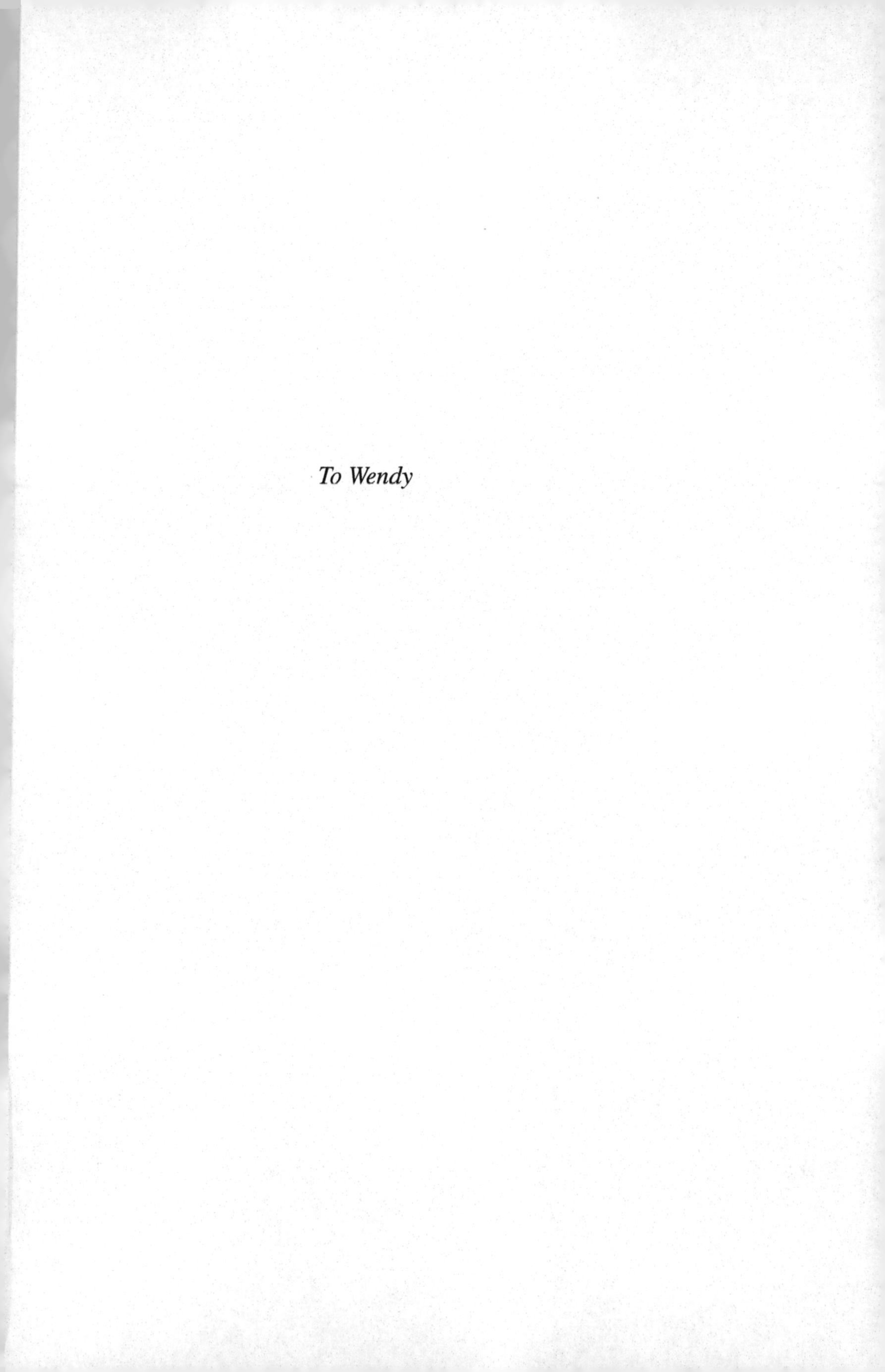

To Wendy

Preface

This is a revised version of my 1984 book of the same name but, because so much time has elapsed since the publication of the first version, it has been considerably modified and enlarged to accommodate all the developments in recursive estimation and time series analysis that have occurred over the last quarter century. Also over this time, the CAPTAIN Toolbox for recursive estimation and time series analysis has been developed by my colleagues and I at Lancaster, for use in the Matlab^{TM} software environment (see Appendix G). Consequently, the present version of the book is able to exploit the many computational routines that are contained in this widely available Toolbox, as well as some of the other routines in Matlab and its other toolboxes.

The book is an introductory one on the topic of recursive estimation and it demonstrates how this approach to estimation, in its various forms, can be an impressive aid to the modelling of stochastic, dynamic systems. It is intended for undergraduate or Masters students who wish to obtain a grounding in this subject; or for practitioners in industry who may have heard of topics dealt with in this book and, while they want to know more about them, may have been deterred by the rather esoteric nature of some books in this challenging area of study. As such, it can also be considered as a primer for the eventual reading of these more advanced theoretical texts on the subject. However it should be emphasized that the book also contains a considerable amount of novel material which does not appear in any other texts on the subject.

There are many people who have influenced my work over many years and who I wish take the opportunity to thank. First, my colleagues in the Environmental Science Department at Lancaster, Keith Beven, Arun Chotai, Wlodek Tych, Andrew Jarvis and Nick Chappell, as well as other colleagues in other Departments: Granville Tunnicliffe-Wilson, Peter Diggle and Jon Tawn in Mathematics and Statistics; James Taylor in Engineering; and Robert Fildes in the Management School. Of these, particular thanks are due to Keith Beven, who has continually encouraged my trespass into the hydrological world and has heavily influenced my research on both *Data-Based Mechanistic* (DBM) modelling and flood forecasting (see 12); Wlodek Tych, one of the major architects of the CAPTAIN Toolbox, who persuaded me to try Matlab out in the first place and who has solved many Matlab problems for me;

and Arun Chotai, who has been a good friend over many years and shared with me the systems and control teaching in the Department.

Whilst at Lancaster I have worked with research students and assistants, all of whom have enriched my life and thinking but are too numerous to mention. Suffice it to say that those who have particularly influenced my time series and modelling research are Cho Ng, Matthew Lees, Stuart Parkinson, Laura Price, Chris Fawcett, Miranda Foster, Paul McKenna, Andrew Jarvis, Renata Romanowcz and, more recently, Dave Leedal and Paul Smith. Particular thanks are due to James Taylor and Diego Pedregal who both worked with me for a number of years and have been of great assistance in the development of the CAPTAIN Toolbox.

Before I came to Lancaster I was a Professorial Fellow at the Australian National University (ANU), in Canberra. It was there that I met and discussed many things with Ted Hannan, one of the all-time greats of time series analysis; and, with him, supervised Vic Solo, whose significant contributions to recursive time series analysis and signal processing are well known. Also I worked with Tony Jakeman, now a long-time and trusted friend, on the development of the recursive *Refined Instrumental Variable* (RIV) approach to time series analysis that figures so strongly in this book (see chapters 6 to 10). I was also influenced by discussions with Mike Osborne, David (Dingle) Smith and Tom Beer, who helped develop the *Aggregated dead Zone* (ADZ) model (see chapters 6, 8 and 12). Being in Australia also allowed me to continue my friendship with Graham Goodwin at Newcastle, NSW, whose contributions to the theory and practice of time series analysis and automatic control have been so impressive for the past forty years, and whose work has influenced me to a great extent. I was also able to continue my friendship, first established in California during the 1960s, with Neville Rees, with whom I have had so many useful discussions on adaptive control and, latterly with Chris Lu, on large simulation model emulation and control (see chapter 12).

I moved to the ANU from Cambridge, where I was a Lecturer in Engineering and Fellow of Clare Hall. It was here that I consolidated my earlier Ph.D research at Cambridge on recursive estimation and self-adaptive control. This Ph.d research was continually encouraged by John Coales, a great friend who was so influential in my career; and Howard Rosenbrock, who excited my interest in Karl Friedrich Gauss (who first developed the recursive least squares algorithm: see Appendix A). Over the six years that I was at Cambridge, I supervised numerous research students who helped to shape my later research interests, such as Karl Neethling, Joseph Kittler, John Naughton and Paul Whitehead. But no one had a more profound influence on my future career than Bruce Beck, whose enthusiasm for pursuing research on environmental modelling and control was so infectious and who made me realize that this would be one focus of my own future research.

During our work on the modelling of dissolved oxygen and biochemical oxygen demand (DO-BOD) variations in the River Cam, Bruce and I realized that the conventional 'hypothetic-deductive' approach to modelling such systems was limited by the difficulty of performing planned experiments. It became clear that more research was required on an alternative 'inductive' approach, where the model structure was not assumed *a priori*, but was inferred statistically from real experimental or monitored data. And it was such thinking that shaped both of our subsequent careers and led, in my case, to the development of the DBM approach to modelling; an approach that has been such a strong motivation for the development of the recursive methods considered in the present book and which is reviewed in its final chapter 12.

Of course, there are many others outside of Lancaster, the ANU and Cambridge who have influenced my work over many years. Just to mention those with whom I have worked and whose friendship I particularly value: John Norton, who I have known for many years and is the author of a well known book on system identification; Howard Wheater, a distinguished hydrologist and long-time friend from the 1970s who, together with Neil Macintyre, Thorsten Wagener and others, has promoted the use of DBM modelling in hydrological applications (see chapters 7, 8, 10 and the Epilogue); Antonio Garcia Ferrer who has strongly supported the application of CAPTAIN tools for unobserved component modelling to economic forecasting (see chapter 5); Keith Burnham who has always been supportive of my research; Barry Croke and Ian Littlewood, with whom I have worked recently (see chapter 8); Marco Ratto and Andrea Pagano, who have been my main collaborators on the development of state dependent parameter regression and large model emulation methods (see chapters 11 and 12); Hugues Garnier and Marion Gilson, who have helped me develop further the RIVC approach to the identification and estimation of continuous-time transfer function models (see chapter 8) and made other important contributions in this area; and Liuping Wang, with whom I have worked on model-based predictive control and state-dependent parameter modelling.

Liuping is a very generous and gifted academic who, with the assistance of Hugues Garnier and Tony Jakeman, was kind enough to put an enormous amount of time into two recent activities that have left me with extremely pleasant memories and provided the stimulus for me to write the present book. First, the organisation of a Workshop in Melbourne, Australia, that marked my 70th birthday in December 2009; and second, the preparation of a book *System Identification, Environmetric Modelling and Control System Design* which collects together 26 contributions to the Workshop and will be published this year. My grateful thanks also to the many other friends and colleagues from all over the World who were kind enough either to attend the Workshop or contribute chapters to the book.

Of course, writing a book is one thing, checking it is quite another. I am immensely grateful to my young friends and excellent research workers, Tomasz Larkowski and Ivan Zajic, who have laboriously checked the draft, noted my many careless errors and made useful suggestions that have considerably improved the book. Of course, I remain solely responsible for any remaining errors.

Finally and most importantly, I wish to thank my wife Wendy, who has been my partner for over fifty years. Without her selfless support and encouragement, I could not have pursued my studies or have written this book.

Lancaster, UK
March, 2011 *Peter C. Young*

Contents

Part II Recursive Estimation of Parameters in Transfer Function Models

Chapter 1
Introduction

Nous traiterous particulièrement le problème suivant, tant à cause de son utilité pratique que de la simplicité de la solution[1].

K.F. Gauss, *Méthode des Moindres Carrés*, translation by J. Bertrand, 1855

1.1 The Historical Context

The concept of least squares estimation is inextricably linked with the name of Karl Friedrick Gauss. Although Legendre was responsible for the first published account of the theory in 1805 and, indeed, first coined the term 'least squares', it was Gauss who developed the method into a statistical tool, embedding it within a statistical framework involving a probabilistic treatment of observational errors. Gauss's first published exposition on least squares appeared in his famous *Theoria Motus Corporum Coelestum* which appeared in 1809 when he was 31 years of age. But, as Sprott (1978) has pointed out in his excellent review of Gauss's contributions to statistics, the basic ideas were most probably formulated while he was still in his twenties.

In the *Theoria Motus*, the discussion on least squares appears in relation to an important practical problem; namely the estimation of the six constant coefficients or 'parameters' that determine the elliptical orbit of a planetary body, on the basis of $n > 6$ observations. His second exposition on the subject was presented in a series of papers (1821, 1823, 1826) which were collected together under the title *Theoria Combinationis Erroribus Minimum Obnoxiae*. Here he abandoned the previous 'inferential' treatment delineated in the *Theoria Motus* and concentrated on a 'decision theoretic' approach, in which he restricted attention to estimates that are a linear function of the observations. And it was here that he presented perhaps his most famous statistical theorem that, '*among all linear error-consistent estimates, the least squares estimate has minimum mean square error*'.

[1] We will deal particularly with the following problem, both due to its practical usefulness as the simplicity of the solution.

But our current interest lies elsewhere in the *Theoria Combinationis*: in a quite short and apparently little known section of some five pages (Bertrand, 1855, p.53-58; Trotter, 1957, p.67-72), Gauss shows how it is possible '*to find the changes which the most likely values of the unknowns undergo when a new equation (observation) is adjoined and to determine the weights of these new determinations*'. In other words, and to use more contemporary terminology, he developed an algorithm for sequentially or recursively updating the least squares parameter estimates on receipt of additional data.

Bertrand's French translation of Gauss's work appeared in 1855 under the appropriate title *Méthode des Moindres Carrés* and was authorized by Gauss himself. This translation, together with my commentary on it, appears in Appendix A of the present book. In addition to its importance in historical terms, Gauss's analysis is interesting because it demonstrates the elegance of his approach and the power of his mind: without the advantages of matrix algebra which, as we shall see, considerably simplifies the derivation, Gauss was able to obtain the *Recursive Least Squares* (RLS) algorithm with consummate ease.

Gauss's analysis represents the birth of recursive least squares theory; a theory so much ahead of its time that it would lie dormant for almost a century and a half before it was rediscovered on two separate occasions: first by the statistician Plackett (1950); and then later and in a more sophisticated form, as the core of the linear filtering and prediction theory evolved by the control and systems theorist Rudolf Kalman (1960).

Not surprisingly, perhaps, Plackett's paper went almost unnoticed in the pre-computer age of the early nineteen fifties. Harking back to Gauss, he re-worked the original results in more elegant vector-matrix terms and developed an algorithm for the general case in which additional observations occur in sets $S > 1$. In the present book, like Gauss, I restrict the analysis mainly to S=1, although the extension to $S > 1$ is straightforward and is discussed directly in chapters 4 and 10, and implicitly when considering hybrid recursive algorithms (see later).

Kalman's results, almost certainly obtained without knowledge of either Gauss's or Plackett's prior contributions, were developed within the context of state variable estimation and filter theory, using an argument based on orthogonal projection. Not only were Kalman's results mathematically elegant in providing a computationally straightforward solution to the optimal filtering problem, which had a number of advantages over the earlier Wiener solution (Wiener, 1949), but they also had good potential for practical application. Not surprisingly, therefore, they caused quite a revolution in the automatic control and systems field at the time; and then provided, during the next fifty years, a rich source of material for research workers in diverse areas of science and social science. Subsequently, the term 'Kalman Filter' has become very widely used and, as Rudolf remarked to me on one occasion, he only wished that he had been able to copyright his name!

The KF is model-based, with a linear, stochastic, dynamic model describing the evolution of the system state variables and providing a predictive estimate of the variables one or more samples ahead; and the Gauss-Plackett recursion providing an update to the state prediction on receipt of the latest observation(s). It is now well known that the KF algorithm can be derived in various ways; via orthogonal projection, as in Kalman's exposition; as well as from the standpoint of maximum likelihood or Bayesian estimation. It can also be developed in various different forms, for application to discrete (Kalman, 1960), continuous (Kalman and Bucy, 1961) and hybrid dynamic systems (i.e. continuous-time system model, discrete-time observations). But, in all its forms, it has had a profound effect on data processing during the last five decades, being used in applications ranging from trajectory and orbit estimation for the Apollo Moon mission, to the forecasting of economic time series.

Sprott (1978) has questioned whether the Kalman filter is really a significant 'development' of the Gauss-Plackett recursive least squares algorithm. While it is true that the Gauss-Plackett recursion formulae are an essential aspect of the Kalman filter equations, it is also clear, as we shall see in this book, that Kalman considerably extended the theory both to allow for the estimation of state variables or time-variable parameters, and to handle the analysis of statistically non-stationary time series. Nevertheless the Gauss-Plackett recursion is undoubtedly the central component of the Kalman filter and the basis of most other recursive least squares algorithms. Thus a good understanding of its function in a data processing sense is an essential pre-requisite for the practical application of the algorithm. It is the provision of such understanding, therefore, which is one of the primary aims in the first part of this book.

As Gauss pointed out so succinctly in the quotation at the beginning of this chapter, recursive least squares theory is both simple and useful. Here I will exploit this simplicity and take the reader gently through the mysteries of the subject, avoiding wherever possible undue rigour and complexity. In the spirit of Gauss, I will concentrate on mathematical analysis which, while it is often algebraic in form, also has sufficient statistical content to ensure that the reader is fully aware of the important statistical aspects of the results. I will, however, allow myself one luxury not available to Gauss and simplify the analysis still further by resort to matrix algebra, assuming that the reader is already acquainted with such analysis; has access to a good text on the subject; or finds that the background notes in Appendix B of the book provide sufficient revision.

Finally, in order to emphasize the practical utility of the various recursive least squares algorithms that emerge during the analysis, I will provide a number of simulation and practical examples, with applications which range from the man-made world of engineering to the more natural world of ecology and the environment. Many other applications in diverse areas, from economics to hydrodynamics, are discussed in a variety of technical papers that my colleagues and I have produced and these are referred to in the text and listed in the bibliography.

The text is divided into three major parts.

1. The first is primarily concerned with the estimation of constant or time-variable parameters in general models which are linear-in-the-parameters. This includes the development of the Kalman Filter and 'Fixed Interval Smoothing' algorithms, as well as the use of these algorithms in the development of 'Unobserved Component' models based on linear regression relationships with time variable parameters.

2. The second part discusses the limitations of recursive least squares estimation and introduces recursive 'Instrumental Variable' estimation that is able to overcome some of these limitations. It shows how the recursive procedures developed in the first part can be modified for use in the identification and estimation of constant and time variable parameters in both discrete and continuous-time transfer function models, including when the dynamic system is enclosed within a closed feedback loop.
3. The third part, which is entirely new in this second edition, first shows how the algorithms for the estimation of time variable parameters in TF models can be modified to allow for the estimation of 'State-Dependent Parameter' models that are able characterize the dynamics of nonlinear stochastic, dynamic systems, including those with chaotic dynamics. It then goes on to discuss the 'Data-Based Mechanistic' modelling philosophy that I have evolved over many years and for which most of the algorithmic tools described in this book were designed to serve.

In sympathy with this introductory nature of the book and its intended readership, there has been a conscious attempt to simplify the mathematical analysis as much as possible, particularly in the early chapters, so as to enhance the readability of the book and avoid an overly esoteric presentation. For the reader unfamiliar with some of the mathematics used in the book, Appendix B provides background notes, not only on matrix algebra but also on probability and statistics, as well as some simple concepts in dynamic systems and discussion of *Monte Carlo Simulation* (MCS) analysis of the kind used in evaluating estimation algorithms and stochastic time series models. In all cases, the results in Appendix B are chosen because of their relevance to the material in the book and are quoted without rigorous proof. They should, however, be sufficient for the reader, either as an aid to memory, or as a guide to reading of standard texts in the various areas: see e.g. Kendall and Stuart (1961); Bryson and Ho (1969); Ljung and Söderström (1983); Söderström and Stoica (1989); Norton (2009); Harvey (1989); Johnston and DiNardo (1997); Ljung (1999).

1.2 The Contents of the Book

Following this Introduction, Part I *Recursive Estimation of Parameters in Linear Regression Models* begins with chapter 2, which provides an introduction to recursive analysis by way of the simplest example, namely the recursive estimation of the constant mean value of a stationary random variable, and goes on to consider both the physical nature of the recursive algorithm and its relationship to the linear least squares regression algorithm for a single unknown parameter (or regression coefficient). Then, from this simple basis, chapter 3 develops the deterministic, recursive least squares (RLS) algorithm for multiple unknown parameters and points out how it can be viewed within the context of the procedure known to mathematicians as *Stochastic Approximation* (SA), which is reviewed in Appendix C. It then shows how, with statistical assumptions about the nature of the stochastic disturbances, it is possible to modify the deterministic RLS algorithm so that it constitutes the recursive version of linear least squares estimation for that stalwart of statistics, the *General Linear Regression* (GLR) model. It also includes an introduction to two important problems associated with statistical regression analysis - namely multi-collinearity and estimation bias caused by *Errors-in-Variables* (EIV) - and discusses briefly one of the simplest solutions to the latter problem which is based on the concept of an *Instrumental Variable* (IV) that is used as an instrument in removing such bias.

Chapter 4 shows how the recursive algorithms are in an ideal form for extension to *Time Variable Parameter* (TVP) estimation, by the introduction of either a fading memory capability or, in the statistical versions of the algorithms, modifications based on stochastic modelling of the parameter variations. This latter approach leads naturally to the development of the Kalman filter and the estimation of time-variable states in stochastic dynamic systems, including the important topic of *Fixed Interval Smoothing* (FIS), a procedure that improves the estimates obtained from the Kalman filter when 'off-line' analysis of the data is possible. The chapter also deals briefly with other related topics: namely simplified TVP estimation algorithms; rapidly variable and 'state dependent' parameters; and *Variance Intervention* (VI), a technique for estimating sudden and discontinuous jumps in the values of the estimated states or parameters.

Chapter 5, which completes the first part of the book, follows naturally from chapter 4 by considering various, practically useful, models and algorithms for estimating parameters in 'dynamic' regression models of various types, where the term 'dynamic' is of historic precedent and slightly misleading: it simply means that the models have time variable parameters, including long term stochastic trends. These *Unobserved Component* (UC) models include: *Dynamic Linear Regression* (DLR), *Dynamic Harmonic Regression* (DHR), which provides a powerful vehicle for advanced time series forecasting and signal extraction; and *Dynamic AutoRegression* (DAR), which can be used for 'time frequency' analysis, i.e. the estimation of time variable spectra of temporal signals.

Chapter 5 also deals with simple but real dynamic models that can be considered in terms of the *Transfer Function* (TF) representations of a dynamic systems but, because of their simplicity, can be estimated by application of RLS estimation. This is important because it leads on rather naturally to the detailed discussion of TF models and the recursive estimation of their parameters in part II of the book.

The Kalman Filter provides a natural approach to the analysis of time series data that are assumed to be generated from stochastic state space equations. Kalman's analysis is concerned particularly with the estimation of the time-variable, but often unmeasurable, state variables of the system; whilst assuming perfect prior knowledge of the parameters that characterize the chosen state space mathematical model of the system. At the beginning of Part II, chapter 6 considers the alternative observation space models, where the dynamic relationships are usually characterized in TF terms. It is on the estimation of parameters in such models that much of the more recent research on recursive analysis has been concentrated and the chapter goes on to discuss why the simple recursive least squares algorithms developed in the previous chapters are not particularly appropriate to the solution of these more complex time series estimation problems. And it shows how the IV method introduced in chapter 3 provides a simple way of correcting some of these deficiences.

Chapters 7 and 8 describe in detail the particular *Refined Instrumental Variable* (RIV) approach to time series model estimation that I suggested in 1976 and which I and my colleagues have developed over the past thirty years. This is based on an optimal recursive-iterative version of the IV method for estimating TF model parameters introduced in chapter 6. Most importantly, RIV estimation is based on the formulation of a *Pseudo-Linear Regression* (PLR) model form and the estimation of this PLR by means of an iterative algorithm that can utilize a recursive solution at each iteration, if this is required. This can be contrasted with more conventional methods of TF model estimation for discrete-time TF models that use gradient methods of numerical optimization and do not have *inherent* recursive solutions (although recursive approximations have been evolved, as discussed in chapter 10).

RIV estimation is chosen as a vehicle for introducing time series model estimation to the reader for two reasons. First, together with the closely related RIV algorithm for continuous-time models (RIVC), it provides what I believe to be *the only unified, time domain approach to the identification and estimation of both discrete and continuous-time models*. Second, because of its inherent exploitation of instrumental variables, it is relatively insensitive to assumptions about the statistical characteristics of the noise on the measured time series data. These chapters contain simulation examples that compare the RIV and RIVC estimation performance with other approaches to TF model estimation, such as the well known *Prediction Error Minimization* (PEM) and *Four Step Instrumental Variable* (IV4) algorithms, the latter of which is different from, but has some relationship with, the RIV algorithm.

Chapter 9 deals with the topic of closed loop TF model estimation that has particular relevance in automatic control systems analysis and design. In particular, it introduces a new, three-stage closed loop estimation algorithm for discrete or continuous-time models that exploits the RIV and RIVC algorithms described in chapters 7 and 8.

Most of the algorithms considered up to Chapter 10 have been designed for 'off-line' use: i.e. they are normally applied to a set of data that is available in the computer and can be analysed as a fixed block of data, albeit using recursive estimation in order to expose aspects of the data that are not clear from non-recursive *en bloc*[2] analysis, or to estimate possible changes in the model parameters. Chapter 10, which is the last chapter in Part II of the book, considers true real-time recursive estimation, starting with the recursive PEM (RPEM) algorithm and showing how it relates to RIV estimation. It then develops the real-time recursive RIV (RRIV) algorithm in a form where iteration is performed at each recursive update, after the latest input-output sampled data are received. A continuous-time RRIVC version of the RIVC algorithm can be implemented in a very similar manner, but this is not discussed in the chapter.

[2] 'as a body or whole' or 'batch' processing.

The rest of chapter 10 describes the ubiquitous *Extended Kalman Filter*; the first recursive parameter estimation algorithm for nonlinear, stochastic, dynamic models to be suggested after the publication of Kalman's seminal paper in 1960 and one that has been widely used in many areas of application but has a number of limitations when used in practice. It also introduces briefly some recent 'computationally intensive' successors to the EKF that exploit stochastic *Monte Carlo Simulation* (MCS) analysis and have received a lot of attention in the past few years: namely, the *Ensemble Kalman Filter* (EnKF), the *Particle Filter* (PF) and the *Unscented Kalman Filter* (UKF).

In the final Part III of the book *Other Topics*, chapter 11 is primarily concerned with the recursive estimation of parameters in the wide class of nonlinear, stochastic, dynamic systems whose behaviour can be represented by a *State-Dependent Parameter* (SDP) transfer function model: i.e. a TF model whose parameters are dependent on dynamic 'state variables', which normally include the input and output variables, as well as their past values. This novel algorithm is intended mainly as a *non-parametric* (graphical) approach to nonlinear model structure identification: i.e. the identification of the location and nature of significant nonlinearities affecting the behaviour of the dynamic system. As such, it is normally a precursor to the *parameterization* of the identified nonlinearities and the final estimation of the resulting parametric model.

One major use of SDP estimation, and all of the other algorithms discussed in the book, is in connection with *Data-Based Mechanistic* (DBM) modelling, a general inductive approach to a modelling that I and my colleagues have developed over many years. The DBM modelling philosophy is built on the principle that a data-based model should not just explain the time series data well, it should also provide a mechanistic description of the system under investigation; a description that further enhances our confidence in its ability to approximate reality in a meaningful manner. The various inter-linked stages of the DBM approach are outlined in chapter 12 of the book, which then goes into more detail on the topic of model reduction and the use of reduced order models in the emulation (or meta-modelling) of large computer models. These latter topics are important because they provide a bridge between large computer simulation models and the parsimonious (parametrically efficient, low order) DBM models produced by the estimation methods described in this book.

Finally, there is an Epilogue, followed by six appendices, of which the first three, *The K.F.Gauss Derivation of Recursive Least Squares*, *Basic Mathematical and Statistical Background*, and *Stochastic Approximation* are the most important.

1.3 Software

I believe that an important requirement of any book that promotes the use of computer algorithms is that these algorithms can be available to the reader so that he or she is able to evaluate and, hopefully, apply them to practical examples with which they are concerned. In this regard, most of the algorithms developed and applied in the present book are available in the *Computer Aided Program for Time series Analysis and Identification of Noisy systems* (CAPTAIN) Toolbox that we have developed for use in the Matlab^{TM} software environment. The content of this Toolbox is reviewed briefly in Appendix G, which notes that a very useful e-Handbook (Pedregal et al., 2005) is provided for users. This includes a comprehensive listing of the computational routines, as well as many examples that help to guide their use. In a very real sense, this handbook can be considered as companion volume to the present book and its availability will enhance the learning process, particularly as regards the use of the recursive algorithms in practical time series analysis.

1.4 The Aims of the Book

Recursive estimation is of undoubted importance in general time series analysis. Indeed, since the first edition of this book, twenty seven years ago, recursive estimation has become commonplace, taking its position as an essential component in most degree courses concerned with control and systems theory, signal and image processing, statistical estimation and econometrics; and it is becoming increasingly important in other applied science courses, such as the earth and atmospheric sciences (e.g. hydrology, oceanography, atmospheric science), as well as some courses in the social sciences (e.g. psychology, sociology). Existing computer packages, such as the CAPTAIN Toolbox, the *Systems Identification* (SID) Toolbox in Matlab and the STAMP toolbox for econometricians, include tools for recursive estimation and have become popular in various areas of science and social science. Recognising this importance, the book is aimed at students and practitioners who want to learn about recursive estimation and its practical relevance in a straightforward manner that steadily enhances understanding. Consequently, the analytical treatment starts at a very basic level, requiring only a knowledge of simple algebra, and it then increases in difficulty with each new chapter, hopefully in a manner which matches the reader's increasing understanding of a subject, until more advanced, state-of-the-art research topics are addressed in the later chapters.

The book also encourages the reader to question theoretical assumptions and not to assume that methodological procedures will always work as expected when applied to real data. In the Epilogue, for instance, I end the book with a cautionary discussion which starts with the question *What is Optimality*? The present book develops numerous recursive algorithms, most often on the basis of some criterion of optimality. But optimality depends upon the nature of the optimization criterion and the assumptions on which it relies. If these assumptions are violated or the algorithm used for optimization has some weaknesses, then the results may not be as good as one would hope and, indeed, a simpler approach may provide results that are more satisfactory to the user. For instance, the statistically inspired optimization cost functions may be too narrow to satisfy the requirements of scientists, engineers and social scientists who will often prefer a model that satisfies a wider, less formal, definition of optimality. A practical example is used to illustrate this dilemma and it is concluded that, while the optimal algorithms, such as RIV and PEM mentioned above, should be the algorithms of first choice, the user should always check the results carefully and try other, possibly simpler algorithms that may well have advantages in practical situations.

Finally, I hope the book will encourage the reader to speculate a little and ask a few questions that may not have obvious answers, given our current level of knowledge. In this regard, one could ask whether recursive estimation has a wider significance than its pragmatic use as a computational tool in science and engineering might suggest? Could it not permeate some of the fundamental processes of the physical and biological world that we live in? I think that its inherent stochastic mechanism of sequential model prediction, followed by correction based on the latest measured data, may be present in many processes that we still do not understand, such as functions that are operative in the human brain or that characterize the stochastic-dynamic aspects of quantum mechanics. Recursive estimation is important because it gives us insight into the uncertain nature of all discrete and continuous time series data that derive from measurement; and it shows how these data can be assimilated into an associated mathematical model of a dynamic system in order to improve our probabilistic knowledge of the system's evolving state.

But does this insight reveal anything about the nature of the World and our concept of 'reality'? The answer to this question has to be subjective. Having worked on aspects of recursive estimation all my life, however, and applied this technique to many real systems, it is has led me to seriously question any deterministic view of reality; a view that there is an underlying set of truths that, if we continue to search, we will eventually discover. I believe that all things observed by humans, from what our eyes perceive to the measured effects of fundamental particles, can never be 'seen' with ultimate clarity; they can only be defined probabilistically in some manner and we should reject the idea for an underlying deterministic system. Rather, we should consider 'reality' in terms of the probability distributions, or other representations of uncertainty, that are estimated from noisy observations by procedures such as the recursive algorithms described in this book. Or, better still, perhaps these algorithms will be succeeded by other, still more powerful ones, that may be developed in the future, hopefully by someone who has been stimulated by the present book.

Whether or not it is of such, more general, importance, the undeniable value of recursive estimation lies not only in its obvious elegance and flexibility, but also in its demonstrable practical utility, as exemplified by the applications discussed in this book and many related publications. In this sense, the book attempts to follow the example of Gauss who says, in the preface to the *Theoria Motus*:

> *The more important problems are, for the most part, illustrated by appropriate examples, taken, wherever it was possible, from actual observations. In this way not only is the efficacy of the methods more fully established and their use more clearly shown, but also, care, I hope, has been taken so that inexperienced computers [analysts] should not be deterred from the study of these subjects, which undoubtedly constitute the richest and most attractive part of [the subject].*[3]

[3] Words in parentheses added by the present author.

Part I
Recursive Estimation of Parameters in Linear Regression Models

Chapter 2
Recursive Estimation: A Simple Tutorial Introduction

Figure 2.1 is a plot of the annual rainfall $y_i, i = 1, 2, \ldots, 132$, for the town of Walgett in New South Wales, Australia, over the period 1879 to 2010. It is a typical example of time-series data and the human eye (which is an extremely good and underrated filter of data) can discern various characteristics and patterns which could be described verbally: for example, the rainfall is extremely variable; it averages about 472 *mm* a year; and there seems to be indications of some changes in the mean and variance over the last 30 to 40 years.

But such a description is largely qualitative; if we are to be more quantitative and precise in our evaluation, and particularly if we wish to compare the rainfall at Walgett with that measured at other stations in N.S.W. or elsewhere, then the data must be compressed in some manner to yield a reduced and hopefully small number of 'statistics' that can be computed easily from the original data and which collectively provide a good description of these data. The most obvious statistics in the case of the Walgett data are the first two statistical moments, in the form of the sample mean or average rainfall $\bar{y}(N)$, the sample variance $\sigma^2(N)$ and the sample standard deviation $\sigma(N)$ about this mean value (the sample covariance is considered in Appendix B), where these are defined as follows:

$$\begin{aligned} \bar{y}(N) &= \frac{1}{N} \sum_{i=1}^{N} y(i) \\ \sigma^2(N) &= \frac{1}{N} \sum_{i=1}^{N} [y(i) - \bar{y}(N)]^2 \\ \sigma(N) &= \sqrt{\sigma^2(N)} \end{aligned} \tag{2.1}$$

and N is the total number of samples available, in this case 132. This yields values of $\bar{y}(N) = 471.9\ mm$, $\bar{\sigma}^2(N) = 26409\ mm^2$ and $\sigma(N) = 162.5\ mm$.

There is an implicit assumption in the formulae (2.1) that the true mean and variance are constant. If we trust the evidence of our eyes, we might suspect that this is not the case and that there are possibly changes in both the mean and variance over

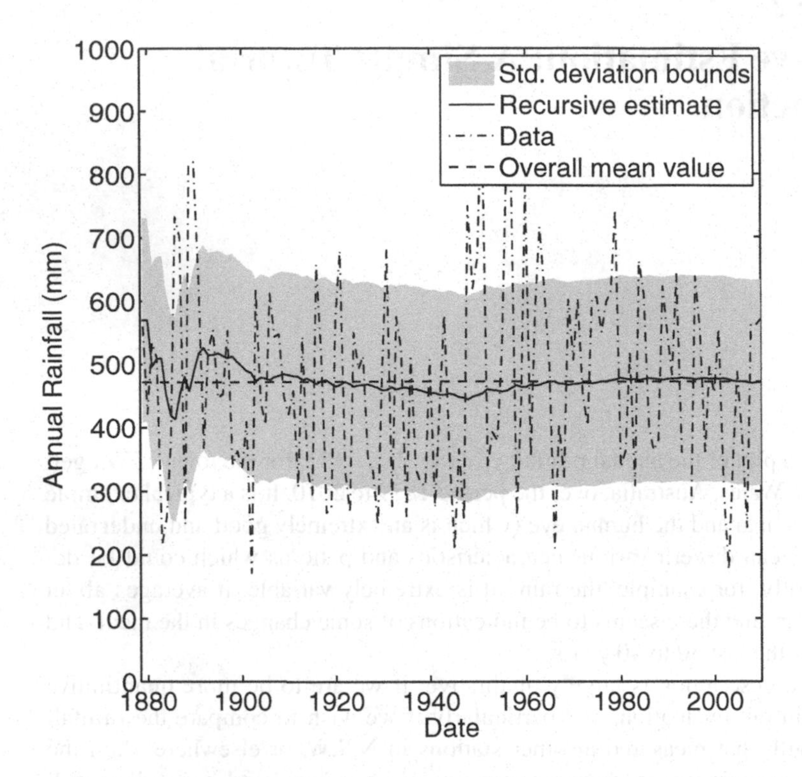

Fig. 2.1 Rainfall at Walgett, N.S.W, 1879 to 2009, showing also the recursive estimate of the mean value generated by the algorithm (2.3).

the observation interval. For the moment, however, let us take on the role of the naïve analyst who either did not bother to look at the graphical plot of the data or, as is quite common these days, merely consulted that most sterile of visual presentations, the computer data listing.

The computations involved in the calculation of $\bar{y}(N)$ and $\sigma^2(N)$ from (2.1) are extremely simple, so that the evaluation of these statistics for any subset of the N samples is straightforward. If the statistics had been computed over some $k-1$ samples within the data set, however, and then we wished to obtain them for k samples, it seems reasonable to assume that there is a simple relationship between $\bar{y}(k-1)$ and $\bar{y}(k)$. Similarly, if the estimates are available for N samples and an additional sample is received, as might be the case in the present example when the total rainfall measurement for 2010 becomes available, then it might be suspected that some combination of the statistic already computed for N samples and the new sample at $N+1$ would yield the new value of the statistic at $N+1$.

2.1 Recursive Estimation of the Mean Value of a Random Variable

The above conjectures are correct and it is quite easy to obtain a sequentially updated or *recursive* relationship for estimating the mean value of a series such as the Walgett data. Since this recursive algorithm will be extended later to the estimation of more general parameters, which are denoted by the letter a, it is convenient in the following analysis to replace $\bar{y}$ by $\bar{a}$. Using this nomenclature, there are basically two ways to develop a recursive relationship between the mean estimate $\bar{a}(k)$ of the true mean value a at any arbitrary k^{th} sample and its prior computed estimate $\bar{a}(k-1)$ at the previous $(k-1)^{th}$ sample: first, by simple manipulation of the *en bloc* solution in (2.1); and second, by considering the problem within the wider context of recursive least squares estimation.

In the former regard, $y(i)$ can be considered as an observation of a at the ith sampling instant i.e.,

$$y(i) = a + e(i);\; i = 1,2,\ldots,N \tag{2.2}$$

where the difference between a and $y(i)$, $e(i) = y(i) - a$, can be considered as the error or 'noise' on the observation. Now the sample estimate of the mean at the kth instant is given by

$$\bar{a}(k) = \frac{1}{k}\sum_{i=1}^{k} y(i) \;\text{ or }\; k\bar{a}(k) = \sum_{i=1}^{k} y(i)$$

so that, at the $(k-1)^{th}$ instant,

$$(k-1)\,\bar{a}(k-1) = \sum_{i=1}^{k-1} y(i)$$

As a result, $\bar{a}(k)$ is related to $\bar{a}(k-1)$ by

$$\bar{a}(k) = \frac{1}{k}[(k-1)\,\bar{a}(k-1) + y(k)] = \frac{k-1}{k}\bar{a}(k-1) + \frac{1}{k}y(k)$$

and so, multiplying by $k/k-1$,

$$\frac{k}{k-1}\bar{a}(k) = \bar{a}(k-1) + \frac{1}{k-1}y(k)$$

or

$$\frac{k}{k-1}\bar{a}(k) = \bar{a}(k-1) + \frac{1}{k-1}[y(k) - \bar{a}(k-1)] + \frac{1}{k-1}\bar{a}(k-1)$$

Finally, collecting terms,

$$\bar{a}(k) = \bar{a}(k-1) + \frac{1}{k}[y(k) - \bar{a}(k-1)] \tag{2.3}$$

This is the simple recursive algorithm for estimating the sample mean value of the observations up to the kth instant from the sample mean computed at the previous $(k-1)^{th}$ instant and the new observation $y(k)$ received at the kth instant. Note that it is computationally efficient, requiring the immediate past value of $\bar{a}(k-1)$ and no past observations to be stored; and it can be designated as an 'on-line' or 'real-time' estimator, since an update of the estimate occurs every time a new sample of $y(k)$ is received. The recursive estimate of the mean value for the Walgett data, as generated by (2.3) with $\bar{a}(0) = y(0)$, is shown as the full line in Figure 2.1.

The diagrammatic representation of recursive estimation in Figure 2.2 shows that the recursive estimates are obtained by working serially through the data, one sample at a time, using the recursion (2.3). As shown on the diagram, such recursive estimation is widely known as 'recursive filtering', because it can be considered as a filtering process, as subsequent analysis will reveal. This can be compared with the more conventional *en bloc* method of analysis, where a single estimate is obtained by operating on the whole set of data in one operation according to the first equation in (2.1). And it can be contrasted with *iterative* data processing which, in the present context, means the sequential processing of a *complete set of data*, as shown in Figure 2.2. Here, at each iterative step, the data base remains the same at N samples and only some estimated variable is modified. Using this terminology, the *en bloc* solution in (2.1) can be considered as a single, non-recursive, estimation operation on the data. Iterative processing involving multiple iterations is not relevant to the simple problem under discussion here but it will assume some significance later when more complex parameter estimation problems are considered.

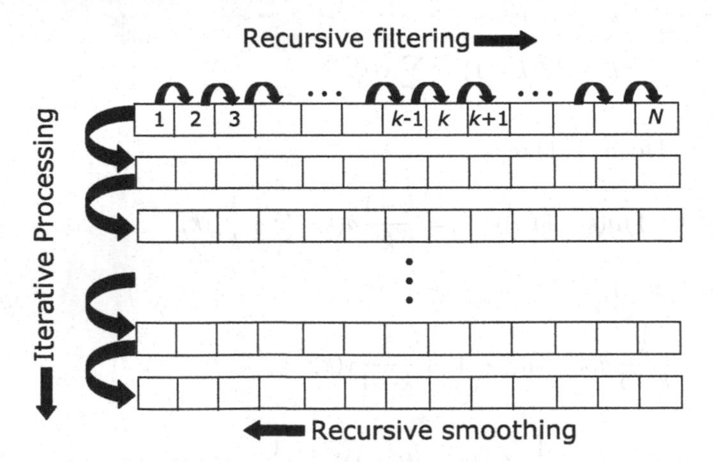

Fig. 2.2 Diagrammatic representation of recursive (forward filtering, backward smoothing) and iterative data processing.

Also shown in Figure 2.2 is the operation known as ‘recursive smoothing’, where the time series data are processed in reverse order in order to improve the estimates obtained in the forward filtering pass. This analysis can be used when a complete block of data has been measured and is available for analysis; and it can be contrasted to ‘on-line’ analysis where the measured data are being received (e.g. from a remote sensor) and analyzed, recursively, often in ‘real-time’. However, FIS analysis is only relevant if the parameter being estimated (here the mean value of the series) is assumed to change over time. The ability to estimate such ‘time variable parameters’ is an important advantage of recursive estimation and it is considered later in chapter 4.

While the *en bloc* (non-recursive or single iteration) method of determining the mean value is well known, the recursive algorithm is not known nearly so well. And yet the algorithm is significant in a number of ways: not only is it elegant and computationally attractive, but it also exposes, in a most vivid manner, the physical nature of the estimate for increasing sample size and so provides insight into a mechanism that, as subsequent chapters in this book will show, is useful in many more general problems.

Referring to equation (2.3), it is clear that the estimate of the mean after k samples is equal to the previous estimate $\bar{a}(k-1)$ obtained after $k-1$ samples, plus a correction term which is the product of $1/k$ and the difference between the new sample observation $y(k)$ and $\bar{a}(k-1)$. In effect, therefore, the previous estimate $\bar{a}(k-1)$ is modified in proportion to the error between the observation of the random variable and the latest estimate of its mean value. However, because this modification is multiplied by $1/k$, its effect becomes progressively smaller as the sample size increases until, for large k its makes very little difference to the estimate. Figure 2.1 reveals this effect well in this case: the full line is the recursive estimate computed by the algorithm (2.3), starting with $\bar{a}(0) = 0$, and we see that it fluctuates quite a lot at first but the changes are becoming very small by the end of the series. Note that, provided the algorithm is initiated in a particular manner (see later), the recursive estimate at any sampling instant k is *precisely* the same as the *en bloc* mean estimate obtained from equation (2.3) using all the data up to that point. Interestingly, because the recursive analysis used here assumes that the mean is constant, the recursive smoothed estimate is constant for all the data and is shown as a dash-dot line. Later examples in chapter 4 and 5 will show that this is not the case if it is assumed that the parameter may vary over the sampling period.

Another way of looking at the problem of estimating the mean value is to consider it as a problem of least squares estimation. Here the least squares estimate $\hat{a}$ of a is simply that value of $\bar{a}$ which minimizes a cost function $\mathfrak{J}_2$, defined as the sum of the squares of the differences between $y(i)$ and a over the observation period, i.e., in mathematical terms:

$$\hat{a}(k) = \arg\min_{a} \mathfrak{J}_2(a) \qquad \mathfrak{J}_2 = \sum_{i=1}^{k} [y(i) - a]^2 \tag{2.4}$$

where arg min simply means the value of a which minimizes the argument, here the least squares cost function $\mathcal{J}_2$. The minimum in this case is obtained in the usual manner by differentiating $\mathcal{J}_2$ with respect to a and equating the result to zero in the usual manner, i.e.

$$\frac{\partial \mathcal{J}_2}{\partial a} = 2\sum_{i=1}^{k}\left\{[y(i)-a]\frac{\partial [y(i)-a]}{\partial a}\right\} = 2[ka - \sum_{i=1}^{k} y(i)] = 0$$

Consequently,

$$\hat{a}(k) = \frac{1}{k}\sum_{i=1}^{k} y(i) \tag{2.5}$$

where $\hat{a}(k)$ denotes the least squares estimate based on an observation size of k samples. This solution is, of course, the arithmetic mean utilized previously. But continuing within the context of least squares estimation, equation (2.5) can be written as

$$\hat{a}(k) = p(k)b(k) \tag{2.6}$$

where

$$p(k) = \frac{1}{k} \quad \text{and} \quad b(k) = \sum_{i=1}^{k} y(i) \tag{2.7}$$

It is easily seen that $p(k)$ and $b(k)$ are related to their previous values $p(k-1)$ and $b(k-1)$ by the equations:

$$\frac{1}{p(k)} = \frac{1}{p(k-1)} + 1 \tag{2.8}$$

and

$$b(k) = b(k-1) + y(k) \tag{2.9}$$

Multiplying throughout equation (2.8) by $p(k)p(k-1)$,

$$p(k-1) = p(k) + p(k)p(k-1) \tag{2.10}$$

so that

$$p(k) = \frac{p(k-1)}{1+p(k-1)} \tag{2.11}$$

and finally substituting (2.11) into (2.10)

$$p(k) = p(k-1) - \frac{p^2(k-1)}{1+p(k-1)} \tag{2.12}$$

This equation and equation (2.9) can now be substituted into equation (2.6) to yield

$$\hat{a}(k) = \left[p(k-1) - \frac{p^2(k-1)}{1+p(k-1)}\right][b(k-1) + y(k)]$$

which, after reduction and using the fact that $\hat{a}(k-1) = p(k-1)b(k-1)$, can be written as

$$\hat{a}(k) = \hat{a}(k-1) + \frac{p(k-1)}{1+p(k-1)}[y(k) - \hat{a}(k-1)] \tag{2.13}$$

Equations (2.12) and (2.13) taken together constitute the recursive least squares algorithm for the estimation of the mean of the random variable $y(i)$, $i = 1,2, \ldots, k$. Not surprisingly, they are exactly equivalent to equation (2.3), as can be seen if it is noted that $p(k-1) = 1/(k-1)$ and then substitute into equation (2.13) to yield

$$\hat{a}(k) = \hat{a}(k-1) + \frac{1}{k}[y(k) - \hat{a}(k-1)] \tag{2.14}$$

or,

$$\hat{a}(k) = \hat{a}(k-1) + p(k)[y(k) - \hat{a}(k-1)] \tag{2.15}$$

Finally, note that similar analysis to the above could be carried out for the problem of estimating the variance of $y(k)$, in which case the recursive algorithm would be of the form

$$\hat{\sigma}^2(k) = \hat{\sigma}^2(k-1) + p(k)[d^2(k) - \hat{\sigma}^2(k-1)] \tag{2.16}$$

where $d(k) = y(k) - \hat{a}(N)$ is the deviation of $y(k)$ from the sample mean $\hat{a}(N)$ and again $p(k) = 1/k$. Note that, at every sampling instant k, this requires the computation of the sample mean value $\hat{a}(N)$, which is based on all the data, so that unless this is known *a priori* ('prior to' or 'before the fact', from the Latin for 'from what comes before'[1]), it is not strictly an 'on-line' algorithm. But we shall have more to say about the recursive estimation of the variance later in the book.

The exercise of obtaining (2.3) again, in what appears to be a rather roundabout fashion, is not wasted, for now we see how the algorithm fits within the context of least squares optimization theory. Moreover, it is now clear that the error term can be interpreted as being proportional to the negative gradient of the instantaneous cost $[y(k) - \hat{a}(k-1)]^2$, i.e.,

$$\frac{\partial}{\partial \hat{a}(k-1)}\{[y(k) - \hat{a}(k-1)]^2\} = -2[y(k) - \hat{a}(k-1)]$$

In other words, the recursive mean value estimation algorithm (2.15) can be viewed as a 'gradient algorithm', in which the estimate $\hat{a}(k-1)$ is updated to $\hat{a}(k)$ in a direction defined by the negative gradient of the instantaneous cost and with a magnitude or step size dictated by the weighting factor $p(k)$.

But $p(k)$ is not constant; it is, in fact, in inverse proportion to the number of observations k. Thus, as the algorithm proceeds and confidence in the estimate increases, so less and less notice is taken of the gradient measure, since it is more likely to arise from the noise $e(k)$ than from error in the estimate of the mean value. And it is the harmonic sequence $1/i$, i.e., $1, 1/2, 1/3, \ldots, 1/k$, which provides the

---

[1] *a priori* and *a posteriori* ('posterior to' or 'after the fact', from the Latin for 'from what comes after') are widely used in recursive estimation, particularly when it is considered in Bayesian estimation terms (see Appendices B and E; and later chapters).

mechanism for the attenuation of the corrective action. Indeed it can be seen that, as the data base becomes very large, new observations have little effect since, under the basic assumption that the mean value is constant, they provide essentially little 'new' information on which to base statistical inference. In effect, the variable weighting factor $p(k)$ acts to smooth or 'filter' out the inaccuracy injected by the observation noise. Needless to say, if the assumption of a stationary mean value is not justified then such a procedure would be dangerous and could lead to poor estimation performance with, for example, heavily biased estimates. This will be discussed further in subsequent sections.

2.1.1 Filtering interpretation of the recursive algorithm

The interpretation of equation (2.15) as a discrete-time filtering algorithm becomes even more transparent if a block diagram of the algorithm is constructed as shown in Figure 2.3, where z^{-1} is the 'backward shift' operator, i.e., $z^{-1}\,\hat{a}(k) = \hat{a}(k-1)$ and, in general[2], $z^{-r}\,\hat{a}(k) = \hat{a}(k-r)$. The reader can easily verify that Figure 2.3(a) is a representation of equation (2.14) or (2.15) and, following the rules of block diagram manipulation (see Appendix B) it is also straightforward to manipulate the diagram into the form of Figure 2.3(b), where z is the 'forward shift' operator, i.e., $z\,\hat{a}(k) = \hat{a}(k+1)$.

It is now clear that the algorithm can be considered as a discrete feedback system or digital filter characterized by a unity feedback loop together with a forward path which includes a time variable gain $p(k) = 1/k$ and a discrete integrator or summation device $1/(z-1)$. The feedback system of Figure 2.3 is, in fact, a variable gain 'low pass' filter mechanism of a type often used in communication and control systems design. Here the term 'low pass' is applied because the filter in its fixed gain form (i.e. $p(k) =$ constant) 'passes' low frequency variations in the input signal $y(k)$ but attenuates or 'filters off' any high frequency changes. When $p(k)$ is made a strictly decreasing function of time by defining it as $1/k$, the effect is simply to reduce sequentially the 'pass-band' of the filter until, at $k = \infty$, it passes no change in signal at all and the output of the integrator remains fixed at the final asymptotic estimate of the mean, $\hat{a}(N)$. However, later in chapter 4, it will be shown how, *in the case of time variable parameter estimation algorithms*, the asymptotic gain $p(k)$ for $k \to \infty$ tends to a constant value, so that the recursive estimator becomes a true low-pass filter and is able to estimate changes in the parameter (here the mean value of $y(i)$ $i = 1, 2, \ldots, k$).

With this filtering interpretation in mind, it also seems fairly clear, by physical intuition, that the averaging effect of the gain $p(k) = 1/k$ will only work successfully if the noise $e(k)$ is, in fact, a zero mean random variable with finite variance. And, as

[2] The z^{-1} notation is used here in preference to alternatives such as B (Box and Jenkins, 1970), in the time series analysis literature, L (Johnston and DiNardo, 1997), in the econometrics literature, where it is called the 'lag' operator, or q^{-1} (Ljung, 1999), in the control and systems literature, inorder to emphasize the close relationship between the forward shift z and the z-transform operator.

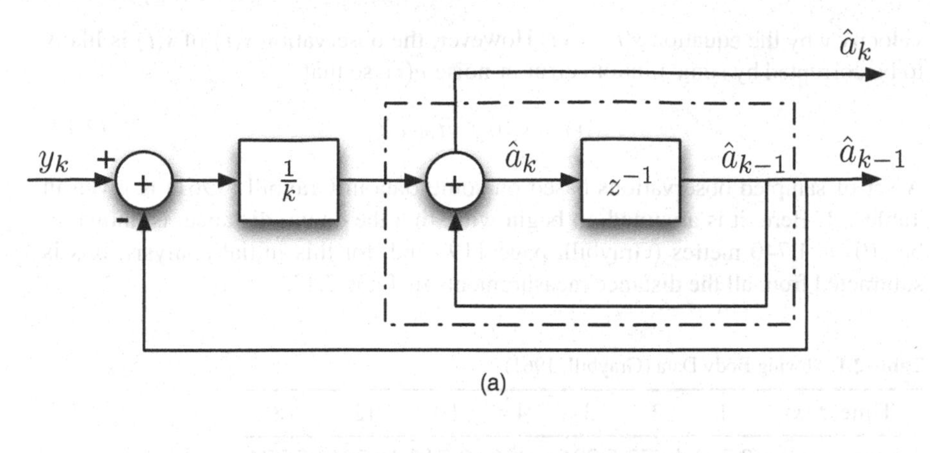

(a)

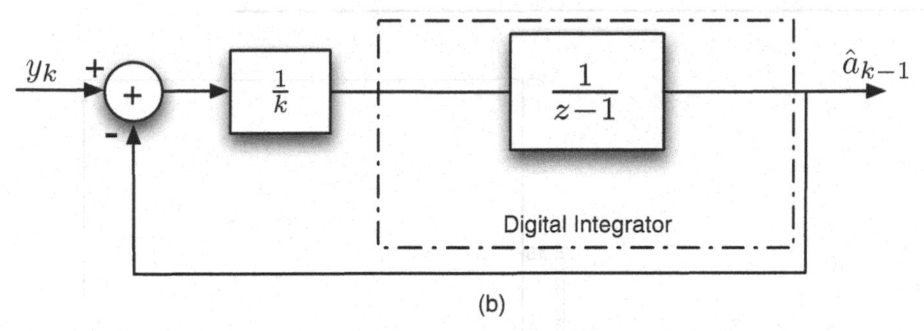

(b)

Fig. 2.3 The recursive algorithm as a first order, digital low-pass filter: (a) basic block diagram; (b) reduced block diagram showing presence of digital integrator.

we shall see later, it is necessary to put even stricter requirements on the statistical properties of $e(k)$ if the estimate is to be relatively efficient in a statistical sense, i.e., if it is to possess low variance when compared with the theoretical minimum variance estimate (see Appendix B).

2.2 Recursive Least Squares Estimation for a Single Unknown Parameter

When viewed within the context of least squares estimation, the recursive algorithm defined by Equations (2.12) and (2.13) or (2.15) for estimating the mean value of a random variable can be considered as a special example of regression analysis for a single unknown parameter (the mean value) in which the regression variable or 'regressor' is unity. Let us consider a somewhat more general situation where the regressor is an exactly known but variable quantity. For example, consider a body moving in a straight line with constant velocity v from some datum position, say $s(0)$. From basic physical principles, the distance $s(t)$ at time t is related to the velocity v by the equation $s(t) = vt$. However, the observation $y(t)$ of $s(t)$ is likely to be corrupted by some random error or noise $e(t)$, so that,

$$y(t) = s(0) + vt + e(t) \tag{2.17}$$

A set of sampled observations based on some data in Graybill (1961) is given in Table 2.1. Here, it is assumed, to begin with, that the datum distance, is known to be $s(0) = 1.740$ metres (Graybill, page 119) and, for this initial analysis, this is subtracted from all the distance measurements in Table 2.1[3].

Table 2.1 Moving Body Data (Graybill, 1961).

Time, t (s)	1	2	3	4	10	12	18
Distance, y (m)	2.743	4.572	5.791	6.096	13.716	16.764	23.774

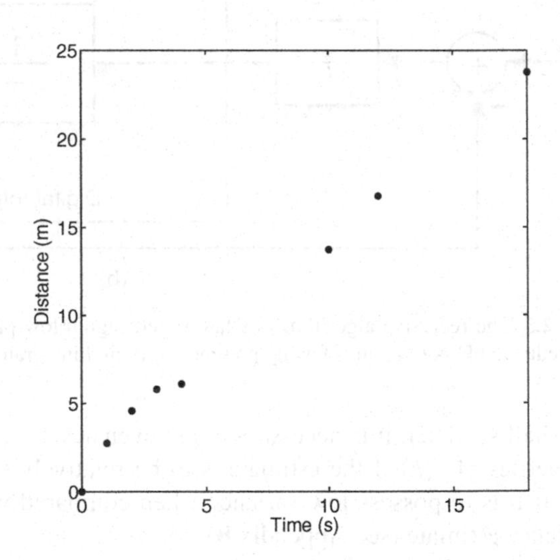

Fig. 2.4 Data for the moving body example: from (Graybill, 1961).

By plotting the measured distance against time, as in Figure 2.4, the observed data should lie, very approximately, on a straight line through the origin, with slope v. No straight line, however, will pass exactly through all the data points because of the measurement errors. Consequently, one has to ask which of the many straight lines that could be drawn through the points should be chosen as being the best representation of the data? To obtain the least squares solution to this problem, the observation equation at the kth sampling instant is written in the sampled data form,

$$y(k) = x(k)a + e(k); \quad k = 1, 2, \ldots, 7 \tag{2.18}$$

[3] The measurements used here are converted to metres from Graybill's original measurements in feet. Also, in the next chapter 3, the data will be re-analyzed assuming that $s(0)$ unknown.

where $x(k)$ denotes the time when the the observation $y(k)$ is made and a is the unknown velocity.

Following the procedure of the previous section, the least squares cost function $\mathfrak{J}_2$ is now formulated in the form

$$\mathfrak{J}_2 = \sum_{k=1}^{N} [y(k) - x(k)a]^2$$

where, in this case, $N = 7$. The condition for minimum of $\mathfrak{J}_2$ with respect to a is then

$$\frac{\partial \mathfrak{J}_2}{\partial a} = -2\sum_{k=1}^{N} [y(k) - x(k)a]x(k) = 0$$

so that,

$$\left[\sum_{k=1}^{N} x^2(k)\right] \hat{a} = \sum_{k=1}^{N} x(k)y(k)$$

and the least squares estimate $\hat{a}(k)$ after k samples is given by

$$\hat{a}(k) = p(k)b(k) \tag{2.19}$$

where

$$p(k) = \left[\sum_{i=1}^{k} x^2(i)\right]^{-1} \quad \text{and} \quad b(k) = \sum_{i=1}^{k} x(i)y(i)$$

Once again, this formulation and solution of the least squares estimation problem is very simple; its simplicity appealed to Gauss and it has appealed to almost everyone concerned with the analysis of observations ever since. Other criterion functions such as the least magnitude (see e.g. Young (1965a))

$$\mathfrak{J}_1 = \sum_{k=1}^{N} |y(k) - x(k)a|$$

could also be utilized but are less convenient in analytical terms. Proceeding in a similar manner to that used in the case of the recursive estimation of the mean, note that here,

$$\begin{aligned} p^{-1}(k) &= \sum_{i=1}^{k} x^2(i) = p^{-1}(k-1) + x^2(k) \qquad (a) \\ b(k) &= \sum_{i=1}^{k} x(i)y(i) = b(k-1) + x(k)y(k) \qquad (b) \end{aligned} \tag{2.20}$$

Rearranging (2.20)(a),

$$p(k-1) = p(k) + p(k)x^2(k)p(k-1) \tag{2.21}$$

therefore,

$$\begin{aligned} p(k-1)x(k) &= p(k)x(k) + p(k)x^3(k)p(k-1) \\ &= p(k)x(k)[1 + p(k-1)x^2(k)] \end{aligned}$$

so that

$$p(k-1)x(k)\,[1+p(k-1)x^2(k)]^{-1} = p(k)x(k)$$

Now, multiplying by $p(k-1)x(k)$ and using (2.21),

$$p^2(k-1)x^2(k)\,[1+p(k-1)x^2(k)]^{-1} = p(k)x^2(k)p(k-1) = p(k-1) - p(k)$$

Consequently,

$$p(k) = p(k-1) - p^2(k-1)x^2(k)\,[1+p(k-1)x^2(k)]^{-1}$$

or,

$$p(k) = p(k-1) - p(k-1)x(k)\,[1+x(k)p(k-1)x(k)]^{-1}x(k)p(k-1) \quad (2.22)$$

The reason for writing the latter expression in an apparently too complex and, in this single unknown parameter case, rather redundant form, will become apparent later when vector-matrix versions of the recursive least squares algorithm are considered. Now, substituting (2.22) and (2.20)(b) into (2.19),

$$\hat{a}(k) = \left[p(k-1) - p^2(k-1)x^2(k)\,\{1+p(k-1)x^2(k)\}^{-1}\right][b(k-1) + x(k)y(k)]$$

Noting that $\hat{a}(k-1) = p(k-1)b(k-1)$, this expression can be expanded to yield

$$\hat{a}(k) = \hat{a}(k-1) + g(k)[y(k) - x(k)\,\hat{a}(k-1)] \quad (2.23)$$

where,

$$g(k) = p(k-1)x(k)\,[1+x(k)p(k-1)x(k)]^{-1} \quad (2.24)$$

Another expression for $g(k)$ can be obtained by writing (2.24) in the form:

$$g(k) = [p(k)p^{-1}(k)]p(k-1)x(k)\,[1+x(k)p(k-1)x(k)]^{-1}$$

and then, substituting for $p^{-1}(k)$ from (2.20a),

$$\begin{aligned} g(k) &= p(k)[p^{-1}(k-1) + x^2(k)]p(k-1)x(k)\,[1+p(k-1)x^2(k)]^{-1} \\ &= p(k)\left[x(k) + x^2(k)p(k-1)x(k)\right]\left[1+p(k-1)x^2(k)\right]^{-1} \\ &= p(k)x(k) \end{aligned}$$

Finally, after collecting together equations (2.23), (2.24) and (2.22), we obtain the following recursive algorithm for the estimate $\hat{a}(k)$ of the parameter a at the kth sample:

$$\hat{a}(k) = \hat{a}(k-1) + g(k)[y(k) - x(k)\,\hat{a}(k-1)] \quad \text{(I-1)}$$
$$g(k) = p(k-1)x(k)\,[1 + x(k)p(k-1)x(k)]^{-1} \quad \text{(I-2a)}$$
$$\text{or} \quad g(k) = p(k)x(k) \quad \text{(I-2b)}$$
$$p(k) = p(k-1) - g(k)x(k)p(k-1) \quad \text{(I-3)}$$

The recursive algorithm I, consisting of the relations (I-1), (I-2a) or (I-2b), and (I-3) taken together, constitutes the scalar recursive least squares algorithm for estimating the single parameter a in the simple single parameter regression relationship (2.18). The similarity with the recursive least squares algorithm for estimating mean value is obvious on inspection: the reader can easily verify that the algorithm is in the form of a discrete-step gradient algorithm, as before, with the weighting factor or gain $p(k)$ being a strictly decreasing function of k. The performance of the algorithm (I) is illustrated in Figure 2.5, which shows the results obtained in the moving body example.

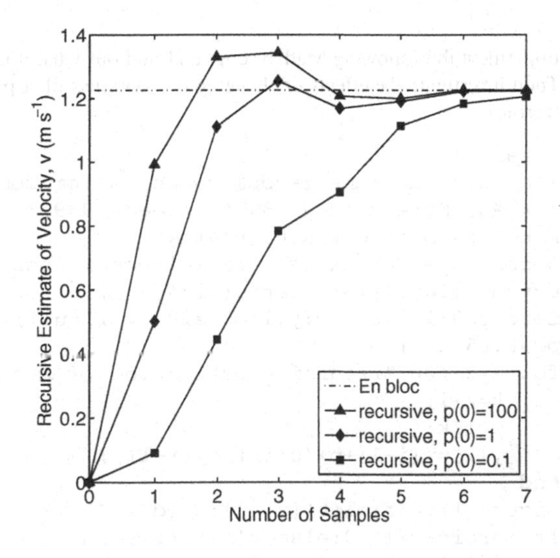

Fig. 2.5 Recursive estimates of the velocity in the moving body example.

The initiation of the recursive algorithm is again rather obvious: $a(1)$ and $p(1)$ can be computed easily from the non-recursive relationships in (2.19) and then the recursive algorithm can be used thereafter. There is, however, an alternative initiation procedure which, while providing no real advantage in the present simple context, will assume major significance when the estimation of multiple unknown parameters in a general multivariate regression relationship is considered later. Here the algorithm is initiated with $\hat{a}(0)$ set to some arbitrary finite value, for example zero, and $p(0)$ set to some large number, say 10^2 in the present example. While this is less obvious from a numerical standpoint, it yields equivalent results provided $p(0)$ is chosen large enough. And, as shown in the next chapter, it can be justified when the algorithm is considered from a statistical standpoint.

Considering Figure 2.5, the recursive estimates are plotted for initial conditions $\{\hat{a}(0) = 0;\ p(0) = 100\}$, $\{\hat{a}(0) = 0;\ p(0) = 1.0\}$ and $\{\hat{a}(0) = 0;\ p(0) = 0.1\}$, in order to show the effect of choosing different $p(0)$ values. Also shown is the overall *en bloc* solution ($\hat{a} = 1.226$, shown as a dash-dot line plotted right across the graph for reference). The results for $p(0) = 100$ are identical with the stage-wise results (i.e. the repeated solution of (2.19) for increasing k), which are not plotted because of this) but those for $p(0) = 1.0$ are quite different, with slower convergence, although the *final* estimates at $k = 7$ are virtually the same and equal to the *en bloc* estimate, as required. However, when $p(0) = 0.1$, the convergence is too slow for this number of samples and the final estimates do not agree with the *en bloc* estimate. This emphasizes the need to select large values of $p(0)$ in order to ensure equivalence with the stage-wise and *en bloc* results. The statistical interpretation of this requirement is discussed in the next chapter 3.

2.3 Exercises

1. Implement the following Matlab script and find out what it is doing (hints in the annotations). Then investigate the effects of changing the low-pass filter parameter pc and initial conditions (zeroic).

```
clear all
% ensure the same random sequence for comparisons (reseed)
s = RandStream('mcg16807', 'Seed',93973);
RandStream.setDefaultStream(s);
% generate data & set initial conditions
u=randn(100,1)+5;y=zeros(size(u));
zeroic=0;if zeroic;y(1)=0;else;y(1)=u(1);end
pc=0.05;k=0;
for i=2:100;%recursive estimation of mean
    k=k+1;
    p=1/k;
    y(i)=y(i-1)+p*(u(i-1)-y(i-1));
end
figure(1);clf;plot([u y]);hold
if zeroic;y(1)=0;else;y(1)=u(1);end
for i=2:100;%Low-pass filter
    k=k+1;
    y(i)=y(i-1)+pc*(u(i-1)-y(i-1));
end
plot(y,'r');hold off
axis([0,100,0,10])
legend('Data','recursive mean','LP filter')
```

2. Write a Matlab m-file script to implement the recursive least squares algorithm (I) for a single unknown parameter and show that this reproduces the results shown in Figure 2.5 when applied to the Graybill data. Then apply the algorithm in exercise 1. to a set of your own data.

2.4 Summary

In this chapter, we have considered at some length the simple problem of recursively estimating a single unknown parameter. At this point the reader may well see little advantage in the recursive procedure since the stage-wise solution of the equivalent *en bloc* problem yields the same results and is computationally no more demanding. But the reason for treating the problem in detail will become clearer as more complicated examples of recursive estimation are considered. It will become apparent that the analytical framework provided in this chapter provides a firm basis for solving these more complex problems and that the physical insight into the nature of the simple recursive algorithms gained here is valuable in understanding the functioning of their more complicated and sophisticated relatives. In particular, the interpretation of the recursive algorithms as simple gradient procedures with time varying gain factors should help the reader to better appreciate the functioning of the recursive algorithms and counteract, to some extent, the superficial complexity of the recursive algorithms in the multiple parameter situation, as discussed in the next and subsequent chapters.

Chapter 3
Recursive Least Squares Estimation

So far, we have considered the least squares solution to a particularly simple estimation problem in a single unknown parameter. A more general problem is the estimation of the n unknown parameters a_j, $j = 1,2,\ldots,n$, appearing in a general n^{th} order *linear regression* relationship of the form,

$$x(k) = a_1x_1(k) + a_2x_2(k) + \cdots + a_nx_n(k) \tag{3.1}$$

where the measurement $y(k)$ of $x(k)$ is assumed to be contaminated by zero mean value noise $e(k)$ of some form, so that the observation equation takes the form

$$y(k) = x(k) + e(k); \quad k = 1,2,\ldots,N \tag{3.2}$$

or,

$$y(k) = \mathbf{x}^T(k)\mathbf{a} + e(k) \tag{3.3}$$

where,

$$\mathbf{x}^T(k) = [x_1(k)\; x_2(k)\; \cdots\; x_n(k)]\,; \;\; \mathbf{a} = [a_1\; a_2\; \cdots\; a_n]^T \tag{3.4}$$

and the superscript T denotes the vector/matrix transpose. Here, $\mathbf{x}^T(k)\mathbf{a}$ is the vector inner product (see Appendix B), while the $x_j(k), j = 1,2,\ldots,n$, are exactly known, linearly independent variables that are also statistically independent of the measurement noise on $y(k)$. In this case, the least squares optimization after k samples have been obtained, is defined as:

$$\hat{\mathbf{a}} = \arg\min_{\mathbf{a}} \mathfrak{J}_2(\mathbf{a}) \qquad \mathfrak{J}_2(\mathbf{a}) = \sum_{i=1}^{k}\left[y(i) - \sum_{j=1}^{n} x_j(i)a_j\right]^2 = \sum_{i=1}^{k}[y(i) - \mathbf{x}^T(i)\mathbf{a}]^2 \tag{3.5}$$

As in the scalar case considered in the previous chapter, this requires that all the partial derivatives of $\mathfrak{J}_2(\mathbf{a})$ with respect to each of the parameter estimates should be set simultaneously to zero. Such a procedure yields a set of n linear, simultaneous algebraic equations that are usually termed the 'normal equations' of linear regression analysis and which can be solved for the parameter estimates $\hat{a}_j(k), j = 1,2,\ldots,n$, all at the kth instant.

Now, using the results in section B.1.12 of Appendix B for the differentiation of vector-matrix expressions,

$$\frac{\partial}{\partial \mathbf{a}}[y(i) - \mathbf{x}^T(i)\mathbf{a}]^2 = -2\mathbf{x}(i)[y(i) - \mathbf{x}^T(i)\mathbf{a}]$$

the normal equations become

$$\frac{1}{2}\nabla_{\mathbf{a}}(\mathfrak{J}_2) = -\sum_{i=1}^{k}\mathbf{x}(i)y(i) + \left[\sum_{i=1}^{k}\mathbf{x}(i)\mathbf{x}^T(i)\right]\mathbf{a}(k) = 0 \tag{3.6}$$

where $\nabla_{\mathbf{a}}(\mathfrak{J}_2)$ denotes the gradient of $\mathfrak{J}_2(\mathbf{a})$ with respect to all the elements of $\mathbf{a}$. As a result,

$$\left[\sum_{i=1}^{k}\mathbf{x}(i)\mathbf{x}^T(i)\right]\hat{\mathbf{a}}(k) = \sum_{i=1}^{k}\mathbf{x}(i)y(i) \tag{3.7}$$

and, provided that the matrix $\sum \mathbf{x}(i)\mathbf{x}^T(i)$ is non-singular and invertible, the solution to the simultaneous equations (3.7) takes the *en bloc* form

$$\hat{\mathbf{a}}(k) = \mathbf{P}(k)\mathbf{b}(k) \tag{3.8}$$

where,

$$\mathbf{P}(k) = \left[\sum_{i=1}^{k}\mathbf{x}(i)\mathbf{x}^T(i)\right]^{-1} \quad ; \quad \mathbf{b}(k) = \sum_{i=1}^{k}\mathbf{x}(i)y(i) \tag{3.9}$$

which is simply the vector-matrix equivalent of equation (2.19) in the previous chapter.

3.1 The Deterministic Recursive Linear Least Squares Algorithm

First of all, it is necessary to consider the 'deterministic' recursive least squares algorithm. Although it is concerned with parameter estimation, the algorithm is considered as a deterministic estimation procedure in the sense that it makes few assumptions about either the statistical nature of the signals or the noise $e(k)$; and it does not provide any statistical information on the nature of the estimates. The stochastic situation, when such assumptions are made, is considered in the next section 3.2.

Following a similar procedure to that used in the single parameter case (see equations (2.8) *et seq*), but noting that we are now dealing with matrices and vectors rather than scalar quantities (see Appendix B), it is clear that $\mathbf{P}(k)$ and $\mathbf{b}(k)$ in equation (3.9) can be defined as follows:

$$\begin{aligned} \mathbf{P}^{-1}(k) &= \mathbf{P}^{-1}(k-1)+\mathbf{x}(k)\mathbf{x}^T(k) \qquad (a) \\ \mathbf{b}(k) &= \mathbf{b}(k-1)+\mathbf{x}(k)y(k) \qquad (b) \end{aligned} \tag{3.10}$$

In order to develop a recursive version of equation (3.8), equation (3.10)(a) is now pre-multiplied by $\mathbf{P}(k)$ and then post-multiplied by $\mathbf{P}(k-1)$, to give

$$\mathbf{P}(k-1) = \mathbf{P}(k) + \mathbf{P}(k)\mathbf{x}(k)\mathbf{x}^T(k)\mathbf{P}(k-1) \tag{3.11}$$

Post multiplying by $\mathbf{x}(k)$ then yields,

$$\begin{aligned} \mathbf{P}(k-1)\mathbf{x}(k) &= \mathbf{P}(k)\mathbf{x}(k) + \mathbf{P}(k)\mathbf{x}(k)\mathbf{x}^T(k)\mathbf{P}(k-1)\mathbf{x}(k) \\ &= \mathbf{P}(k)\mathbf{x}(k)[1+\mathbf{x}^T(k)\mathbf{P}(k-1)\mathbf{x}(k)] \end{aligned}$$

Then, post-multiplying by $[1+\mathbf{x}^T(k)\mathbf{P}(k-1)\mathbf{x}(k)]^{-1}\mathbf{x}^T(k)\mathbf{P}(k-1)$,

$$\mathbf{P}(k-1)\mathbf{x}(k)[1+\mathbf{x}^T(k)\mathbf{P}(k-1)\mathbf{x}(k)]^{-1}\mathbf{x}^T(k)\mathbf{P}(k-1) = \mathbf{P}(k)\mathbf{x}(k)\mathbf{x}^T(k)\mathbf{P}(k-1)$$

Substituting from (3.11),

$$\mathbf{P}(k) = \mathbf{P}(k-1) - \mathbf{P}(k-1)\mathbf{x}(k)[1+\mathbf{x}^T(k)\mathbf{P}(k-1)\mathbf{x}(k)]^{-1}\mathbf{x}^T(k)\mathbf{P}(k-1) \tag{3.12}$$

which is termed the 'matrix inversion lemma' (Bodewig, 1956; Ho, 1962) because it is an alternative to the matrix inversion required for the *en bloc* solution (3.8).

It is now straightforward to obtain the equivalent recursive equation for recursively updating the estimate of the parameter vector $\hat{\mathbf{a}}(k)$:

$$\begin{aligned} \hat{\mathbf{a}}(k) &= \{\mathbf{P}(k-1) - \mathbf{P}(k-1)\mathbf{x}(k)[1+\mathbf{x}^T(k)\mathbf{P}(k-1)\mathbf{x}(k)]^{-1} \\ &\qquad \mathbf{x}^T(k)\mathbf{P}(k-1)\}\{\mathbf{b}(k-1)+\mathbf{x}(k)y(k)\} \\ &= \hat{\mathbf{a}}(k-1) - \mathbf{P}(k-1)\mathbf{x}(k)[1+\mathbf{x}^T(k)\mathbf{P}(k-1)\mathbf{x}(k)]^{-1}\mathbf{x}^T(k)\hat{\mathbf{a}}(k-1) \\ &\quad + \mathbf{P}(k-1)\mathbf{x}(k)y(k) - \mathbf{P}(k-1)\mathbf{x}(k)[1+\mathbf{x}^T(k)\mathbf{P}(k-1)\mathbf{x}(k)]^{-1} \\ &\qquad \mathbf{x}^T(k)\mathbf{P}(k-1)\mathbf{x}(k)y(k) \\ &= \hat{\mathbf{a}}(k-1) - \mathbf{P}(k-1)\mathbf{x}(k)[1+\mathbf{x}^T(k)\mathbf{P}(k-1)\mathbf{x}(k)]^{-1} \\ &\quad \{\mathbf{x}^T(k)\hat{\mathbf{a}}(k-1) + \mathbf{x}^T(k)\mathbf{P}(k-1)\mathbf{x}(k)y(k) - [1+\mathbf{x}^T(k)\mathbf{P}(k-1)\mathbf{x}(k)]y(k)\} \end{aligned}$$

so that finally,

$$\hat{\mathbf{a}}(k) = \hat{\mathbf{a}}(k-1) + \mathbf{g}(k)[y(k) - \mathbf{x}^T(k)\hat{\mathbf{a}}(k-1)] \tag{3.13}$$

where

$$\mathbf{g}(k) = \mathbf{P}(k-1)\mathbf{x}(k)[1+\mathbf{x}^T(k)\mathbf{P}(k-1)\mathbf{x}(k)]^{-1} \tag{3.14}$$

An alternative expression for $\mathbf{g}(k)$ can be obtained straightforwardly by the following manipulation:

$$\begin{aligned}\mathbf{g}(k) &= \left[\mathbf{P}(k)\mathbf{P}^{-1}(k)\right]\mathbf{P}(k-1)\mathbf{x}(k)\left[1+\mathbf{x}^T(k)\mathbf{P}(k-1)\mathbf{x}(k)\right]^{-1} \\ &= \mathbf{P}(k)\left[\mathbf{P}^{-1}(k-1)+\mathbf{x}(k)\mathbf{x}^T(k)\right]\mathbf{P}(k-1)\mathbf{x}(k)\left[1+\mathbf{x}^T(k)\mathbf{P}(k-1)\mathbf{x}(k)\right]^{-1} \\ &= \mathbf{P}(k)\left[\mathbf{x}(k)+\mathbf{x}(k)\mathbf{x}^T(k)\mathbf{P}(k-1)\mathbf{x}(k)\right]\left[1+\mathbf{x}^T(k)\mathbf{P}(k-1)\mathbf{x}(k)\right]^{-1} \\ &= \mathbf{P}(k)\mathbf{x}(k)\left[1+\mathbf{x}^T(k)\mathbf{P}(k-1)\mathbf{x}(k)\right]\left[1+\mathbf{x}^T(k)\mathbf{P}(k-1)\mathbf{x}(k)\right]^{-1}\end{aligned}$$

so that

$$\mathbf{g}(k) = \mathbf{P}(k)\mathbf{x}(k) \tag{3.15}$$

Collecting together equations (3.12), (3.13) (3.14) and (3.15) then yields the following general *Recursive Least Squares* (RLS) algorithm:

$$\hat{\mathbf{a}}(k) = \hat{\mathbf{a}}(k-1) + \mathbf{g}(k)[y(k) - \mathbf{x}^T(k)\hat{\mathbf{a}}(k-1)] \qquad \text{(RLS-1a)}$$

$$\mathbf{g}(k) = \mathbf{P}(k-1)\mathbf{x}(k)[1+\mathbf{x}^T(k)\mathbf{P}(k-1)\mathbf{x}(k)]^{-1} \qquad \text{(RLS-2)}$$

$$\mathbf{P}(k) = \mathbf{P}(k-1) - \mathbf{g}(k)\mathbf{x}^T(k)\mathbf{P}(k-1) \qquad \text{(RLS-3)}$$

Alternatively, using the definition of $\mathbf{g}(k)$ in (3.15), the recursive update equation (RLS-1a) can be written in the form

$$\hat{\mathbf{a}}(k) = \hat{\mathbf{a}}(k-1) + \mathbf{P}(k)\mathbf{x}(k)[y(k) - \mathbf{x}^T(k)\hat{\mathbf{a}}(k-1)] \qquad \text{(RLS-1b)}$$

Although (RLS-1a) is normally preferred in computational terms, this alternative expression will be useful in the subsequent analysis.

In contrast to the scalar algorithm (I) in chapter 2, the RLS algorithm provides some considerable advantage over the stage-wise solution of (3.8). In addition to the now convenient recursive form, which provides for a minimum of computer storage, note that the term

$$[1+\mathbf{x}^T(k)\mathbf{P}(k-1)\mathbf{x}(k)]$$

is simply a scalar quantity. As a result, there is no requirement for direct matrix inversion even though the repeated solution of the equivalent classical *en bloc* solution (3.8) entails inverting an $n \times n$ matrix for each solution.

Finally as regards the implementation of this RLS algorithm, it is necessary once again to specify starting values $\hat{\mathbf{a}}(0)$ and $\mathbf{P}(0)$ for the estimated parameter vector $\mathbf{a}(k)$ and the matrix $\mathbf{P}(k)$, respectively. As in the scalar case, however, this presents no real problem; it can be shown that the criterion function-parameter hypersurface is unimodal and that an arbitrary finite $\hat{\mathbf{a}}(0)$ (normally a zero vector) coupled with $\mathbf{P}(0)$ having large diagonal elements (say 10^6, in general) will yield convergence and performance commensurate with the stage-wise solution of the same problem.

Let us now explore the nature of the RLS algorithm further. Note that the term $\mathbf{x}^T(k)\hat{\mathbf{a}}(k-1)$ in (RLS-1a) and (RLS-1b) can be considered as a prediction of $y(k)$ at the kth sampling instant based on the estimate $\hat{\mathbf{a}}(k-1)$ at the previous $(k-1)^{th}$ instant, so that the recursive residual

$$\varepsilon(k) = y(k) - \mathbf{x}^T(k)\hat{\mathbf{a}}(k-1) \tag{3.16}$$

can also be considered as the one-step-ahead *prediction error*. These one-step-ahead prediction errors $\varepsilon(k)$, $k = 1,2, \ldots, N$ are also termed the *innovations* sequence because they represent that part of $y(k)$ that cannot be predicted from the past data and introduce the latest new information on the adequacy of the parameter estimates. It is important to realize that the $\varepsilon(k)$ are different from the least squares residuals $\hat{e}(k)$ obtained in the same manner but with $\hat{\mathbf{a}}(k-1)$ replaced by $\hat{\mathbf{a}}(N)$: see later, equation (3.29). In fact, $\varepsilon(k)$ can be interpreted as the *a priori* prediction of $e(k)$ at sample k, given the latest measurement $y(k)$ and the previous estimate $\hat{\mathbf{a}}(k-1)$.

The following alternative expression for $\hat{\mathbf{a}}(k)$ follows directly from the second form of the recursive update equation (RLS-1b), i.e.,

$$\hat{\mathbf{a}}(k) \;=\; \hat{\mathbf{a}}(k-1) + \mathbf{P}(k)\{\mathbf{x}(k)y(k) - \mathbf{x}(k)\mathbf{x}^T(k)\hat{\mathbf{a}}(k-1)\} \tag{3.17}$$

Consequently, another similarity with the scalar case is that the RLS algorithm can be interpreted as a special form of gradient procedure. In particular, the gradient of the *instantaneous* part J_I of the cost function $\mathcal{J}_2$ at the kth sampling instant, i.e.,

$$J_I = [y(k) - \mathbf{x}^T(k)\hat{\mathbf{a}}(k-1)]^2 = \varepsilon^2(k) \tag{3.18}$$

is an $n \times 1$ vector that takes the form (see section B.1.12 of Appendix B again),

$$-\frac{1}{2}\nabla_{\hat{\mathbf{a}}}J_I = -\frac{\partial\varepsilon(k)}{\partial\hat{\mathbf{a}}}\varepsilon(k) = \mathbf{x}(k)\varepsilon(k) \tag{3.19}$$

and the scalar weighting factor is replaced by the time variable weighting matrix $\mathbf{P}(k)$ in (3.17). However, the strictly decreasing nature and consequent smoothing effect of this matrix is directly analogous to that of $p(k)$ in equation (I-2) of the single parameter case, as pointed out by Ho (1962).

It is interesting to note that the instantaneous gradient in (3.19) can be presented in the form

$$\frac{1}{2}\nabla_{\hat{\mathbf{a}}}[\varepsilon^2(k)] = \nabla_{\hat{\mathbf{a}}}[\varepsilon(k)]\varepsilon(k)$$

revealing that $\nabla_{\hat{\mathbf{a}}}[\varepsilon(k)]^2$ is the product of the derivative $\nabla_{\hat{\mathbf{a}}}[\varepsilon(k)]$ and the recursive error $\varepsilon(k)$. As a result, noting equations (3.17) and (3.19), the recursive estimation equation (RLS-1b) can be written as

$$\begin{aligned} \hat{\mathbf{a}}(k) &= \hat{\mathbf{a}}(k-1) + \mathbf{P}(k)\mathbf{x}(k)\varepsilon(k) \\ \mathbf{x}(k) &= -\frac{\partial\varepsilon(k)}{\partial\hat{\mathbf{a}}} \end{aligned} \tag{3.20}$$

This is, some ways, a more general form of the recursive algorithm, one which has relevance to situations where the prediction error variable $\varepsilon(k)$ is not, as here, a simple linear function of the unknown parameters. This interpretation is fundamental to the *Prediction Error Minimization* (PEM) methods of estimation: see e.g. Ljung (1987) and the later chapter 10.

3.2 The Stochastic Recursive Linear Least Squares Algorithm

As pointed out in the previous section, the RLS algorithm makes no statistical assumptions about the nature of the signals and noise. To illustrate how even the minimum of statistical information can help to improve this aspect of the algorithm, let us make certain mild assumptions about the nature of the measurement noise $e(k)$ in equation (3.3), which is repeated below for convenience:

$$y(k) = \mathbf{x}^T(k)\mathbf{a} + e(k) \tag{3.21}$$

In particular let us assume that the $e(k)$ are:

1. a zero mean sequence of random variables: i.e. $E\{e(k)\} = 0$;
2. serially uncorrelated and have a constant variance: i.e. $E\{e(k)e(j)\} = \sigma^2\delta_{kj}$ where δ_{kj} is the Kronecker delta function:

$$\delta_{kj} = \begin{Bmatrix} 1 & k = j \\ 0 & k \neq j \end{Bmatrix} \quad \forall k, j;$$

3. independent of the variables $x_j(k)$ that compose the vector $\mathbf{x}(k)$; i.e.

$$E\{\mathbf{x}(k)e(j)\} = 0, \ \forall k, j.$$

In this situation, the model (3.21), is known as the *General Linear Regression Model* and the least squares equations (3.7) are termed the *Normal Equations of Least Squares Regression Analysis* (Kendall and Stuart, 1961; Priestley, 1981). In order to examine the effect of these statistical assumptions, equation (3.21) is written in the alternative vector-matrix form

$$\mathbf{y}(k) = \mathbf{X}(k)\mathbf{a}(k) + \mathbf{e}(k) \tag{3.22}$$

where $\mathbf{y}(k) = [y(1), y(2), \ldots, y(k)]^T$ and $\mathbf{X}(k)$ is a $k \times n$ matrix with rows $\mathbf{x}^T(i)$, $i = 1, \ldots, k$. Consequently, the normal equations (3.7) can be written concisely as

$$\mathbf{X}^T(k)\mathbf{X}(k)\hat{\mathbf{a}}(k) - \mathbf{X}^T(k)\mathbf{y}(k) = 0 \tag{3.23}$$

so that

$$\hat{\mathbf{a}}(k) = [\mathbf{X}^T(k)\mathbf{X}(k)]^{-1}\mathbf{X}^T(k)\mathbf{y}(k) \tag{3.24}$$

Substituting for $\mathbf{y}(k)$ from (3.22),

$$\hat{\mathbf{a}}(k) = \mathbf{a}(k) + [\mathbf{X}^T(k)\mathbf{X}(k)]^{-1}\mathbf{X}^T(k)\mathbf{e}(k)$$

or

$$\tilde{\mathbf{a}}(k) = \hat{\mathbf{a}}(k) - \mathbf{a}(k) = [\mathbf{X}^T(k)\mathbf{X}(k)]^{-1}\mathbf{X}^T(k)\mathbf{e}(k) \tag{3.25}$$

where $\tilde{\mathbf{a}}(k) = \hat{\mathbf{a}}(k) - \mathbf{a}(k)$ is the estimation error after k samples.

It is now a simple matter to determine the statistics of the estimation error. The expected value of $\tilde{\mathbf{a}}(k)$ is given by

$$E\{\tilde{\mathbf{a}}(k)\} = E\{[\mathbf{X}^T(k)\mathbf{X}(k)]^{-1}\mathbf{X}^T(k)\mathbf{e}(k)\}$$

where E is the expected value operator. But since $\mathbf{e}(k)$ is independent of $\mathbf{X}(k)$ from assumption 3., above, and has zero mean from assumption 1., it is clear that this expression can be simplified in the following manner:

$$E\{\tilde{\mathbf{a}}(k)\} = [\mathbf{X}^T(k)\mathbf{X}(k)]^{-1}\mathbf{X}^T(k)E\{\mathbf{e}(k)\}$$

because $\mathbf{X}(k)$ is a deterministic matrix composed of exactly known elements. As a result,

$$E\{\tilde{\mathbf{a}}(k)\} = 0 \quad \text{or} \quad E\{\hat{\mathbf{a}}(k)\} = \mathbf{a} \tag{3.26}$$

so that that the estimate $\hat{\mathbf{a}}(k)$ is unbiased. The covariance matrix of the estimation errors, $\mathbf{P}^*(k)$, is defined as

$$\mathbf{P}^*(k) = E\{\tilde{\mathbf{a}}(k)\tilde{\mathbf{a}}^T(k)\} = \begin{bmatrix} E\{\tilde{a}_1^2\} & E\{\tilde{a}_1\tilde{a}_2\} & \cdots & E\{\tilde{a}_1\tilde{a}_n\} \\ E\{\tilde{a}_2\tilde{a}_1\} & E\{\tilde{a}_2^2\} & \cdots & E\{\tilde{a}_2\tilde{a}_n\} \\ \cdots & \cdots & \cdots & \cdots \\ E\{\tilde{a}_n\tilde{a}_1\} & E\{\tilde{a}_n\tilde{a}_2\} & \cdots & E\{\tilde{a}_n^2\} \end{bmatrix}$$

Now, by using the matrix product rule $[\mathbf{PQR}]^T = \mathbf{R}^T\mathbf{Q}^T\mathbf{P}^T$ (see Appendix B) and noting that $[\mathbf{X}^T(k)\mathbf{X}(k)]^{-1}$ is symmetric,

$$\begin{aligned} \mathbf{P}^*(k) &= E\{\tilde{\mathbf{a}}(k)\tilde{\mathbf{a}}^T(k)\} \\ &= E\{[\mathbf{X}^T(k)\mathbf{X}(k)]^{-1}\mathbf{X}^T(k)\mathbf{e}(k)\mathbf{e}^T(k)\mathbf{X}(k)[\mathbf{X}^T(k)\mathbf{X}(k)]^{-1}\} \end{aligned} \tag{3.27}$$

Once again this expression can be simplified by noting assumption 3., to yield

$$\mathbf{P}^*(k) = [\mathbf{X}^T(k)\mathbf{X}(k)]^{-1}\mathbf{X}^T(k)E\{\mathbf{e}(k)\mathbf{e}^T(k)\}\mathbf{X}(k)[\mathbf{X}^T(k)\mathbf{X}(k)]^{-1}$$

But from assumption 2., the covariance matrix of the noise vector $\mathbf{e}(k)$, i.e.,

$$E\{\mathbf{e}(k)\mathbf{e}^T(k)\} = \begin{bmatrix} E\{e^2(1)\} & E\{e(1)e(2)\} & \cdots & E\{e(1)e(k)\} \\ E\{e(2)e(1)\} & E\{e^2(2)\} & \cdots & E\{e(2)e(k)\} \\ \cdots & \cdots & \cdots & \cdots \\ E\{e(k)e(1)\} & E\{e(k)e(2)\} & \cdots & E\{e^2(k)\} \end{bmatrix}$$

assumes the special form,

$$E\{\mathbf{e}(k)\mathbf{e}^T(k)\} = \sigma^2 \mathbf{I}_k$$

where $\mathbf{I}_k$ is the kth order identity matrix. As a result,

$$\mathbf{P}^*(k) = \sigma^2[\mathbf{X}^T(k)\mathbf{X}(k)]^{-1}\mathbf{X}^T(k)\mathbf{X}(k)[\mathbf{X}^T(k)\mathbf{X}(k)]^{-1} = \sigma^2[\mathbf{X}^T(k)\mathbf{X}(k)]^{-1}$$

so that,

$$\mathbf{P}^*(k) = \sigma^2 \mathbf{P}(k) \tag{3.28}$$

where $\mathbf{P}(k) = [\mathbf{X}^T(k)\mathbf{X}(k)]^{-1} = [\sum_{i=1}^{k} \mathbf{x}(i)\mathbf{x}^T(i)]^{-1}$, and the matrix $\mathbf{P}(k)$ is obtained from the recursive least squares analysis via the matrix inversion lemma, (3.12). Finally, an estimate $\hat{\sigma}^2$ of σ^2 can be obtained by defining the vector of least squares residuals $\hat{\mathbf{e}}(k)$, after k samples, as[1]

$$\hat{\mathbf{e}}(k) = [\hat{e}(1), \ldots, \hat{e}(k)]^T, \text{ where } \hat{e}(i) = y(i) - \mathbf{x}^T(i)\hat{\mathbf{a}}(k),\ i = 1, 2, \ldots, k \tag{3.29}$$

Substituting from (3.22) and (3.24),

$$\begin{aligned}
\hat{\mathbf{e}}(k) &= \mathbf{X}(k)\mathbf{a}(k) + \mathbf{e}(k) - \mathbf{X}(k)\{[\mathbf{X}^T(k)\mathbf{X}(k)]^{-1}\mathbf{X}^T(k)[\mathbf{X}(k)\mathbf{a}(k) + \mathbf{e}(k)]\} \\
&= \mathbf{e}(k) - \mathbf{X}(k)\{[\mathbf{X}^T(k)\mathbf{X}(k)]^{-1}\mathbf{X}^T(k)\mathbf{e}(k) \\
&= [\mathbf{I}_k - \mathbf{X}(k)[\mathbf{X}^T(k)\mathbf{X}(k)]^{-1}\mathbf{X}^T(k)]\mathbf{e}(k)
\end{aligned}$$

so that,

$$\hat{\mathbf{e}}(k) = \mathbf{B}(k)\mathbf{e}(k) \tag{3.30}$$

where $\mathbf{B}(k) = [\mathbf{I}_k - \mathbf{X}(k)[\mathbf{X}^T(k)\mathbf{X}(k)]^{-1}\mathbf{X}^T(k)]$ is a symmetric idempotent matrix (see Appendix B), i.e., $\mathbf{B}^T = \mathbf{B}$; $\mathbf{B}^2 = \mathbf{B}$. Now referring to (3.30), the sum of the squares of the least squares residuals is given by

$$\sum_{i=1}^{k} \hat{e}^2(i) = \hat{\mathbf{e}}^T(k)\hat{\mathbf{e}}(k) = \mathbf{e}^T(k)\mathbf{B}^T(k)\mathbf{B}(k)\mathbf{e}(k) = \mathbf{e}^T(k)\mathbf{B}(k)\mathbf{e}(k).$$

Consequently,

$$\begin{aligned}
E\{\hat{\mathbf{e}}^T(k)\hat{\mathbf{e}}(k)\} &= E\{\mathbf{e}^T(k)[\mathbf{I}_k - \mathbf{X}(k)[\mathbf{X}^T(k)\mathbf{X}(k)]^{-1}\mathbf{X}^T(k)]\mathbf{e}(k)\} \\
&= E\{\mathbf{e}^T(k)\mathbf{e}(k) - \mathbf{e}^T(k)\mathbf{X}(k)[\mathbf{X}^T(k)\mathbf{X}(k)]^{-1}\mathbf{X}^T(k)\mathbf{e}(k)\}
\end{aligned} \tag{3.31}$$

But from assumption 2.,

$$E\{\mathbf{e}^T(k)\mathbf{e}(k)\} = E\{e^2(1) + \ldots + e^2(k)\} = \sigma^2 k$$

and

$$E\{\mathbf{e}^T(k)\mathbf{X}(k)[\mathbf{X}^T(k)\mathbf{X}(k)]^{-1}\mathbf{X}^T(k)\mathbf{e}(k)\} = \sigma^2 \text{trace}\{\mathbf{X}(k)[\mathbf{X}^T(k)\mathbf{X}(k)]^{-1}\mathbf{X}^T(k)\}$$

[1] Note that the final estimate of $\hat{\mathbf{a}}(k)$ at the sample k is used here to define the $\hat{e}(i)$ $i = 1, 2, \ldots, k$.

where the trace of matrix is defined as the sum of its diagonal elements. Thus (3.31) can be simplified to the following

$$E\{\hat{\mathbf{e}}^T(k)\hat{\mathbf{e}}(k)\} = \sigma^2[k - \text{trace}\{[\mathbf{X}^T(k)\mathbf{X}(k)]^{-1}\mathbf{X}^T(k)\mathbf{X}(k)\} = \sigma^2[k - \text{trace}\,\mathbf{I}_n]$$

$$E\{\hat{\mathbf{e}}^T(k)\hat{\mathbf{e}}(k)\} = \sigma^2[k-n] \tag{3.32}$$

since $\mathbf{X}^T(k)\mathbf{X}(k)$ is of order n. From this result we can now see that an unbiased estimator $\hat{\sigma}^2$ of σ^2 can be obtained from

$$\hat{\sigma}^2 = \frac{1}{k-n}\hat{\mathbf{e}}^T(k)\hat{\mathbf{e}}(k) = \frac{1}{k-n}\sum_{i=1}^{k}\hat{e}^2(i) \tag{3.33}$$

which makes sense since, as the sample size k increases, this quickly approaches the average of the squared values of $\hat{e}(i)$, $i = 1, 2, \ldots, k$. It is now a simple matter to obtain the following stochastic recursive least squares regression algorithm by substituting $\mathbf{P}^*(k)/\hat{\sigma}^2$ for $\mathbf{P}(k)$ into the basic RLS algorithm, i.e.,

$$\hat{\mathbf{a}}(k) = \hat{\mathbf{a}}(k-1) + \mathbf{g}(k)[y(k) - \mathbf{x}^T(k)\hat{\mathbf{a}}(k-1)] \qquad \text{(SRLS-1)}$$

$$\mathbf{g}(k) = \mathbf{P}^*(k-1)\mathbf{x}(k)[\hat{\sigma}^2 + \mathbf{x}^T(k)\mathbf{P}^*(k-1)\mathbf{x}(k)]^{-1} \qquad \text{(SRLS-2a)}$$

$$\text{or} \quad \mathbf{g}(k) = \frac{\mathbf{P}^*(k)}{\hat{\sigma}^2}\mathbf{x}(k) \qquad \text{(SRLS-2b)}$$

$$\mathbf{P}^*(k) = \mathbf{P}^*(k-1) - \mathbf{g}(k)\mathbf{x}^T(k)\mathbf{P}^*(k-1) \qquad \text{(SRLS-3)}$$

In this algorithm $\hat{\sigma}^2$ is either an *a priori* estimate of the noise variance σ^2 or an estimate obtained from the data and updated at each sample (see the later exercises and chapter 4). Based on our previous discussion of the RLS algorithm, $\hat{y}(k|k-1) = \mathbf{x}^T(k)\hat{\mathbf{a}}(k-1)$ is the one-step-ahead prediction of $y(k)$ and the innovations sequence $\varepsilon(k) = y(k) - \hat{y}(k|k-1)$ is the associated prediction error. It is interesting, therefore, to consider the variance of $\varepsilon(k)$, which is given by

$$\begin{aligned} E\{\varepsilon^2(k)\} &= E\{[\mathbf{x}^T(k)\mathbf{a} - \mathbf{x}^T(k)\hat{\mathbf{a}}(k-1) + e(k)]^2\} \\ &= E\{[\mathbf{x}^T(k)\tilde{\mathbf{a}} + e(k)]^2\} \\ &= \mathbf{x}^T(k)E\{\tilde{\mathbf{a}}\,\tilde{\mathbf{a}}^T\}\mathbf{x}(k) + E\{e^2(k)\}]\} \end{aligned}$$

because the parametric estimation error $\tilde{\mathbf{a}}(k) = \mathbf{a} - \hat{\mathbf{a}}(k-1)$ is independent of $e(k)$, so that the expected values of the cross products in the expansion of $[\mathbf{x}^T(k)\tilde{\mathbf{a}} + e(k)]^2$ are zero. So, finally, the variance of the innovations sequence at any sample k is given by:

$$\text{var}\{\varepsilon(k)\} = \sigma^2 v(k) = 1 + \mathbf{x}^T(k)\mathbf{P}(k)\mathbf{x}(k) \tag{3.34}$$

since, by the definition (3.27), $E\{\tilde{\mathbf{a}}\,\tilde{\mathbf{a}}^T\} = \mathbf{P}^*(k)$ and $\mathbf{P}^*(k) = \sigma^2\mathbf{P}(k)$, from (3.28). This is an important relationship because it shows how the the variance of the innovations sequence, which is such a critical aspect of recursive estimation, changes as the recursive estimation proceeds. In statistics, this changing variance is referred to as 'heteroscedasticity' and $v(k)$ can be used to produce a normalized sequence

$$\varepsilon_n(k) = \varepsilon(k)/\sqrt{v}(k) = \frac{\varepsilon(k)}{[1+\mathbf{x}^T(k)\mathbf{P}(k)\mathbf{x}(k)]^{0.5}} \tag{3.35}$$

such that, if the statistical assumptions are satisfied, the sample variance of $\varepsilon_n(k)$ is constant over the observation interval (i.e. it is converted to a 'homoscedastic' sequence). This device is used later when the important topic of *Prediction Error Decomposition* is discussed in the next Chapter 4.

The SRLS algorithm is very powerful and elegant: it not only supplies the parameter estimates at each sampling instant but also an indication of the accuracy of these estimates through the error-covariance matrix $\mathbf{P}^*(k)$. Obviously, $\mathbf{P}^*(k)$ behaves like $\mathbf{P}(k)$ except for the scale factor $\hat{\sigma}^2$ and is a strictly decreasing function of the sample size. This property is one manifestation of the statistical property of consistency (see Appendix B) because it demonstrates how the error-covariance is a strictly decreasing function of sample size and, since the estimate is unbiased, the accuracy of the estimate increases as more data are utilized.

It is well known that, with certain statistical assumptions about the noise and signal characteristics, least squares estimation can be interpreted from both the maximum likelihood and Bayesian standpoints. These same interpretations can of course also be applied to the RLS regression algorithm. For instance, suppose it is assumed that, in addition to its statistical properties in (3.21), $e(k)$ has a Gaussian amplitude distribution. It is then possible to argue that the parameter estimate $\hat{\mathbf{a}}(k)$ and the estimation errors $\tilde{\mathbf{a}}(k)$ also have Gaussian normal distributions of the form (see Appendix B).

$$\hat{\mathbf{a}}(k) \;=\; \mathcal{N}[\mathbf{a},\mathbf{P}^*(k)]; \quad \tilde{\mathbf{a}}(k) \;=\; \mathcal{N}[\mathbf{0},\mathbf{P}^*(k)]$$

so that the statistical characteristics of the parameter estimates are completely determined by the first two statistical moments; namely the mean $\mathbf{a}$, the true value of $\mathbf{a}$, and the error covariance matrix $\mathbf{P}^*(k)$. In this case, the SRLS algorithm can be considered as a *Recursive Maximum Likelihood* (RML) estimator for the linear regression model parameters. If it is further assumed that prior information is available on the mean and covariance matrix, in the form of the initial estimates $\hat{\mathbf{a}}(0)$ and $\mathbf{P}^*(0)$, then the algorithm can be considered as a Bayesian estimator, since it can be obtained from application of the Bayes Rule linking *a priori* and *a posteriori* probability statements (see e.g. Appendix B). Indeed, the algorithm can be viewed as the physical embodiment of Bayesian estimation: at each recursive step from the first, the *a priori* estimates are modified with the help of the new (innovatory) information imparted by the latest data sample $\{y(k),\mathbf{x}(k)\}$ and the resultant calculation of the kth recursive residual (innovation) to yield *a posteriori* estimates which, in turn, become the *a priori* estimates for the next recursive step. In this latter sense, the rule of thumb introduced previously for initialising the $\mathbf{P}(0)$ matrix in the RLS algorithm as a diagonal matrix with large diagonal elements makes obvious sense. If little is known about $\mathbf{a}$ and $\hat{\mathbf{a}}(0)$ is set to zero, then the initial $\mathbf{P}(0)$ matrix, interpreted in these covariance terms via equation (3.28), should reflect this uncertainty: the large diagonal elements indicate high initial error variance and little confidence in the initial estimate $\hat{\mathbf{a}}(0)$; with the zero off-diagonal elements showing that there is no prior information on the covariance properties and the best estimate (or guess?) is that they are all zero. In statistics, this initial setting is termed the 'diffuse prior' assumption.

But the direct interpretation of $\mathbf{P}^*(0)$ as an *a priori* covariance matrix in the SRLS algorithm introduces more flexibility since, in general, $\hat{\mathbf{a}}(0)$ and $\mathbf{P}^*(0)$ can be selected to reflect quantitatively our prior knowledge of the model parameters. For instance, a previous estimation experiment using N_p samples could have supplied $\hat{\mathbf{a}}(N_p)$ and $\mathbf{P}^*(N_p)$ and these could be used to define $\hat{\mathbf{a}}(0)$ and $\mathbf{P}^*(0)$ for any new experiment. Or alternatively, $\hat{\mathbf{a}}(0)$ and $\mathbf{P}^*(0)$ could be provided from other physical considerations, as when the coefficients are physically meaningful and are known to lie within certain definable confidence bounds. Appendix E shows how the SRLS algorithm can be obtained directly from minimization of an instantaneous cost function: this demonstrates more clearly what the optimization means at the initiation of the algorithm and further justifies *a priori* selection of $\hat{\mathbf{a}}(0)$ and $\mathbf{P}^*(0)$ in these terms.

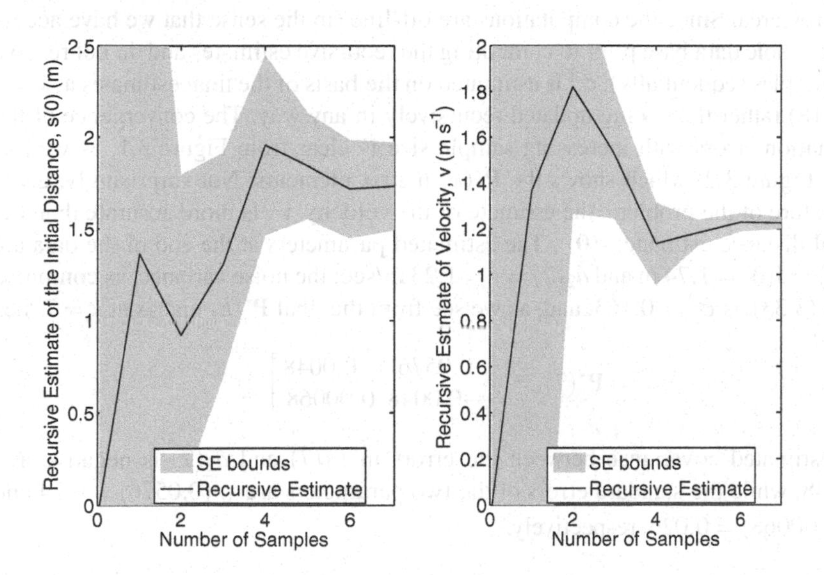

Fig. 3.1 SRLS estimation results for the moving body example with two unknown parameters: the recursive parameter estimates and their estimated standard error bounds.

A simple application of the SRLS algorithm is the estimation of the initial distance $s(0)$ and velocity v in the moving body problem introduced in section 2.2 of chapter 2. The model is now given in equation (2.17) and the results obtained by applying the SRLS algorithm are illustrated in Figures 3.1. Here, the estimates of the

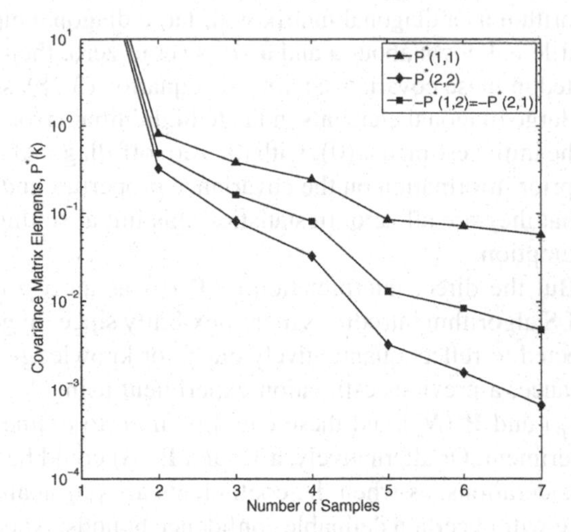

Fig. 3.2 SRLS estimation results for the moving body example with two unknown parameters: the $\mathbf{P}^*(k)$ matrix elements.

two parameters are shown in the two panels, with the *Standard Error* (SE) bounds, computed as the square root of the diagonal elements of the $\mathbf{P}^*(k)$ matrix, shown as the grey area. Since the computations are off-line (in the sense that we have access to the whole data base prior to computing the recursive estimates and do not receive the samples sequentially), $\hat{\sigma}^2$ is estimated on the basis of the final estimates at $k = 7$ ($t = 18$) rather than being updated recursively in any way. The convergence of the estimation errors with increasing sample size is clear from Figure 3.1, as well as from Figure 3.2, which shows the $\mathbf{P}^*(k)$ matrix elements. Not surprisingly, given the nature of the problem, the estimate of the velocity, v , is more accurate than the initial distance estimate, $s(0)$. The estimated parameters at the end of the data are $\hat{a}_1(7) = \hat{s}(0) = 1.74$ m and $\hat{a}_2(7) = \hat{v} = 1.23$ m/sec; the noise variance, as computed from (3.33), is $\hat{\sigma}^2 = 0.163$; and, as we see from the final $\mathbf{P}^*(k)$ matrix at $k = 7$, i.e.

$$\mathbf{P}^*(7) = \begin{bmatrix} 0.0576 & -0.0048 \\ -0.0048 & 0.00068 \end{bmatrix}$$

the estimated covariance between the errors in $\hat{a}_1(7)$ and $\hat{a}_2(7)$ is negative at -0.0048, while the standard errors of the two parameters are $\sqrt(0.0576) = 0.24$ and $\sqrt(0.00068) = 0.026$, respectively.

3.3 Some Cautionary Comments: Multiple Collinearity and Errors-in-Variables

Regression analysis is probably the most widely used yet, at the same time, the most widely abused statistical tool in both the physical and social sciences. Of the many caveats that should be issued to the unwary user of regression analysis, only two will be mentioned here because they are particularly pertinent to the subject of the present book and become important in later chapters concerned with recursive time series analysis.

3.3.1 Multiple collinearity

Although, as pointed out, the regressors or 'independent variables' that compose the vector $\mathbf{x}(k)$ should be linearly independent, it will often be impossible to pose the estimation problem such that there is no correlation at all between the regressors. And if it happens that this correlation is high, so that there is a tendency towards at least 'partial linear dependence'[2], then it can be shown (e.g. Brownlee, 1965) that the estimates will have high estimation error variance, even though the sum of squares of the residuals $\varepsilon(k)$ may be small and indicative of a good fit or 'explanation' of the data. The reader should be wary of this problem, particularly in general practical applications of regression analysis, and should note that a check on the existence of problems of this type is available by using *Multiple Correlation Analysis*: see e.g. Brownlee (1965); Young (1970a).

It is worth noting that the physical effect of multiple correlation problems in RLS and SRLS analysis is for the various recursive estimates to apparently converge slowly and 'wander about', often in sympathy with each other. This is a consequence of the high correlation between regressors tending to drive the data covariance matrix $\mathbf{C}(k) = \mathbf{X}^T(k)\mathbf{X}(k)$ towards singularity, so that the elements of its inverse $\mathbf{P}(k)$ assume correspondingly hiqh values. Given the relationship between $\mathbf{P}(k)$ and the error covariance matrix $\mathbf{P}^*(k)$, we can see that this will imply high parametric estimation error variance and poorly defined parameter estimates.

From a geometric standpoint (see e.g.Young, 1965a) multiple collinearity can also be interpreted as a tendency towards the development of valley-like phenomena in the hypersurface associated with the cost function in the parameter-cost function hyperspace. Thus there is no clearly defined minimum in the surface and the estimates wander along the elongated, valley-like bottom in some partial relationship with each other: this leads to low residual errors but some ambiguity about the parameter values, as indicated by the high estimation error variance. In the case of exact linear dependence there will, of course, be an exact linear relationship between the regressors, a true valley in the cost function hypersurface and complete ambiguity. In theory, the $\mathbf{C}(k)$ matrix will then be singular and $\mathbf{P}(k)$ indeterminate: in practice, however, it turns out that the recursive matrix inversion lemma (3.12) (see section 3.1) will yield an inverse with very large elements indicative of the high level of uncertainty about the resulting parameter estimates. Note that this arises because the matrix inversion lemma can be considered as a 'pseudo-inverse' in the sense of Penrose (1955).

[2] There are, of course, no relative measures of linear dependence and we use this phrase purely for illustrative purposes.

3.3.2 Errors-in-variables, the structural model and instrumental variable estimation

Up to this point, we have only considered a particular form of linear estimation model, namely the regression model in which the variables associated with the unknown parameters are exactly known quantities. In practice, a number of different estimation models are encountered and the techniques described here may need to be modified in some manner. This whole topic is discussed fully in a book by Graybill (1961), in which he treats all of the major possibilities in considerable detail. In the present context, it will suffice to mention one of the most important models met in practical situations, the *structural model*.

In the structural model, the basic relationship between the parameters is still in the form shown in (3.21) but the elements of $\mathbf{x}(k)$ are no longer exactly known quantities and can only be observed in error. In other words, the observed value of $\mathbf{x}(k)$ is $\mathbf{z}(k)$ where

$$\mathbf{z}(k) = \mathbf{x}(k) + \mathbf{e}_z(k) \tag{3.36}$$

and $\mathbf{e}_z(k)$ is an n vector of measurement noise associated with the observation of $\mathbf{x}(k)$. The estimation problems introduced by such 'errors-in-variables' are discussed later in section 7.4 of chapter 7, where it is shown how they affect the estimation of parameters in transfer function models of stochastic, dynamic systems. In the context of regression analysis, however, it is worth noting that the presence of errors-in-variables of the above type induces an asymptotic bias on the parameter estimates which is a function of the signal/noise ratio on the signals and is zero only when there is no noise, i.e., $\mathbf{e}_z(k) = 0 \; \forall \; k$. In other words, no matter how many data are utilized to estimate the parameters in $\mathbf{a}(k)$, the resultant estimates will always be biased away from the true values i.e.,

$$\underset{k\to\infty}{\mathrm{p.lim}}\; \hat{\mathbf{a}}(k) \neq \mathbf{a}(k)$$

where p. lim is the probability in the limit (see Appendix B).

The importance of the asymptotic bias on the parameter estimates will depend upon the application of the regression analysis. If it is being used merely for forecasting the 'dependent variable' $y(k)$ on the basis of the noisy 'independent variables' $\mathbf{z}(k)$, then the results may be acceptable, since it can be shown that the forecast $\hat{y}(k) = \mathbf{z}^T(k)\hat{\mathbf{a}}(k)$ is unbiased in most situations, provided the noise is stationary and retains those statistical characteristics it possessed in the data used to obtain the estimate: see e.g. (Johnston, 1963). But if the parameters or regression coefficients in $\mathbf{a}$ are important in their own right and, for example, have physical significance, then the problem can assume major proportions. Indeed one wonders how many exercises of regression analysis in the social sciences have, unknowingly, been affected by the problem of errors-in-variables and how this, in turn, may have affected the conclusions of the analyst.

There are a number of ways of solving the structural model problem. If the noise statistics are known *a priori*, for example, then there is no real problem since it is possible to compensate directly for the bias, although the estimates can be highly sensitive to errors in the *a priori* assumptions: see e.g. Levin (1964) and the later discussion of errors-in-variables in chapter 7. If non-linear estimation is utilized, then it is possible to exploit more sophisticated procedures, such as maximum likelihood, which provide a good but usually more complicated general approach to the problem. If the simplicity of linear least squares estimation is to be retained, however, there seems to be only one real solution to the structural model problem: this is the method of *Instrumental Variables* (IV).

The IV approach has a rich history in the many scientific disciplines, including statistics, econometrics, control and systems, with the earliest algorithms of this type dating back to the 1950s (Durbin, 1954; Kendall and Stuart, 1961; Young, 1965b, 1966; Mayne, 1967; Wong and Polak, 1967; Young, 1968b, 1970a; Dhrymes, 1970; Young, 1970b, 1976b; Young and Jakeman, 1979b; Jakeman and Young, 1979b; Young and Jakeman, 1980a; Söderström and Stoica, 1983). The basic IV method is an extremely simple technique which retains the single iteration, linear least squares-like solution and is particularly attractive because it does not require detailed *a priori* information on the noise statistics to yield consistent, asymptotically unbiased estimates.

In relation to the linear regression problem, the basic IV approach involves first noting that, in a structural model situation, the normal equations (3.5) must be formulated as

$$\left[\sum_{i=1}^{k} \mathbf{z}(i)\mathbf{z}^T(i)\right]\hat{\mathbf{a}} - \sum_{i=1}^{k} \mathbf{z}(i)y(i) = 0 \tag{3.37}$$

since $\mathbf{x}$ has to be replaced by its noisily measured value $\mathbf{z}$. This solution is then modified, for the purposes of estimation, to yield the following IV normal equations

$$\left[\sum_{i=1}^{k} \hat{\mathbf{x}}(i)\mathbf{z}^T(i)\right]\hat{\mathbf{a}} - \sum_{i=1}^{k} \hat{\mathbf{x}}(i)y(i) = 0 \tag{3.38}$$

so that the IV estimate is obtained as

$$\hat{\mathbf{a}}(k) = \left[\sum_{i=1}^{k} \hat{\mathbf{x}}(i)\mathbf{z}^T(i)\right]^{-1} \sum_{i=1}^{k} \hat{\mathbf{x}}(i)y(i) \tag{3.39}$$

or, in relation to equation (3.24),

$$\hat{\mathbf{a}}(k) = \left[\hat{\mathbf{X}}^T(k)\mathbf{Z}(k)\right]^{-1} \hat{\mathbf{X}}^T(k)\mathbf{y}(k) \tag{3.40}$$

where $\hat{\mathbf{X}}(k)$ is an $k \times n$ matrix with rows $\hat{\mathbf{x}}^T(i)$, $i = 1,2,\ldots,k$; and $\mathbf{Z}(k)$ a similarly dimensioned matrix with elements $\mathbf{z}^T(i)$, $i = 1,2,\ldots,k$.

It is clear that the IV estimate $\hat{\mathbf{a}}(k)$ in (3.40) only exists if the matrix $\hat{\mathbf{X}}^T(k)\mathbf{Z}(k)$ is invertible, so that the instrumental variable vector $\hat{\mathbf{x}}(k)$ must be chosen to ensure that this is the case. Moreover, in order that the IV estimate is 'good' in some sense, it must possess other properties. As might be expected on intuitive grounds, the main requirement, in this regard, is that $\hat{\mathbf{x}}(k)$ should be as highly correlated as possible with the equivalent variables in the 'noise free' vector $\mathbf{x}(i)$; but totally statistically independent of the noise $\mathbf{e}_z(k)$, i.e.,

$$E\{\hat{\mathbf{x}}(i)\mathbf{x}(j)\} \neq 0; \text{ and } E\{\mathbf{x}(i)\mathbf{e}_z(j)\} = 0, \ \forall\, i, j$$

We shall see the reason why the introduction of the IV vector is so effective in combating the effects of the errors-in-variables in chapters 6 to 8, which also show how it is possible to select instrumental variables that are optimal in statistical terms. For the moment, however, it will suffice to note that the *en bloc* solution (3.39) can be written in the form

$$\hat{\mathbf{a}}(k) = \hat{\mathbf{P}}(k)\hat{\mathbf{b}}(k) \tag{3.41}$$

where

$$\hat{\mathbf{P}}(k) = \hat{\mathbf{C}}(k)^{-1} = \left[\sum_{i=1}^{k} \hat{\mathbf{x}}(i)\mathbf{z}^T(i)\right]^{-1} \quad \text{and} \quad \hat{\mathbf{b}}(k) = \sum_{i=1}^{k} \hat{\mathbf{x}}(i)y(i) \tag{3.42}$$

As a result, it is a simple matter to obtain a recursive IV algorithm using an approach similar to that employed in the RLS case discussed in the previous chapter. We will leave the reader to derive these equations, which take the following form,

$$\hat{\mathbf{a}}(k) = \hat{\mathbf{a}}(k-1) + \mathbf{g}(k)[y(k) - \mathbf{z}^T(k)\hat{\mathbf{a}}(k-1)] \tag{IV-1}$$

$$\mathbf{g}(k) = \hat{\mathbf{P}}(k-1)\hat{\mathbf{x}}(k)[1 + \mathbf{z}^T(k)\hat{\mathbf{P}}(k-1)\hat{\mathbf{x}}(k)]^{-1} \tag{IV-2a}$$

$$\text{or} \quad \mathbf{g}(k) = \hat{\mathbf{P}}(k)\hat{\mathbf{x}}(k) \tag{IV-2b}$$

$$\hat{\mathbf{P}}(k) = \hat{\mathbf{P}}(k-1) - \mathbf{g}(k)\mathbf{z}^T(k)\hat{\mathbf{P}}(k-1) \tag{IV-3}$$

It is clear that this basic IV algorithm is very similar to the RLS algorithm, except that alternate $\mathbf{z}(k)$ vectors are replaced by the IV vector $\hat{\mathbf{x}}(k)$

The major problem with the general application of the IV algorithm is that of choosing the instrumental variables themselves: how can we obtain or generate variables with such specific statistical properties? The difficulty of answering this question in any general manner has acted as a strong deterrent to the widespread use of the IV approach in statistics: see Kendall and Stuart (1961). But the question is surprisingly easy to answer in the context of estimating transfer function models for stochastic, dynamic systems, as considered in chapters 6 to 8, which show how the IV algorithm has assumed major significance in time series analysis.

Another question that immediately arises concerns optimality. The RLS algorithm is clearly optimal in the sense of the least squares optimization cost function in (3.5). But, since the IV estimate is obtained simply by a *modification* of the least squares solution to this optimization problem, it is not clear whether or not it is optimal and, if so, in what sense. This question will also be pursued later in chapters 6 to 8.

3.4 Connection with Stochastic Approximation

The scalar recursive algorithm (I) discussed in the last chapter and the multidimensional recursive RLS, SRLS and IV algorithms in the previous sections of this chapter are examples of a more general technique known to mathematicians as a *Stochastic Approximation* (SA) or 'stochastic gradient' procedure. The procedure is approximate in the sense that the noisy data are used recursively to generate better and better 'approximations' or estimates of the unknown parameters; it is stochastic because it is able to handle the random character of the additive noise effects; and finally, the alternative 'stochastic gradient' terminology applies because, as we have seen, it represents the stochastic equivalent of the well known, deterministic, hill climbing procedures, in which the search direction at any point is chosen by reference to the measured gradient of the cost function at that point.

This latter remark is significant because such deterministic gradient techniques have been used both for the optimization of functions in a parameter space (see e.g. Wilde (1964)) and the optimization of functionals in a function space (see any book on mathematical optimization or Wikipedia[3]): they are, in other words, of wide practical significance. Some further background and information on stochastic approximation is given in Appendix C.

3.5 Exercises

1. Write a Matlab m-file script to implement the RLS algorithm: apply this to the moving body example and use the results to obtain an estimate $\hat{\sigma}^2$ of the noise variance σ^2.
2. Modify the RLS m-file script to the SRLS form, using $\hat{\sigma}^2$ from exercise 1 as an *a priori* estimate of the noise variance. Apply this to the moving body example and show that it reproduces the results shown in Figures 3.1 and 3.2.

3.6 Summary

This chapter has shown how the deterministic, recursive least squares algorithm can be derived and converted easily into a stochastic form by making certain simple statistical assumptions about the nature of the observational errors. The resultant

---

[3] e.g. `http://en.wikipedia.org/wiki/Optimization`
then select `Optimization(mathematics)`.

stochastic, recursive least squares regression algorithm, SRLS, is of wide general significance, providing a recursive method for estimating the parameters in the 'general linear regression model' of statistics. But the algorithm, like its *en bloc* counterpart, can be abused if applied outside the range of restrictions inherent in its derivation. In this chapter, two such problem areas have been mentioned briefly: multiple collinearity and errors-in-variables. In the next chapter, we will see how it is possible to resolve another potential problem; namely the difficulty introduced if the parameters, contrary to the assumptions, are time-variable (non-stationary); i.e. they tend to vary in a statistically significant manner over the observation interval.

Chapter 4
Recursive Estimation of Time Variable Parameters in Regression Models

In time series analysis, a series is said to be stationary in the wide sense or stationary to second order if the first and second statistical moments, i.e. the mean, variance and covariance (see Appendix B), are not functions of time. In this book, such a series will be termed simply stationary (although this term is often reserved for processes obeying stricter conditions than these).

A useful property of stationary series is that they are ergodic; in other words, the mean, variance and covariance can be estimated by averages over time, as in the case of the Walgett rainfall data in Chapter 2, rather than by defining them in terms of ensemble averages over a number of random realizations of the series. But if, as in the case of the Walgett data, the first two statistical moments may not be constant and the series is *non-stationary*, then these ordinary temporal averages are not appropriate and alternative measures are required.

Non-stationarity, in this general sense, can arise for many different reasons and can be explained in many different ways. In standard statistical texts on time series analysis however, discussion is often restricted to non-stationarity in the form of stochastically variable means or 'trends' in data. A well known example is the book by Box and Jenkins (1970), where such phenomena are ascribed to a form of non-stationarity described by Yaglom (1955) where the non-stationary characteristics are assumed to arise from the presence of one or more integrators in the stochastic part of signal generation process. This will be discussed in more detail later, but the simplest discrete-time example of such a process is the *Random Walk* (RW) sequence,

$$\xi(k) = \xi(k-1) + e(k-1) \tag{4.1}$$

or in z transfer function terms (cf. the transfer function of the digital integrator in Figure 2.3, Chapter 2, and Appendix B)

$$\xi(k) = \frac{z^{-1}}{1-z^{-1}}e(k) = \frac{1}{z-1}e(k) \tag{4.2}$$

where $e(k)$ is a zero mean, serially uncorrelated white noise sequence. This is sometimes termed an integrated process of order unity, or I(1), with higher order random walk models defined accordingly. For example the *Integrated Random Walk* (IRW) or I(2) process takes the form,

$$\begin{aligned} \xi(k) &= \xi(k-1) + \nabla\xi(k-1) \\ \nabla\xi(k) &= \nabla\xi(k-1) + e(k-1) \end{aligned} \tag{4.3}$$

This can be considered in the transfer function form,

$$\xi(k) = \frac{z^{-2}}{(1-z^{-1})^2} e(k) \tag{4.4}$$

or the state space form (see Appendix B),

$$\begin{bmatrix} \xi(k) \\ \nabla\xi(k) \end{bmatrix} = \begin{bmatrix} 1 & 1 \\ 0 & 1 \end{bmatrix} \begin{bmatrix} \xi(k-1) \\ \nabla\xi(k-1) \end{bmatrix} + \begin{bmatrix} 0 \\ 1 \end{bmatrix} e(k) \tag{4.5}$$

Note here that the second state variable in (4.5) is denoted by $\nabla\xi(k)$ because, from the first equation in the IRW model, $\nabla\xi(k) = \xi(k+1) - \xi(k)$ is defined as the change in $\xi(k)$ over one sampling instant Δt (or Δs if the data are measurements at spatial locations), so that,

$$\frac{\nabla\xi(k)}{\Delta t} = \frac{\xi(k+1) - \xi(k)}{\Delta t}$$

measures the discrete rate of change of $\xi(k)$. Such random walk models are considered further in section 4.2.

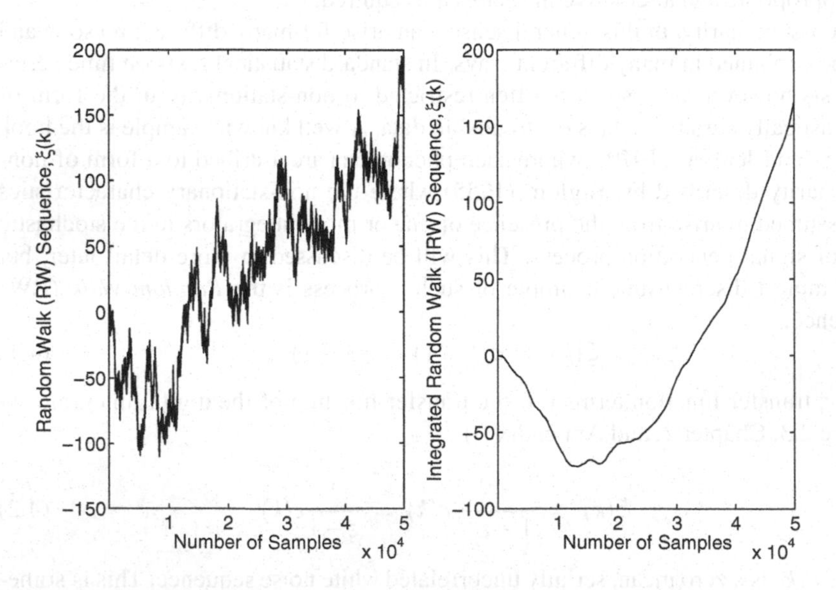

Fig. 4.1 Typical RW (left panel) and IRW (right panel) sequences.

The transfer function (4.2) will be recognized from Figure 2.3 as a discrete summer or integrator. Clearly sequences such as $\xi(k)$ will 'wander' with stochastically variable mean value, as shown by the typical examples in Figure 4.1, but the IRW sequence is much smoother than the RW sequence because of the additional integrator (summer) in the second equation for $\nabla\xi(k)$. As a result, the above models are good candidates for the description of such behaviour in time series and this is the motivation for their use by Box and Jenkins. However, their method of handling such phenomena in time series analysis, namely to difference the series a requisite number of times, is questionable because of the noise amplification that is inherent in the differencing operation. This is considered further in section 4.2, where an alternative approach is proposed that exploits random walk models such as (4.1) and (4.3).

But Yaglom-type non-stationarity is not, in itself, an entirely appropriate description for all forms of non-stationary behaviour. One embodiment of non-stationarity which is particularly interesting and relevant in many areas of science and engineering is parametric time-variability. This applies in those cases where the model of the time series data can only be characterized adequately by parameters which vary over time in some significant manner. For instance, linear, small perturbation, dynamic models are used very successfully in the design of autostabilization systems for aircraft provided due allowance is made for systematic changes in the dynamic characteristics, as evidenced by the variation in the model parameters as a function of the flight condition (Young, 1979c). And in various industrial processes, model parameters are known to vary because of factors such as catalyst ageing in chemical processes, changing carbon content in steel when using the basic oxygen furnace, changing characteristics in paper making plant, and the changing nature of a river catchment area in hydrological modelling. From the theoretical standpoint, we might expect simple models of complex, nonlinear phenomena such as these to exhibit parametric non-stationarity: it is well known, for instance, that the linearization of nonlinear models often results in linear, time-varying model structures. The systems analyst is well aware of these problems and sees in the estimates of parametric variation possible information which could lead to the identification of the nonlinearities (e.g., Young, 1978). This is discussed further in Chapter 11.

These latter remarks apply particularly in the case of 'badly defined' systems (e.g., Young, 1978; Young et al., 1978; Kalman, 1979) such as those encountered in environmental and socio-economic systems analysis. In the socio-economic case, for example, the possibility of behavioural characteristics changing through time is often a topic of discussion; as, for instance, when it was suggested that the so-called 'Phillips Curve' for wage inflation as a function of unemployment (Phillips, 1958) was not applicable during periods of strong union bargaining power. In such cases, the inference is that the (possibly nonlinear) relationships are themselves changing through the passage of time as a function of the changing socio-economic attitudes and conditions.

In all of these examples, the hypothesis of parametric non-stationarity (or its inverse) needs to be tested in some manner; and it is here that recursive estimation has a potentially important role to play. Conventional statistical hypothesis testing can be utilized, as we will discuss later. But the recursive parameter estimates themselves are clearly of primary utility in providing information not only on the existence of non-stationarity, but also on the possible nature of the parametric variations (Young, 1975). In this Chapter, therefore, we will explore how the recursive algorithms of the previous Chapter can be extended in various ways to handle parametric variation; and how they can, in this manner, provide a valuable additional weapon in the armoury of the statistician, econometrician, environmental scientist and systems analyst.

4.1 Shaping the Memory of the Estimator

If a common factor $1/k$ is introduced into equation (3.7), we obtain the following least squares normal equation,

$$\left[\frac{1}{k}\sum_{i=1}^{k}\mathbf{x}(i)\mathbf{x}^T(i)\right]\hat{\mathbf{a}} - \frac{1}{k}\sum_{i=1}^{k}\mathbf{x}(i)y(i) = 0 \tag{4.6}$$

It is now clear that the data processing can be considered as the finite-time averaging of the various products and cross products in $\mathbf{x}(i)\mathbf{x}^T(i)$ and $\mathbf{x}(i)y(i)$. This means that, in arriving at the estimates, all data are weighted equally over the observation interval of k samples and so there is an implicit assumption that the parameters remain constant during this period. In order to allow for possible parametric variation, it is necessary to remove the effects of 'obsolete' data in some manner. Since the conventional, time-invariant parameter solution (4.6) has an unrestricted memory of k samples, then such a procedure can be visualized as restricting or 'shaping' the memory of the estimator.

The two most obvious procedures for shaping the memory of an estimation scheme are: first, to base an estimation on only the most recent portion of the data, say s samples; and second to weight the data exponentially into the past with an exponential 'fading memory', characterized by a decay time constant T_e (see later). Both have been employed in conventional *en bloc* data processing and estimation, and can also be applied to recursive algorithms. They can best be visualized as moving window or weighting functions, as shown in Figure 4.2.

4.1.1 The moving Rectangular Window (RW)

From the top panel of Figure 4.2, it is clear that the procedure in this case is to choose an observation period of s samples which, on the basis of the conditions applying in the particular application under consideration, will provide estimates of the desired accuracy: typically this might require, for example, that the error covariance matrix obtained for s samples, i.e. $P_s^* = \sigma^2 P_s$, is acceptable in some sense. The estimation

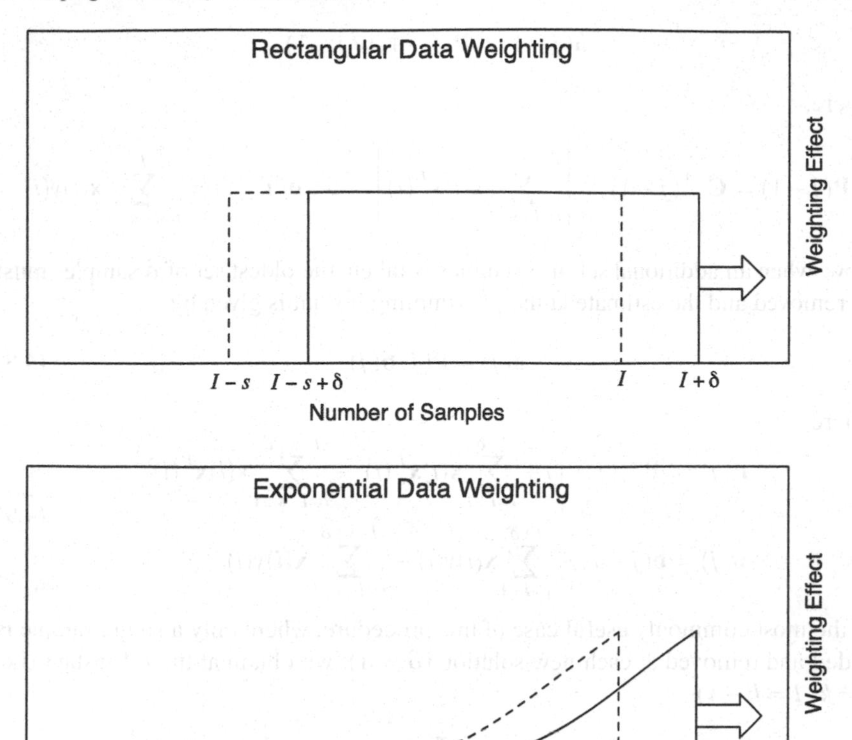

Fig. 4.2 Moving rectangular and exponentially-weighted-past weighting functions.

equations can then be solved for overlapping or adjoining intervals of this length. In other words, after each solution based on s samples is completed, δ new samples are taken, the oldest δ samples are removed and a new solution is obtained.

If the least squares estimation problem is considered in this manner then, after $j-1$ estimation intervals each of s samples, the normal equations can be written (cf. equation (3.8))

$$\left[\sum_{i=I-s+1}^{I} \mathbf{x}(i)\mathbf{x}^T(i)\right] \hat{\mathbf{a}} = \sum_{i=I-s+1}^{I} \mathbf{x}(i)y(i) \tag{4.7}$$

where $I = s + (j-2)\,\delta$ is the total number of samples up to the $(j-1)^{th}$ interval. After $j-1$ estimation intervals, the solution to (4.7) can be written as

$$\hat{\mathbf{a}}(j-1) = \mathbf{P}(j-1)\,\mathbf{b}(j-1)$$

where,

$$\mathbf{P}(j-1) = \mathbf{C}^{-1}(j-1) = \left[\sum_{i=I-s+1}^{I} \mathbf{x}(i)\mathbf{x}^T(i)\right]^{-1} \quad ; \quad \mathbf{b}(j-1) = \sum_{i=I-s+1}^{I} \mathbf{x}(i)y(i)$$

Now, when an additional set of δ samples is taken, the oldest set of δ samples must be removed and the estimate at the jth sampling instant is given by

$$\hat{\mathbf{a}}(j) = \mathbf{P}(j)\,\mathbf{b}(j) \tag{4.8}$$

where

$$\begin{aligned} \mathbf{P}(j) &= [\mathbf{P}^{-1}(j-1) + \sum_{i=I+1}^{I+\delta} \mathbf{x}(i)\mathbf{x}^T(i) - \sum_{i=I-s+1}^{I-s+\delta} \mathbf{x}(i)\mathbf{x}^T(i)]^{-1} \\ \mathbf{b}(j) &= \mathbf{b}(j-1) + \sum_{i=I+1}^{I+\delta} \mathbf{x}(i)y(i) - \sum_{i=I-s+1}^{I-s+\delta} \mathbf{x}(i)y(i) \end{aligned} \tag{4.9}$$

In the most commonly useful case of this procedure, where only a single sample is added and removed at each new solution ($\delta = 1$), we obtain at the kth instant (i.e. $j = k$; $I = k-1$)

$$\begin{aligned} \mathbf{P}(k) &= [\mathbf{P}(k-1) + \mathbf{x}(k)\mathbf{x}^T(k) - \mathbf{x}(k-s)\mathbf{x}^T(k-s)]^{-1} \\ \mathbf{b}(k) &= \mathbf{b}(k-1) + \mathbf{x}(k)y(k) - \mathbf{x}(k-s)y(k-s) \end{aligned} \tag{4.10}$$

Employing a similar approach to that used in previous Chapters, it is straightforward to obtain from these relationships and (4.8) a recursive, *Rectangularly-Weighted-Past* (RWP) algorithm, which consists of the following two steps at each recursion,

Receipt of new data at k^{th} instant

$$\hat{\mathbf{a}}'(k) = \hat{\mathbf{a}}(k-1) + \mathbf{g}(k)\,\{y(k) - \mathbf{x}^T(k)\,\hat{\mathbf{a}}(k-1)\} \tag{RWP-1}$$

$$\mathbf{g}(k) = \mathbf{P}(k-1)\mathbf{x}(k)[1 + \mathbf{x}^T(k)\mathbf{P}(k-1)\mathbf{x}(k)]^{-1} \tag{RWP-2}$$

$$\mathbf{P}'(k) = \mathbf{P}(k-1) - \mathbf{g}(k)\,\mathbf{x}^T(k)\,\mathbf{P}(k-1) \tag{RWP-3}$$

Removal of data received at $(k-s)^{th}$ instant

$$\hat{\mathbf{a}}(k) = \hat{\mathbf{a}}'(k) + \mathbf{g}(k)\,\{y(k-s) - \mathbf{x}^T(k-s)\,\hat{\mathbf{a}}'(k-1)\} \tag{RWP-4}$$

$$\mathbf{g}(k) = -\mathbf{P}'(k)\mathbf{x}(k-s)[1 + \mathbf{x}^T(k-s)\mathbf{P}'(k)\mathbf{x}(k-s)]^{-1} \tag{RWP-5}$$

$$\mathbf{P}(k) = \mathbf{P}'(k) - \mathbf{g}(k)\,\mathbf{x}^T(k-s)\,\mathbf{P}'(k) \tag{RWP-6}$$

4.1.2 The moving Exponentially-Weighted-Past (EWP) window

Exponential-Weighting-into-the-Past can be introduced into the least squares problem formulation by replacing the finite-time averaging operations in the moving rectangular window algorithm by EWP averaging. This is equivalent to considering an EWP least squares cost function $\mathfrak{J}_{EWP}$ of the form

$$\mathfrak{J}_{EWP} = \sum_{i=1}^{k} [y(i) - \mathbf{x}^T(i)\hat{\mathbf{a}}]^2 \alpha(k-i) \tag{4.11}$$

where $0 < \alpha < 1.0$ is an exponential 'forgetting' factor related to the time constant T_e of the exponential weighting by the expression $\alpha = \exp(-\Delta t/T_e)$, where Δt is the sampling interval in time units appropriate to the application. Two typical exponential decay functions with $\alpha = 0.7\ (T_e = 2.8\Delta t)$ and $\alpha = 0.9\ (T_e = 9.5\Delta t)$ are shown in Figure 4.3. With $\alpha = 1.0$, $\mathfrak{J}_{EWP}$ becomes, of course, the usual least squares cost function $\mathfrak{J}$ in (3.5).

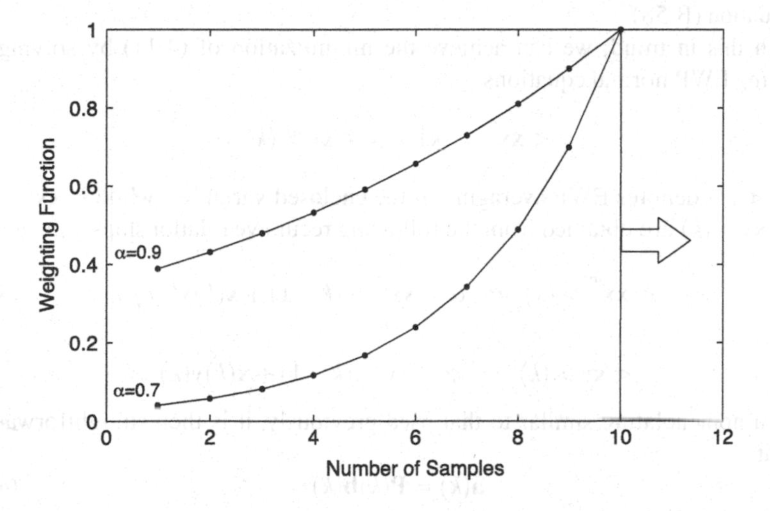

Fig. 4.3 Typical EWP weighting functions for $\alpha = 0.9$ and $\alpha = 0.7$ parameter values.

In a recursive context, the easiest way of incorporating these modifications is to note that a first order, discrete-time, dynamic system has an exponential unit impulse response and, therefore, an exponential weighting effect on past inputs. This can be demonstrated quite simply by considering the following discrete-time filter with input $u(k)$ and output $x(k)$

$$x(k) = \alpha x(k-1) + u(k) \tag{4.12}$$

In z^{-1} transform transfer function terms (see Chapter 3 and Appendix 1) this can be written,

$$x(k) = \frac{1}{1-\alpha z^{-1}} u(k)$$

On long division of 1 by $1 - \alpha z^{-1}$, this yields

$$x(k) = (1 + \alpha z^{-1} + \alpha^2 z^{-2} + \cdots + \alpha^i z^{-i} + \cdots + \alpha^\infty z^{-\infty})\, u(k)$$
$$x(k) = u(k) + \alpha u(k-1) + \alpha^2 u(k-2) + \ldots + \alpha^i u(k-i) + \ldots + \alpha^\infty u(k-\infty)$$

If $u(k)$ is a unit impulse, i.e. $u(k) = 1.0$ for $k = 0$ and $u(k) = 0$ for $k > 0$, then it is clear that $x(0) = 1.0$; $x(1) = \alpha^2$; ... ; $x(\infty) = \alpha^\infty$, as required. Referring to section B.3.7.1 of Appendix B, the time constant for this exponential decay is defined by $T_e = -\Delta t / \log_e \alpha$, where Δt is the sampling interval of the time series data: see equation (B.59). Considering the time constant in terms of the number of samples (i.e. when $\Delta t = 1.0$), a first order approximation is $T_e = (1-\alpha)^{-1}$ samples: see equation (B.58).

With this in mind, we can achieve the minimization of (4.11) by solving the following EWP normal equations

$$< \mathbf{x}\mathbf{x}^T > (k)\, \hat{\mathbf{a}} = < \mathbf{x}y > (k)$$

where $< . >$ denotes EWP averaging of the enclosed variables, while $< \mathbf{x}\mathbf{x}^T > (k)$ and $< \mathbf{x}y > (k)$ are obtained from the following recursive relationships: *cf.* equation (4.12):

$$< \mathbf{x}\mathbf{x}^T > (k) = \alpha < \mathbf{x}\mathbf{x}^T > (k-1) + \mathbf{x}(k)\mathbf{x}^T(k) \tag{4.13}$$

and

$$< \mathbf{x}y > (k) = \alpha < \mathbf{x}y > (k-1) + \mathbf{x}(k)y(k)$$

Using a nomenclature similar to that used previously, it is then straightforward to see that

$$\hat{\mathbf{a}}(k) = \mathbf{P}(k)\mathbf{b}(k) \tag{4.14}$$

where

$$\mathbf{P}(k) = [< \mathbf{x}\mathbf{x}^\mathbf{T} > (k)]^{-1};\ \mathbf{b}(k) = < \mathbf{x}y > (k)$$

so that

$$\begin{aligned} \mathbf{P}(k) &= [\alpha \mathbf{P}^{-1}(k-1)] + \mathbf{x}(k)\mathbf{x}^\mathbf{T}(k)]^{-1} \\ \mathbf{b}(k) &= [\alpha \mathbf{b}(k-1)] + \mathbf{x}(k)y(k)] \end{aligned} \tag{4.15}$$

The relationships (4.14) and (4.15) can be converted into the fully recursive form using an identical approach to that employed for the RLS algorithm of Chapter 3, but taking care to carry the scalar α through the calculations. These recursive equations, as presented below, are one form of the EWP algorithm and other forms are possible, depending upon the exact nature of the filter algorithm. But while superficially dissimilar, these other forms tend to perform in a similar manner: for example, the 'unity steady state gain' filter or its equivalent is sometimes preferred (Bray et al., 1965; Young, 1969a,b, 1970a).

$$\hat{\mathbf{a}}(k) = \hat{\mathbf{a}}(k-1) + \mathbf{g}(k)\{y(k) - \mathbf{x}^T(k)\hat{\mathbf{a}}(k-1)\} \quad \text{(EWP-1)}$$

$$\mathbf{g}(k) = \mathbf{P}(k-1)\mathbf{x}(k)[\alpha + \mathbf{x}^T(k)\mathbf{P}(k-1)\mathbf{x}(k)]^{-1} \quad \text{(EWP-2a)}$$

$$\text{or } \mathbf{g}(k) = \mathbf{P}(k)\,\mathbf{x}(k) \quad \text{(EWP-2b)}$$

$$\mathbf{P}(k) = \frac{1}{\alpha}\{\mathbf{P}(k-1) - \mathbf{g}(k)\mathbf{x}^T(k)\mathbf{P}(k-1)\} \quad \text{(EWP-3)}$$

The EWP weighting introduced by α leads to an algorithmic memory that dies away into the past in an exponential fashion, as defined by α and the associated time constant T_e. In other words, the EWP algorithm progressively reduces or 'discounts'[1] the importance attached to old data. Unlike the RWP algorithm, therefore, the memory is not a defined number of samples: the data are 'forgotten' gradually and it is normally the practice to define the data memory by T_e or, in terms of the number of samples $(1-\alpha)^{-1}$ (see above).

In practice, it has been found advisable to replace the constant α by a variable $\alpha(k)$ which is initially smaller than α but approaches it asymptotically. In this manner, the exponential decay time constant and, therefore, the fading memory is initially small and grows towards the specified length $T_e = -\Delta t/\log_e \alpha$ as time progresses and more samples are processed. This aids initial convergence by ensuring that the effects of initial conditions on the filters (4.13) are quickly removed: when α is constant, these effects themselves decay with a time constant T_e and this can slow convergence considerably.

One possible definition of $\alpha(k)$ is the following,

$$\alpha(k) = \lambda_0\,\alpha(k-1) + (1-\lambda_0)\,\alpha \quad (4.16)$$

with typical values of $\alpha(0)$ and λ_0 being 0.95 and 0.99 respectively. This is simply a 'unity steady state gain' filter with time constant $T_i = -\Delta t/\log_e \lambda_0$ and input α: the reader can verify that the output $\alpha(k)$ rises exponentially with a time constant T_i from the initial condition $\alpha(0)$ to a steady state level of α (the solution of (4.16) when $\alpha(k) = \alpha(k-1)$. An alternative is to use the following algorithm for $\alpha(k)$,

$$\alpha(k) = 1 - \gamma(k)$$
$$\gamma(k) = \frac{1-\alpha}{1-\alpha^{k+1}} \quad (4.17)$$

[1] 'Discounting' is the term used in the statistics and econometrics literatures: see e.g. West and Harrison (1989), page 59 *et seq*, where it is discussed in a Bayesian context; and Harvey (1989), page 25 *et seq*.

This is a derivation of the scheme used successfully by Bray et al. (1965) and Young (1970a) and is rather simpler to use than (4.16) since it does not require the specification of any additional parameters, such as $\alpha(0)$ and λ_0 in (4.16). In this case $\alpha(k)$ approaches α with increasing k starting from an initial condition $\alpha/(1+\alpha)$, although it approaches it rather slowly.

One EWP algorithm that avoids certain 'wind-up' problems that are sometimes encountered with high noise/signal ratios is the *Directional Forgetting* (DF) algorithm suggested by Kulhavy (1987), which is used in comparison with the EWP algorithm in a later simulation example. This algorithm takes the following form:

$$\hat{\mathbf{a}}(k) = \hat{\mathbf{a}}(k-1) + \mathbf{g}(k)\{y(k) - \mathbf{x}^T(k)\hat{\mathbf{a}}(k-1)\} \qquad \text{(DF-1)}$$

$$\mathbf{g}(k) = \mathbf{P}(k-1)\mathbf{x}(k)[r(k-1)^{-1} + \mathbf{x}^T(k)\mathbf{P}(k-1)\mathbf{x}(k)]^{-1} \qquad \text{(DF-2)}$$

$$\mathbf{P}(k) = \mathbf{P}(k-1) - \mathbf{g}(k)\mathbf{x}^T(k)\mathbf{P}(k-1) \qquad \text{(DF-3)}$$

$$r(k) = \alpha - \left[\mathbf{x}^T(k+1)\mathbf{P}(k)\mathbf{x}(k+1)\right]^{-1}(1-\alpha) \qquad \text{(DF-4)}$$

Here, the scalar α is similar to the forgetting factor used above: if $\alpha = 1.0$ then the normal infinite memory RLS algorithm is operative.

The EWP and DF algorithms only differ from the RLS algorithm by their inclusion of the forgetting factor α, which must be specified by the user. This is our first encounter of what will be termed a 'hyper-parameter', to differentiate it from the parameters being estimated in the model. But it is, nonetheless, a parameter that has to be specified or estimated in some manner. How this may be accomplished is discussed later in section 4.5.3 of this Chapter.

It is easy to see that EWP algorithms with time varying exponential weighting, such as those obtained with the definitions (4.16) and (4.17) for $\alpha(k)$, can be very useful in practice. A case in question is the recursive estimation of the observational error variance $\hat{\sigma}^2$ mentioned in earlier Chapters. If this estimation is based on the recursive residual or innovation sequence $\varepsilon(k) = y(k) - \mathbf{x}^T(k)\hat{\mathbf{a}}(k-1)$, then it is clear that, since the initial $\varepsilon(k)$ values will be influenced by the relatively poor estimates obtained when little data have been processed, the estimate may not be particularly good in comparison with the *en bloc* estimate (3.33). But, if the following EWP algorithm is used in place of (2.16), with $\alpha(k)$ defined by (4.17), for example, then the 'poor' early estimates $\varepsilon(k)$, for small k, will be forgotten exponentially,

$$\begin{aligned}\hat{\sigma}^2(k) &= \hat{\sigma}^2(k-1) + p(k)\left[\varepsilon^2(k) - \hat{\sigma}^2(k-1)\right] \\ p(k) &= \frac{1}{\alpha(k)}\left[p(k-1) - \frac{p^2(k-1)}{\alpha(k)+p(k-1)}\right]\end{aligned} \qquad (4.18)$$

And if α is chosen as unity, then this algorithm will forget the poor initial estimates but progressively utilize later data with equal weighting; i.e. it will convert to the normal least squares algorithm. In this case, the estimate will not be as good as that obtained off-line, but we would at least have a reasonable on-line estimator.

The left hand panel of figure 4.4 shows the results obtained by applying the above recursive EWP algorithm, with $\alpha = 0.9$, to the Walgett time series (full line). Also shown are the standard RLS estimates obtained in Chapter 2, as well as the memory parameter $\alpha(k)$. The recursive estimate of the residual variance obtained from equation (4.18) is shown in the right hand panel (full line), where it is compared with the squared residuals $\varepsilon^2(k)$ on which it is based.

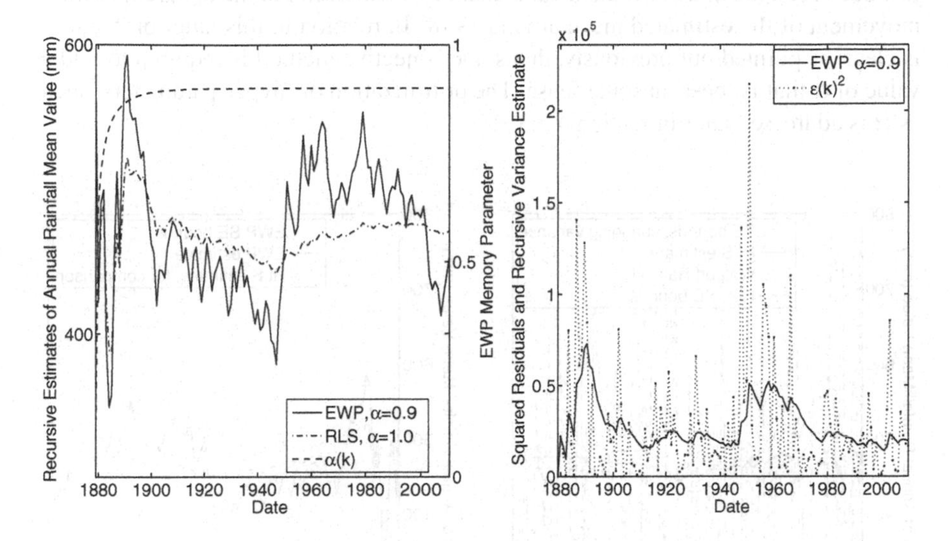

Fig. 4.4 Walgett rainfall mean value estimation: results obtained by RLS and recursive EWP estimation of the mean, together with the EWP memory parameter $\alpha(k)$ (left panel); the squared residuals and recursive EWP estimate of the variance are shown in the right panel.

The results shown in Figure 4.4 are deterministic because we do not have estimates of the uncertainty associated with them. However, following from the analysis in the previous Chapter 3, it is clear that, *if the statistical assumptions about the measurement noise $e(k)$ are satisfied*, then the covariance matrix $\mathbf{P}(k)^*$ associated with the recursive EWP estimate can be obtained from (3.28) of Chapter 3, with $\mathbf{P}(k)$ defined in the EWP algorithm (EWP-3) where, in this case, $\mathbf{P}^*(k) = p^*(k)$ is a scalar. Then, the standard error on the EWP estimate of the mean is given by $\sqrt{p^*(k)}$. The results of this analysis are shown in Figure 4.5, where the left hand panel is a plot of the recursive RLS estimates; and those in the right hand panel are the recursive EWP estimates for $\alpha = 0.9$. In the left hand panel, the standard error bounds, computed as $\pm\sqrt{p^*(k)}$, are shown as the dashed lines for the RLS estimate assuming a constant variance; and as the grey area when the variance is estimated recursively using (4.18). In the right hand panel, only the latter standard error bounds are plotted in grey but the recursive RLS estimate is given as the dash-dot line for comparison.

Of course, the adequacy of these statistical results needs to be checked in relation to the assumptions on which they are based. The autocorrelation function confirms that the residuals are serially uncorrelated; and given the standard deviation of the recursive residuals is 162.9, the mean value of -2.04 is very small, with the Student t-test (http://en.wikipedia.org/wiki/Student's-t-distribution) suggesting that it is insignificantly different from zero. But, there are two problems with generating these standard statistical results: first, a histogram of the residuals suggests that they are *not* normally distributed and this is confirmed, for example, by the Jarque-Bera (Jarque and Bera., 1987) test (http://en.wikipedia.org/wiki/Jarque-Bera); second, the recursive EWP results depend upon the value of the forgetting factor α and this has just been selected at a convenient value that draws attention to the apparent upward movement of the estimated mean around 1948. In relation to this latter problem, it is clear, as pointed out previously, that some objective method is required to find a value of α that is 'best' in some sense. The optimization of 'hyper-parameters' such as α is addressed later in section 4.5.3.

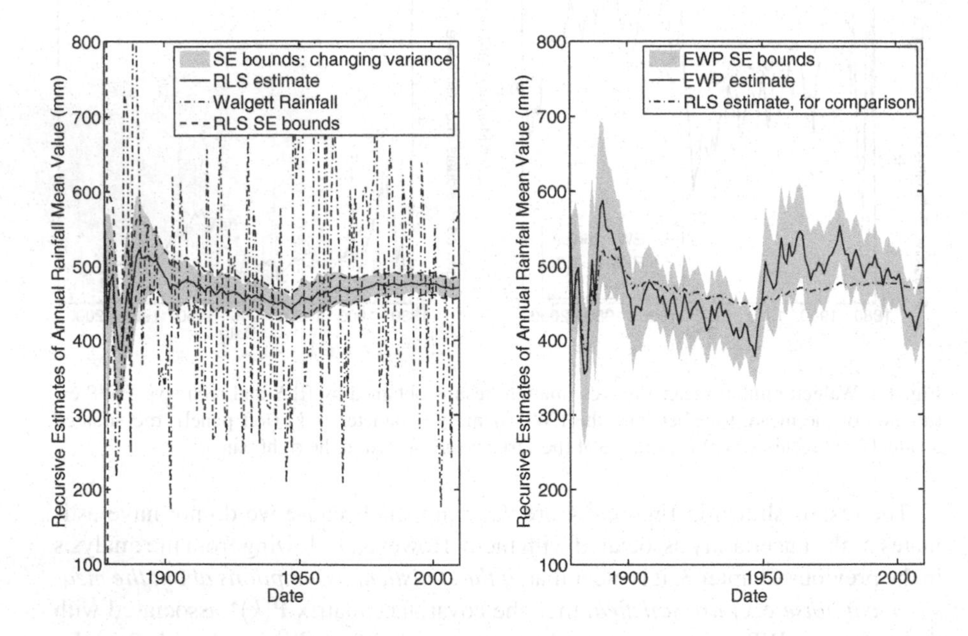

Fig. 4.5 Walgett rainfall mean value estimation: recursive stochastic estimation results obtained with RLS, incorporating changing residual variance estimation (left panel); and the heteroscedastic recursive EWP estimate, incorporating changing residual variance estimation (right panel).

Finally, it is interesting to investigate the effect of the fading memory modifications on the RLS estimator. In the case of rectangular data weighting, the effect is immediately obvious: $p(k)$ is strictly decreasing during the period when the first s samples are being processed but, thereafter, it remains constant since the additions to $p(k)$ from (RWP-6) compensate exactly for the reductions arising from (RWP-3); indeed, in this simple case, the former is redundant, since (RWP-3) could be used for the first s samples and then $p(k)$ could be frozen at its $p(s)$ value, thereafter.

In the EWP situation, the recursive relationship for $p(k)$ is given in (4.18). Once again $p(k)$ is no longer a strictly decreasing function of sample size: in the steady state $(p(k) = p(k-1) = p(\infty))$ the solution of (4.18) is given by

$$p(\infty) = \frac{1}{\alpha}\left[p(\infty) - \frac{p^2(\infty)}{\alpha + p(\infty)}\right]$$

so that

$$p(\infty)\left\{p(\infty) + (\alpha - 1)\right\} = 0 \tag{4.19}$$

and the non zero solution is $p(\infty) = 1 - \alpha$. In other words, for $p(0) > 1 - \alpha$, $p(k)$ reduces from its initial condition to a steady state solution of $1 - \alpha$ which, as might be expected, is zero only for $\alpha = 1$ corresponding to the 'infinite' time constant and unlimited memory. For $1 - \alpha > p(0) > 0$, $p(k)$ increases slowly to again reach a steady state value of $1 - \alpha$.

4.2 Modelling the Parameter Variations

In practice, shaping the memory of the estimation algorithm can prove to be a somewhat inflexible approach to the problem of estimating time variable parameters. In the RWP and EWP algorithms, for example, the analyst can only specify a single parameter (the memory length s or the exponential forgetting factor α) in order to 'match' the algorithm to the specific time-variable parameter model under investigation. If there happened to be *a priori* information available which suggested that certain of the parameters varied at different rates or, indeed, if it was known that only some of the parameters varied while others were constant, then it would be necessary to modify the algorithm. While this is possible, it is not a very elegant solution. This sub-section describes a much more convenient approach that has the advantage of allowing for the maximum likelihood optimization of certain 'hyper-parameters' that define the expected rate of change and, therefore, the memory of the algorithm, for each model parameter.

The statistical interpretation of the least squares equations discussed earlier suggests a more flexible alternative to the RWP and EWP approaches to the problem of detecting parameter variations which involves modelling the parameter variations as simple stochastic processes, such as the random walk. Implicit in the RLS algorithm for the estimation of a regression model with assumed constant parameters is the equation

$$\mathbf{a}(k) = \mathbf{a}(k-1) \quad \forall\, k \tag{4.20}$$

This suggests that one alternative approach to TVP estimation is to relax this assumption in some manner. The simplest possibility in this regard is when it is assumed that each element of the parameter vector may change by some small, random amount between samples. For example

$$\mathbf{a}(k) = \mathbf{a}(k-1) + \boldsymbol{\eta}(k-1) \qquad \boldsymbol{\eta}(k) = \mathcal{N}(\mathbf{0}, \mathbf{Q}_a) \tag{4.21}$$

where $\boldsymbol{\eta}(k) = [\eta_1\ \eta_2, \ldots, \eta_n]^T$ is an $n \times 1$ vector of serially-independent and normally distributed random variables with zero mean and covariance matrix $\mathbf{Q}_a$ (from hereon, such a noise vector will be referred to simply as a 'white noise vector'), i.e.

$$E\{\boldsymbol{\eta}(k)\} = \mathbf{0}; \quad E\{\boldsymbol{\eta}(k)\boldsymbol{\eta}(j)^T\} = \mathbf{Q}_a\delta_{kj}$$

where δ_{kj} is the Kronecker delta function (see section B.3, Appendix B) This will be recognized as the vector equivalent of the scalar random walk (4.1) model introduced earlier.

Another, fairly simple assumption is that the parameter vector may change in a random manner about some specified mean value $\bar{\mathbf{a}}$, i.e.,

$$[\mathbf{a}(k) - \bar{\mathbf{a}}] = \boldsymbol{\beta}[\mathbf{a}(k-1) - \bar{\mathbf{a}}] + \boldsymbol{\eta}(k-1) \tag{4.22}$$

where, $\bar{\mathbf{a}}$ is the mean value about which the parameter is assumed to vary and

$$\boldsymbol{\beta} = \begin{bmatrix} \beta_1 & 0 & 0 & 0 \\ 0 & \beta_2 & 0 & 0 \\ \cdots & \cdots & \cdots & \cdots \\ 0 & 0 & 0 & \beta_n \end{bmatrix}$$

and, for stability, $0 < \beta_i < 1$ for $i = 1, 2, \ldots, n$.

The additive random component equations (4.21) and (4.22) are simple examples of a general, linear *Gauss-Markov* (GM) process: a stochastic difference equation that has the following form (see Appendix B and Bryson and Ho (1969))

$$\mathbf{a}(k) = \mathbf{A}\mathbf{a}(k-1) + \mathbf{D}\boldsymbol{\eta}(k-1) \qquad \boldsymbol{\eta}(k) = \mathcal{N}(\mathbf{0}, \mathbf{Q}_a) \tag{4.23}$$

where $\mathbf{A}$ is an $n \times n$ transition matrix, $\mathbf{D}$ is a $n \times m$ input matrix, both of which may be time variable but are assumed constant in the present context. If a parameter variation model of this general form (4.23) is available, it clearly provides additional *a priori* information which can be used to advantage when attempting to estimate the parameter vector $\mathbf{a}(k)$. The purpose of this section is to demonstrate how this *a priori* assumed information may be used to construct 'dynamic' least squares algorithms capable of tracking unknown parametric variations. The very real problem of obtaining prior information in this form is discussed later.

With the above assumptions in mind, consider the problem of estimating the parameter vector $\mathbf{a}(k)$ in a nth order linear regression model of the form

$$y(k) = \mathbf{x}^T(k)\mathbf{a}(k) + e(k) \qquad e(k) = \mathcal{N}(0, \sigma^2) \tag{4.24}$$

where now $\mathbf{a}(k)$ is a $n \times 1$, time variable parameter vector which can be described by the stochastic difference equation (4.23). This model allows us to make *a priori* updates $\hat{\mathbf{a}}(k|k-1)$ and $\mathbf{P}^*(k|k-1)$ to the estimate $\hat{\mathbf{a}}(k-1)$ and covariance matrix $\mathbf{P}^*(k-1)$ obtained at the previous $(k-1)^{th}$ instant from the recursive regression algorithm (RLS) of Chapter 3. Here, $\hat{\mathbf{a}}(k|k-1)$ denotes the *a priori* estimate of the parameter vector at the kth sampling instant given information only to the $(k-1)^{th}$ instant.

In order to show how these a priori updates can be obtained, note that, based on the statistical properties of $\boldsymbol{\eta}(k)$,

$$E\{\mathbf{a}(k)\} = \mathbf{A}\mathbf{a}(k-1) \tag{4.25}$$

It is intuitively reasonable to assume that, given a least squares estimate $\hat{\mathbf{a}}(k-1)$ at the $(k-1)^{th}$ instant, the best *a priori* estimate of $\mathbf{a}(k)$ at the kth instant (i.e. prior to the receipt of new data at the kth instant) can be generated by

$$\hat{\mathbf{a}}(k|k-1) = \mathbf{A}\hat{\mathbf{a}}(k-1) \tag{4.26}$$

which is the estimate at k based on the previous estimate at $k-1$ and *a priori* knowledge of the parameters variation law (4.23). Now, if the *a priori* estimation error is defined as

$$\tilde{\mathbf{a}}(k|k-1) = \hat{\mathbf{a}}(k|k-1) - \mathbf{a}(k)$$

then, from (4.23) and (4.25),

$$\tilde{\mathbf{a}}(k|k-1) = \mathbf{A}\hat{\mathbf{a}}(k-1) - \mathbf{A}\mathbf{a}(k-1) - \mathbf{D}\boldsymbol{\eta}(k-1) = \mathbf{A}\tilde{\mathbf{a}}(k-1) - \mathbf{D}\boldsymbol{\eta}(k-1)$$

where $\tilde{\mathbf{a}}(k-1) \triangleq \hat{\mathbf{a}}(k-1) - \mathbf{a}(k-1)$ is the estimation error at the $(k-1)^{th}$ instant.

The covariance matrix $\mathbf{P}^*(k|k-1)$ of the *a priori* estimation errors $\tilde{\mathbf{a}}(k|k-1)$, can now be obtained from

$$\begin{aligned}
\mathbf{P}^*(k|k-1) &= E\{\tilde{\mathbf{a}}(k|k-1)\tilde{\mathbf{a}}^T(k|k-1)\} \\
&= E\{[\mathbf{A}\tilde{\mathbf{a}}(k-1) - \mathbf{D}\boldsymbol{\eta}(k-1)][\mathbf{A}\tilde{\mathbf{a}}(k-1) - \mathbf{D}\boldsymbol{\eta}^T(k-1)]\} \\
&= \mathbf{A}E\{\tilde{\mathbf{a}}(k-1)\tilde{\mathbf{a}}^T(k-1)\}\mathbf{A}^T + \mathbf{D}E\{\boldsymbol{\eta}(k-1)\boldsymbol{\eta}^T(k-1)\}\mathbf{D}^T \\
&\quad - \mathbf{A}E\{\tilde{\mathbf{a}}(k-1)\boldsymbol{\eta}^T(k-1)\}\mathbf{D}^T - \mathbf{D}E\{\boldsymbol{\eta}(k-1)\tilde{\mathbf{a}}^T(k-1)\}\mathbf{A}^T
\end{aligned}$$

But, from equation (4.23), $\boldsymbol{\eta}(k-1)$ only affects $\mathbf{a}(k)$, so that

$$E\{\tilde{\mathbf{a}}(k-1)\boldsymbol{\eta}^T(k-1)\} = 0; \; E\{\boldsymbol{\eta}(k-1)\tilde{\mathbf{a}}^T(k-1)\} = 0$$

and so, finally,

$$\mathbf{P}^*(k|k-1) = \mathbf{A}\mathbf{P}^*(k-1)\mathbf{A}^T + \mathbf{D}\mathbf{Q}_a\mathbf{D}^T \tag{4.27}$$

where $\mathbf{P}^*(k-1) = E\{\tilde{\mathbf{a}}(k-1)\tilde{\mathbf{a}}^T(k-1)$ and $\mathbf{Q}_a$ is the covariance matrix of the parameter variation disturbance vector in equation (4.23).

Equations (4.26) and (4.27) provide the required *a priori* predictions $\hat{\mathbf{a}}(k|k-1)$ and $\mathbf{P}^*(k|k-1)$ of $\hat{\mathbf{a}}(k)$ and $\mathbf{P}^*(k)$, respectively. Bearing in mind the basic recursive least squares regression algorithm in the time invariant parameter case, these equations can be used to construct the following least squares prediction-correction algorithm for estimating the time variable parameter $\mathbf{a}(k)$ defined by equation (4.23) i.e.

Prediction:

$$\hat{\mathbf{a}}(k|k-1) = \mathbf{A}\hat{\mathbf{a}}(k-1) \quad \text{(STVP-1)}$$

$$\mathbf{P}^*(k|k-1) = \mathbf{A}\mathbf{P}^*(k-1)\mathbf{A}^T + \mathbf{D}\mathbf{Q}_a\mathbf{D}^T \quad \text{(STVP-2)}$$

Correction:

$$\hat{\mathbf{a}}(k) = \hat{\mathbf{a}}(k|k-1) + \mathbf{g}(k)\{y(k) - \mathbf{x}^T(k)\hat{\mathbf{a}}(k|k-1)\} \quad \text{(STVP-3)}$$

$$\mathbf{g}(k) = \mathbf{P}^*(k|k-1)\mathbf{x}(k)[\hat{\sigma}^2 + \mathbf{x}^T(k)\mathbf{P}^*(k|k-1)\mathbf{x}(k)]^{-1} \quad \text{(STVP-4)}$$

$$\mathbf{P}^*(k) = \mathbf{P}^*(k|k-1) - \mathbf{g}(k)\,\mathbf{x}^T(k)\mathbf{P}^*(k|k-1) \quad \text{(STVP-5)}$$

Notice that this since, from (3.28), $\mathbf{P}^*(k) = \sigma^2\mathbf{P}(k)$, algorithm can be written down in terms of $\mathbf{P}(k)$ rather than $\mathbf{P}^*(k)$, where it now takes the form:

Prediction:

$$\hat{\mathbf{a}}(k|k-1) = \mathbf{A}\hat{\mathbf{a}}(k-1) \quad \text{(TVP-1)}$$

$$\mathbf{P}(k|k-1) = \mathbf{A}\mathbf{P}(k-1)\mathbf{A}^T + \mathbf{D}\mathbf{Q}_{nvr}\mathbf{D}^T \quad \text{(TVP-2)}$$

Correction:

$$\hat{\mathbf{a}}(k) = \hat{\mathbf{a}}(k|k-1) + \mathbf{g}(k)\{y(k) - \mathbf{x}^T(k)\hat{\mathbf{a}}(k|k-1)\} \quad \text{(TVP-3)}$$

$$\mathbf{g}(k) = \mathbf{P}(k|k-1)\mathbf{x}(k)[1 + \mathbf{x}^T(k)\mathbf{P}(k|k-1)\mathbf{x}(k)]^{-1} \quad \text{(TVP-4)}$$

$$\mathbf{P}(k) = \mathbf{P}(k|k-1) - \mathbf{g}(k)\,\mathbf{x}^T(k)\mathbf{P}(k|k-1) \quad \text{(TVP-5)}$$

where $\mathbf{Q}_{nvr} = \mathbf{Q}_a/\sigma^2$ is the *Noise-Variance ratio* (NVR) matrix. This form of the algorithm is useful in practical terms because it does not require prior estimation of σ^2. Note that the only changes to the RLS algorithm required to accommodate this approach to TVP estimation are the addition of the prediction equations: the correction step is simply the RLS algorithm with the past estimates $\hat{\mathbf{a}}(k-1)\}$ replaced by their predictions $\hat{\mathbf{a}}(k|k-1)\}$.

At this point, the reader may well feel that, while the TVP algorithms have theoretical appeal, it may be exceedingly difficult to specify a parameter variation model such as (4.23). Fortunately, such understandable pessimism is not fully justified in this case. For example, consider the situation when the RW model (4.21) is used to describe parameter variations. In this case $\mathbf{A} = \mathbf{D} = \mathbf{I}$, so that the prediction equations become simply

$$\begin{aligned} \hat{\mathbf{a}}(k|k-1) &= \hat{\mathbf{a}}(k-1) \\ \mathbf{P}^*(k|k-1) &= \mathbf{P}^*(k-1) + \mathbf{Q}_a \end{aligned} \tag{4.28}$$

or, alternatively, the covariance matrix update can be replaced by

$$\mathbf{P}(k|k-1) = \mathbf{P}(k-1) + \mathbf{Q}_{nvr} \tag{4.29}$$

In order to utilize this simplified algorithm, it is only necessary to specify $\hat{\sigma}^2$ and the covariance matrix $\mathbf{Q}_a$ of the random parameter variations between samples; or, equivalently, the NVR matrix $\mathbf{Q}_{nvr}$. In practice, this is a relatively straightforward task

Specifying $\mathbf{Q}_a$ Experimental evidence over many years, with simulated and real data, has shown that the TVP algorithm with (STVP-1) and (STVP-2) in the form of equations (4.28) provides a reasonable approach to the estimation of slowly variable parameters (i.e. parameters whose percentage change per sampling instant is small, say $\leq 5\%$). In such cases, the $\mathbf{Q}_a$ matrix can be chosen to be diagonal in form (so implying a simplifying *a priori* assumption that the parameter variations are not correlated), with the diagonal elements selected to reflect the expected rate of variation between samples of each parameter in the parameter vector. For example, if the ith element a_i of $\mathbf{a}$ is expected to have changes with a standard deviation σ_{a_i} between samples, then the ith diagonal element $q_{a(i,i)}$ of $\mathbf{Q}_a$ should be set to $\sigma^2_{a_i}$. This is inherently more flexible than the equivalent EWP approach because different expected rates of change can be specified for different parameters: for instance, any known time-invariant parameters can be accommodated simply by setting the appropriate diagonal element of $\mathbf{Q}_a$ to zero.

Specifying $\mathbf{Q}_{nvr}$ Here, the user can specify $\mathbf{Q}_{nvr}$, noting that $\mathbf{Q}_{nvr} = \mathbf{Q}_a/\sigma^2$. However, we will see later that an associated optimization routine can be devised that provides the user with a more objectively computed diagonal $\mathbf{Q}_{nvr}$ that is optimized by either maximum likelihood or by minimizing the single or multi-step-ahead forecasting errors.

The latter specification is the one utilized by the TVP estimation routines in the CAPTAIN Toolbox, as discussed in the next Chapter 5.

In the single parameter case, letting $q_a(1,1) = q$, for simplicity, we can see that the algorithm (TVP) with the RW model for the parameter variations performs quite similarly to the EWP algorithm. For example, in the case of the Walgett mean value estimation problem, the recursive relationship for $p(k)$ is given by

$$\begin{aligned} p^*(k|k-1) &= p^*(k-1) + q \\ p^*(k) &= p^*(k|k-1) - p^{*2}(k|k-1)/[1+p^*(k|k-1)] \end{aligned}$$

where $\hat{\sigma}^2$ is assumed to be unity to facilitate comparison with the EWP equivalent equation. The steady state solution of these equations is obtained when $p^*(k) = p^*(k-1) = p(\infty)$ and is given by the solution of

$$p^2(\infty) + p(\infty)q - q = 0 \qquad (4.30)$$

i.e.,

$$p(\infty) = \frac{1}{2}[-q \pm \sqrt{q^2 + 4q}]$$

Equating $p(\infty)$ in this equation with $p(\infty)$ in (4.19), we find that, for similar steady state solutions,

$$q = \frac{1}{\alpha}(1 - 2\alpha + \alpha^2)$$

so that, for instance, when $\alpha = 0.9$ $q = 0.0111$. A comparison of the recursive estimation results using these settings is given in Figure 4.6 and it is clear that, after about 20 years, the two algorithms are producing virtually identical results, both as regards the recursive estimates and the estimated SE bounds.

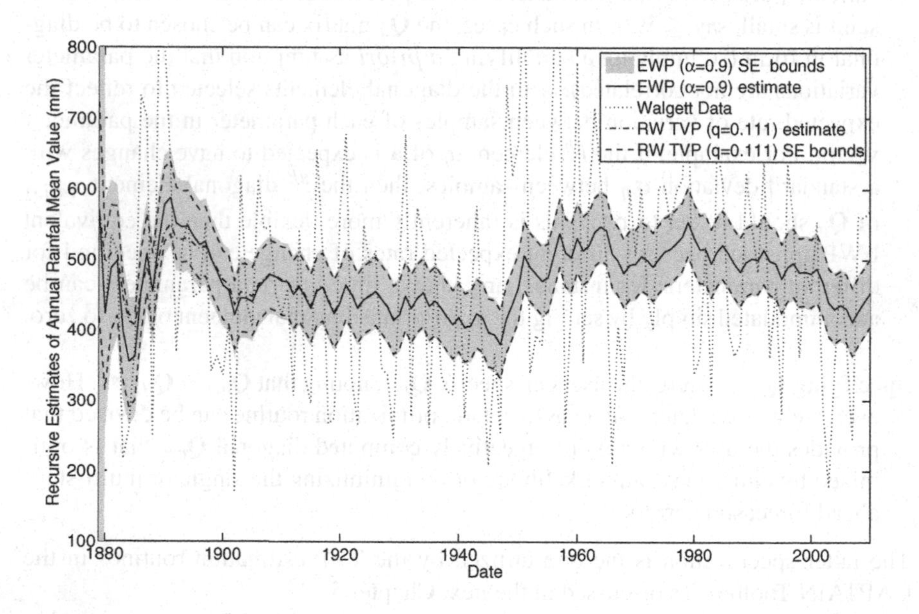

Fig. 4.6 Walgett rainfall mean value estimation: recursive stochastic estimation results comparing the recursive EWP estimates ($\alpha = 0.9$) and the 'equivalent' recursive TVP estimates based on a RW model for the parameter variations ($q = 0.0111$).

The first order Gauss-Markov model (4.22) can also be useful in time variable parameter estimation, but it is clearly only appropriate if the parameter variations are around a given constant mean value $\bar{\mathbf{a}}$. The smoothness of the expected variation can be specified by the selection of the β_i: by analogy with (4.12) and the forgetting factor α in recursive EWP estimation, the time constant T_{a_i} of the smoothing for the ith parameter is defined by

$$T_{a_i} = -\Delta t / \log_e \beta_i$$

so that for $\beta_i = 0.9$, $T_{a_i} = 9.49$ time units; and the closer β_i is to unity, the greater the smoothing effect.

The above analysis for the RW model of the parameter variations could be repeated with each parameter a_i in the parameter vector $\mathbf{a}$ modelled as an IRW process (4.5). Another model that has been found useful is the *Smoothed Random Walk* (SRW), where each parameter is modelled as follows:

$$\begin{bmatrix} a_i(k) \\ \nabla a_i(k) \end{bmatrix} = \begin{bmatrix} \alpha & 1 \\ 0 & 1 \end{bmatrix} \begin{bmatrix} a_i(k-1) \\ \nabla a_i(k-1) \end{bmatrix} + \begin{bmatrix} 0 \\ 1 \end{bmatrix} e(k-1) \tag{4.31}$$

Intuitively, one would expect the IRW model in (4.5) to be useful when there are expected to be large and smooth variations in the parameters (Norton, 1975); while (4.31) is a compromize between the IRW and RW models, which requires specification of the smoothing hyper-parameter α. In fact, when parameters modelled in this SRW manner are combined into a model of the whole parameter vector $\mathbf{a}(k)$, the resulting state space model can be considered as the equivalent of the model (4.22) when the mean value of the parameter is time variable, rather than constant.

In order to obtain a better idea of the way in which the assumptions (4.1), (4.5) and (4.31) affect the performance of the estimator, it is instructive to consider the nature of the prediction equations (STVP-1) for the various simple random walk processes. In the RW case, we have seen already that the *a priori* prediction for each parameter $\hat{a}(k|k-1)$ is simply the estimate obtained at the previous recursive instant $k-1$. For the IRW and SRW model, however, an additional unknown parameter state $\nabla a(k)$ is introduced and estimated. In order to see the implications of this on the prediction equations, the $\nabla \hat{a}(k)$ equation can be substituted into the $\hat{a}(k)$ equation to yield the following results:

$$\begin{aligned} \text{IRW:} \quad \hat{a}(k|k-1) &= \hat{a}(k-1) + [\hat{a}(k-1) - \hat{a}(k-2)] \\ &= \hat{a}(k-1) + \nabla \hat{a}(k-1) \\ \text{SRW:} \quad \hat{a}(k|k-1) &= \hat{a}(k-1) + \alpha [\hat{a}(k-1) - \hat{a}(k-2)] \\ &= \hat{a}(k-1) + \alpha \nabla \hat{a}(k-1) \end{aligned}$$

where $\nabla \hat{a}(k-i)$ is used here the define rate of change of the estimate between the $k-i-1$ and $k-i$ recursions. In other words, the *a priori* prediction is obtained by projecting the *a priori* estimate obtained at the previous $(k-1)^{th}$ recursion in proportion to its immediate past rate of change: in the IRW case the projection is equal to the past rate of change; in the SRW case it is made a fraction, α, of this rate of change. And if higher order integrated random walks are considered, then it is clear that more past information on rate of change would be used for *a priori* prediction. In the case of the *Double Integrated Random Walk* (DIRW), for example,

$$\begin{bmatrix} a(k) \\ \nabla a(k) \\ \nabla^2 a(k) \end{bmatrix} = \begin{bmatrix} 1 & 1 & 0 \\ 0 & 1 & 1 \\ 0 & 0 & 1 \end{bmatrix} \begin{bmatrix} a(k-1) \\ \nabla a(k-1) \\ \nabla^2 a(k-1) \end{bmatrix} + \begin{bmatrix} 0 \\ 0 \\ 1 \end{bmatrix} e(k-1\} \tag{4.32}$$

where $\nabla^2 a(k)$ is the second difference of $a(k)$ (a measure of its 'acceleration'). As a result, in this case,

$$\hat{a}(k|k-1) = \hat{a}(k-1) + \nabla\hat{a}(k-1) + \nabla^2\hat{a}(k-1)$$

Simple illustrative examples of the RW, SRW, IRW and DIRW *a priori* predictions are given in Figure 4.7 for different past $\hat{a}(k)$ histories.

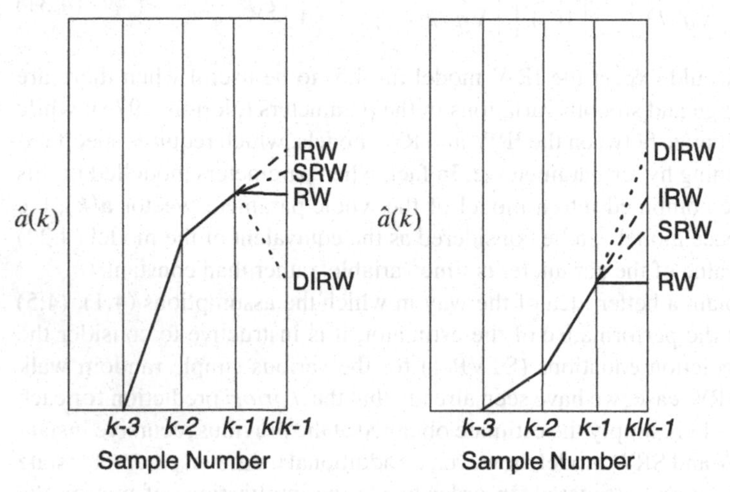

Fig. 4.7 Typical predictions for different random walk models.

It is interesting to note that, in the IRW case, this predictive procedure is analogous to that used in a hybrid (analog-digital) estimation algorithm that I used long ago (Young, 1966), where the past rates of change of the estimates were generated using purely analog equipment. It is also clear from the examples in Figure (4.7) that, although the applicability of the different random walk models will depend to some extent upon the nature of the parametric variation, all of the models (except perhaps the DIRW) seem to provide reasonable extrapolations when it is remembered that these extrapolations are automatically corrected by the update equations (STVP-3) to (STVP-5) on receipt of the next data sample.

Finally, it is useful to bring the various random walk-type models discussed above into a *Generalized Random Walk* (GRW) form where, in terms of practical utility, attention will be restricted to second-order, sub-vector models. Here, the stochastic evolution of each parameter sub-vector $\mathbf{a}_i(k)$, where,

$$\mathbf{a}_i(k) = \begin{bmatrix} a_i(k) \\ \nabla a_i(k) \end{bmatrix}$$

is assumed to follow the following state space model:

$$\mathbf{a}_i(k) = \mathbf{A}_i\mathbf{a}_i(k-1) + \mathbf{D}_i\boldsymbol{\eta}_i(k-1) \qquad i = 1,2,\ldots,n, \tag{4.33}$$

where

$$\mathbf{A}_i = \begin{bmatrix} \alpha & \beta \\ 0 & \gamma \end{bmatrix}, \quad \mathbf{D}_i = \begin{bmatrix} \delta & 0 \\ 0 & \varepsilon \end{bmatrix}$$

and $\boldsymbol{\eta}_i(k) = [\eta_{1i}(k)\ \eta_{2i}(k)]^T$ is a 2×1, zero mean, white noise vector that allows for stochastic variability in the parameters and is assumed to be characterized by a (normally diagonal) covariance matrix $\mathbf{Q}_{\eta_i}$. This general model comprises as special cases the *Integrated Random Walk* (IRW: $\alpha = \beta = \gamma = \varepsilon = 1$; $\delta = 0$); the scalar *Random Walk* (RW: scalar but equivalent to (4.33) with $\beta = \gamma = \varepsilon = 0$; $\alpha = \delta = 1$: i.e. just the first equation in (4.33)); the intermediate case of *Smoothed Random Walk* (SRW: $0 < \alpha < 1$; $\beta = \gamma = \varepsilon = 1$; and $\delta = 0$); the first order autoregressive process (AR(1): again scalar with $\beta = \gamma = \varepsilon = 0$; $0 < \alpha < 1$; $\delta = 1$); and, finally, both the *Local Linear Trend* (LLT: $\alpha = \beta = \gamma = \varepsilon = \delta = 1$)[2]; and *Damped Trend* (DT: $\alpha = \beta = \delta = \varepsilon = 1$; $0 < \gamma < 1$): see Harvey (1984, 1989). The hyper-parameters (see Appendix B) in this case are the various, normally constant, coefficients in this GRW model (α, β, γ, δ, ε, as well as the elements of $\mathbf{Q}_{\eta_i}$ or the NVR matrix $\mathbf{Q}_{nvr} = \mathbf{Q}_{\eta_i}/\sigma^2$). These are also assumed to be unknown *a priori* and need to be specified by the user or estimated from the data, as discussed later.

4.2.1 The complete TVP regression model

Having introduced the GRW models for the parameter variations, an overall state space model can be constructed straightforwardly by the aggregation of the subsystem matrices defined by (4.33) for each parameter vector $\mathbf{a}_i(k)$, with the 'observation' equation, which is actually the model equation expressed in a linear regression form. For example, in the case of the linear regression model (4.24), the state equations take the form:

$$\begin{aligned} \text{State Equations: } & \mathbf{x}(k) = \mathbf{A}\mathbf{x}(k-1) + \mathbf{D}\boldsymbol{\eta}(k-1) \quad \text{(i).} \\ \text{Observation Equation: } & y(k) = \mathbf{h}^T(k)\mathbf{x}(k) + e(k) \quad \text{(ii)} \end{aligned} \tag{4.34}$$

Here, the $p \times 1$ dimensional state vector $\mathbf{x}(k)$ is composed of the parameter subvectors $\mathbf{a}_i(k)$, $i = 1,2,\ \ldots,\ n$, each defined by the GRW model (4.33), i.e.,

$$\mathbf{x}(k) = \left[\mathbf{a}_1^T(k)\ \mathbf{a}_2^T(k)\ \ldots\ \mathbf{a}_n^T(k)\right]^T. \tag{4.35}$$

[2] Interestingly, the LLT model can be considered simply as the combination of the simpler RW and IRW models.

Note that, in order to conform with standard nomenclature, $\mathbf{x}(k)$ is used here to denote the state vector; it should not be confused with the regressor vector $\mathbf{x}(k)$ used previously in this Chapter. In general, $p = 2n$ because of the two dimensional nature of the GRW state vector. The state transition matrix $\mathbf{A}$ and the input matrix $\mathbf{D}$ are $p \times p$ block diagonal matrices with blocks defined by the corresponding sub-system matrices $\mathbf{A}_i$ and $\mathbf{D}_i$, respectively, in the GRW equation (4.33), i.e.,

$$\mathbf{A} = \begin{bmatrix} \mathbf{A}_1 & \mathbf{0} & \mathbf{0} & \mathbf{0} \\ \mathbf{0} & \mathbf{A}_2 & \mathbf{0} & \mathbf{0} \\ \mathbf{0} & \mathbf{0} & \ddots & \mathbf{0} \\ \mathbf{0} & \mathbf{0} & \mathbf{0} & \mathbf{A}_n \end{bmatrix}; \quad \mathbf{D} = \begin{bmatrix} \mathbf{D}_1 & \mathbf{0} & \mathbf{0} & \mathbf{0} \\ \mathbf{0} & \mathbf{D}_2 & \mathbf{0} & \mathbf{0} \\ \mathbf{0} & \mathbf{0} & \ddots & \mathbf{0} \\ \mathbf{0} & \mathbf{0} & \mathbf{0} & \mathbf{D}_n \end{bmatrix}$$

The stochastic input vector $\boldsymbol{\eta}(k)$ is a p-dimensional vector containing, in appropriate locations, the white noise input vectors $\boldsymbol{\eta}_i(k)$ ('system disturbances' in normal SS terminology) to each of the GRW models in (4.33), i.e.,

$$\boldsymbol{\eta}(k) = \left[\boldsymbol{\eta}_1^T(k)\ \boldsymbol{\eta}_2^T(k) \cdots \boldsymbol{\eta}_n^T(k)\right]^T$$

These white noise inputs, which provide the stochastic stimulus for parametric change in the model, are assumed to be independent of the 'observation' noise $e(k)$ and $\boldsymbol{\eta}(k)$ has a covariance matrix $\mathbf{Q}$ formed from the combination of the individual covariance matrices $\mathbf{Q}_{\eta_i}$.

Finally, in the 'observation' equation (4.34)(ii), $\mathbf{c}(k)$ is, in general, a $p \times 1$ vector of the following form,

$$\mathbf{h}^T(k) = [x_1(k)\ 0\ x_2(k)\ 0\ x_3(k)\ 0\ \ldots\ 0\ x_n(k)\ 0] \tag{4.36}$$

that relates the scalar observation $y(k)$ to the parametric state variables defined by (4.34)(i) and (4.35), so that the observation equation represents the linear regression model (4.24), with each parameter defined as a GRW process. In the case of the scalar RW and AR(1) models, $p = n$ and the alternate zeros are simply omitted.

Within the context of time variable parameters modelled specifically by GRW processes, the STVP and TVP algorithms now need to be modified to reflect the above state space model (4.34). In the practically more useful TVP case, the resulting GRWTVP algorithm takes the following form:

Prediction:

$$\hat{\mathbf{x}}(k|k-1) = \mathbf{A}\hat{\mathbf{x}}(k-1) \tag{GRWTVP-1}$$

$$\mathbf{P}(k|k-1) = \mathbf{A}\mathbf{P}(k-1)\mathbf{A}^T + \mathbf{D}\mathbf{Q}_{nvr}\mathbf{D}^T \tag{GRWTVP-2}$$

Correction:

$$\hat{\mathbf{x}}(k) = \hat{\mathbf{x}}(k|k-1) + \mathbf{g}(k)\{y(k) - \mathbf{h}^T(k)\hat{\mathbf{x}}(k|k-1)\} \quad \text{(GRWTVP-3)}$$
$$\mathbf{g}(k) = \mathbf{P}(k|k-1)\mathbf{h}(k)[1 + \mathbf{h}^T(k)\mathbf{P}(k|k-1)\mathbf{h}(k)]^{-1} \quad \text{(GRWTVP-4)}$$
$$\mathbf{P}(k) = \mathbf{P}(k|k-1) - \mathbf{g}(k)\,\mathbf{h}^T(k)\mathbf{P}(k|k-1) \quad \text{(GRWTVP-5)}$$

4.3 Vector Measurements

The analysis in Chapters 2, 3 and Section 4.2 can be extended, as could the RWP and EWP algorithms, to the case where the equation (4.24) is in terms of a p vector of observations $\mathbf{y}(k)$ at each sampling instant, i.e.,

$$\mathbf{y}(k) = \mathbf{H}(k)\mathbf{a}(k) + \mathbf{e}(k) \quad (4.37)$$

where $\mathbf{y}(k) = [y_1(k)\; y_2(k)\; \ldots\; y_p(k)]^T$ and $\mathbf{e}(k)$ is a $p \times 1$ white noise vector of observational errors whose elements are assumed to be serially uncorrelated but which may, however, be instantaneously mutually correlated with covariance matrix $\mathbf{R}(k)$, which may be time variable, i.e.

$$E\{\mathbf{e}(k)\} = 0; \; E\{\mathbf{e}(k)\mathbf{e}^T(j)\} = \mathbf{R}(k)\delta_{kj} \quad (4.38)$$

where δ_{kj} is, once again, Kronecker delta function. Finally, $\mathbf{H}(k)$ is a $p \times n$ matrix of exactly known but possibly time-variable 'regressors' assumed to have the same properties as the analogous regression variables in the scalar observation case. Further discussion on the nature of $\mathbf{H}(k)$ and how it may be defined by the user will be delayed until later.

In order to develop a recursive least squares algorithm in this vector case, it is first necessary to define a suitable cost function. Clearly a simple least squares function such as (3.5) with its scalar norm

$$\mathcal{J}_2(\mathbf{a}) = \sum_{i=1}^{k} [y(i) - \mathbf{x}^T(i)\mathbf{a}]^2$$

is no longer appropriate and it is necessary to consider a vector quadratic form of some kind (see Appendix B). The most general quadratic form in the present context is as follows:

$$||\mathbf{y}(i) - \mathbf{H}(i)\mathbf{a}(i)||_W^2 = [\mathbf{y}(i) - \mathbf{H}(i)\mathbf{a}(i)]^T\mathbf{W}[\mathbf{y}(i) - \mathbf{H}(i)\mathbf{a}(i)] = \mathbf{e}^T(i)\mathbf{W}\mathbf{e}(i) \quad (4.39)$$

where $\mathbf{e}(i) = \mathbf{y}(i) - \mathbf{H}(i)\mathbf{a}(i)$ and $\mathbf{W}$ is a $p \times p$, symmetric, positive definite weighting matrix with elements $w_{k,j}, k, j = 1, 2, \ldots, p$. By expansion of (4.39), the reader can verify that it is equivalent to the following summation

$$\mathbf{e}^T(i)\,\mathbf{W}\mathbf{e}(i) \;=\; \sum_{k,j=1}^{p} w_{kj} e_k(i) e_j(i)$$

where $e_k(i), k = 1, 2, \ldots, p$ are the elements of $\mathbf{e}(i)$. Thus the norm (4.39) is a general quadratic form in the elements of the error vector with weightings $w_{k,j}$ to be specified by the analyst.

How the weighting matrix $\mathbf{W}$ should be specified will be considered subsequently but, for the moment, the estimation problem can be posed in terms of the norm (4.39) as one of minimizing the cost function $\mathcal{J}(\mathbf{a})$ with respect to $\mathbf{a}$, i.e.,

$$\hat{\mathbf{a}} = \arg\min_{\mathbf{a}} \mathcal{J}(\mathbf{a}) \qquad \mathcal{J}(\mathbf{a}) = \sum_{i=1}^{k} \|\mathbf{y}(i) - \mathbf{H}(i)\mathbf{a}(i)\|^2_{\mathbf{W}} \tag{4.40}$$

In the case of constant parameters, i.e. $\mathbf{a}(k) = \mathbf{a}\ \forall\ k$, setting the gradient of $\mathcal{J}(\mathbf{a})$ to zero in the normal manner, yields the following normal equations (see Appendix B):

$$\sum_{i=1}^{k} [\mathbf{H}^T(i)\,\mathbf{W}\mathbf{H}(i)]\hat{\mathbf{a}}(k) \;=\; \sum_{i=1}^{k} \mathbf{H}^T(i)\mathbf{W}\mathbf{y}(i)$$

As a result, the recursive expressions (3.10) in the scalar error case now take the form,

$$\begin{aligned} \mathbf{P}^{-1}(k) &= \mathbf{P}^{-1}(k-1) + \mathbf{H}^T(k)\,\mathbf{W}\mathbf{H}(k) \\ \mathbf{b}(k) &= \mathbf{b}(k-1) + \mathbf{H}^T(k)\,\mathbf{W}\mathbf{y}(k) \end{aligned} \tag{4.41}$$

and the reader will find that the recursive estimation algorithms for $\mathbf{a}(k)$ and $\mathbf{P}(k)$ can be derived quite straightforwardly in the following form,

$$\hat{\mathbf{a}}(k) \;=\; \hat{\mathbf{a}}(k-1) + \mathbf{G}(k)\{\mathbf{y}(k) - \mathbf{H}(k)\hat{\mathbf{a}}(k-1)\} \tag{VRLS-1}$$

$$\mathbf{G}(k) \;=\; \mathbf{P}(k-1)\mathbf{H}^T(k)[\mathbf{W}^{-1} + \mathbf{H}(k)\mathbf{P}(k-1)\mathbf{H}^T(k)]^{-1} \tag{VRLS-2}$$

$$\mathbf{P}(k) \;=\; \mathbf{P}(k-1) + \mathbf{G}(k)\mathbf{H}(k)\mathbf{P}(k-1) \tag{VRLS-3}$$

where it will be noted that the expression $\mathbf{W}^{-1} + \mathbf{H}(k)\mathbf{P}(k-1)\mathbf{H}^T(k)$ now denotes a $p \times p$ matrix and so $[\mathbf{W}^{-1} + \mathbf{H}(k)P(k-1)\mathbf{H}^T(k)]^{-1}$ requires matrix inversion at every recursive update, in contrast to the previous scalar observation situation.

Before proceeding to develop the time-variable parameter equivalent of algorithm (TVP), it is first necessary to consider how the weighting matrix $\mathbf{W}$ should be chosen and what the statistical properties of the parameter estimates will be with this choice of $\mathbf{W}$. It is well known (Young, 1968a; Johnston and DiNardo, 1997) that if the observational errors $\mathbf{e}(k)$ have a Gaussian amplitude distribution (Appendix B), i.e. $\mathbf{e}(k) = \mathcal{N}(\mathbf{0},\mathbf{R})$, then the cost function (4.40) is equivalent to the maximum likelihood cost function when $\mathbf{W} = \mathbf{R}^{-1}$. And, in this situation, it can be shown, by an argument similar to that used in Chapter 3, that the parameter estimates obtained from the recursive algorithm (VRLS) are zero mean and have an error covariance matrix $\mathbf{P}^*(k)$ defined by (cf. equation (3.28) in Chapter 3):

$$\mathbf{P}^*(k) = \left[\sum_{i=1}^{k} \mathbf{H}^T(i)\mathbf{R}^{-1}\mathbf{H}(i)\right]^{-1} \tag{4.42}$$

Given this information and pursuing the same approach used to develop the STVP algorithm, it is now possible to derive the following prediction-correction algorithm for the case where $\mathbf{a}(k)$ is stochastically time-variable and can be modelled by equation (4.23):

Prediction:

$$\hat{\mathbf{a}}(k|k-1) = \mathbf{A}\hat{\mathbf{a}}(k-1) \tag{VTVP-1}$$

$$\mathbf{P}^*(k|k-1) = \mathbf{A}\mathbf{P}^*(k-1)\mathbf{A}^T + \mathbf{D}\mathbf{Q}_a\mathbf{D}^T \tag{VTVP-2}$$

Correction:

$$\hat{\mathbf{a}}(k) = \hat{\mathbf{a}}(k|k-1) + \mathbf{G}(k)\{\mathbf{y}(k) - \mathbf{H}(k)\hat{\mathbf{a}}(k|k-1)\} \tag{VTVP-3}$$

$$\mathbf{G}(k) = \mathbf{P}^*(k|k-1)\mathbf{H}^T(k)[\mathbf{R} + \mathbf{H}(k)\mathbf{P}^*(k|k-1)\mathbf{H}^T(k)]^{-1} \tag{VTVP-4}$$

$$\mathbf{P}^*(k) = \mathbf{P}^*(k|k-1) - \mathbf{G}(k)\,\mathbf{H}(k)\mathbf{P}^*(k|k-1) \tag{VTVP-5}$$

Given the assumptions about the observational errors and the stochastic nature of the parameter variation model, this VTVP algorithm is optimal in a maximum likelihood sense. Moreover, given *a priori* information on the initial estimate $\hat{\mathbf{a}}(0)$ and its associated initial covariance matrix $\mathbf{P}^*(0)$, the estimates are also optimal in a Bayesian sense. In fact, as shown in the next section 4.4, although the VTVP algorithm is intended for TVP estimation, it is algorithmically very similar to the famous optimal filter-estimation algorithm for state variable estimation, first proposed by Kalman (1960), which is considered in the next sub-section 4.4.

4.4 The Kalman Filter

In his seminal paper (Kalman, 1960), Kalman considered the problem of estimating a state vector $\mathbf{x}(k)$ associated with a stochastic dynamic system modelled by the simple Gauss-Markov process:

$$\begin{aligned} \text{State equation}: \quad & \mathbf{x}(k) = \mathbf{A}(k)\mathbf{x}(k-1) + \boldsymbol{\eta}(k-1) \\ \text{Output measurement}: \quad & \mathbf{y}(k) = \mathbf{C}(k)\mathbf{x}(k) \end{aligned} \tag{4.43}$$

For generality, however, let us consider the more complex stochastic system below (see Appendix B), where the vector of output measurements are contaminated by noise $\boldsymbol{\xi}(k)$:

$$\begin{aligned} \mathbf{x}(k) &= \mathbf{A}(k)\mathbf{x}(k-1) + \mathbf{D}(k)\boldsymbol{\eta}(k-1) \\ \mathbf{y}(k) &= \mathbf{C}(k)\mathbf{x}(k) + \boldsymbol{\xi}(k) \end{aligned} \tag{4.44}$$

where $\mathbf{y}(k)$ is now a $p \times 1$ dimensional vector of observations that are linearly related to the state vector $\mathbf{x}(k)$ by the matrix $\mathbf{C}$; while $\boldsymbol{\eta}(k)$ and $\boldsymbol{\xi}(k)$ are zero mean, statistically independent, white noise disturbance vectors with possibly time-variable covariance matrices $\mathbf{Q}(k)$ and $\mathbf{R}(k)$, respectively: i.e.

$$\begin{aligned} E\{\boldsymbol{\eta}(k)\} = E\{\boldsymbol{\xi}(k)\} &= \mathbf{0}; \quad E\{\boldsymbol{\eta}(k)\boldsymbol{\xi}^T(j)\} = \mathbf{0}, \\ E\{\boldsymbol{\eta}(k)\boldsymbol{\eta}^T(j)\} = \mathbf{Q}(k)\delta_{kj}; &\quad E\{\boldsymbol{\xi}(k)\boldsymbol{\xi}^T(j)\} = \mathbf{R}(k)\delta_{kj} \end{aligned} \tag{4.45}$$

By an argument based on orthogonal projection, Kalman showed that the optimal estimate $\hat{\mathbf{x}}(k)$ of $\mathbf{x}(k)$ in (4.44) can be obtained from the following KF algorithm:

Prediction:

$$\hat{\mathbf{x}}(k|k-1) = \mathbf{A}(k)\hat{\mathbf{x}}(k-1) \tag{KF-1}$$

$$\mathbf{P}^*(k|k-1) = \mathbf{A}(k)\mathbf{P}^*(k-1)\mathbf{A}^T(k) + \mathbf{D}(k)\mathbf{Q}(k)\mathbf{D}^T(k) \tag{KF-2}$$

Correction:

$$\hat{\mathbf{x}}(k) = \hat{\mathbf{x}}(k|k-1) + \mathbf{G}(k)\{\mathbf{y}(k) - \mathbf{C}(k)\hat{\mathbf{x}}(k|k-1)\} \tag{KF-3}$$

$$\mathbf{G}(k) = \mathbf{P}^*(k|k-1)\mathbf{C}^T(k)[\mathbf{R}(k) + \mathbf{C}(k)\mathbf{P}^*(k|k-1)\mathbf{C}^T(k)]^{-1} \tag{KF-4}$$

$$\mathbf{P}^*(k) = \mathbf{P}^*(k|k-1) - \mathbf{G}(k)\,\mathbf{C}(k)\mathbf{P}^*(k|k-1) \tag{KF-5}$$

This is algebraically identical to the VTVP algorithm except that, here, the model matrices are $\mathbf{A}(k)$, $\mathbf{D}(k)$ and $\mathbf{C}(k)$; and the covariance matrices $\mathbf{Q}(k)$, $\mathbf{R}(k)$ are considered as possibly time-variable. So we see that this KF algorithm is quite complex and flexible: not only are the basic state variables $\mathbf{x}(k)$ time-variable, but they can also be generated from a non-stationary dynamic system excited by non-stationary stochastic disturbances. Bearing this result in mind, it is clearly reasonable to extend the VTVP parameter estimation algorithm in a similar manner, if required.

Note that the Kalman filter is often derived in terms of a model with a vector of external *deterministic* inputs or exogenous inputs $\mathbf{u}(k)$ affecting the state equations,

$$
\begin{aligned}
\mathbf{x}(k) &= \mathbf{A}(k)\mathbf{x}(k-1) + \mathbf{B}(k)\mathbf{u}(k-1) + \mathbf{D}(k)\boldsymbol{\eta}(k-1) \\
\mathbf{y}(k) &= \mathbf{C}(k)\mathbf{x}(k) + \mathbf{B}_I(k)\mathbf{u}(k) + \boldsymbol{\xi}(k)
\end{aligned} \tag{4.46}
$$

where the term $\mathbf{B}_I(k)\mathbf{u}(k)$ in the output observation equation allows the input vector $\mathbf{u}(k)$ to affect the output instantaneously (see section B.3.3 of Appendix B). The effect of these modifications is to change the KF equations as follows:

Prediction:

$$\hat{\mathbf{x}}(k|k-1) = \mathbf{A}(k)\hat{\mathbf{x}}(k-1) + \mathbf{B}(k)\mathbf{u}(k-1) \quad \text{(KFX-1)}$$

$$\mathbf{P}^*(k|k-1) = \mathbf{A}(k)\mathbf{P}^*(k-1)\mathbf{A}^T(k) + \mathbf{D}(k)\mathbf{Q}(k)\mathbf{D}^T(k) \quad \text{(KFX-2)}$$

Correction:

$$\hat{\mathbf{x}}(k) = \hat{\mathbf{x}}(k|k-1) + \mathbf{G}(k)\{\mathbf{y}(k) - \mathbf{C}(k)\hat{\mathbf{x}}(k|k-1) - \mathbf{B}_I(k)\mathbf{u}(k)\} \quad \text{(KFX-3)}$$

$$\mathbf{G}(k) = \mathbf{P}^*(k|k-1)\mathbf{C}^T(k)[\mathbf{R}(k) + \mathbf{C}(k)\mathbf{P}^*(k|k-1)\mathbf{C}^T(k)]^{-1} \quad \text{(KFX-4)}$$

$$\mathbf{P}^*(k) = \mathbf{P}^*(k|k-1) - \mathbf{G}(k)\,\mathbf{C}(k)\mathbf{P}^*(k|k-1) \quad \text{(KFX-5)}$$

so allowing for the effect of the input vector $\mathbf{u}(k)$ on the evolution of the state to be incorporated into the KF. The covariance prediction and update equations remain unchanged since the elements of $\mathbf{u}(k)$ are assumed purely deterministic variables and do not affect the stochastic properties of the estimates. The reader can verify that these equations make sense by introducing a similar vector of deterministic input disturbance into the Gauss Markov equation (4.23) and repeating the previous analysis.

Many different derivations of the Kalman filter algorithm have been suggested since the appearance of Kalman's seminal paper in 1960 (e.g. early contributions by Rauch et al. (1965); Duncan and Horn (1972)). The treatment of Bryson and Ho (1969) is, however, of particular merit because of its simple derivation and clarity in statistical terms. Using a Bayesian argument under the assumption that the probability densities $p[x(0)]$, $p[\boldsymbol{\xi}(k)]$ and $p[\boldsymbol{\eta}(k)], k = 1,2, \ldots, N$ are all Gaussian, they point out that the conditional probability $p[\mathbf{x}(0)|\mathbf{Y}(k)]$ is also Gaussian (where $\mathbf{Y}(k)$ denotes all data up to sample instant k) and is thus characterized completely by its mean and covariance matrix, with the mean giving the maximum of the density. They then derive equations for the conditional mean $\hat{\mathbf{x}}(k)$ and covariance $\mathbf{P}^*(k)$ of the densities $p[\boldsymbol{\xi}(k), \mathbf{Y}(k)]$, which are precisely the Kalman filter equations. Rauch et al. (1965) use a similar approach from a maximum likelihood standpoint, but concentrate on the 'smoothing solution' (i.e. the estimation of $\hat{\mathbf{x}}(k|N)$ where $k < N$, as considered in the next section 4.5.

At this point, it is interesting to note that, from a starting point of time-invariant parameter estimation, we have progressed via algorithms for slowly-variable parameter estimation to procedures for estimating the possibly rapid variations in the state of a stochastic dynamic system. This link between state and parameter estimation, which was first pointed out by Lee (1964) and Rauch et al. (1965) (see also Young, 1969a,b), raises certain philosophical questions regarding the relationship between state variables and parameters in dynamic system models. Although it is inappropriate to enter this kind of discussion in the present context, it is interesting to note how the similarities between states and parameters in dynamic systems are emphasized in the simple moving body example discussed earlier. There, the unknown 'parameters' can actually be interpreted as the states of a dynamic system; it is possible to consider them as stationary parameters simply because they do not vary over the observation interval. In the light of the above discussion, the problem could equally well have been considered as one of state estimation and solved by application of the Kalman filter-estimation algorithm (KF). In the case of the single unknown velocity problem, for example, the reader may easily verify that the state equations are as follows,

$$\begin{bmatrix} x_1(k) \\ x_2(k) \end{bmatrix} = \begin{bmatrix} 1 & a_{12} \\ 0 & 1 \end{bmatrix} \begin{bmatrix} x_1(k-1) \\ x_2(k-1) \end{bmatrix} ; \; y(k) = \begin{bmatrix} 1 & 0 \end{bmatrix} \begin{bmatrix} x_1(k) \\ x_2(k) \end{bmatrix} + \xi(k) \tag{4.47}$$

where $x_1(k) = s(k) - s_0$, $x_2(k) = v(k)$ and a_{12} is the sampling interval. Consequently, estimates $\hat{x}_1(k)$ and $\hat{x}_2(k)$ of $x_1(k)$ and $x_2(k)$, respectively, could be obtained from algorithm (KF) simply by making the following substitutions:

$$\mathbf{A}(k) = \begin{bmatrix} 1 & a_{12} \\ 0 & 1 \end{bmatrix} ; \; \mathbf{D}(k) = [0] ; \; \mathbf{Q} = [0] ; \; \mathbf{C}(k) = \begin{bmatrix} 1 & 0 \end{bmatrix} ; \; \mathbf{R} = \sigma^2$$

for all k. Then, the results obtained in this manner, as shown in Figure 4.8 to 4.10, are directly equivalent to those obtained via linear regression analysis.

Figure 4.8 shows the recursive estimate of the distance $\hat{x}_1(k)$ compared with the measured data $y(k)$. Since the analysis is now being carried out in real-time (seconds) there are missing $y(k)$ samples so, in the analysis, these are set to the Matlab 'Not-a-Number' (NaN) when they occur. When this is detected at any sample, only the predictive step is carried out, and such prediction continues for subsequent samples until the next measurement of $y(k)$ occurs. Note that over these missing data periods, the corrections are shown equal to predictions because no corrections are made.

The positions of the predictive and corrective steps are identified by circular and cross points, respectively. Note how the associated diagonal element of the $\mathbf{P}^*(k)$ matrix widens because no new data have been received while this is happening and the uncertainty is increasing. At the occurrence of the next measurement, however, it decreases because the new information provided by this measurement reduces the uncertainty; and, at the same time, the estimated distance jumps, with the corrective step, to a value that has taken into account this latest measurement. Figure 4.9 shows similar results for the recursive estimate of the velocity and its estimated variance. In this case, of course, there are no measurements and the velocity estimate is being inferred on the basis of the distance estimate and the assumed state space model, via the KF equations. In this example, the underlying model is continuous-time,

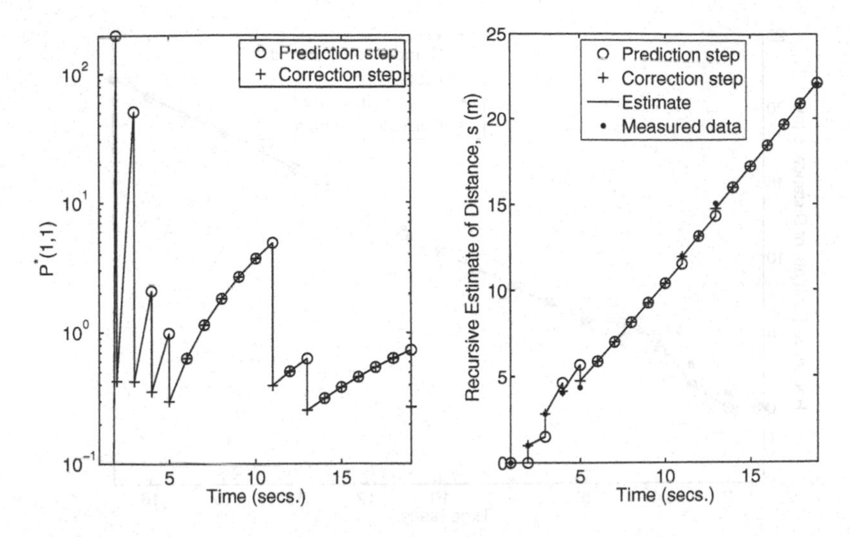

Fig. 4.8 Moving body example: Kalman filter recursive estimate of the distance (right panel) and the associated diagonal element of the $\mathbf{P}^*(k)$ matrix (left panel).

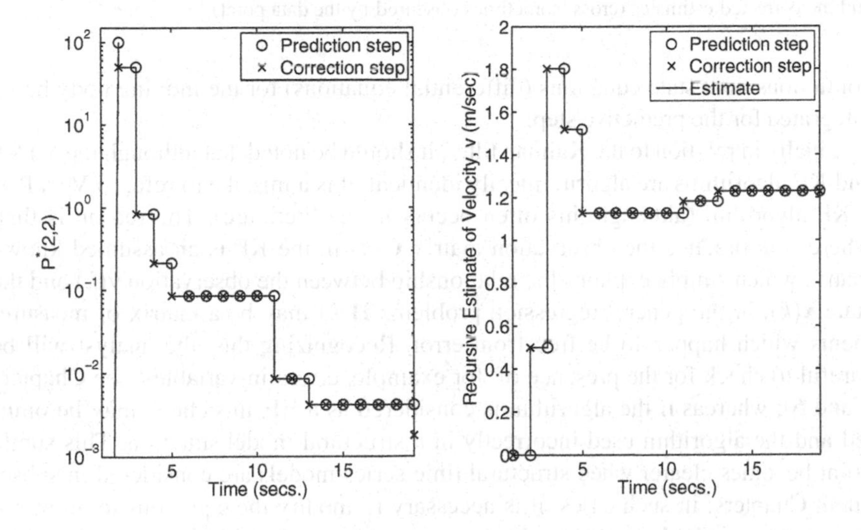

Fig. 4.9 Moving body example: Kalman filter recursive estimate of the velocity (right panel) and the associated diagonal element of the $\mathbf{P}^*(k)$ matrix (left panel).

so Figure 4.10 presents the predictions as a continuous line with forecasts of up to 6 seconds ahead (between 5 and 11 seconds), with the associated standard error bounds shown in grey, revealing again how the bounds expand over a gap such as this, where no measurements are available, and then contract suddenly once a measurement is made and assimilated by the KF. In fact, a CD version of the KF (see section 10.3 of Chapter 10) could have been used in this example, with the

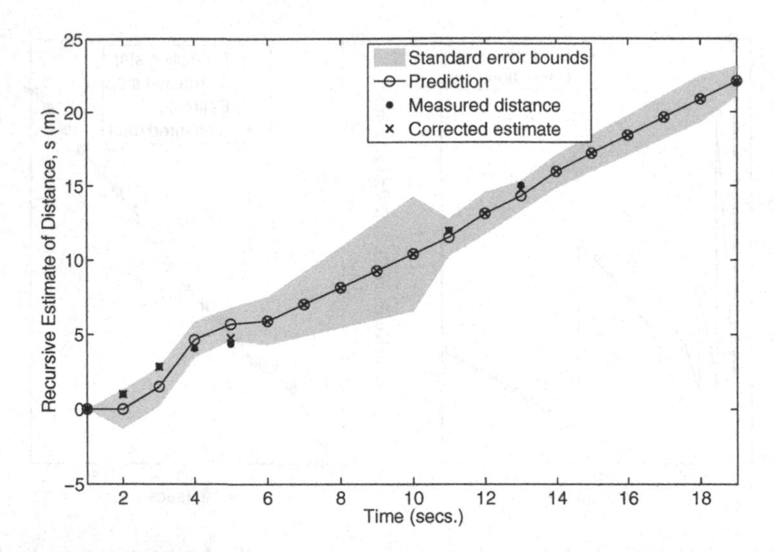

Fig. 4.10 Moving body example: Kalman filter recursive estimate of the distance showing the measured distance (solid dots); the one-second-ahead predictions (full-line with circular points); and the corrected estimates (cross, sometimes obscured by the data point).

continuous time state equations (differential equations) for the moving body being integrated for the predictive step.

Finally in relation to the Kalman filter, it should be noted that although the VTVP and KF algorithms are algorithmically identical, it is a mistake to refer to VTVP as a KF algorithm (although this often occurs in the literature). The reason is that, whereas in practice the observation matrix $\mathbf{C}(k)$ in the KF is an assumed known matrix which simply explains the relationship between the observation $\mathbf{y}(k)$ and the state $\mathbf{x}(k)$, in the general regression problems $\mathbf{H}(k)$ may be a matrix of measurements which happen to be free from error. Recognizing this, the analyst will be careful to check for the presence of, for example, errors-in-variables (see Chapters 2 and 6); whereas if the algorithm is considered as a KF, this check may be omitted and the algorithm used incorrectly in a structural model situation. This subtle point becomes clearer when structural time series models are considered in subsequent Chapters: in such cases, it is necessary to modify the equations to allow for the errors-in-variables and to avoid asymptotic bias on the estimates. They are then, however, significantly different from the KF equations.

4.5 Recursive Fixed-Interval Smoothing

Recursive estimation is not only useful in an 'on-line' situation, where data are being received continuously from some source and are being processed or 'assimilated' as they are received. It can also be of great utility in the off-line situation, where blocks of the time series data are available for analysis. However, in this situation, there are some advantages if the recursive 'filtering' equations, in all the algorithms from STVP to KFX, are accompanied by optimal recursive 'smoothing' equations.

At any instant k, the recursive 'filtered' estimate $\hat{\mathbf{a}}(k)$ obtained from the TVP algorithm is based on the data up to and including the kth sample, but it is clearly not a function of the future data $k+l,\ k+2,$ etc. But if a block of N samples is being processed (which could, for example, have been collected during some prior experiment or monitoring exercise), then it is tempting to consider how an 'off-line' recursive estimate at sample k could be generated that is a function of all the available data, i.e., using previous terminology, the estimate given all the data, or $\hat{\mathbf{a}}(k|N)$, where N is the total sample size or 'fixed interval'[3]. Note that the recursive FIS estimate is only relevant in the time-variable parameter (or state) estimation context: clearly if the parameters are constant, then $\hat{\mathbf{a}}(k|N) = \hat{\mathbf{a}}(N)$, for all k.

4.5.1 Simple FIS estimation

A good introductory treatment of recursive smoothing is given in Gelb et al (1974) who point out that the smoothed estimate at time k can be considered as the optimal combination of the recursive estimate obtained up to k, starting at the beginning of the data ($k = 1$), and the estimate obtained working backwards through the data from sample N to k. Optimal recursive FIS estimation will be discussed in the next sub-section but it is instructive to consider how a sub-optimal, but useful, FIS estimate can be obtained in a rather simple manner in the scalar case. Here, the recursive filtering algorithm, as used in the *forward pass* through the data, is used to process the filtered estimate $\hat{a}(k)$ in a *backward pass*, starting with the final estimate $\hat{a}(N)$ and covariance $p^*(N)$ obtained in the forward filtering pass: i.e. the 'data' in this backward pass are not the observations $y(k)$ but the recursive filtered estimate $\hat{a}(k)$ for $k = 1, 2, \ldots, N$, processed recursively *in reverse order*.

This achieves two objectives. First, the filtering effect of the backward pass is in addition to the filtering effect of the forward pass; and so, as shown in the example below, the parameter estimates are doubly filtered and so 'smoothed', as required. Second, and in some respects more importantly, the lag effect (see Chapter 2) of the forward pass recursive filtering is effectively negated by the backward pass because the forward pass estimates are filtered again *in reverse order*: effectively, the lag effect in one direction becomes a lead in the other and they cancel each other out (see later, exercise 3.).

Figure 4.11 illustrates the effect of FIS smoothing in the case of the Walgett data. The recursive filtering estimate of the mean value, as obtained by the application of the STVP algorithm with $q = 0.0111$, is shown as the dash-dot line; the full line is

[3] On-line 'Fixed Lag Smoothing' is also possible (Moore, 1973), where the recursive estimation works in a forward-pass, filtering mode but with smoothed estimates provided at every sampling instant k over a finite interval of l samples into the past (i.e. over the interval $k-l$ to k), but this is not discussed here.

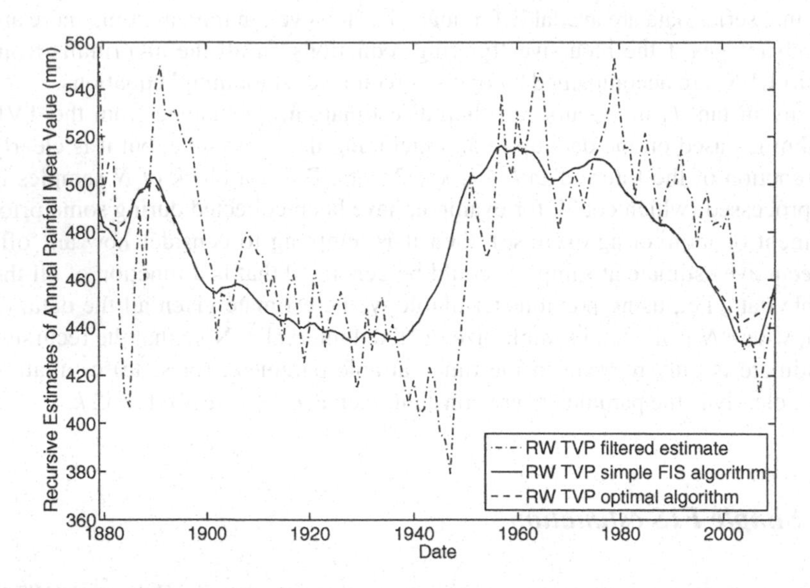

Fig. 4.11 Walgett rainfall mean value estimation: comparison of 'filtered' estimate (dash-dot); simple backward pass FIS estimate (full line); and the FIS estimate obtained using the optimal FIS algorithm (dashed line), all using sub-optimal $q = 0.0111$.

the estimate obtained by the simple FIS algorithm and, for comparison, the dashed line is the FIS estimate generated by the optimal FIS algorithm described in the next sub-section. It is clear that, except for the beginning and end of the data, the two FIS estimates are identical.

4.5.2 Optimal FIS algorithms

The simple recursive smoothing illustrated in Figure 4.11 is easy in the scalar case but not when estimating more than one parameter. The problem is then much more difficult and its solution attracted a lot of attention in the 1960s. Rauch et al. (1965) and Bryson and Ho (1969) consider the development of the optimal FIS algorithm from a Bayesian viewpoint. As pointed out previously, recursive estimation can be considered as the embodiment of Bayesian estimation and the Kalman filter, for example, can be obtained in a rigorous fashion using Bayesian concepts. In the smoothing case, however, the solution is not so well defined as in the previous recursive (or filtering) context. Norton (1975) has discussed this matter and compared various solutions to the problem which differ in computational and numerical stability terms. In the context of the GRWTVP estimation algorithm, his conclusions suggest that the most useful forms of the smoothing algorithm are obtained using either of the following equations:

$$\hat{\mathbf{x}}(k|N) = \hat{\mathbf{x}}(k) - \mathbf{P}^*(k)\mathbf{A}^T\boldsymbol{\lambda}(k) \tag{4.48}$$

or

$$\hat{\mathbf{x}}(k|N) = \mathbf{A}^{-1}[\hat{\mathbf{x}}(k+1|N)] - \mathbf{DQD}^T\boldsymbol{\lambda}(k) \tag{4.49}$$

where $\hat{\mathbf{x}}(k)$ and $\mathbf{P}^*(k)$ are obtained from the forward (filtering) pass through the data using the GRWTVP algorithm. Note that (4.49) involves the inverse of the state transition matrix $\mathbf{A}$, so that this matrix must be non-singular when utilizing this form of the smoothing update. The associated error covariance matrix $\mathbf{P}^*(k|N)$ can be generated in a number of ways; for example, Rauch et al. use:

$$\begin{aligned}\mathbf{P}^*(k|N) = \mathbf{P}^*(k) + \mathbf{P}^*(k)\mathbf{A}\mathbf{P}^*(k+1|k)^{-1} \\ [\mathbf{P}^*(k+1|N) - \mathbf{P}^*(k+1|k)]\mathbf{P}^*(k+1|k)^{-1}\mathbf{A}\mathbf{P}^*(k)\end{aligned} \tag{4.50}$$

where $\mathbf{P}^*(k)$ and $\mathbf{P}^*(k+1|k)$ are the covariance matrices generated by either of the recursive STVP or TVP algorithms (where strictly, using this nomenclature, $\mathbf{P}^*(k)$ should be denoted by $\mathbf{P}^*(k|k)$ to avoid ambiguity). The vector $\boldsymbol{\lambda}$ in algorithm (4.48) and (4.49) is a Lagrange multiplier term required in the solution of the optimization problem in order to ensure that the estimate is obtained under the constraint that the parameter vector evolves according to the state equations (4.34). It is obtained recursively from the equations

$$\begin{aligned}\boldsymbol{\lambda}(k-1) =& [\mathbf{I}_p - \mathbf{P}^*(k)\frac{\mathbf{h}(k)\mathbf{h}^T(k)}{\hat{\sigma}^2}]^T \\ & \{\mathbf{A}^T\boldsymbol{\lambda}(k) - \frac{\mathbf{h}(k)}{\hat{\sigma}^2}[y(k) - \mathbf{h}^T(k)\mathbf{A}\hat{\mathbf{x}}(k-1)]\}\end{aligned} \tag{4.51}$$

with $\boldsymbol{\lambda}(N) = 0$

where p is the number of parameters in the parameter vector $\mathbf{x}$ (see section 4.2.1). Other derivations are considered by Norton but are not favoured because of numerical and computational aspects. Note from the form of the above smoothing algorithm that, in contrast to the forward filtering pass, the covariance matrix equation (4.50) is not essential to the computation of the estimate and need only be computed if $\mathbf{P}^*(k|N)$ is required to assess the error covariance properties.

The combination of the recursive GRWTVP algorithm with the above recursive smoothing equations constitutes the *Kalman Filtering and Fixed Interval Smoothing* (KALMSMO) algorithm for estimating the parametric state $\mathbf{x}(k)$ defined by (4.34). Note that, as the name implies, the same general algorithmic form applies to more standard state equations (4.44) considered in section 4.4, except that, in general, the state equations will then relate to a physical system (such as the moving body example). As a result, the state variables will be defined in relation to this physical system (distance travelled and velocity in the moving body example), rather than the parametric states in the case of recursive TVP estimation. Moreover, in this more general context, the state equations could take any observable form and would not have the special form used in the present, more limited, TVP context. Since the KALMSMO algorithm, used within the TVP context, is the main algorithm exploited in the next Chapter 5, these filtering and smoothing equations are gathered together below in the alternative, and practically more useful form where, *in the initial forward filtering pass*, the NVR matrix $\mathbf{Q}_{nvr}$ is used, rather than $\mathbf{Q}_a$:

Prediction:

$$\hat{\mathbf{x}}(k|k-1) = \mathbf{A}\hat{\mathbf{x}}(k-1) \qquad \text{(KALMSMO-1)}$$

$$\mathbf{P}(k|k-1) = \mathbf{A}\mathbf{P}(k-1)\mathbf{A}^T + \mathbf{D}\mathbf{Q}_{nvr}\mathbf{D}^T \qquad \text{(KALMSMO-2)}$$

Correction:

$$\hat{\mathbf{x}}(k) = \hat{\mathbf{x}}(k|k-1) + \mathbf{g}(k)\{y(k) - \mathbf{h}^T(k)\hat{\mathbf{x}}(k|k-1)\} \qquad \text{(KALMSMO-3)}$$

$$\mathbf{g}(k) = \mathbf{P}(k|k-1)\mathbf{h}(k)[1 + \mathbf{h}^T(k)\mathbf{P}(k|k-1)\mathbf{h}(k)]^{-1} \qquad \text{(KALMSMO-4)}$$

$$\mathbf{P}(k) = \mathbf{P}(k|k-1) - \mathbf{g}(k)\,\mathbf{h}^T(k)\mathbf{P}(k|k-1) \qquad \text{(KALMSMO-5)}$$

$$\mathbf{P}^*(k) = \hat{\sigma}^2(k)\mathbf{P}(k) \qquad \text{(KALMSMO-6)}$$

Smoothing:

$$\hat{\mathbf{x}}(k|N) = \hat{\mathbf{x}}(k) - \mathbf{P}^*(k)\mathbf{A}^T\boldsymbol{\lambda}(k) \qquad \text{(KALMSMO-7)}$$

$$\text{or } \hat{\mathbf{x}}(k|N) = \mathbf{A}^{-1}[\hat{\mathbf{x}}(k+1|N) - \mathbf{D}\mathbf{Q}_a\mathbf{D}^T\boldsymbol{\lambda}(k)] \qquad \text{(KALMSMO-8)}$$

$$\begin{aligned}\boldsymbol{\lambda}(k-1) &= [\mathbf{I}_p - \mathbf{P}^*(k)\frac{\mathbf{h}(k)\mathbf{h}^T(k)}{\hat{\sigma}^2}]^T \\ &\quad \{\mathbf{A}^T\boldsymbol{\lambda}(k) - \frac{\mathbf{h}(k)}{\hat{\sigma}^2}[y(k) - \mathbf{h}^T(k)\mathbf{A}\hat{\mathbf{x}}(k-1)]\} \\ \text{with } \boldsymbol{\lambda}(N) &= 0 \qquad \text{(KALMSMO-9)}\end{aligned}$$

$$\begin{aligned}\mathbf{P}^*(k|N) &= \mathbf{P}^*(k) + \mathbf{P}^*(k)\mathbf{A}\mathbf{P}^*(k+1|k)^{-1} \\ &\quad [\mathbf{P}^*(k+1|N) - \mathbf{P}^*(k+1|k)]\,\mathbf{P}^*(k+1|k)^{-1}\mathbf{A}\mathbf{P}^*(k)\end{aligned} \qquad \text{(KALMSMO-10)}$$

Here, it will be noted that an estimate $\hat{\sigma}^2(k)$ is required to compute the covariance matrix $\mathbf{P}^*(k)$. However, in this off-line analysis, $\hat{\sigma}^2$ can be evaluated and then $\mathbf{Q}_a$ can be evaluated from $\mathbf{Q}_a = \hat{\sigma}^2\mathbf{Q}_{nvr}$ (see later, sub-section 4.5.4).

As pointed out earlier in section 4.4, the idea of modelling time variable parameters as simple stochastic processes and estimating their evolution in time using the above filtering and smoothing algorithms has its origin in the 1960s when control engineers realized that recursive estimation and, in particular the Kalman filter, could be applied to the problem of estimating time variable parameters in regression-type models. However, the full implications of this approach and its further development occurred later. Relevant publications are Norton (1975); West and Harrison (1989); Harvey (1989); Young et al. (1989); Young and Ng (1989); Young (1989); Ng and Young (1990); Young (1994) and Young (2010b), which show how this approach can be extended, in various ways, to problems of forecasting, backcasting, smoothing and signal extraction, as discussed in the next Chapter 5. An important aspect of these more sophisticated approaches relates to the optimization of the hyper-parameters, as discussed in the next section 4.5.3.

4.5.3 Optimization of hyper-parameters

In order to utilize the KALMSMO algorithm, it is necessary to specify the hyper-parameters associated with the stochastic model chosen to represent the parameter variations (e.g. the NVR parameters or α in the GRW model (4.33)). Let these unknown hyper-parameters be collected together in a hyper-parameter vector $\boldsymbol{\gamma}$. In the CAPTAIN Toolbox, this is accomplished by two methods: (i) maximum likelihood optimization based on *Prediction Error Decomposition*; and (ii) optimization of the single- or multiple-step-ahead prediction errors.

The former ML optimization method, which is more satisfactory in theoretical terms, derives originally from the work of Schweppe (1965), who showed how to generate likelihood functions for Gaussian signals based on the innovations term in the KALMSMO-3 equation, $\varepsilon(k) = y(k) - \mathbf{h}^T(k)\hat{\mathbf{x}}(k|k-1)$. Under the assumption that the stochastic white noise inputs to the system have a normal distribution, $\varepsilon(k)$ should constitute a zero mean, normally distributed, white noise sequence if the KF parameters are defined optimally: see also Bryson and Ho (1969), page 389. Its importance in the present TVP context was first recognized by Harvey (1981) and Kitagawa (1981). Since then, it has become one of the two standard approaches to the problem, the other being the *Expectation and Minimization* (EM) algorithm: see Dempster et al. (1977).

In order to define the likelihood function, note first that equations (KALMSMO-3) and (KALMSMO-4) can be combined to the following form:

$$\hat{\mathbf{x}}(k) = \hat{\mathbf{x}}(k|k-1) + \mathbf{P}(k|k-1)\mathbf{h}(k)\frac{\varepsilon(k)}{[1+\mathbf{h}^T(k)\mathbf{P}(k|k-1)\mathbf{h}(k)]}$$

because $v(k) = 1 + \mathbf{h}^T(k)\mathbf{P}(k|k-1)\mathbf{h}(k)$ is a scalar quantity. Now, if p is the number of TVPs being estimated and the first p observations are regarded as fixed, then the log-likelihood function of the series $y(k+1), \ldots, y(N)$ can be defined as follows in terms of the prediction error decomposition:

$$\begin{aligned}\log L = \frac{-(N-p)}{2}\log(2\pi) - \frac{1}{2}\log(\sigma^2) - \frac{1}{2}\sum_{k=p+1}^{k=N}\log(v(k)) \\ - \frac{1}{2\sigma^2}\sum_{k=p+1}^{k=N}\frac{\varepsilon^2(k)}{v(k)}\end{aligned} \quad (4.52)$$

Referring to equation (3.34) in the previous Chapter 3, $\sigma^2 v(k)$ is the variance of $\varepsilon(k)$, so that the last term in (4.52) is based on the sum of squares of the normalized one-step-ahead prediction errors.

Now the ML estimate of σ^2, conditional on the hyper-parameters, is given by,

$$\hat{\sigma}^2 = \frac{1}{N-p} \sum_{k=p+1}^{k=N} \frac{\varepsilon^2(k)}{v(k)} \tag{4.53}$$

so that it can be estimated in this manner and concentrated out of the expression (4.52) by substituting (4.53) into (4.52), to yield the following expression for this 'concentrated likelihood':

$$\mathscr{L}_c(\boldsymbol{\gamma}) = -\frac{N-p}{2}\log(2\pi+1) - \frac{1}{2}\sum_{k=p+1}^{k=N} \log(v(k)) - \frac{N-p}{2}\log(\hat{\sigma}^2) \tag{4.54}$$

which needs to be maximized with respect to the unknown elements of the hyper-parameter vector $\boldsymbol{\gamma}$ in order to obtain their ML estimates.

Since (4.54) is nonlinear in the hyper-parameters, the likelihood maximization needs to be carried out numerically. Consequently, it is more convenient to remove the constant term (since it will play no part in the optimization) and multiply (4.54) by -2, to yield the following optimization problem:

$$\underset{\boldsymbol{\gamma}}{\arg\min}\, \mathscr{L}_c(\boldsymbol{\gamma}) = \sum_{k=p+1}^{k=N} \log(v(k)) + (N-p)\log(\hat{\sigma}^2) \tag{4.55}$$

This minimization is accomplished by initiating the optimization with the hyper-parameter estimates either selected by the user or set to some default values (in both cases, ensuring that the resulting optimization does not converge on a local minimum). The recursive filtering part of the KALMSMO algorithm is used repeatedly to generate the one step ahead prediction errors $\varepsilon(k)$ and, thence, the value of $\mathscr{L}_c(\boldsymbol{\gamma})$ in (4.55) associated with the latest selection of hyper-parameter vector $\boldsymbol{\gamma}$ made by the optimization algorithm. The optimization algorithm then adjusts its selection of hyper-parameter estimates in order to converge on those estimates which minimize this concentrated likelihood. Further details of this and alternative ML optimization procedures are given, for example, in Harvey and Peters (1990). In the CAPTAIN Toolbox, the Matlab fmins and fminu functions are used for optimization and the initiation of optimization is discussed in the next Chapter 5.

4.5.4 Implementation of the KALMSMO algorithm

In order to implement the KALMSMO algorithm, it is necessary to specify the estimate of the measurement noise variance, $\hat{\sigma}^2(k)$. In the case where this is constant for all k, it is straightforward to obtain an estimate at the end of the filtering pass through the data, which uses the 'normalized' form of the recursion with $\hat{\sigma}^2 = 1.0$ and $\mathbf{Q}_{nvr}$ optimized as shown in the previous section 4.5.3. The $\hat{\sigma}^2$ estimate is then obtained from (4.53) and used to compute the covariance matrix $\mathbf{P}^*(k) = \hat{\sigma}^2\mathbf{P}(k)$ and $\mathbf{Q}_a = \hat{\sigma}^2\mathbf{Q}_{nvr}$, prior to running the recursive FIS pass through the data. This is the approach used in the standard versions of the 'dynamic' TVP routines in the CAPTAIN Toolbox (see next Chapter 5). Note that there are two routines in CAPTAIN that implement the KALMSMO algorithm: kalmsmo is the special implementation, as described above, which is called by other routines, in particular those associated with the 'unobserved components' model-type considered in the next chapter 5; while kalmanfis is a more general implementation, where the user has to supply the information on the state space model and parameters required for implementation of the Kalman filter in more general applications.

If the measurement noise is *heteroscedastic*, that is it has changing variance, and this changing variance is unknown *a priori*, then the procedure is not nearly as simple, since the estimate $\hat{\sigma}^2(k)$ has to be computed for all k. There are various ways of approaching this problem (see e.g. Kitagawa and Takanami, 1985, 2003) but the simplest approach is probably the three-stage procedure that is available as an option in the CAPTAIN Toolbox. This routine is an implementation of the KALMSMO algorithm for the simplest case of a single parameter linear regression,

$$y(k) = x(k) + e(k) \tag{4.56}$$

where $x(k)$ is being considered normally as a low frequency component or 'trend' in $y(k)$ (see the next Chapter 5) modelled as an RW, IRW or DIRW process. For example, in the case of the IRW model, the stochastic state equations take the following form (cf equation (4.5)):

$$\begin{aligned}
\begin{bmatrix} x(k) \\ \nabla x(k) \end{bmatrix} &= \begin{bmatrix} 1 & 1 \\ 0 & 1 \end{bmatrix} \begin{bmatrix} x(k-1) \\ \nabla x(k-1) \end{bmatrix} + \begin{bmatrix} 0 \\ 1 \end{bmatrix} \eta(k) \qquad \eta(k) = \mathcal{N}(0, \sigma_\eta^2) \\
y(k) &= \begin{bmatrix} 1 & 0 \end{bmatrix} \begin{bmatrix} x(k) \\ \nabla x(k) \end{bmatrix} + e(k) \qquad e(k) = \mathcal{N}(0, \sigma^2(k))
\end{aligned} \tag{4.57}$$

where, by definition, $\nabla x(k)$ is the change in $x(k)$ over the sampling interval Δt and the discrete-time derivative in time units is $\nabla x(k)/\Delta t$. Since this IRW model is normally the most effective in this role, the very simple version of the KALMSMO algorithm in this case is then often referred to as the 'IRWSMOOTH' and the CAPTAIN routine is called irwsm. The associated irwsmopt routine is also available in CAPTAIN to optimize the NVR hyperparameter using either ML optimization, as described in the above section 4.5.3, or by simply minimizing the sum of squares of the f-step-ahead predictions, where f is specified by the user.

Given the above model, the three-stage procedure used by irwsm in the heteroscedastic situation is outlined below, but the same basic approach could be used for the general TVP regression in n parameters.

1. Assume initially that the observation noise is homoscedastic, so that $\sigma^2(k) = \sigma^2 \; \forall \, k$ and use the irwsmopt routine to obtain an optimal estimate of the Noise

Variance Ratio , $NVR = \sigma_\eta^2/\sigma^2$. Now, apply the standard IRWSMOOTH routine irwsm to obtain an initial estimate $\hat{x}_0(k)$ of the trend $x(k)$ conditional on this optimized NVR value, and subtract this estimated trend from the data to provide an initial estimate $\hat{e}_0(k) = y(k) - \hat{x}_0(k)$ of the heteroscedastic noise $e(k)$.

2. Compute the squared residuals $\hat{e}_0^2(k)$ and repeat the procedure in stage 1, with $y(k)$ replaced by $\hat{e}_0^2(k)$, in order to obtain an estimate $\hat{\sigma}^2(k|N)$ of the changing variance $\sigma^2(k)$.
3. Run the KALMSMO algorithm with $\hat{\sigma}^2(k) = \hat{\sigma}^2(k|N)$ and, employing the same approach to NVR optimization as in steps 1. and 2., use this optimized NVR and the estimate of the heteroscedastic noise variance $\hat{\sigma}^2(k|N)$ to generate filtering and smoothing estimates of the state in (4.57). This state estimate provides the smooth trend as its first element $\hat{x}(k)$, with the second element $\nabla\hat{x}(k)$ providing the discrete derivative of the trend. The FIS estimate $\hat{e}(k)$ of the heteroscedastic noise $e(k)$ can then be obtained as $\hat{e}(k) = y(k) - \hat{x}(k|N)$.

If necessary, this procedure above could be iterated over steps 2. and 3. to further improve the estimates. However, experience suggests that this normally introduces no discernible change and is not essential. It is interesting to note that step 1. alone (i.e simple application of the IRWSMOOTH algorithm) often yields good estimates of the trend and its derivative, and steps 2. and 3. may not change the estimates too much (see the example in section 6.10.2 of the next Chapter 5). However, these additional steps should provide much improved estimates of the standard error bounds on the estimates.

Although the above approach to the estimation of heteroscedastic noise variance in step 2. is simple and probably sufficient for most practical purposes, it is not strictly optimal because the squared residuals $\hat{e}^2(k)$ do not have a normal distribution. One approach that attempts to correct this was suggested by Kitagawa and Gersch (1996) and it has been used in Young (2002a, 2006d) to estimate the heteroscedastic noise associated with models of hydrological and volcanic data. It is based on the fact that, under fairly non-restrictive conditions, the stochastic process $\chi^2(m)$ defined by

$$\chi^2(m) = \frac{\hat{e}^2(2m-1) + \hat{e}^2(2m)}{2}, \qquad m = 1, 2, \cdots, N/2$$

constitutes an independent random sequence of chi-squared random variables with two degrees of freedom. Then the transformation $t(m) = log\chi^2(m) + \gamma$, where $\gamma =$ 0.5722 is the Euler constant, yields $t(m)$ with an almost normal distribution. As a result, the FIS algorithm can be applied to $t(m)$ in order to provide a smoothed estimate $\hat{t}_s(m)$ of the low frequency changes in the variance (the 'envelope' function) of $y(k)$ over time. This estimate can then be inverse transformed by $\exp(\hat{t}_s(m) -$ 0.57722) and used to normalize y_k.

4.5.5 The physical nature of FIS estimation

Although the FIS estimates are computed by the recursive KALMSMO algorithm, the implicit, physical nature of the estimation can be compared with the well known *Centralized Moving Average* (CMA) approach to smoothing, where the estimate at any sample k is based on a simple average of the sampled values in a rectangular 'window' centralized on the sample at k. The rectangular window is particularly simple and other, more complex, window functions can be used to provide different types of smoothed estimates: see e.g. (Priestley, 1981), page 432 *et seq*, who discusses smoothing in relation to the spectral estimation in the frequency domain.

The nature of FIS in this CMA context can be investigated by smoothing $x(k)$ in the linear regression model (4.56), using a time series $x(k)$ that consists of $N+1$ values, all of which are zero, except for a single, unity value at sample $N/2+1$. This yields the impulse response of the FIS algorithm and reveals the shape of the smoothing window or 'kernel' at sample $N/2+1$ in the center of the series. Figure 4.12 shows the resulting weighting kernels associated with the RW, IRW DIRW and SRW parameter variation models.

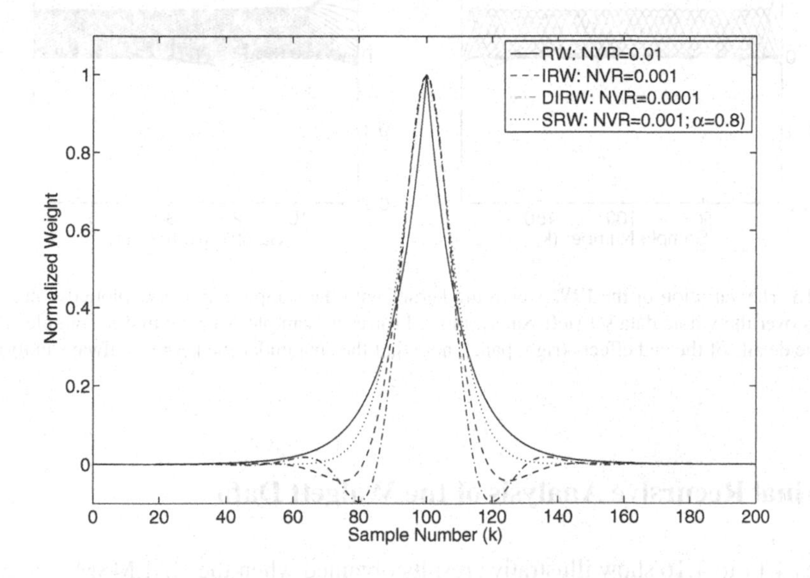

Fig. 4.12 Weighting kernels associated with the various GRW processes.

These show clearly that, in each case, the smoothing is based on averaging the data around any sample in a CMA manner with the weights defined by the shape of the kernel function (with the area under the kernel, as defined by the sum of the weights, always maintained at unity), rather than the simple equal weights of the rectangular kernel used in classical CMA. In this connection, note how the weights are all positive in the RW model case but have some negative values in the case of the other models: for instance, the IRW model exhibits the well known 'mexican hat' shape and reveals the connection of FIS estimation and 'wavelet' analysis: see e.g. Strang and Nguyen (1996).

Of course, the weighting kernels shown in Figure 4.12 are not static, they perform moving averages as they sweep through the data. And, in contrast to the simple CMA process, they yield estimates right to the edges of the data set, as shown in Figure 4.13, where it will be noted again that the area under all the weighting kernels is always maintained at unity.

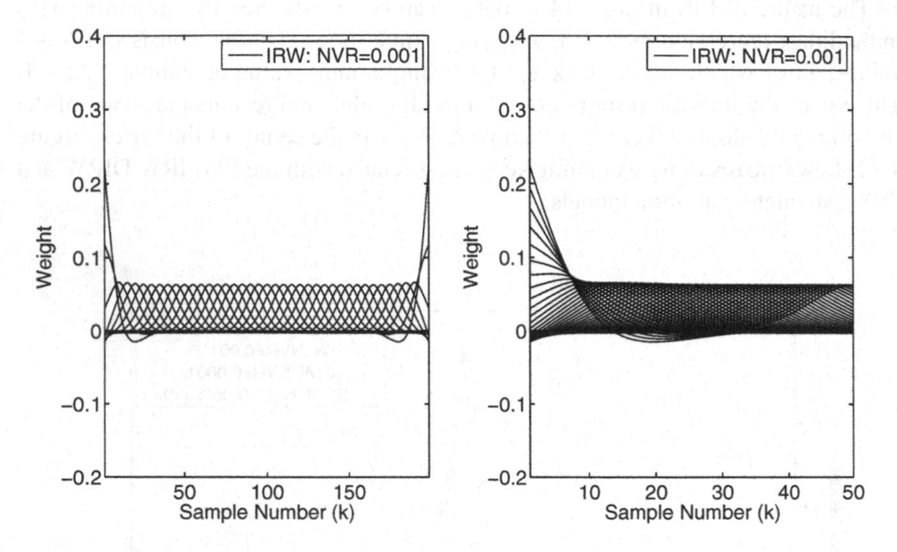

Fig. 4.13 The variation of the IRW weighting kernel with the sample number k: plotted every 5 samples over the whole data set (left panel); plotted for every sample over the first 50 samples to show the details of the end effects (right panel: note that the area under the kernel is always unity).

4.6 Final Recursive Analysis of the Walgett Data

Figures 4.14 to 4.16 show illustrative results obtained when the KALMSMO algorithm is used to estimate the possibly changing mean value of the Walgett rainfall data series. This is, of course, the simplest possible application of this general parameter estimation algorithm. First, Figure 4.14 compares various recursive filtering (left panel) and smoothing (right panel) estimation results based on IRW and SRW parameter variation models with a range of NVR and α hyper-parameter values. These were obtained using the dlr routine in CAPTAIN which is based on the KALMSMO algorithm and is described in more detail in the next Chapter 5. It is clear that a range of results can be obtained and those resulting from the larger values of the NVR and α hyper-parameters exhibit more variation, suggesting possibly significant changes in the mean value over the series. But on what basis should we be selecting these hyper-parameter values and are these estimated changes really significant?

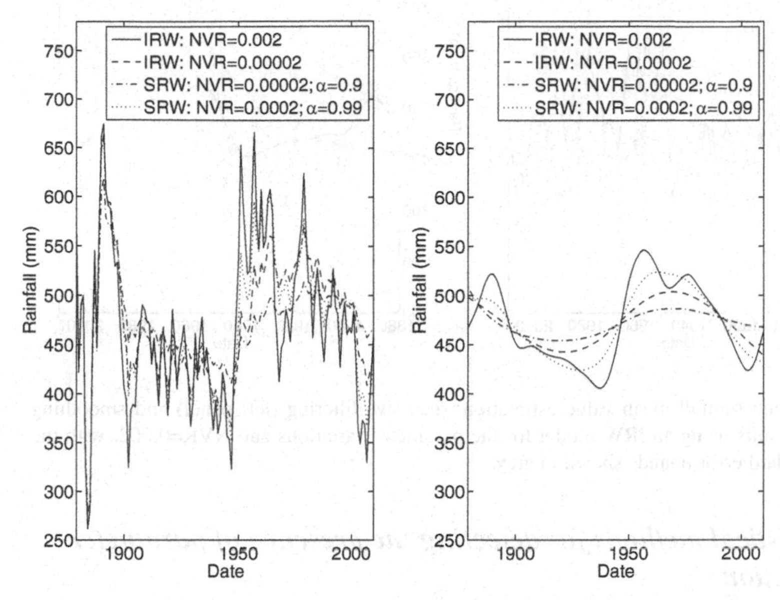

Fig. 4.14 Walgett rainfall mean value estimation: various recursive filtering (left panel) and smoothing (right panel) results for different TVP models and NVR values.

These questions can be addressed in various ways. First, the standard deviation confidence bounds could be considered, as shown in Figure 4.15, where only one set of results, with an IRW model for the parameter variations and NVR=0.002, is shown to avoid confusion. Now we see that, although the estimated changes in the mean value are quite large, so are the estimated SD bounds; as a result, it is very difficult to conclude that, within this level of estimated uncertainty, the changes are significant. Indeed, the FIS estimates, in particular, suggest that the changes around 1948 are similar to those that occurred at the end of the 19th Century; or even, stretching ones imagination perhaps too far given that there are only two peaks separated by about 70 years, that the changes may be part of a quasi-periodic pattern. But this is pure speculation and we are dealing with statistical algorithms, so surely it is much better if some form of statistical test can be applied.

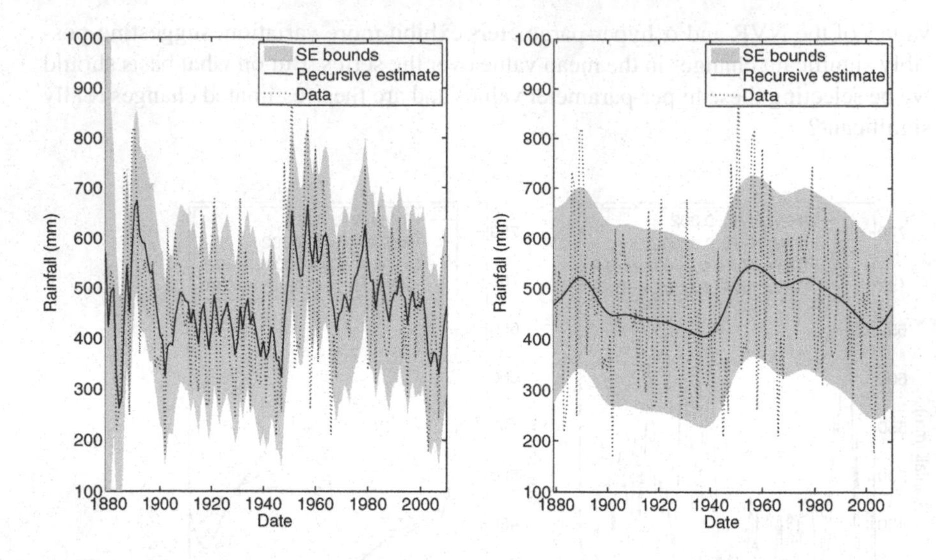

Fig. 4.15 Walgett rainfall mean value estimation: recursive filtering (left panel) and smoothing (right panel) results using an IRW model for the parameter variations and NVR=0.002, with the estimated standard error bounds shown in grey.

4.6.1 Statistical methods for detecting the presence of parameter variation

En bloc procedures for testing whether statistically significant parameter variation occurs over a set of observations have received a considerable amount of attention. Usually the data set is split into sub-sections and tests are applied to see if there is a significant difference in the estimates obtained from the different sub-sections. There appears to be three main sets of test statistics available: those based on the 't' test of residual means (Quandt, 1960); those based on analysis of variance (e.g. the F test of Quandt or its extension by Chow (1960) as a test of so-called 'structural change' - see also Harvey (1976); and those based on analysis of variance in the frequency domain using 'evolutionary spectra' Priestley (1981). More recently Tsay (1988) considers detection of outliers, level shifts and variance changes in time series.

Recursive methods for testing for parameter variation are also available and the best known is that suggested by Brown et al. (1975), which will help to exemplify the kind of approach that is possible in the recursive situation. As pointed out previously, the innovations sequence $\varepsilon(k) = y(k) - \mathbf{h}^T(k)\hat{\mathbf{x}}(k|k-1)$ should, if the assumptions regarding the regression problem are satisfied, be a zero mean, serially uncorrelated sequence of random variables with changing variance. Furthermore, if it is assumed that the errors of observation $e(k)$ have a normal Gaussian amplitude distribution then the innovations will also be Gaussian. Brown et al. build on the result developed earlier in Chapter 3 and show that the normalized innovations sequence $\varepsilon_n(k)$ in (3.35), i.e.,

$$\varepsilon_n(k) = \frac{\varepsilon(k)}{[1+\mathbf{h}^T(k)\mathbf{P}(k)\mathbf{h}(k)]^{0.5}} \tag{4.58}$$

is, under the assumption of parametric invariance, distributed as $\mathcal{N}(0,\sigma^2)$. As a result, they suggest that a test for detecting a departure from the zero expectation can function as a test for parametric change. Such a test can be derived from the cumulative sum or CUSUM, $W(k)$, where

$$W(k) = \frac{1}{\hat{\sigma}} \sum_{i=h+1}^{k} \varepsilon_n(i) \text{ with } \sigma^2 = \frac{1}{N-h} \sum_{i=h+1}^{N} \varepsilon_n^2(i) = \frac{S(N)}{N-h}$$

Here $\varepsilon_n(i)$ are obtained from the normal constant parameter RLS algorithm for each point i; h is the number of estimated parameters and N is the sample size. As might be expected, a plot of $W(k)$ against k will yield a graph which varies about zero if the parameters are indeed time-invariant, but which moves away from zero if permanent parametric change occurs. The significance of these movements is tested by the application of a procedure in which the probability limits are given by the lines joining the points h, $\pm a(N-h)^{0.5}$ and N, $\pm 3a(N-h)^{0.5}$, where a is a coefficient given by Brown et al for various significance levels β, as shown below:

$$\beta = 0.01,\ a = 1.143\ :\ \beta = 0.05,\ a = 0.948\ :\ \beta = 0.10,\ a = 0.850$$

Brown et al. also suggest a CUSUM-squared test (as well as other tests based on likelihood ratios). In this case, it is argued that the cumulative sum of squares $S(k)$, where

$$S(k) = \frac{1}{S(N)} \sum_{i=h+1}^{k} \varepsilon_n^2(i)$$

is distributed as a beta distribution in the time-invariant parameter case, with a mean of $(k-h)/(N-h)$ for h parameters. This follows a straight line on the plot of $S(k)$ against k and probability limits are given by drawing lines parallel to the mean value line at distances $\pm c_0 + (k-h)/(N-h)$, where c_0 is a parameter given in Table 4.1 for different significance levels β and sample sizes N.

In practice the CUSUM and CUSUM-squared graphs are plotted with associated significance bands chosen in accordance with the requirements of the analyst. If the test statistic crosses or comes close to the probability limits, then the hypothesis of parametric invariance is rejected and the possibility of parametric change needs to be investigated further. Brown et al. point out that the CUSUM-squared test is likely to be more sensitive to noise effects and changes in variance. They also stress that the significance bands are only approximate guides since the innovations are estimated model residuals, and tests such as those of Kolmogorov-Smirnov may be affected by this factor.

If the CUSUM and CUSUM-squared tests are applied to the Walgett data then the results are a little ambiguous. Even with $\beta = 0.1$ the CUSUM plot does not

Table 4.1 Probability of crossing CUSUM-squared test lines at significance levels (Brown et al., 1975).

N	$\beta = 0.1$	$\beta = 0.05$	$\beta = 0.01$
10	0.259	0.302	0.385
15	0.224	0.261	0.331
20	0.201	0.233	0.295
25	0.184	0.213	0.269
30	0.170	0.197	0.249
40	0.151	0.174	0.219
50	0.134	0.158	0.198
60	0.126	0.145	0.182
70	0.118	0.135	0.170
100	0.100	0.155	0.144

transgress the significance band; on the other hand, the CUSUM-squared plot does comes very close to the lower boundary. So, once again it is difficult to say that any statistically significant change in the mean value is occurring.

Finally, the ML optimization of the NVR parameter can be accomplished, as discussed in sub-section 4.5.3. Here, the results are relatively clear if we consider the FIS estimates, which are the most important because the estimates at any sample are based on all the data. Figure 4.16 shows the optimum estimation results for the IRW model of parameter variations, with an optimized NVR=0.00001, in the left hand panel; and for the RW model, with an optimized NVR that is virtually zero (2.5×10^{-17}), in the right hand panel. So the IRW model assumption suggests some possibly small, but not really significant changes in the FIS estimate; while the RW assumption suggests no change at all in the FIS estimate. Moreover, if we look at the computed *Autocorrelation* (ACF) and *Partial Autocorrelation* (PACF) functions for the recursive residuals in the RW case, as shown in Figure 4.17, they indicate that these are serially uncorrelated, as required, with no correlations exceeding the confidence bounds.

So we must conclude from the above analysis that there is very little evidence to support the supposition that the mean level of the rainfall changed in a statistically significant manner around 1948, at least on the basis of the available data. There are clearly long-term deviations in the mean and the variance of the series, which is consistent with the climate of Australia but, on the present evidence, they appear to be changes around a long-term constant mean value of 471.82 ± 13.938 cm. It is interesting to note that, in the first edition of this book, 36 years ago, the position was not quite so clear and it is the additional data acquired since then that has clarified the situation. So, when dealing with time series data that exhibit long-term fluctuations, it is obviously better to make sure one has sufficient data before reaching any firm conclusions, even when this means waiting a long time. Perhaps there is a message

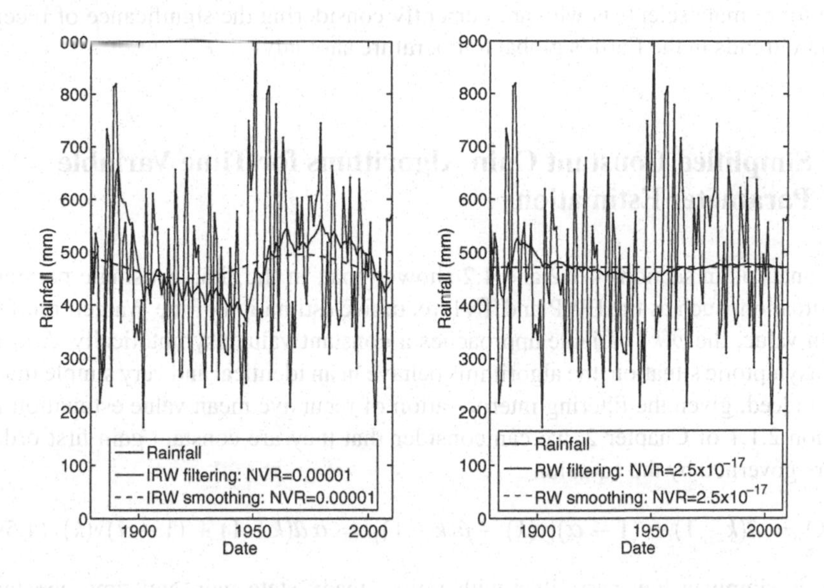

Fig. 4.16 Walgett rainfall mean value estimation: optimum estimation results for IRW model (left panel) and RW model (right panel).

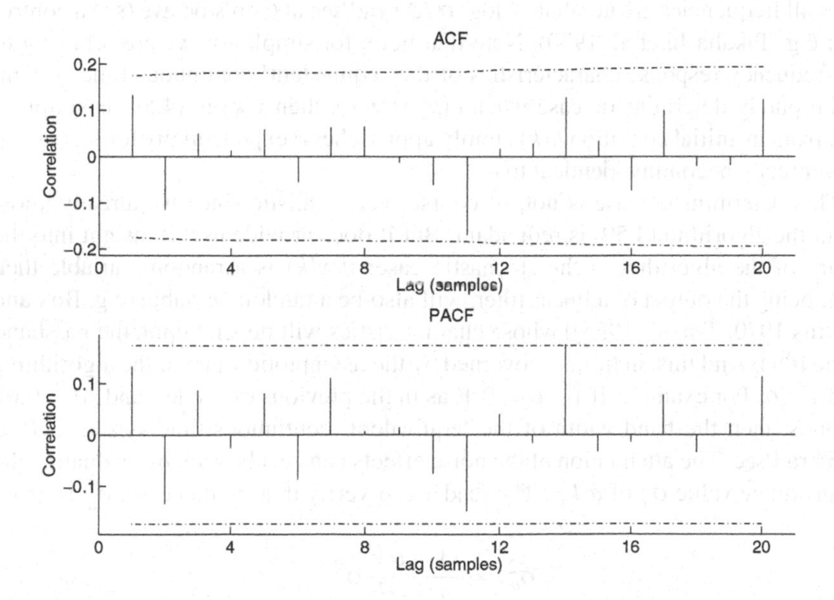

Fig. 4.17 Walgett rainfall mean value estimation: autocorrelation (ACF) and partial autocorrelation (PACF) analysis of the residuals for the optimum RW model case.

here for climate scientists who are currently considering the significance of recent upward trends in the Earth's global temperature anomaly?

4.7 Simplified Constant Gain Algorithms for Time Variable Parameter Estimation

The analysis in sections 4.1.2 and 4.2 showed that, in the simplest single parameter situation, such as the EWP and TVP recursive estimation of the Walgett rainfall mean value, the $p(k)$ variable approaches a constant value asymptotically. And in this asymptotic situation, the algorithms behave in an identical and very simple manner. Indeed, given the filtering interpretation of recursive mean value estimation in section 2.1.1 of Chapter 2, we can consider that they are constant gain first order filters governed by the equation

$$\hat{a}(k) = \hat{a}(k-1) + (1-\alpha)[y(k) - \hat{a}(k-1)] = \alpha\hat{a}(k-1) + (1-\alpha)y(k) \quad (4.59)$$

This is simply a low pass filter with unity steady state gain and time constant $-\Delta t/\log_e \alpha$; i.e. is the output of a first order, linear, discrete time filter with equivalent continuous-time bandwidth 0 to $-\log_e \alpha/\Delta t$ rad/sec which progressively attenuates all frequencies greater than $-\log_e \alpha/\Delta t$ rad/sec at 6 db's/octave (see a control text; e.g. Takahashi et al.,1970). Note that here, for simplicity, we are referring to the frequency response characteristics of the 'equivalent' continuous-time system. In the purely deterministic case when $e(k)$ is zero, then $y(k)$ in (4.59) is simply $\bar{a}$ and, from its initial condition $\hat{a}(k)$ simply approaches $\bar{a}$ exponentially for increasing k, eventually becoming identical to $\bar{a}$.

This deterministic case is not, of course, very realistic since we already know $\bar{a}$ and the algorithm (4.59) is redundant. But it does provide useful insight into the nature of the algorithm in the stochastic case: if $y(k)$ is a random variable then $\hat{a}(k)$, being the output of a linear filter, will also be a random variable (e.g. Box and Jenkins 1970; Truxal (1955)) whose characteristics will depend upon the passband of the filter. And this, in turn, is governed by the asymptotic value of the algorithmic gain $1-\alpha$. For example, if $1-\alpha = 0.1$, as in the previous examples, and $\Delta t = 0.01$ seconds, then the band width of the 'equivalent' continuous time system is 0 to 10.54 rad/sec. The attenuation of the noise effects can best be seen by evaluating the mean square value $\sigma_{\hat{a}}^2$ of $\hat{a}(k)$. The reader can verify that, in this case, $\sigma_{\hat{a}}^2$ is given by[4]

$$\sigma_{\hat{a}}^2 = \frac{(1-\alpha)^2}{1-\alpha^2}\sigma^2$$

[4] Hint:

$$E\{\hat{a}^2(k)\} = E\{[\alpha\hat{a}(k-1) + (1-\alpha)y(k)]^2\} \text{ and } E\{\hat{a}(k-1)y(k)\} = 0$$

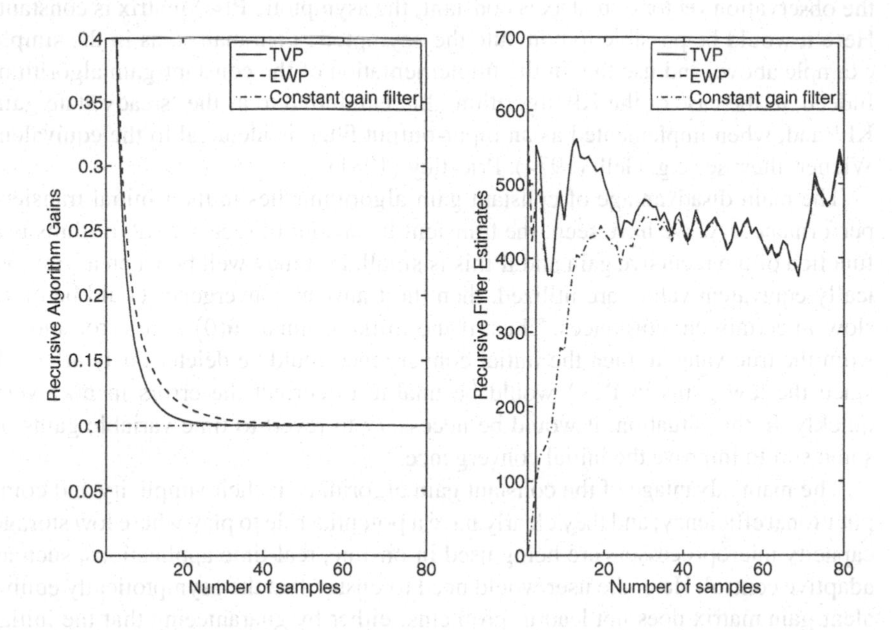

Fig. 4.18 Walgett rainfall mean value estimation: comparison of TVP, EWP and constant gain recursive filter performance showing filter gains (left panel) and recursive parameter estimates (right panel).

where σ^2 is the variance of $y(k)$. For $\alpha = 0.9$, therefore, $\sigma_{\hat{a}}^2 = 0.053\ \sigma^2$ and the residual variance of the estimate $\hat{a}^2(k)$ about the mean $\bar{a}$ (the true value) is quite small in relation to the 'observation' noise variance σ^2. And, at the same time, the convergence from an initial large error in the estimate $\tilde{a}(0) = \hat{a}(0) - \bar{a}$ will take place with a reasonable time constant of $\tau = -1/\log_e \alpha = 9.49$ sampling intervals (0.095secs)

Of course, the whole point of this filtering mechanism in terms of time variable parameter estimation is that, by passing some of the noise in the above manner, the algorithm is also able to detect variations in $\bar{a}$. And these variations should be apparent to the analyst provided they are of a different nature to, and larger than, the noise effects. In general, this will be the case, with the noise tending to have wide band characteristics with power at the higher frequencies while the changes in the parameter (here $\bar{a}$) will usually be in the form of low frequency 'trends' or slow variation as in the Walgett rainfall example.

The situation is, of course, much more complicated in the multi-parameter case with a time varying observation vector or matrix ($\mathbf{x}(k)$, $\mathbf{h}(k)$, $\mathbf{H}(k)$ or $\mathbf{C}(k)$, depending on which algorithm is being used). It is still reasonable to assume, however, that a constant gain matrix $\mathbf{P}(k)$ can be selected that, in some sense, 'approximates' the $\mathbf{P}(k)$ matrix for large k and which yields asymptotic results similar to those of the algorithm with time varying $\mathbf{P}(k)$. In one important case, namely the situation where the observation vector or matrix is constant, the asymptotic $\mathbf{P}(\infty)$ matrix is constant. Here it would be possible to compute the asymptotic gain matrix, as in the simple example above, and use this in the implementation of the constant-gain algorithm. Indeed, in the case of the KF algorithm, this is referred to as the 'steady state gain KF' and, when implemented as an input-output filter, is identical to the equivalent Wiener filter: see e.g. Gelb (1974); Priestley (1981).

The main disadvantage of constant gain algorithms lies in their initial transient performance. As we have seen, the transient behaviour of recursive algorithms is a function of the recursive gain and if this is small, as it may well be when asymptotically equivalent values are utilized, then the transient convergence could be quite slow in certain circumstances. Thus, if the initial estimate $\hat{\mathbf{a}}(0)$ is in error and far from the true value $\mathbf{a}$, then the initial convergence could be deleteriously affected, since the low gains in $\mathbf{P}(\infty)$ would be unable to correct the errors in $\hat{\mathbf{a}}(k)$ very quickly. In this situation, it would be necessary to revert to time variable gains of some sort to improve the initial convergence.

The main advantage of the constant gain algorithms is their simplicity and computational efficiency; and they clearly have a potential role to play where low storage capacity microprocessors are being used in on-line, real-time applications, such as adaptive control. Here, the user would need to ensure that the asymptotically equivalent gain matrix does not lead to problems: either by guaranteeing that the initial error $\tilde{\mathbf{a}}(0)$ is small enough to be corrected adequately in a reasonable time; or by employing some simple variable gain mechanism, similar to that used in the EWP algorithm (section 4.1.2).

4.8 The Estimation of Rapidly Varying Parameters

The various possibilities for the stochastic modelling of parameter variations provided by the GRW model (4.33) are useful in practice simply because they require the specification of so little *a priori* information on the nature of parameter variations. But, if it is likely that the parameters will vary rapidly between samples, then we cannot expect algorithms based on these simple stochastic parameter variation laws to necessarily function satisfactorily. In this situation, we need to look for procedures whereby additional information on the nature of the parameter variations can, if available, be incorporated into the algorithm to aid its 'parameter tracking' performance; in other words, procedures which utilize this information to give the algorithm prior warning of the direction and magnitude of the parametric change.

The most obvious approach to the problem is to make use of the more complicated Gauss-Markov model of parameter variation than (4.33). But, as pointed out previously, this would require the specification of the $\mathbf{A}$ and $\mathbf{D}$ matrices and there is no general procedure for doing this: its efficacy in practice will depend upon the problem at hand and the nature of the *a priori* knowledge available on parameter variation. An example where this idea has been exploited is described in Young (1981b) (see also Mendel and Fu (1970), page 230 *et seq.*, and the first edition of this book) where a 'self-adaptive' autostabilization system for an airborne vehicle is described in which the model parameter vector $\mathbf{a}(k)$ is considered to be a simple linear function of on-board sensor measurements that provide information on 'external' variables (here dynamic pressure, mass and Mach number) that define the flight conditions of the vehicle. These is then used to define a suitable Gauss-Markov model for the parameter variations that reflect the changing flight conditions of the vehicle.

In particular the model parameter vector $\mathbf{a}(k)$ is defined as follows:

$$\mathbf{a}(k) = \mathbf{T}(k)\mathbf{a}^*(k); \quad \mathbf{a}^*(k) = \mathbf{a}^*(k-1) + \boldsymbol{\eta}(k-1) \tag{4.60}$$

where $\mathbf{T}(k)$ is a time variable, diagonal transformation matrix with the sensor measurements defining its elements; and $\mathbf{a}^*(k)$ is a slowly variable parameter that allows for any changes in $\mathbf{a}(k)$ not accounted for by $\mathbf{T}(k)$. As shown, this is described by a vector random walk process, which is simply the vector equivalent of the RW model considered previously in this chapter (see also section 5.4 of the next chapter 5). It is easy to see that these equations can be combined to produce a Gauss-Markov model for the variations in $\mathbf{a}(k)$ of the same form as the state equations in (4.44), i.e.,

$$\begin{aligned} &\mathbf{a}(k) = \mathbf{A}(k)\mathbf{a}(k-1) + \mathbf{D}(k)\boldsymbol{\eta}(k-1) \\ &\text{where,} \quad \mathbf{A} = \mathbf{T}(k)\mathbf{T}^{-1}(k-1); \quad \mathbf{D}(k) = \mathbf{T}(k) \end{aligned} \tag{4.61}$$

and that, if the changes in the $\mathbf{T}(k)$ matrix elements reflect the major changes in the model parameters, then this will considerably enhance the TVP estimation. This is important because the estimated TVPs are used to update the control gains in the autostabilization control system.

Within this control and systems context, an alternative to self-adaptive control is 'gain scheduling', where measured variables, such as the measurements from the on-board sensors mentioned above, which provide information about the changing dynamic characteristics of the controlled system, are used to update the control gains and so help to ensure that good control is maintained despite these changes. One approach to gain scheduling is the use of *Linear Parameter Varying* (LPV) models (Shamma and Athans, 1991), which are TVP models where the parameters are assumed to be functions of the gain scheduling variables that can be parameterized in some manner: for example as a polynomial functions characterized by constant coefficients: see e.g. Laurain et al. (2010) and the prior references therein. Clearly, this LPV approach is conceptually quite similar to the self-adaptive approach outlined above. The main differences are that the LPV approach considers a wider class of parameterizations than the simple linear relationship in (4.60); and the coefficients that define the functions of the gain scheduling variables are not often assumed to change, so they are not normally updated recursively on-line. Rather these parameters are estimated off-line using some suitable parameter estimation algorithm in which the parameters are normally assumed to be time invariant. For instance, Laurain et al. (2010) use a 'refined instrumental variable' estimation algorithm of the kind described later in chapters 7 and 8. Two examples that do use recursive estimation are Giarre et al. (2006) and Wei et al. (2005), but here it is used for initial estimation of the LPV function coefficients, again assuming that they are constant.

There are many publications on LPV modelling, to numerous to review here given the tenuous link with recursive estimation, but some recent research (Previdi and Lovera, 2004) in this area on *Nonlinear Parametrically Varying* (NLPV) models is of relevance in this book. However, such models can be considered as members of a wider *State-Dependent Parameter* (SDP) class of models that are considered later in Chapter 11. Here, the 'state' includes any variables on which the parameters may depend, including 'internal' variables *within* the system equations, such as the input, output and their past values, so the resultant model is clearly nonlinear.

4.9 Variance Intervention

It cannot be assumed that all changes in time variable parameters will be smooth: for instance, some time series can exhibit sharp changes in the trend and its slope at some locations that will be referred to here as 'intervention' points. There have been numerous papers published on this general topic of 'Intervention Analysis': see e.g. Harvey and Durbin (1986); Harvey (1989); Tsay (1988); Weston and Norton (1997) and the prior references therein.

It is not possible to review the various approaches to intervention analysis and only one of the simplest methods, *Variance Intervention* (VI) will be considered and used in this book (Young and Ng, 1989). This is based on exploitation of the NVR hyper-parameter's ability to control the rate of change of the associated time variable parameters, so that rapid change can be specified by making these large. Consequently, if the NVR is transiently increased in value at any user-specified or automatically detected sampling instant, then the estimated parameter will change significantly at this sampling instant *provided there is information in the data to suggest a significant change in value*. However if, at the same time, the NVR is maintained at the normal optimized value for the samples either side of this intervention point, then this transient increase in the NVR will simply signal to the recursive algorithm the need for a sudden change in the parameter only at the intervention point (or points).

Considering the simplest case of the IRW model within the setting of the single parameter linear regression model,

$$y(k) = x(k) + e(k) \qquad e(k) = \mathcal{N}(0, \sigma^2) \tag{4.62}$$

the IRW model for $x(k)$ is given by a modified version of equation (4.57):

$$\begin{bmatrix} x(k) \\ \nabla x(k) \end{bmatrix} = \begin{bmatrix} 1 & 1 \\ 0 & 1 \end{bmatrix} \begin{bmatrix} x(k-1) \\ \nabla x(k-1) \end{bmatrix} + \begin{bmatrix} 1 & 0 \\ 0 & 1 \end{bmatrix} \begin{bmatrix} \eta_1(k-1) \\ \eta_2(k-1) \end{bmatrix} \tag{4.63}$$

where now,

$$\boldsymbol{\eta}(k) = \begin{bmatrix} \eta_1(k) \\ \eta_2(k) \end{bmatrix} = \mathcal{N}(\begin{bmatrix} 0\,0 \\ 0\,0 \end{bmatrix}, \begin{bmatrix} \sigma^2_{\eta_1}(k) & 0 \\ 0 & \sigma^2_{\eta_2}(k) \end{bmatrix})$$

The variances of $\sigma^2_{\eta_1}(k)$ and $\sigma^2_{\eta_2}(k)$, as well as the associated $NVR_1(k) = \sigma^2_{\eta_1}(k)/\sigma^2$ and $NVR_2(k) = \sigma^2_{\eta_2}(k)/\sigma^2$ are now time variable to allow for intantaneous changes in both the level $x(k)$ and the rate of change $\nabla x(k)$ of the parameter. Typically, if there are I interventions at $k = r_1, r_2, \ldots, r_I$, then the $NVR_i(k), i = 1,2$, will be constant at their optimized values for all k except at the r intervention samples, where they will be set to a larger value, NVR_I. Clearly, continuity in the slope over the intervention can be obtained by maintaining $\sigma^2_{\eta_2}$ at a constant optimized value, with no intervention changes,

Variance intervention is facilitated by the fact that the estimation is not very sensitive to the value of the NVR_I at the intervention point, so that any large value will suffice to allow for the estimation of any discontinuous change in the parameter that is justified at the specified intervention sample. For example, all the TVP estimation algorithms in the CAPTAIN Toolbox allow for VI and specify a default value of 100. This concept can be applied in any recursive TVP estimation algorithm that is using a Gauss-Markov process, such as the GRW model (4.33) or the DIRW process (4.32), to model the parameter variations; and it can be applied to any of the level or the slope states associated with the model. An example of VI applied to the estimation of discontinuous-change in a time variable 'trend' parameter is described in section 5.2.3 of the next Chapter 5.

4.10 Exercises

1. By reference to the derivation of the RLS algorithm in Chapter 3, derive the EWP algorithm.
2. Implement the simple scalar FIS algorithm described briefly in the text of section 4.5.1. Then apply this to the Walgett data and check the result using the following Matlab m-file script, which exploits the dlr routine in CAPTAIN, where z is the Walgett data vector:

```
%Forward path filtering
[fit,fitse,par,parse,comp,e,y0]=dlr(z,ones(size(z)),0,qa,...
[],570,100,0);
zr=flipud(par);%reverse order of par filtered estimate
%Simple backward path smoothing of z
[fitb,fitseb,parb,parseb,compb,eb,y0b]=dlr(zr,ones(size(zz)),...
0,qa,[],pard(end),parsed(end)^2,0);
%Optimal FIS for comparison
[fitd,fitsed,pards,parsed,compd,ed,y0d]=dlr(z,ones(size(z)),...
0,qa,[],570,100,1);
```

3. Below is the Matlab m-file script to implement the Kalman filter algorithm for the moving body example, with the state equations specified in (4.47) but with added system noise inputs $\eta_1(k)$ and $\eta_2(k)$.

```
N=30;%number of samples
```

```
a=1;A=[1 a;0 1]; D=[1 0;0 1];C=[1 0];%model parameters
x=zeros(2,N);y=zeros(N,1);xh=zeros(2,N);%define variables
veta1=0.0;veta2=0.0;ve=0.43;%variance of noise variables
eta1=sqrt(veta1)*randn(N,1);eta2=sqrt(veta2)*randn(N,1);
e=sqrt(ve)*randn(N,1);%generate noise variables
x(:,1)=[0 1.22]';%initial conditions
for i=2:N%simulation
x(:,i)=A*x(:,i-1)+D*[eta1(i) eta2(i)]';
y(i)=C*x(:,i)+e(i);
end
Q=[veta1 0;0 veta2];%system input covariance matrix
xh(:,1)=[0 1.22]';P=[100 0;0 100];%initial conditions
for i=2:N%KF estimation
    xh(:,i)=A*xh(:,i-1);
    P=A*P*A'+Q;
    g=P*C'./(ve+C*P*C');
    inn(i)=y(i)-C*xh(:,i);
    xh(:,i)=xh(:,i)+g*inn(i);
    P=P-g*C*P;
end
```

Check and run the script; evaluate the results; and then modify the noise variance parameter veta2 (the variance of η_2) in order to get a feeling for the operation of the Kalman filter when there are random changes in the velocity.

4. Write a Matlab script to implement the simple IRWSMOOTH algorithm (see section 4.5.4) and use the irwsm routine in CAPTAIN to check that it is working satisfactorily. This will provide valuable insight into the nature of the filtering and smoothing recursions.

4.11 Summary

This long Chapter has explored the many aspects of recursive time variable estimation from modifications of the simple RLS algorithm to the optimal state estimation procedures of R.E. Kalman. The objective has not just been simply to derive algorithms, but more to provide the reader with an insight into why and how they work. In this manner, it is hoped that both the synthesis and the use of the various algorithms will become more transparent.

But the discussion has been limited in one important practical respect: we have restricted the analysis to the 'general linear regression model' and have not addressed the important problem of errors-in-variables. This will be corrected later in Chapters 6 and 7, which show how it is possible to modify the regression algorithms to handle the very real errors-in-variables problems that arise naturally when attempting to estimate parameters in certain important transfer function models of stochastic dynamic systems. Before this, however, the next Chapter 5 describes how the KALMSMO algorithm developed in sub-section 4.5.2 of the present Chapter provides the basis for the estimation of the more general and practically important *Unobserved Component Model* (UCM).

Chapter 5
Unobserved Component Models

For many years, the problem of filtering and the extraction of component signals from noisy time series data has occupied the minds of mathematicians and statisticians from a variety of disciplines. Indeed, three of the most notable mathematicians of the Twentieth Century - Wiener (1941), Kolmogorov (1941), and Kalman (1960) - have all made central contributions to the development of filter and signal extraction theory. A useful concept in filtering, smoothing and signal extraction is the *Unobserved Component* (UC) model, where the observed variable $y(k)$ is related to a number of components which represent different perceived features of the data, usually differentiated by their characteristic spectral properties. For example, consider the two well known time series in Figure 5.1: the monthly atmospheric carbon dioxide (CO_2) measurements in Mauna Loa in Hawaii, over the period 1958-2000; and the monthly airline passenger data over the period 1949-1960. Both of these time series have two rather obvious characteristics: both are 'nonstationary' time series, exhibiting a pronounced upward 'trend' and an 'annual cycle'; and, in the airline passenger case, this annual cycle is growing in amplitude over the observed time period. Note also that atmospheric CO_2 series has some missing monthly measurements which, as we shall see, present no real difficulty when these data are analysed using an appropriate UC model.

This Chapter shows that the UC model can be considered as a direct extension of the TVP and STVP time variable parameter regression models presented in the previous Chapter 4. Within a macro-economic context, unobserved components, such as the business cycle, have been termed 'stylized facts' (Agénor et al., 2000) and one of the best known methods used for extracting such behaviour is the 'Hodrick-Prescott (HP) filter' (Hodrick and Prescott, 1980, 1997; Pedregal and Young, 2001). This 'filter' is actually a fixed interval smoothing algorithm and it can yield exactly the same smoothing results as the simple IRWSMOOTH smoothing (CAPTAIN routine is irwsm) discussed in section 4.5.4 of the previous Chapter 4. However, the HP filter is not as flexible as IRWSMOOTH because it is derived using an alternative, non-recursive 'deterministic regularization' approach (see Appendix D and Jakeman and Young (1984)).

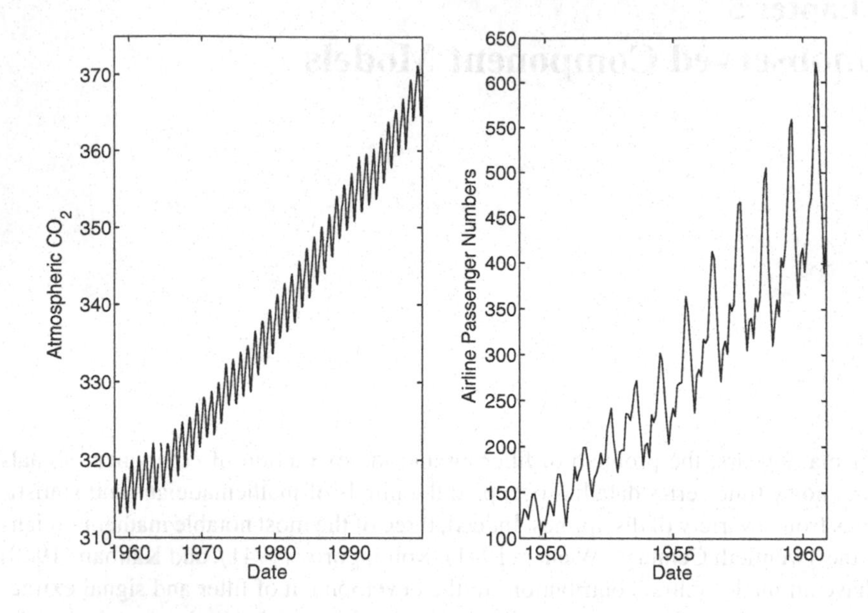

Fig. 5.1 Two typical, nonstationary time series: atmospheric carbon dioxide (CO_2) measurements at Mauna Loa, Hawaii, 1958-2000; and airline passenger numbers 1949–1960 (Box and Jenkins, 1970).

Although the UC model can be formulated in full multivariable (multi-input, multi-output) terms, the discussion here will be restricted to the case of a single output variable $y(k)$, since multivariable formulations are complicated and the single output case is much easier to use and better for the tutorial purposes of the present book. A typical and fairly general discrete-time UC model for $y(k)$ takes the form:

$$y(k) = T(k) + S(k) + C(k) + \mathcal{F}\{\mathbf{u}(k)\} + \xi(k) + e(k) \qquad e(k) = \mathcal{N}(0, \sigma^2) \quad (5.1)$$

Here, $y(k)$ is the observed output time series; $T(k)$ is a trend or low frequency component, such as that seen in the CO_2 and airline passenger series; $S(k)$ is a seasonal component, such as the annual cycle that is also so obvious in both of these series; $C(k)$ is a sustained cyclical or quasi-cyclical component, such as the business cycle in economic data or the 'El-Nino' variations in climate data, with a period different from that of any seasonality in the data; $\mathcal{F}\{\mathbf{u}(k)\}$ is a term that captures the influence of an input or 'exogenous' variable vector $\mathbf{u}(k)$. The effects of such exogenous variables are potentially multifarious since the component $\mathcal{F}\{\mathbf{u}(k)\}$ is included to model any non-random behaviour in $y(k)$ that is not accounted for by the other components but can be explained by the effects of other variables. For example, even if these other variables are not themselves periodic, they may inject periodic or quasi-periodic components in $y(k)$ through the dynamic properties of any transfer functions or nonlinear dynamic processes that are identified in $\mathcal{F}\{\mathbf{u}(k)\}$. Not all oscillatory behaviour is sustained, however, and this is the reason for the inclusion of the stochastic perturbation component $\xi(k)$, which is introduced to explain any decaying oscillations in $y(k)$ that are not accounted for by the effects of any exogenous variables. Finally, the irregular component $e(k)$ represents any stochastic and totally unpredictable temporal variations in $y(k)$ that have not been explained by all the other components.

In order to allow for nonstationarity in the time series $y(k)$, it is assumed that the various components in (5.1), including the trend $T(k)$, can be characterized by time variable parameters. Most often, the nature of such parametric time variability will not be known prior to the analysis and so each TVP introduced into the model is defined as a nonstationary stochastic variable, as discussed in the last Chapter 4. This adds a statistical degree of freedom to the estimation problem, so allowing for the estimation of any slow parameter variations that may result from slow physical changes in the process: for example, in the case of environmental systems, deforestation of a catchment, the silting of a reservoir, and climate change; or in an economic context, the possible influence of social and political changes over long periods of time.

In practice, not all the components in the UC model (5.1) are necessary: indeed, the simultaneous presence of all these components can induce identifiability problems in which it is not possible to unambiguously estimate all the components. As a result, use of the complete model (5.1) is not advisable in practice unless special precautions are taken to avoid these problems. In this Chapter, therefore, four special examples of the general UC model (5.1) are considered that can prove particularly useful in practical terms: *Dynamic Linear Regression* (DLR); Dynamic Harmonic Regression (DHR); *Dynamic Auto-Regression* (DAR) and *Dynamic Auto-Regression with Exogenous Variables* (DARX).

The adjective 'dynamic' used in these names is rather misleading: it derives from historical precedent, primarily through the work of Harrison and West (Harrison and Stevens, 1976; West and Harrison, 1989) on the *Dynamic Linear Model* (DLM). These authors exploit the Kalman filter for time variable parameter estimation, exactly as we have done in the previous Chapter but, consequent upon their background, they stress its interpretation in Bayesian estimation terms that were not, as we have pointed out, the original stimulus for Kalman's derivation of his filter. Moreover, this Bayesian analysis is not as simple as the alternative approach used in the present book and it is not essential to the understanding of UC models, their estimation and use. Consequently it will not be considered further here and the interested reader should consult West and Harrison (1989) which deals with the subject in considerable depth.

In general, the term 'dynamic' simply denotes that these DLM models, which are normally formulated within a linear regression framework, have time varying parameters that are modelled as stochastic dynamic processes, as discussed in the previous Chapter 4. As a result, models such as DLR and DHR are not inherently dynamic at all in a systems sense: rather they are models that are able to adapt their parameters to changes in the characteristics of the time series being analysed. On the other hand, the DAR and DARX models consist of dynamic difference equations and are truly dynamic although, fortuitously, they can also be written and analysed in simple linear regression terms.

But the most important characteristic of all these 'dynamic' models is undoubtedly the fact that they have time variable parameters that can be estimated from the time series data in an optimal manner provided the assumptions of the UC model are satisfied. For this reason, they are inherently self-adaptive: namely, they change their parameters automatically in an optimal manner to reflect changes in the statistical nature of the time series being analyzed. For this reason, they can be exploited in the 'off-line' analysis of time series, where recursive FIS estimation allows for signal extraction and the investigation of the reasons for the parameter variation; or 'on-line' in applications such as self-adaptive forecasting, operational control and management, where recursive filtering provides the main algorithmic tool.

5.1 The Dynamic Linear Regression (DLR) Model

The simplest model within the UC context is the DLR model, which is merely a straightforward extension of the TVP algorithm introduced in the previous Chapter 4 to include a trend component that allows for changes in the mean value of the underling regression relationship, i.e.,

$$y(k) = T(k) + \mathbf{x}^T(k)\mathbf{a}(k) + e(k) \tag{5.2}$$

where $\mathbf{a}(k)$ are either constant parameters (the normal regression model) or they may vary over the observation interval to reflect possible changes in the regression relationship and are modeled by one of the GRW family (4.33) of simple random walk processes, as discussed fully in the previous Chapter 4. In this DLR setting, $T(k)$ is effectively another TVP with an associated and implicit regression variable of unity. In the standard regression terminology, therefore, it can be considered as a stochastically variable 'intercept' parameter. And, as such, it is assumed to be described by one of the GRW class of stochastic processes introduced in the previous section.

Of course, the regression variables in the vector $\mathbf{x}(k)$ can be defined in any manner that is appropriate to the modelling of variations in $y(k)$ and this will be dependent upon the nature of the problem. However, it is assumed that they will represent relationships that are changing slowly over time in an *a priori* unknown manner. For instance, they could be defined as input variables that are causing instantaneous changes in $y(k)$, rather than the dynamic changes induced by inputs that enter through the $\mathcal{F}\{\mathbf{u}(k)\}$ component. The final result of the UC analysis in this case is the DLR estimated output $\hat{y}(k|N)$ in the form:

$$\hat{y}(k|N) = \hat{T}(k|N) + \mathbf{x}^T(k)\hat{\mathbf{a}}(k|N) \qquad k = 1, 2, \ldots, N \tag{5.3}$$

which is obtained rather obviously by inserting the estimated TVPs into the first two terms of equation (5.2) and omitting the noise term $e(k)$.

Forecasting, interpolation and backcasting are an inherent part of the filtering and smoothing algorithms (see e.g. Young, 1984; Harvey, 1989): if missing samples anywhere within, or immediately outside, the series are detected, then the filtering and smoothing algorithms simply replace the missing samples by their expectations, based on the model and the exogenous variables. In the case of the filtering algorithm, the insertion of f missing samples at the end of the series yields the adaptive f-step ahead forecasts (which, in this DLR model case, requires knowledge or forecasts of the exogenous regression variables in $\mathbf{x}(k)$, over the same period); while the smoothing equations, provide the interpolation over gaps within the data set, or backcasts at the beginning of the data (again requiring knowledge or backcasts of the variables in $\mathbf{x}(k)$ over the backcasting period).

Finally, the FIS estimate of the irregular component $e(k)$ for any UC model can be obtained by subtracting $\hat{y}(k|N)$ from $y(k)$: i.e.,

$$\hat{e}(k|N) = y(k) - \hat{y}(k|N) \tag{5.4}$$

However, these have slight serial correlation and so, as we saw earlier in section 4.6.1, any statistical tests on the estimation residuals are better applied to the normalized innovations process (4.58) formed by the KALMSMO algorithm.

5.1.1 Example: DLR analysis of LIDAR data

In this environmental example, DLR is utilized in the analysis of LIDAR (laser-radar) data obtained for a study investigating the concentration of atmospheric atomic mercury in an Italian geothermal field (Holst et al., 1996; Young and Pedregal, 1996). The LIDAR measurements $y(k)$ are plotted against the range k (distance) in Figure 5.2, where the uniform sampling interval is $\Delta s = 1.5\ m$. Within the present DLR context, it can be considered as the following combination of a smooth IRW trend and additive heteroscedastic measurement noise whose variance increases with the range:

$$\begin{gathered} y(k) = x(k) + e(k) \quad e(k) = \mathcal{N}\{0, \sigma(k)^2\} \\ \begin{bmatrix} x(k) \\ \nabla x(k) \end{bmatrix} = \begin{bmatrix} 1 & 1 \\ 0 & 1 \end{bmatrix} \begin{bmatrix} x(k-1) \\ \nabla x(k-1) \end{bmatrix} + \begin{bmatrix} 0 \\ 1 \end{bmatrix} \eta(k); \quad \eta(k) = N\{0, \sigma_\eta^2) \end{gathered} \tag{5.5}$$

As a result, it is possible to use the three-stage approach to estimation outlined in section 4.5.4 of the previous Chapter 4, with $x(k)$ in the model (4.57) representing the trend $T(k)$ and $\nabla x(k)$, its discrete-time derivative $\nabla T(k)$, where now the argument k represents the range sampling index, rather than the time index.

The three-stage procedure, as implemented in the heteroscedastic option of the irwsm routine in the CAPTAIN Toolbox, yields ML optimized values of $NVR = 0.0051$ and $\hat{\sigma}_\eta^2 = 1.259 \times 10^6$ in the first stage of the procedure. Figure 5.3 is a plot of the trend estimate $\hat{T}(k|N)$ (upper panel) and the discrete derivative estimate

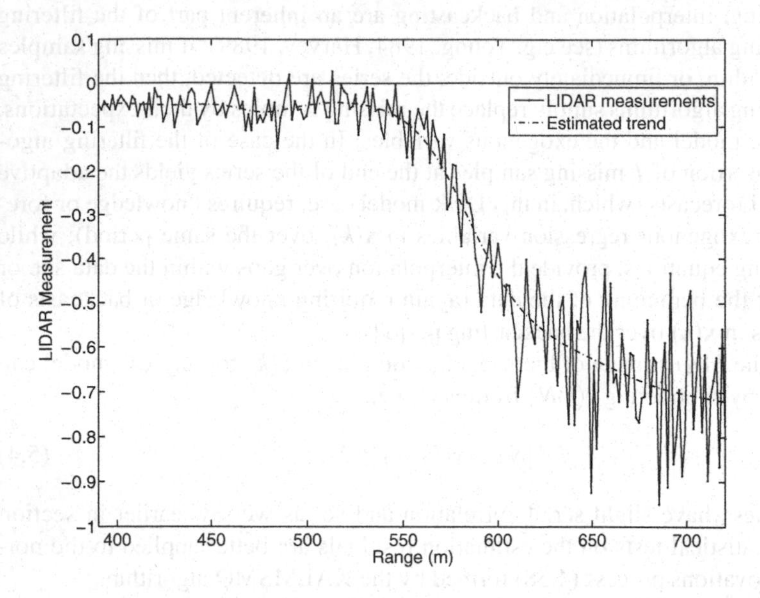

Fig. 5.2 LIDAR example: LIDAR measurements plotted against range (distance) with the finally estimated trend shown as a dash-dot line.

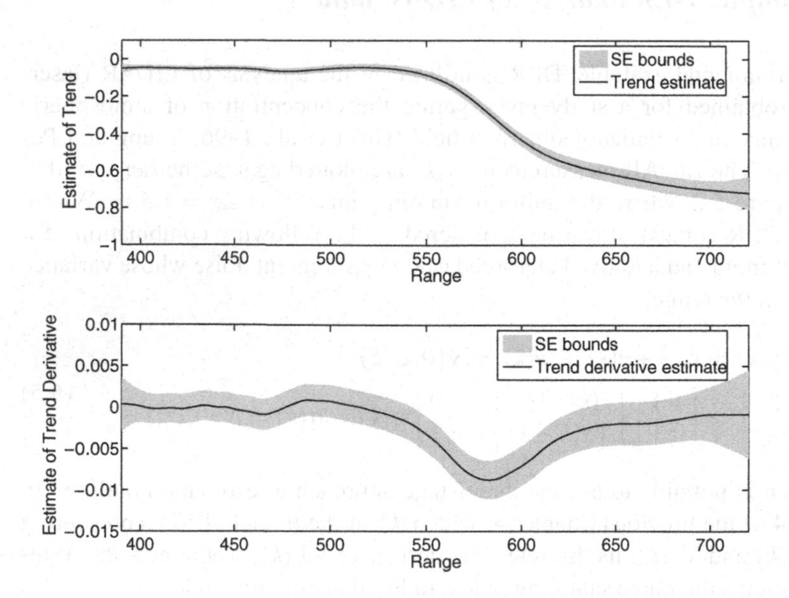

Fig. 5.3 LIDAR example: DLR estimated trend (upper panel) and trend derivative (lower panel).

$\nabla\hat{T}(k|N)$, (lower panel): in each case, the estimated standard errors are shown as the grey areas. These results are quite similar to those of Holst et al. (1996), except that the latter's derivative estimates show some signs of an end effect (a rather sharp turn down at the end of the data set, which does not appear to be entirely consistent with the data). As can be seen, this effect is not present in the equivalent DLR estimate.

Figure 5.4 compares the estimates at the first stage in the estimation process (i.e. the optimized DLR estimates assuming constant variance observational noise) with the final estimates allowing for the heteroscedasticity in Figure 5.4. The trend estimates are very similar but the derivative estimates show some small differences: in particular, the final estimates are less smooth than the initial estimates, probably reflecting the changing levels of uncertainty injected by the assumption of heteroscedasticity. It is clear that, for most practical purposes, the first stage estimates would be quite acceptable in this case (indeed, they are a little smoother so this may add to their attraction). However, as emphasized in the previous section, the standard error bounds on the final stage estimates should be more accurate and this could well be important in some applications .

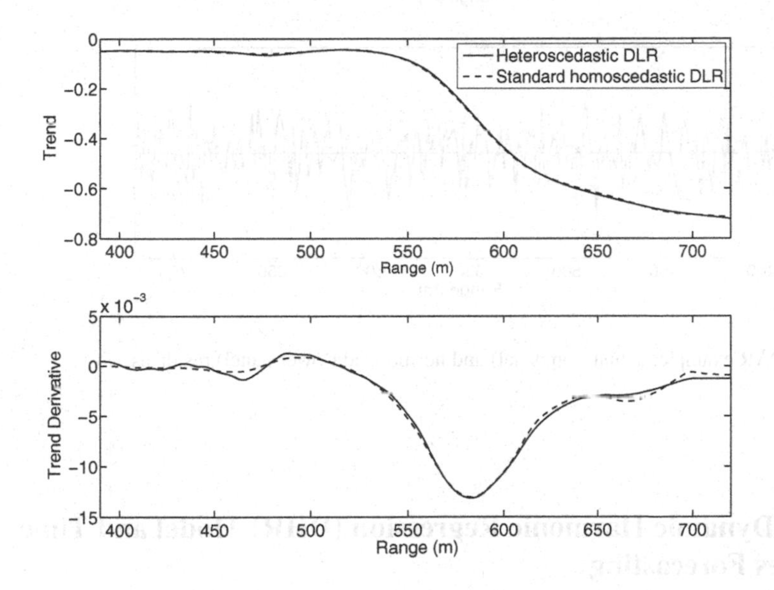

Fig. 5.4 LIDAR example: comparison of standard, homoscedastic DLR (dashed line) and heteroscedastic DLR (full line) estimates of the trend (upper panel) and trend derivative (lower panel).

Finally, the top panel of Figure 5.5 shows the DLR estimate $\hat{e}(k)$ of the irregular component $e(k)$, with the 95% confidence band computed from the estimated changing variance $\hat{\sigma}^2(k/N)$. Below this, in the lower panel, is the normalized irregular component $\hat{e}_n(k)$, where,

$$\hat{e}_n(k) = SF \frac{\hat{e}(k)}{\hat{\sigma}^2(k/N)}; \quad \text{where } SF = \frac{std\{\hat{e}(k)\}}{std\{\frac{\hat{e}(k)}{\hat{\sigma}^2(k/N)}\}}$$

Here the normalization takes into account of this changing variance by simple division, followed by re-scaling SF, based on the appropriate standard deviations. The autocorrelation Function (ACF) of these normalized residuals confirms that the assumption of heteroscedastic white noise is a reasonable one in this case.

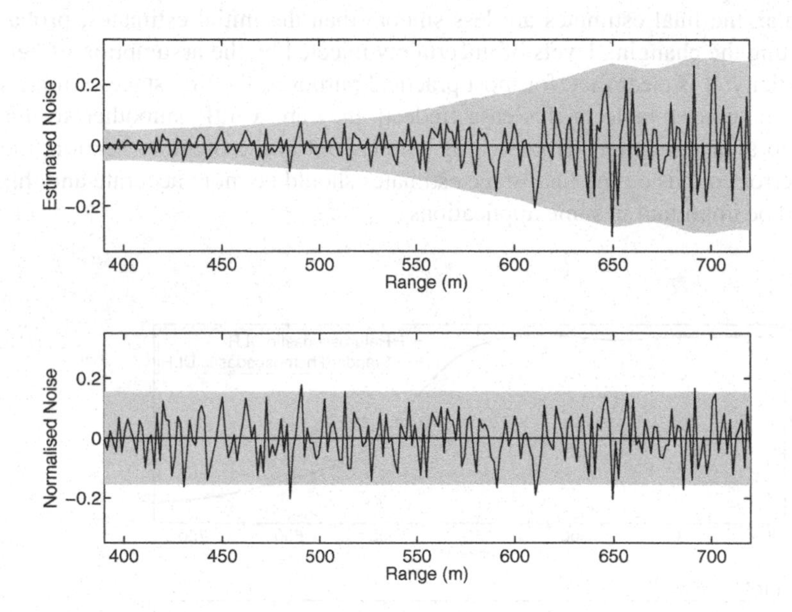

Fig. 5.5 LIDAR example: actual (top panel) and normalized (lower panel) residuals.

5.2 The Dynamic Harmonic Regression (DHR) Model and Time Series Forecasting

Many socio-economic and environmental time series exhibit periodic, seasonal or cyclical effects of various kinds. For instance, both climate and socio-economic phenomena are influenced by annual seasonal variations; life cycles display both diurnal and annual rhythms; and quasi-periodic business cycles influence economic observations. In order to investigate such periodic or quasi-periodic phenomena, as well as performing functions such as forecasting, interpolation over gaps or seasonal adjustment, one needs to estimate the periodic components in some manner. This estimation problem is complicated by the fact that the periodic variations which affect real time series are often nonstationary, in the sense that their amplitude and phase tend to change over the observation interval. Various identification and estimation procedures have been suggested for handling such nonstationary periodicity but some of the most powerful originate in the statistical and econometrics literature, where the estimation problem is often termed *signal extraction*.

One of the oldest and best known techniques for signal extraction is the Census X-11 method and its later developments, X-11 ARIMA and X-12 ARIMA (see Findley et al. (1992, 1996) and the prior references therein). However, three other important approaches to signal extraction have been developed in recent years and have gained favour because of their greater inherent flexibility. These methods are based on: (a) the decomposition of *AutoRegressive Integrated Moving Average* (ARIMA) or 'reduced form' models (see e.g. Burman (1980); Maravall (1993); Gómez and Maravall (1996)); (b) optimal regularization (see e.g. Akaike (1980); Jakeman and Young (1984); Young (1991); Young and Pedregal (1999b)); or (c) formulation of the problem within the stochastic state space setting introduced in section 4.2 of Chapter 4. This latter approach provides the most obvious formulation of UC models of the kind discussed in the present Chapter.

The *Dynamic Harmonic Regression* (DHR) model considered here is formulated within the UC framework and considers the estimation, forecasting and extraction of periodic phenomena in time series within the associated stochastic SS setting. This is not only a more natural and satisfying setting for UC models than the other methods mentioned above, but it is also more attractive computationally because it so nicely integrates the processes of forecasting, interpolation and seasonal adjustment into a single recursive framework. To paraphrase Wiener (1949), it seems to provide a rather natural approach to the "extrapolation, interpolation and smoothing of nonstationary time series". As we shall see, the main difference between the DHR approach and related SS techniques, such as Harvey's *Structural Modelling* method (Harvey, 1989), lies in the formulation of the UC model for the periodic components and the method of optimising the hyper-parameters in this model.

The basic DHR model contains only the trend, cyclical, seasonal and irregular components: i.e.,

$$y(k) = T(k) + C(k) + S(k) + e(k) \tag{5.6}$$

Here, the most important unobserved components are the seasonal term $S(k)$ and the cyclical term $C(k)$. Although these are both modelled in the same manner, it is convenient to define them separately because of their potentially different physical interpretations, i.e.,

$$\text{Seasonal}: \quad S(k) = \sum_{i=1}^{R_s} \{a(i,k)\cos(\omega_i k) + b(i,k)\sin(\omega_i k)\} \tag{5.7}$$

where $a(i,k)$ and $b(i,k)$ are stochastic TVPs and ω_i, $i = 1,2,...,R_s$, are the fundamental and harmonic frequencies associated with the seasonality in the series; and

$$\text{Cyclical}: \quad C(k) = \sum_{i=1}^{R_c} \{\alpha(i,k)\cos(f_i k) + \beta(i,k)\sin(f_i k)\} \tag{5.8}$$

where $\alpha(i,k)$ and $\beta(i,k)$ are stochastic TVPs and $f_i,\ i = 1,2,...,R_c$, are the frequencies associated with the (normally longer period) cyclical component. In both cases, the frequency values are chosen by reference to the spectral properties of the time series, as discussed later. The trend component can also be considered as a stochastic, time variable intercept parameter in the DHR and so it could be incorporated, if so desired, into the cyclical or seasonal components as a zero frequency term (i.e. by introducing an extra term in the summations at $i = 0$ with w_0 or f_0 set equal to zero). This DHR model can be considered as a straightforward extension of the classical, constant parameter, Harmonic Regression (or Fourier series) model, in which the gain and phase of the harmonic components can vary as a result of estimated temporal changes in the parameters $a(i,k)$, $b(i,k)$, $\alpha(i,k)$ and $\beta(i,k)$.

In general, each of these $a(i,k)$, $b(i,k)$, $\alpha(i,k)$ and $\beta(i,k)$, as well as the trend $T(k)$, are modelled as GRW processes (4.33) and the subsequent recursive TVP estimation procedures are exactly the same as for the DLR model, except that the NVR values (and any other hyper-parameters) in the GRW models associated with the parameters of *each component* are constrained to be equal. However, the ML method used for hyper-parameter optimization in the DLR case does not work so well in this DHR context and a novel, frequency domain, optimization algorithm has been developed to replace it. This leads to a much better defined optimum in the objective function-hyper-parameter space, with consequent advantages to DHR modelling, both as regards convergence time and the number of parameters that can be optimized simultaneously.

In order to specify an appropriate form for the model (5.6), it is necessary to identify the number and values of the fundamental and harmonic frequencies associated with the seasonality and cyclicity. This is accomplished by referring to the empirical spectral properties of $y(k)$ using standard methods of spectral analysis: see e.g. Priestley (1981). There are, of course, numerous methods for computing the empirical spectrum but one of the most useful is the *Auto Regressive* (AR) spectrum of the data, with the AR order identified from the *Akaike Information Criterion* (AIC): see Akaike (1974); Priestley (1981). The AR spectrum (see later section 5.3 and Appendix B, section B.3.7.2) has a number of advantages in the present context: it is particularly smooth since it derives directly from the AR model rather than the periodogram; it resolves the modal peaks very well because of its method of computation; and it plays an important role in the method of optimizing the hyper-parameters in the DHR model. These factors will become clearer in the subsequent practical example.

As in the previous DLR modelling, the final result of the above analysis is the estimated output $\hat{y}(k|N)$ of the DHR model in the form (cf. (5.6)):

$$\hat{y}(k|N) = \hat{T}(k|N) + \hat{S}(k|N) + \hat{C}(k|N) \quad k = 1,2,...,N \tag{5.9}$$

where $\hat{T}(k|N)$ is the FIS estimate of the trend; while $\hat{S}(k|N)$ and $\hat{C}(k|N)$ are the estimated seasonal and cyclical components defined by the FIS estimates of the TVPs $a(i,k)$, $b(i,k)$, $\alpha(i,k)$ and $\beta(i,k)$, and their associated sine and cosine functions, via the equations,

$$\begin{aligned}\hat{S}(k|N) &= \sum_{i=1}^{R_s} \{\hat{a}(i,k|N)\cos(\omega_i k) + \hat{b}(i,k|N)\sin(\omega_i k)\} \\ \hat{C}(k|N) &= \sum_{i=1}^{R_c} \left\{\hat{\alpha}(i,k|N)\cos(f_i k) + \hat{\beta}(i,k|N)\sin(f_i k)\right\}\end{aligned} \tag{5.10}$$

However, adaptive forecasting, interpolation and backcasting are much more straightforward than in the DLR case because the regression variables in the DHR model (the sine and cosine variables) are all known functions of the sampling index (time) and can be specified easily outside the data sample when using the model for forecasting and backcasting purposes.

One useful by-product of DHR modelling is an objective approach to 'adaptive seasonal adjustment'; that is the removal of the estimated changing seasonal effects from the data. The FIS estimate of the seasonally adjusted (SA) series is obtained by simply subtracting the FIS estimated seasonal component $\hat{S}(k|N)$ from $y(k)$, i.e.,

$$\hat{y}_{SA}(k|N) = y(k) - \hat{S}(k|N) \tag{5.11}$$

The estimated sub-components of $\hat{S}(k|N)$, given by the individual terms in the summation of (5.10), can also be of practical use since they provide 'band-pass' estimates of the spectral component in the localized region of the harmonic frequencies. Typical practical examples are the DHR analysis of the airline passenger numbers data plotted in Figure 5.1 (Young et al., 1999) and the later practical example considered in section 5.2.3).

5.2.1 Spectral analysis of the DHR model

The method for optimising the hyper-parameters in the DHR model is formulated in the frequency domain and based upon analytical expressions for the power spectra of the DHR model class. In order to obtain these analytical expressions, it is necessary to determine the spectral properties of both the TVPs, as defined by their assumed stochastic models, and the products of such parameters with the sine and cosine functions that characterize the DHR model. As mentioned earlier, these TVPs are normally described by any of the models belonging to the GRW family.

5.2.1.1 The pseudo-spectra of Generalized Random Walk models

Consider first a UC model consisting of only an IRW trend (i.e. equation (4.33) with $\alpha = \beta = \gamma = 1; \delta = 0$) plus a white noise irregular component. The resultant model can be expressed in transfer function terms as,

$$y(k) = T(k) + e(k) = \frac{1}{(1-z^{-1})^2}\eta_{k-1} + e(k)$$

Initially, the inherent non-stationarity of the IRW process can be dealt with by taking second differences of $y(k)$ to yield,

$$\nabla^2 y(k) = \eta_{k-1} + \nabla^2 e(k)$$

where $\nabla = 1 - z^{-1}$ is the difference operator. For this stationary process the power spectrum takes the form,

$$f_{\nabla^2 y}(\omega) = \frac{1}{2\pi}\left[\sigma_\eta^2 + \{2 - 2\cos(\omega)\}^2\sigma^2\right] \tag{5.12}$$

and it is straightforward to see that the pseudo-spectrum of $y(k)$ is given by,

$$f_y(\omega) = \frac{1}{2\pi}\left[\frac{\sigma_\eta^2}{\{2-2\cos(\omega)\}^2} + \sigma^2\right] \tag{5.13}$$

Similarly, the pseudo-spectrum in the case of the simpler RW trend is given by,

$$f_y(\omega) = \frac{1}{2\pi}\left[\frac{\sigma_\eta^2}{2-2\cos(\omega)} + \sigma^2\right] \tag{5.14}$$

The resulting spectral properties of the IRW and RW-type filters are shown graphically in Figure 5.6, for a selection of different NVR values σ_η^2/σ^2.

In the case where the trend is modelled by an SRW process, the pseudo spectrum is somewhat more complex and takes the form

$$f_y(\omega) = \frac{1}{2\pi}\left[\frac{\sigma_\eta^2}{\{1+\alpha^2 - 2\alpha\cos(\omega)\}\{2-2\cos(\omega)\}} + \sigma^2\right] \tag{5.15}$$

Note how the introduction of the smoothing parameter α allows for a continuous transition between the RW model ($\alpha = 0$) and the IRW model ($\alpha = 1.0$). This advantageous property is illustrated in Figure 5.7. From the filter design point of view, the additional parameter allows for varying the shape (width or bandpass) of the filters frequency response, while the NVR parameter shifts the response vertically. Clearly, the pseudo-spectra of other more general GM processes can be analysed in a similar manner.

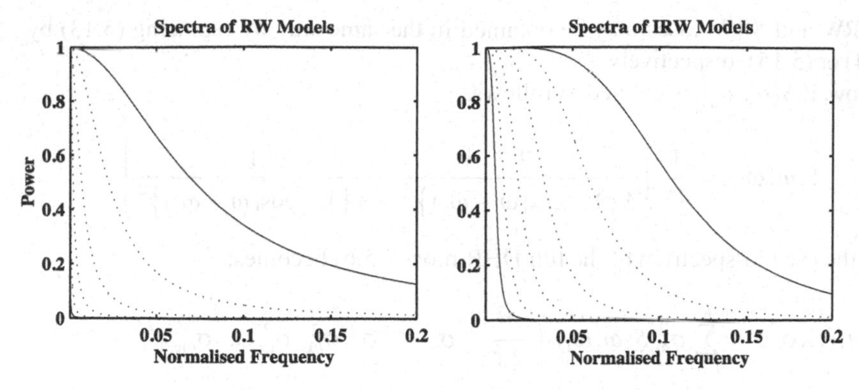

Fig. 5.6 Spectral characteristic of RW and IRW filters for different values of the NVR parameter between 10^{-6} and 10^{-1}.

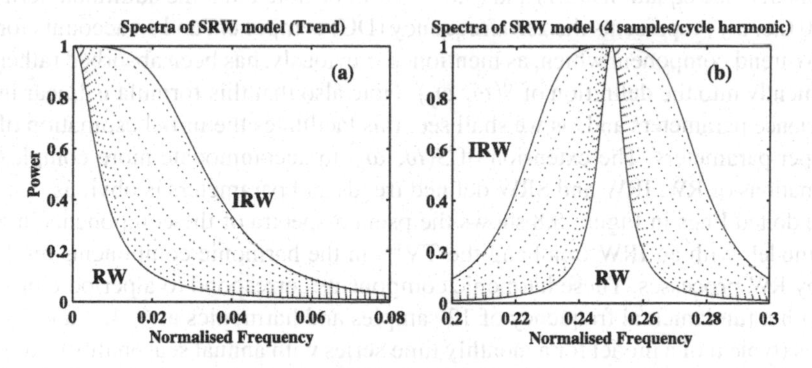

Fig. 5.7 Spectral characteristic of the SRW filter for different values of α between 0 and 1, in: (a) a trend described by an SRW process; and (b) a harmonic with period 4 samples/cycle modulated by a TVP parameter described by an SRW process.

5.2.1.2 The pseudo-spectra of the full DHR model

From the basic Fourier transform properties, the frequency response of amplitude modulated signals of the form $S(k) = a(k)\cos(\omega_j k)$, is known to be:

$$f_s(\omega) = \frac{1}{2}\left[f_A(\omega - \omega_j) + f_A(\omega + \omega_j)\right]$$

where $f_A(\omega)$ is the frequency response of $a(k)$. Consider the case of a single frequency DHR term, $S(k) = a(k)\cos(\omega_j k) + b(k)\sin(\omega_j k)$, in which the TVP associated with the sine and cosine terms are modelled as two IRW processes with equal variance parameters σ_ω^2. The pseudo-spectrum of $S(k)$ then takes the form:

$$f_{\omega_j}(\omega) = \frac{1}{2\pi}\left[\frac{\sigma_{\omega_j}^2}{4\left\{1-\cos(\omega-\omega_j)\right\}^2} + \frac{\sigma_{\omega_j}^2}{4\left\{1-\cos(\omega+\omega_j)\right\}^2}\right] \quad (5.16)$$

The RW and SRW cases may be obtained in the same way by replacing (5.13) by (5.14) or (5.15), respectively.

Now, if $S(\omega,\ \omega_j)$ is defined as follows,

$$S(\omega,\omega_j) = \frac{1}{2\pi}\left[\frac{1}{4\left\{1-\cos(\omega-\omega_j)\right\}^2} + \frac{1}{4\left\{1-\cos(\omega+\omega_j)\right\}^2}\right]$$

then the pseudo-spectrum of the full DHR model (5.6) becomes:

$$f_y(\omega,\boldsymbol{\sigma}^2) = \sum_{j=0}^{R} \sigma^2_{\omega_j} S(\omega,\omega_j) + \frac{\sigma^2}{2\pi} \qquad \boldsymbol{\sigma}^2 = \left[\sigma^2\ \sigma^2_{\omega_0}\ \sigma^2_{\omega_1}\ \ldots\ \sigma^2_{\omega_R}\right]^T \tag{5.17}$$

where R is the total number of number of different frequency components included in the model (see equations (5.7) and (5.8)). It will be noted that the additional term at $j=0$ (i.e. ω_0) represents the zero frequency (DC) component and so accounts for the IRW trend component which, as mentioned previously, has been absorbed rather conveniently into the definition of $S(\omega,\ \omega_j)$. Note also that this formula is linear in the variance parameters and, as we shall see, this facilitates the initial estimation of the hyper-parameters. The extension of $S(\omega,\ \omega_j)$ to accommodate more complex combinations of RW, IRW and SRW defined trends and parameters is obvious.

The dotted lines in Figure 5.8 shows the pseudo-spectra of the components in a DHR model, with an IRW trend and the TVPs in the harmonic components modelled by RW processes. These harmonic components correspond to a periodic process with a fundamental frequency of 12 samples and harmonics at 6, 4, 3 and 2.4 samples (typical of a model for a monthly time series with annual seasonality). Note that, on the linear scale, the components are well separated spectrally and the ratio of maximum power to the minimum is of the order 10^3.

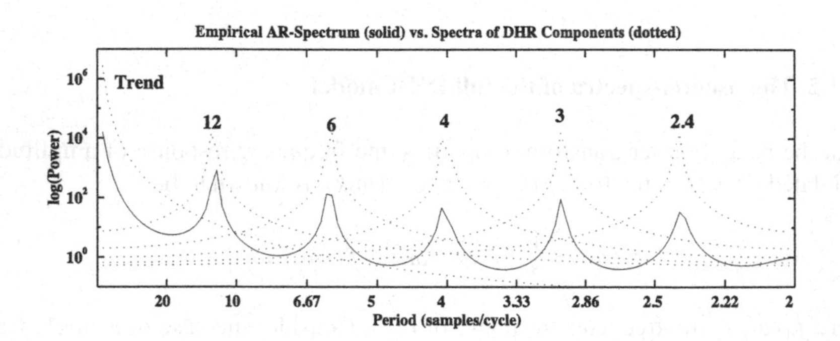

Fig. 5.8 Empirical 14th order AR(14) spectrum (solid) and pseudo-spectra of DHR components (dotted) for the Airline Passenger series.

The close relationship of the DHR to *Discrete Fourier Transform* (DFT) or, equivalently, the constant parameter *Harmonic Regression* (HR) model is clearly apparent from Figure 5.8. As the $\sigma^2_{\omega_j}$ of the jth spectral peak (or more particularly the associated NVR value) becomes smaller, which is equivalent to decreasing the potential variability of the related sine and cosine amplitude parameters, the spectral peak of the jth frequency component approaches the shape of spectral line, as in the case of the constant parameters HR model. In other words, the DHR model can be considered as a logical stochastic TVP (or nonstationary) version of the constant parameter HR and DFT models.

5.2.1.3 Estimation in the frequency domain

Equation (5.17) shows that the spectrum of the DHR process can be considered as a linear combination of known spectral terms, with the unknown variance parameters appearing as associated regression coefficients. Clearly, therefore, one way of estimating the variance parameter vector $\boldsymbol{\sigma}^2 = \left[\hat{\sigma}^2 \ \hat{\sigma}^2_{\omega_0} \ \hat{\sigma}^2_{\omega_1} \ ... \ \hat{\sigma}^2_{\omega_R}\right]^T$ is to solve the following a least squares optimization problem,

$$\hat{\boldsymbol{\sigma}}^2 = \arg\min_{\boldsymbol{\sigma}^2} \mathfrak{J}(f_y, \hat{f}_y, \boldsymbol{\sigma}^2) \quad \mathfrak{J}(f_y, \hat{f}_y, \boldsymbol{\sigma}^2) = \sum_{i=0}^{T-1} \left[f_y(\omega_i) - \hat{f}_y(\omega_i, \boldsymbol{\sigma}^2)\right]^2 \tag{5.18}$$

over T distinct frequencies, ω_i, $i = 0, 1, 2, ..., T-1$, in the range between zero frequency (DC) and the Nyquist frequency (see section 8.7 of Chapter 8): i.e. 0 to 0.5 cycles/sampling interval (or periods of ∞ to 2 samples/cycle). Here $f_y(\omega)$ is the empirical spectrum, while the model pseudo-spectrum $\hat{f}_y(\omega, \hat{\boldsymbol{\sigma}}^2)$ is defined as the following linear function of the $R+2$ unknown variance parameter estimates,

$$\hat{f}_y(\omega, \hat{\boldsymbol{\sigma}}^2) = \sum_{j=0}^{R} \hat{\sigma}^2_{\omega_j} S(\omega, \omega_j) + \frac{\hat{\sigma}^2}{2\pi}$$

A visual illustration of the estimation problem posed in these terms can be obtained by referring again to Figure 5.8, where the spectra of the IRW trend and the 5 harmonic components with RW parameters (dotted lines) can clearly be adjusted by the above optimization approach to fit the AIC identified AR(14) spectrum (full line) of the airline passenger number series (Figure 5.1).

While this linear least squares solution is attractive in its simplicity, practical experience has shown that an alternative objective function,

$$\mathfrak{J}_{NL}(f_y, \hat{f}_y) = \sum_{i=0}^{T-1} \left[\log\left\{f_y(\omega_i)\right\} - \log\left\{\hat{f}_y(\omega_i, \hat{\boldsymbol{\sigma}}^2)\right\}\right]^2 \tag{5.19}$$

defined in logarithmic terms, yields improved estimation results, with a more clearly located optimum and better defined estimates of the variance parameters. This arises predominantly because the logarithmic spectra have clearer information on the shape of the spectral signatures, particularly the shoulders that are so important in defining the magnitude of the variance parameters. This logarithmic measure is also intuitively rather natural in the signal processing area since power spectra are usually presented and measured on logarithmic scales such as decibels. The disadvantage of this approach is, of course, that $\mathcal{J}_{NL}$ is a nonlinear function of the unknown variance parameters and so they must be estimated by nonlinear optimization, as in the conventional ML optimization case.

Fortunately, this nonlinear optimization is quite straightforward and presents no difficulties. Not only is the location of the optimum normally well defined, but the linear least squares solution (5.18) quickly and reliably provides very good initial conditions for the nonlinear minimization problem (5.19). Consequently optimization algorithms with quadratic convergence, that are otherwise sensitive to the initial conditions, can be used with confidence. For this purpose, the objective functions have to fulfil certain conditions (Deutsch, 1965) in order that the parameter estimates are mathematically acceptable, but it is easy to show that the objective function (5.19) satisfies these conditions and is both continuous and convex.

One final simplification is possible: if $\hat{f}_y(\omega, \hat{\sigma}^2)$ is based on the AR spectrum, then the estimate of the residual white noise (based on the one step ahead prediction errors from the estimated AR model) can be used to concentrate out the observation noise variance σ^2 from the model pseudo-spectrum, i.e.,

$$f_y(\omega, \boldsymbol{\sigma}^2) = \sigma^2 \left\{ \frac{f_y(\omega, \sigma^2)}{\sigma^2} \right\} = \sigma^2 \left\{ \sum_{j=0}^{R} \frac{\sigma^2_{\omega_j}}{\sigma^2} S(\omega, \omega_j) + \frac{1}{2\pi} \right\}$$

or equivalently

$$f_y(\omega, \boldsymbol{\theta}_{nvr}) = \sigma^2 \left\{ \sum_{j=0}^{R} NVR_j S(\omega, \omega_j) + \frac{1}{2\pi} \right\} \tag{5.20}$$

where $\boldsymbol{\theta}_{nvr} = [NVR_0 \; NVR_1 \; \dots \; NVR_R]^T$ is the NVR vector. In this manner, the model pseudo-spectrum is defined directly in terms of the unknown NVR values. This not only removes one parameter from the estimation problem but it also introduces a common scale for the remaining unknown parameters which has numerical advantages in the subsequent numerical optimization. For instance, practical experience over many years with the spectral forms (5.17) and (5.20) has demonstrated that the latter formulation provides better defined solutions over a wide range of data series.

5.2.2 The complete DHR estimation algorithm

In the case where a cyclical component is not present, the complete DHR estimation algorithm consists of the following four steps.

1. Estimate an AR(n) spectrum $f_y(\omega)$ of the observed time series $y(k), k = 1, 2, \dots, N$ and its associated residual variance $\hat{\sigma}^2$, with the AR order n normally identified

by reference to the AIC. Note the R significant peaks that characterize the spectrum (these will normally include a fundamental frequency and several of its associated harmonics).

2. Find the linear least squares estimate of the **NVR** parameter vector $\boldsymbol{\theta}_h$ which minimizes the linear least squares objective function: i.e.,

$$\hat{\boldsymbol{\theta}}_h = \arg\min_{\boldsymbol{\theta}_h} \mathfrak{J}(\boldsymbol{\theta}_h) \qquad \mathfrak{J}(\boldsymbol{\theta}_h) = \sum_{i=0}^{T-1} \left[f_y(\omega_i) - \hat{f}_y(\omega_i, \boldsymbol{\theta}_h)\right]^2$$

where $\hat{f}(\omega, \boldsymbol{\theta}_h) = \sigma^2 \sum_{j=0}^{R} \boldsymbol{\theta}_j S(\omega, \omega_j)$; ω_0 and $\boldsymbol{\theta}_0$ refer to the trend term; and the ω_j, $j = 1,2,...,R$, are the R significant frequencies identified from the AR(n) spectrum in step 1.

3. Find the nonlinear least squares estimate of the parameter vector which minimizes nonlinear least squares objective function: i.e.,

$$\hat{\boldsymbol{\theta}}_h = \arg\min_{\boldsymbol{\theta}_h} \mathfrak{J}(\boldsymbol{\theta}_h) \qquad \mathfrak{J}(\boldsymbol{\theta}_h) = \sum_{i=0}^{T-1} \left[\log(f_y(\omega_i)) - \log(\hat{f}_y(\omega_i, \boldsymbol{\theta}_h))\right]^2$$

using result from step 2. to define the initial conditions.

4. Use the NVR estimates $\hat{\boldsymbol{\theta}}_h$ from step 3. to obtain the recursive forward pass (Kalman filter) and backward pass (FIS algorithm) smoothed estimates of the components in the DHR model: i.e. the trend; the total cyclical and seasonal; the fundamental/harmonic components; and the irregulars. In this form, the algorithm can be used for forecasting, backcasting, and signal extraction (e.g. seasonal or cyclical adjustment). Allowance for sharp discontinuities or 'interventions' (see next section 4.9), as well as interpolation over gaps can be introduced, as required.

If a cyclical component is identified then the above procedure is simply expanded to accommodate this additional component.

5.2.3 Practical example: Signals Passed At Danger (SPAD) data

Here, we consider an example where the seasonal pattern changes quite a lot in fairly noisy data. These *Signals Passed At Danger* (SPAD) data, as plotted in Figure 5.9, show the number of railway trains that passed danger signals in the UK for adjacent four week periods over the time 1990-2001. Visually, this figure reveals an underlying and mainly downward trend, as well as considerable volatility around this trend arising, in part, from annual seasonal variations. The bold line on figure 5.9 is an estimate of the long-trend based on conventional 13 sample (one year) CMA analysis (see section 4.5.5) using the Matlab routine `convn`.

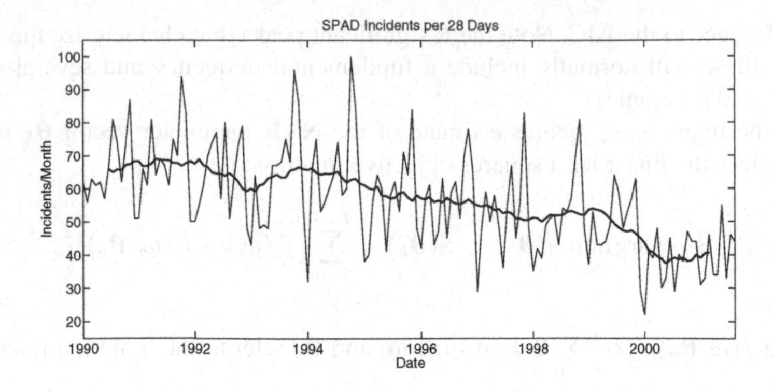

Fig. 5.9 Daily SPAD totals $y(k)$ accumulated over 4 weeks and a standard 13 sample (one year) centralized moving average (CMA) trend line shown in bold.

The patterns of behaviour observed in Figure 5.9 are confirmed by standard graphical time series methods in the form of spectral and temporal (correlation) analysis. For example, the estimated AR(14) model spectrum shown in Figure 5.10 is computed using the arspec routine in CAPTAIN and is typical of the results obtained from such analysis. It indicates that there is power in the spectrum at low frequencies, associated with the long term trend, as well as at various prominent peaks, located at frequencies of 0.0722, 0.1507, 0.2292, 0.3120 and 0.4152 cycles/sample. (i.e. associated with periodic behaviour of 13.85, 6.64, 4.36, 3.21 and 2.41 samples/cycle). Given the uncertainty in the analysis, this spectral signature is consistent with an annual cycle of 52 weeks (13 samples, each of 4 weeks) and harmonics at periods of $13/j$, $j = 2,...,6$, respectively (i.e. 6.5, 4.33, 3.25 and 2.6 samples).

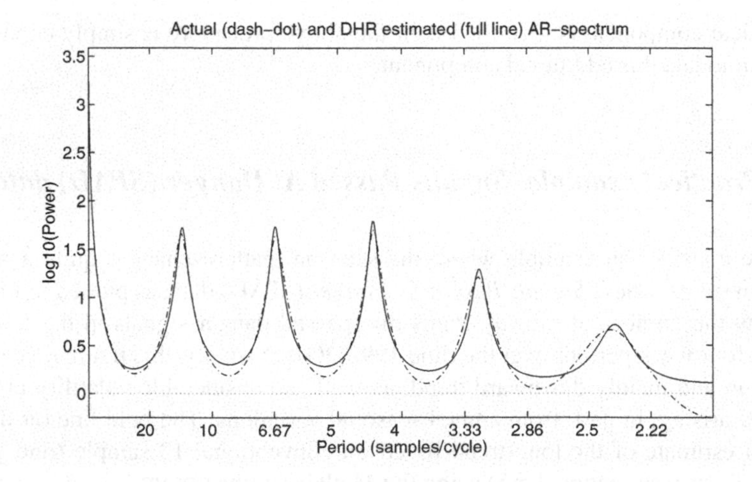

Fig. 5.10 SPAD data: estimated DHR spectrum compared with the AR(14) spectrum.

The CMA estimated trend in Figure 5.9 reveals that the underlying trend is mainly downward, although, after 1999, there appears to be a sharper fall, possibly followed by a small upward movement. The problems with such conventional trend analysis are that:

1. it requires the analyst to provide the averaging interval (here 13 samples or 52 weeks) and, as a result, the subjectively defined level of smoothing is constant, leading to the possibility of under or over-smoothing;
2. The nature of the CMA computation means that the trend estimate is not produced at the ends of the data (here the first and last 6 data points)
3. the analysis takes no real account of the obvious annual seasonality of the data, other than the choice of the averaging interval, so that the trend estimate can be biased;
4. there is no measure of uncertainty associated with the trend estimate, so that confidence in any deductions based on the analysis is difficult to quantify objectively.

In these circumstances, there is a clear need for more objective, non-stationary time series analysis such as that provided by DHR analysis. The DHR model is selected to have an SRW trend component and AR(1) periodic components at the periods of 13.85, 6.64, 4.36, 3.21 and 2.41 samples/cycle, as identified by the empirical AR(14) spectrum in Figure 5.10, since this produces the best match to this spectrum. The optimization of the hyper-parameters using the `dhropt` routine in CAPTAIN results in the following NVR and α values:

Table 5.1 DHR Hyperparameter Optimization for the SPAD data.

Period	T(k)	13.85	6.64	4.36	3.21	2.41
NVR	0.0899	0.0403	0.0172	0.0140	0.0199	0.0145
α	0.9809	0.9589	0.9747	0.9849	0.9619	0.9345

and the DHR estimated spectrum is shown in Figure 5.10, where it is compared with the empirical AR(14) spectrum (dash-dot line).

One statistic that allows us to assess the performance of the DHR model is the well known *Coefficient of Determination*. This is discussed later in section 6.9 of Chapter 6 but, in the present context, it is denoted by R_T^2 and defined as follows:

$$R_T^2 = 1 - \frac{\sigma^2}{\sigma_y^2}; \; \sigma^2 = \frac{1}{N}\sum_{k=1}^{k=N}(\hat{e}(k) - \bar{\hat{e}}(k))^2; \; \sigma_y^2 = \frac{1}{N}\sum_{k=1}^{k=N}(y(k) - \bar{y}(k))^2 \quad (5.21)$$

Here, $\hat{e}(k)$ is the estimated irregular component, as defined in equation (5.4), while σ^2 and σ_y^2 are, respectively, the variances of $e(k)$ and $y(k)$ about their mean values $\bar{y}(k)$ and $\bar{\hat{e}}(k)$ (see equations (2.1) of chapter 2 and section B.2.1 of Appendix B), as computed by the `var` and `cov` routines in Matlab. In these terms, The DHR model explains the data quite well, with $R_T^2 = 0.918$ (i.e. 91.8% of the SPAD variance is explained by the model); the associated variance of $\hat{e}(k)$ is 18.45. The normalized innovations series $\varepsilon_n(k)$ associated with the model, as defined by equation (4.58) has a variance 66.37 and ACF analysis confirms that they constitute a zero mean, white noise process.

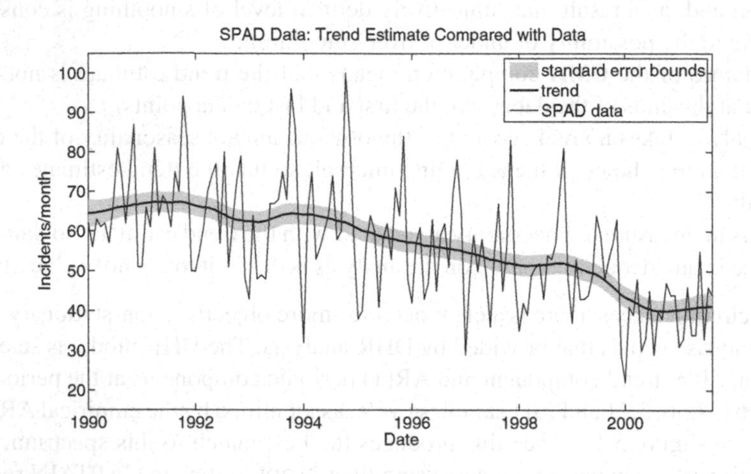

Fig. 5.11 Initially estimated smooth trend for the SPAD data.

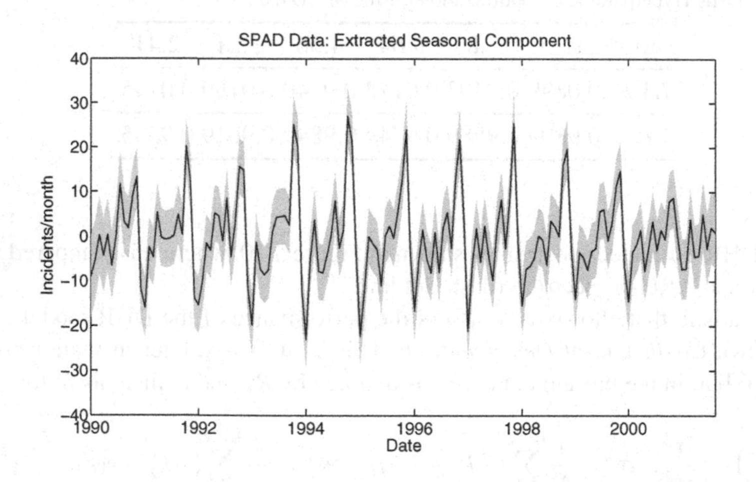

Fig. 5.12 Initially estimated seasonal component for the SPAD data.

Figure 5.11 compares the estimated trend (bold line) and its associated Standard Error (SE) bounds (grey area) with the SPAD data (full line). Note how the standard error bounds expand a little at the beginning and end of the data to reflect the increased uncertainty: this is because the 'centralized' FIS estimates are, of course, computed only on the basis of the available data, with no knowledge of the time series preceding and following the data set. The estimated trend can be compared with the CMA estimated trend in figure 1. We see that the overall behaviour of the two estimated trends is similar but the DHR estimate encompasses the whole of the data, it is smoother and its uncertainty is quantified by the SE bounds. Moreover, the DHR model also provides an estimate of the annual seasonal component, as shown in Figure 5.12. Here, we see that the pattern of the annual seasonality changes quite a lot over the series with considerable variation in the amplitude and some changes in the pattern of the three peaks that normally characterize the seasonal behavioural pattern.

If Figure 5.11 is considered further, we see some indication of sharper changes in the level of the estimated trend at various times over the observation interval: in particular, the sharpest such reduction occurs at the end of 1999. In order to allow the smoothed estimates to accommodate such a level change, the constant hyper-parameters have been optimized so that the estimated level change at each sample (as defined by the NVR values for the trend: 0.0899) is sufficiently large to allow for the estimated change at this point in time. However, this clearly constitutes a limitation in the analysis: if the rate of change at this point is larger than that at most other locations in the series, then the smoothing will not be optimum over most of the series. In particular, there will be insufficient smoothing most of the time and the SE bounds will be larger than necessary.

The possibility of the sharp change in level at the end of 1999, can be evaluated using variance intervention (section 4.9 of the previous Chapter 4), where both the level and slope of the trend are assumed to change significantly at the intervention sample. In this case, the optimization of the hyper-parameters using dhropt results in the following NVR and α values:

Table 5.2 DHR Hyperparameter Optimization with Variance Intervention for the SPAD data.

Period	T(k)	13.85	6.64	4.36	3.21	2.41
NVR	0.0007	0.0242	0.0223	0.0223	0.0211	0.0213
α	0.8602	0.9763	0.9750	0.9777	0.9662	0.9321

Note that the estimate of the intervention sample point was obtained by carrying out the complete analysis over a range of locations and selecting the value that yielded the lowest variance of the residual $\hat{e}(k)$. Note also that, while the NVR values for the periodic component models are comparable with those obtained for the standard DHR model, the trend NVR is much smaller (0.0013 compared with 0.089). This extended DHR model explains the SPAD data a little better than the standard DHR model, with $R_T^2 = 0.921$ and a residual variance of 17.99. Figure 5.13 shows the graphical results in this case, which can be compared with those in Figure 5.11: this clearly reveals the large downward movement at the identified intervention point, coupled with the smoother trend either side of this.

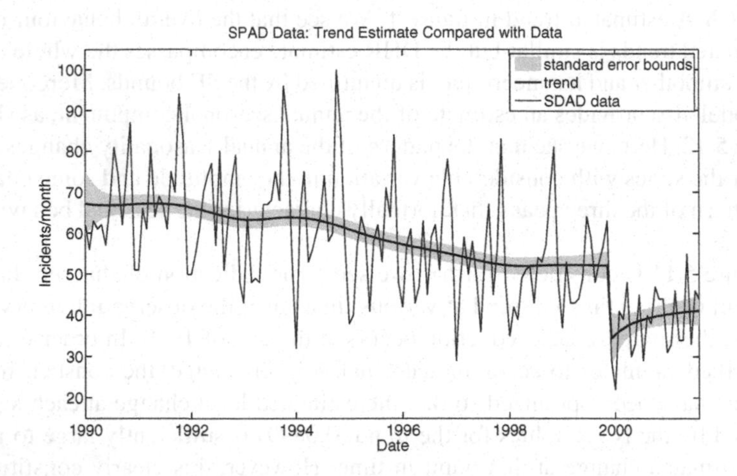

Fig. 5.13 Estimated trend for the SPAD data, with intervention.

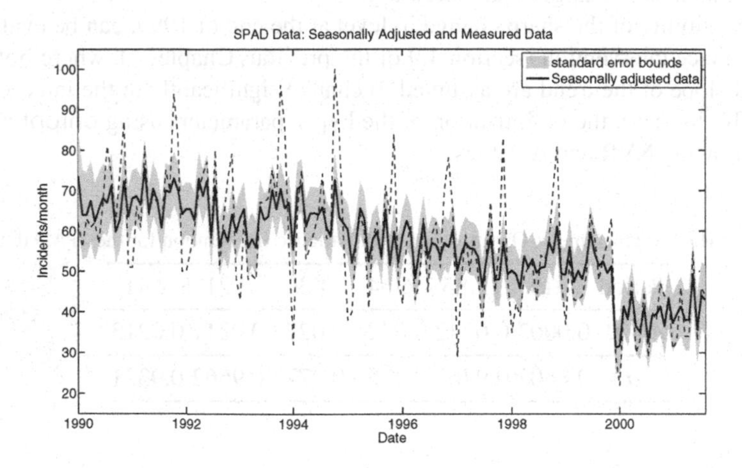

Fig. 5.14 Seasonal adjustment for the SPAD data, with intervention.

Another use of DHR estimation results is to seasonally adjust the data by removing the effects of the estimated seasonal component and revealing the underlying movements in the data. The seasonally adjusted series is obtained by simply subtracting the estimated annual seasonal component $\hat{S}(k)$ from $y(k)$, i.e. $y^{sa}(k) = y(k) - \hat{S}(k)$. Figure 5.14 compares $y^{sa}(k)$ (bold line) with the SPAD data (full line), with the standard error bounds shown as the grey area.

5.2.4 Extensions of DHR model analysis

Various modifications and extensions to the standard DHR algorithm have been proposed.

1. Tych et al. (2002) propose a method of enhancing the low frequency resolution of the AR spectrum, in order to identify any 'hidden' but significant low frequency periodicity, by concatenating a higher order AR spectrum for the low frequencies, with the standard AIC-identified lower order spectrum that specifies the higher frequency seasonal behaviour. The hyper-parameter optimization is then based on this modified spectrum (see the later example in section 5.5).
2. Pedregal and Young (2006) show that, by including the periodic behaviour in both the observation and the state equations, it is possible to generate a UC model that accounts for different periods simultaneously. In particular, this introduces multiplicative cycles, since periodic functions of a given frequency ω_j in the DHR model are multiplied by parameters that are themselves periodic functions of a different set of frequencies ω_i, $i = 1,2, \ldots, R$ defined by the state equations. These are known as 'modulated signals' in the signal processing literature. The advantage of this new model is that the amplitude of the shortest period cycle is modulated by the longer one, so giving a different pattern for each realization of the short period cycle, thereby reducing the dimension of the state vector.
3. Finally, Bujosa et al. (2007) show that it is possible to use an alternative quadratic objective function that allows for linear estimation of the hyper-parameters while retaining the better resolution that is a feature of the nonlinear optimization of the standard DHR algorithm. However, estimation by unrestricted linear least squares can yield estimates that have negative values so, when negative values are found, the algorithm uses non-negative least squares estimation (Lawson and Hanson, 1974). So far, this alternative linear DHR algorithm has not been evaluated sufficiently to assess its practical potential in relation to the well-used and practically very successful standard DHR algorithm.

5.3 The Dynamic AutoRegression (DAR) Model and Time-Frequency Analysis

The basic DAR model is obtained from the UC model (5.1) by considering only the stochastic terms $\xi(k)$ and $e(k)$ as components, and then combining these into a single DAR(p) model of the form:

$$y(k) = \frac{1}{A(k,z^{-1})} e(k) \tag{5.22}$$

in which $A(k,z^{-1}) = 1 + a_1(k)z^{-1} + a_2(k)z^{-2} + \cdots + a_p(k)z^{-p}$ is a time variable parameter polynomial in the backward shift operator z^{-1}. On multiplying throughout equation (5.22) by $A(z^{-1})$, so that it operates on $y(k)$, we obtain the DAR(p) model in the discrete-time (difference equation) equation form:

$$y(k) = -a_1(k)y(k-1) - a_2(k)y(k-2) - \cdots - a_p(k)y(k-p) + e(k) \qquad (5.23)$$

This DAR model can be used for adaptive forecasting but it is not as useful as the DHR model in this regard when the series contains a trend component. This is because the DAR model is not fully identifiable (see later, section 6.6 of Chapter 6) if it contains a low frequency trend component and so it is less flexible than the DHR model in this context.

The most useful aspect of the DAR model is in estimating any significant changes in the spectral properties of $y(k)$: so-called 'time-frequency' or 'evolutionary spectral' analysis. Here, at the kth time instant, the FIS estimated parameters $\hat{a}_i(k|N)$, $i = 1,2,...,p$, of the DAR(p) model can be used to compute the instantaneous AR spectrum at this time from the well known relationship (see e.g. Priestley, 1981 and section B.3.7.2 Appendix B):

$$\hat{h}(\omega,k) = \frac{\hat{\sigma}^2}{2\pi} \frac{1}{\left|1 + a_1(k|N)\exp(-j\omega) + \cdots + a_p(k|N)\exp(-j\omega)\right|^2} \qquad (5.24)$$

for $k = 1, 2, \ldots ,N$, where $\hat{\sigma}^2$ is the estimated variance of the model residuals and can itself be estimated recursively if this is warranted (see section 4.5.4 of Chapter 4). The order p is selected either by the user on the basis of prior knowledge, or by use of the AIC (see section 5.2.1.1). Then, for each user-selected value of ω over the range 0 (zero frequency) to 0.5 (the Nyquist frequency), $\hat{h}(\omega,k)$ or its logarithm, is evaluated with $\exp(-j\omega) = \cos(\omega) + j\sin(\omega)$. The set of all these instantaneous but smoothly changing spectra over the interval $k = 1, 2, \ldots ,N$ then provides an indication of the changing spectral properties of the series $y(k)$ over this time interval.

An illustrative example is shown in Figure 5.15 which is the time-frequency plot produced by the CAPTAIN routine dar for an estimated DAR(2) model of a 'chirp' signal: i.e. a linear swept-frequency cosine signal (as generated by the Matlab routine chirp)[1]. The DAR model has clearly been able to track the changes in the period of the signal, as revealed well in the combined colour 3D surface and contour plot. However, as shown in the example below, time-frequency spectra such as this can be presented in various other ways.

5.3.1 Practical example: palaeoclimatic data analysis

Young and Pedregal (1999b) have applied DHR analysis to the well known SPECtral MAPping series (SPECMAP: see Imbrie et al, 1992), which has been obtained from the analysis of oxygen isotope variations in deep ocean cores. This earlier analysis

[1] This is a command line demonstration example for dar in CAPTAIN.

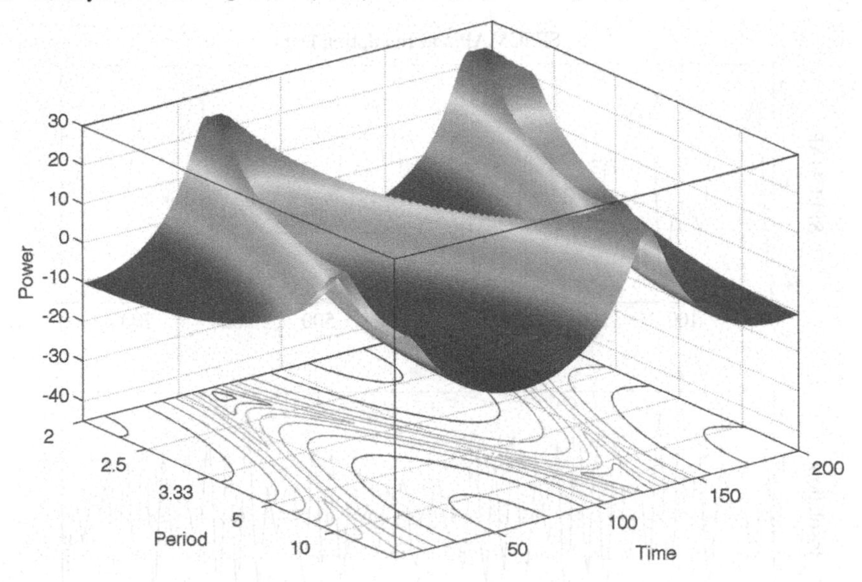

Fig. 5.15 Time-frequency plot showing how the estimated DAR(2) spectrum of a 'chirp' signal changes with time.

is concerned with the possible relationship between this series and solar insolation, in order to evaluate whether there is any evidence in the ocean core record that the astronomical cycles of precession, obliquity, and eccentricity may be important influences on climate, as suggested by Milankovitch theory (Milankovitch, 1920): see e.g. `http://en.wikipedia.org/wiki/Milankovitch_cycles`. Here, the alternative DAR analysis of the SPECMAP series is used to look more closely at the spectral properties of the series and see if there is any evidence that these properties are changing over the 800 kilo-annum (ka) interval.

For the purposes of the DAR analysis, the original SPECMAP series (Imbrie et al., 1992), which has a sampling interval of one ka, is standardised (transformed to zero mean and unity variance) and sub-sampled every 5 ka by 'decimation' (combined filtering and re-sampling). The resulting series is shown in the upper plot of Figure 5.16. The sampling interval of 5 ka is selected for two related reasons: first, it ensures that the AIC identified AR model order is reduced, so simplifying the DAR analysis; second, the important spectral peaks, which are at fairly long periods of *circa* 40, 23 and 19 ka, are effectively moved towards the higher frequency part of the resulting spectrum, where they become more clearly resolved.

The AIC criterion is rather unclear about the best constant parameter AR order but the AR(15) spectrum shown in Figure 5.17 seems to reasonably represent the spectral properties when a range of different AR orders are considered. The four largest peaks in this spectrum are located at periods of 19.69, 8.0, 4.65 and 3.82, in units of 5 ka; in other words, periods of 98.46, 40, 23.27 and 19.10 ka, respectively

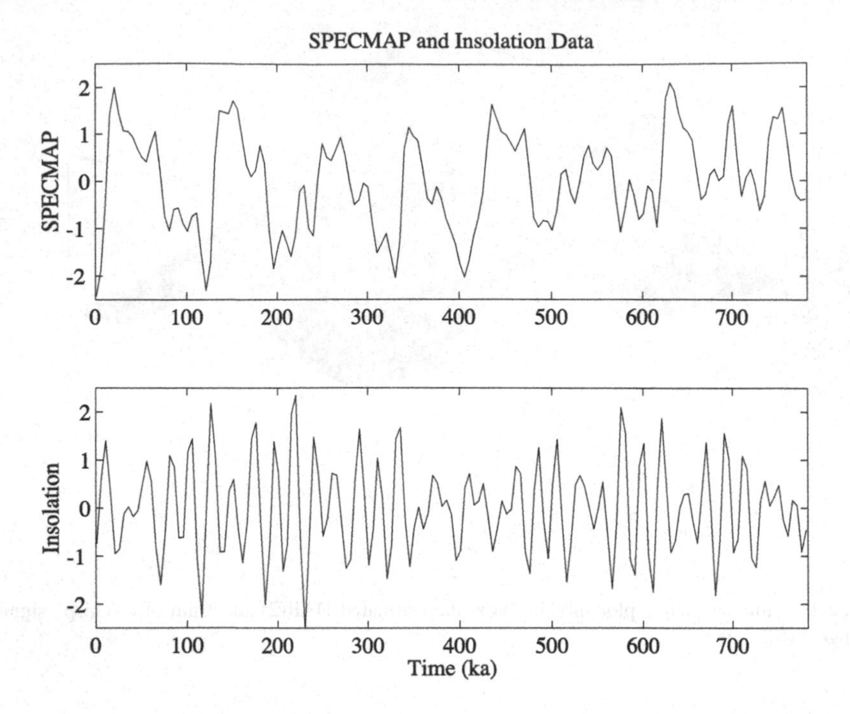

Fig. 5.16 Palaeoclimatic data example: SPECMAP series (upper plot); Insolation series (lower plot)

(frequencies of 0.0102, 0.025, 0.043 and 0.0523 cycles/ka). On this basis, the DAR model is also chosen as 15th order and the parameters are assumed to evolve as simple RW processes. ML optimization of this DAR(15) model then yields the values for the NVR hyper-parameters reported in Table 5.3.

Table 5.3 DAR Hyperparameter Optimization for the SPECMAP data.

Parameter	$a_1(k)$	$a_2(k)$	$a_3(k)$	$a_4(k)$	$a_5(k)$	$a_6(k)$	$a_7(k)$	$a_8(k)$
NVR	0.0023	0.0051	0.0049	0.0047	0.0035	0.0041	0.0025	0.0094

Parameter	$a_9(k)$	$a_{10}(k)$	$a_{11}(k)$	$a_{12}(k)$	$a_{13}(k)$	$a_{14}(k)$	$a_{15}(k)$
NVR	0.0106	0.0233	0.0129	0.0020	0.0017	0.0019	0.0049

The FIS estimates of the 15 DAR parameters obtained with these optimum hyper-parameters define the AR spectrum at every sampling instant and the resulting DAR spectra, as they evolve in time, are presented in three different forms in Figures 5.18 to 5.20. Figure 5.18 is a contour plot with frequency on the vertical axis and time, in ka, along the horizontal axis; Figure 5.19, the spectra at every sampling instant are

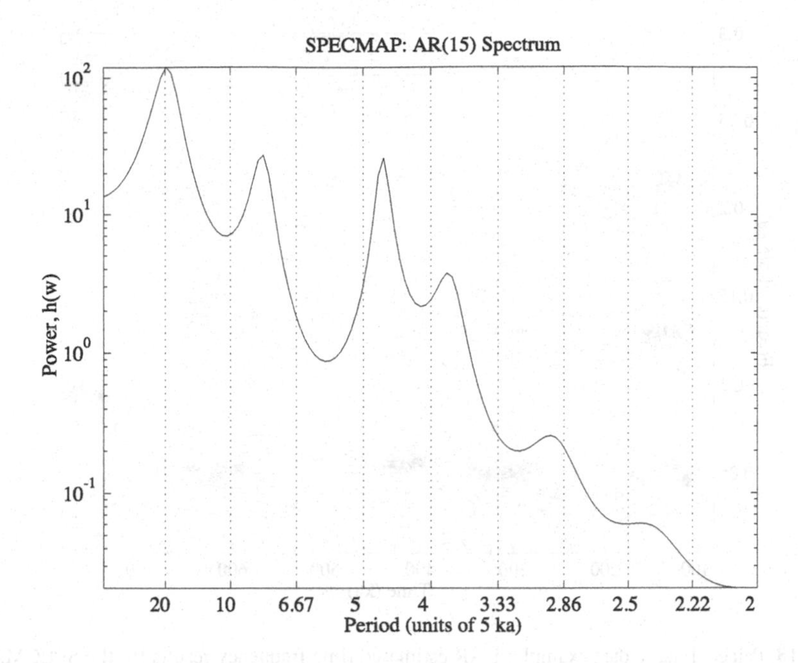

Fig. 5.17 AR(15) spectrum of the SPECMAP series.

'stacked' to show their temporal evolution; and Figure 5.20 shows multiple plots of the temporally evolving spectrum at various times over the 800 ka interval. Other presentations of results such as these are possible, such as the 3D plot in Figure 5.15 and the changing spectra can even be plotted repeatedly as a power-frequency graph to produce a cinematic effect.

One algorithmic factor that should be noted here concerns the optimization of the hyper-parameters. For illustration, all of the NVRs above have been optimized in an unconstrained manner. However, as in the DHR case, the ML optimization does not appear to be too well defined in this situation, with the possibility of many local maxima and the final optimized values being dependent on the user-defined initial values. An alternative approach is to constrain all the NVRs to be equal. This can work well, although it should be carried out with care since the best AR model may be a subset AR (i.e. some of the intermediate parameters should be zero) and so the uniform constraint may not be appopriate. It is also theoretically reasonable to constrain the NVRs in this manner since, as we know from the discussion in section 4.5.5 of Chapter 4, they affect the localized weighting of the data in relation to the estimation of each parameter. Unlike the DLR and DHR cases, where the regressors are not functions of the 'dependent' variable $y(k)$, the regression variables in the DAR case are the past values of $y(k)$ and there seems to be no reason why the weighting should not be uniform across all the parameters (except in special cases such as the subset AR).

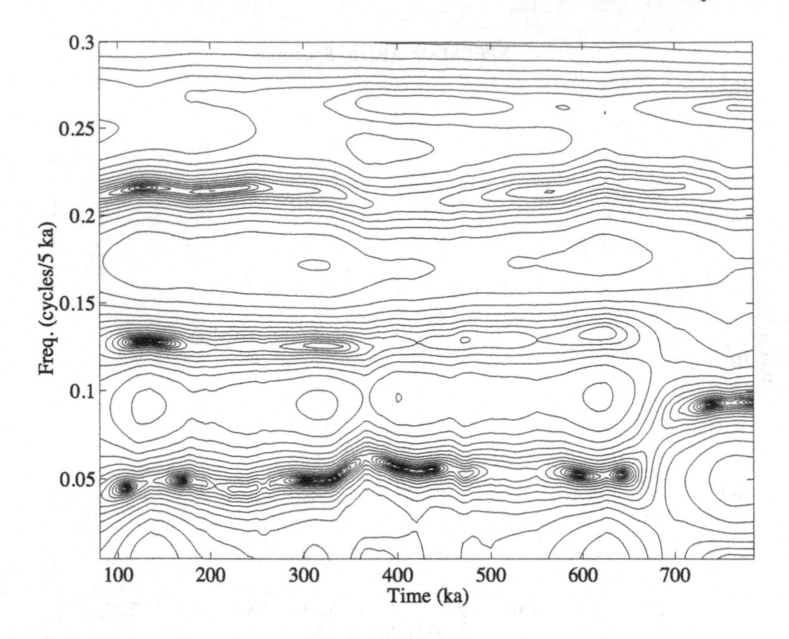

Fig. 5.18 Palaeoclimatic data example: DAR estimated time-frequency results for the SPECMAP series shown as a contour plot.

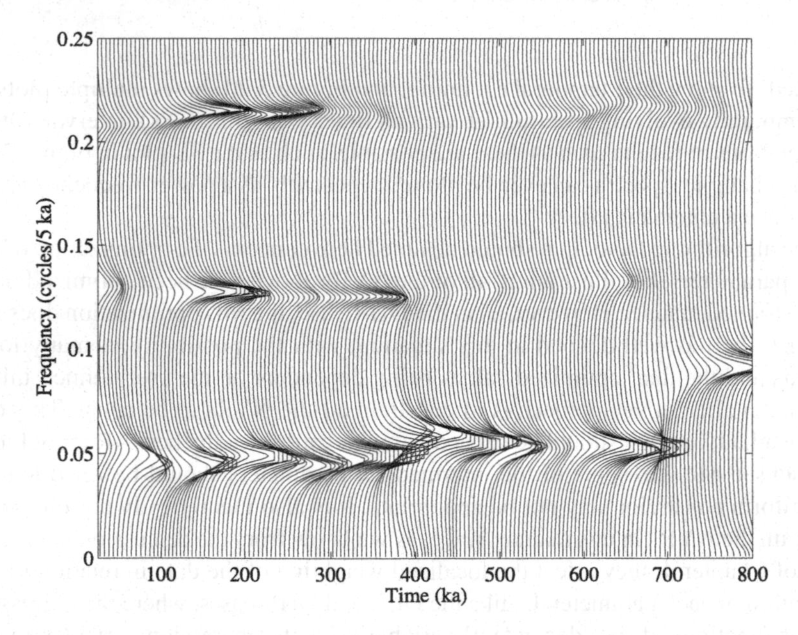

Fig. 5.19 Palaeoclimatic data example: DAR estimated time-frequency results for the SPECMAP series shown as a stacked plot.

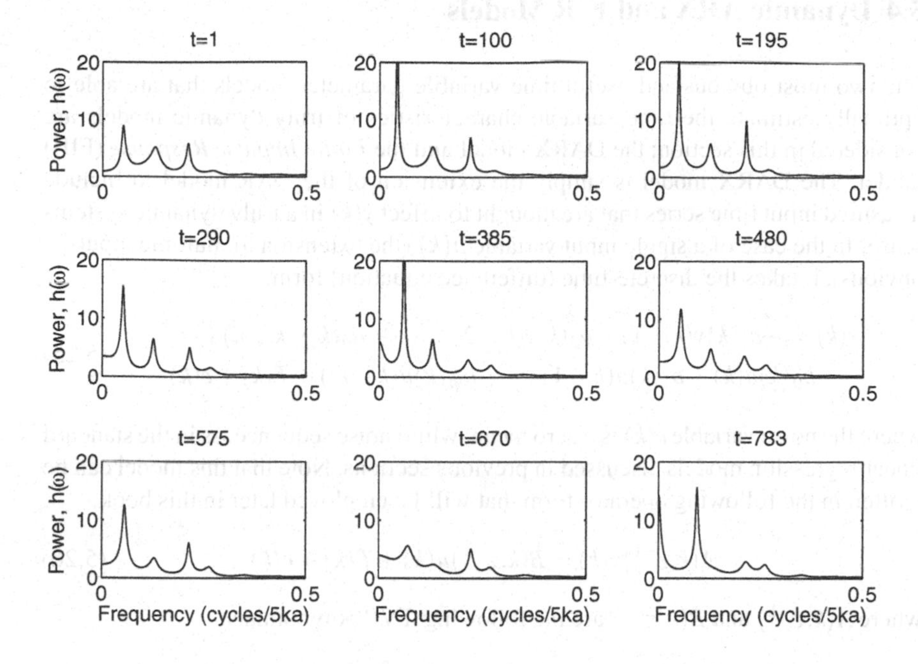

Fig. 5.20 Palaeoclimatic data example: DAR estimated time-frequency results for the SPECMAP series shown as a sequence of spectra at various times.

Bearing these latter observations in mind, the user can also benefit from the ability to 'manually' specify a range of values for the single NVR value in this constrained situation and so evaluate the effects of the different localized weighting effects so induced. This can be very useful and instructive in practical terms. For instance, in the present example, the user is quickly led to the conclusion that the estimated spectral evolution is not too sensitive to the choice of the NVR and a constrained NVR=0.01 for all parameters provides a reasonable compromize between under (e.g. NVR=1) and over (e.g. NVR=0.0001) smoothing, with results very similar to those obtained by unconstrained optimization in Figures 5.18 to 5.20.

These results show that, while the estimated frequencies associated with the three major spectral peaks (or 'modes') are relatively stable over most of the 800 ka, the two lowest frequency modes (periods of 40 and 98.6 ka) appear to coalesce at around 650-700 ka. Also, the estimated power in all three modes varies considerably, with that of the highest frequency mode (23.27 ka period) becoming very low at the beginning, middle and end of the time interval, with higher power in between. This latter result is consistent with the DHR results discussed by Young and Pedregal (1999b) which shows that the three modes at periods of 40, 23.27 and 19.10 ka are also a feature of the AR(15) spectrum of the insolation series shown in the lower plot of Figure 5.16. DAR analysis of this series is also given in this paper.

5.4 Dynamic ARX and FIR Models

The two most obvious and useful time variable parameter models that are able to optimally estimate the time variable characteristics of truly dynamic models are considered in this section: the DARX model and the *Finite Impulse Response* (FIR) model. The DARX model is simply the extension of the DAR model to include measured input time series that are thought to affect $y(k)$ in a truly dynamic systems sense. In the case of a single input variable $u(k)$ (the extension to multiple inputs is obvious), it takes the discrete-time (difference equation) form,

$$\begin{aligned} y(k) = -a_1(k)y(k-1) - a_2(k)y(k-2) - \cdots - a_n(k)y(k-n) + \\ b_0(k)u(k) + b_1(k)u(k-1) \cdots + b_m(k)u(k-m) + T(k) + e(k) \end{aligned} \tag{5.25}$$

where the noise variable $e(k)$ is a zero mean, white noise sequence, as in the standard linear regression models discussed in previous sections. Note that this model can be written in the following operator form that will be employed later in this book.

$$A(k,z^{-1})y(k) = B(k,z^{-1})u(k) + T(k) + e(k) \tag{5.26}$$

where $A(k,z^{-1})$ and $B(k,z^{-1})$ are the following TVP polynomials:

$$A(k,z^{-1}) = 1 + a_1(k)z^{-1} + a_2(k)z^{-2} + \cdots + a_n(k)z^{-n} \tag{5.27a}$$

$$B(k,z^{-1}) = b_0(k) + b_1(k)z^{-1} + \cdots + b_m(k)z^{-m} \tag{5.27b}$$

in which z^{-r} is the backward shift operator, i.e. $z^{-r}y(k) = y(k-r)$. Dividing through this equation by $A(k,z^{-1})$ gives the following *Transfer Function* (TF) form of the model:

$$y(k) = \frac{B(k,z^{-1})}{A(k,z^{-1})}u(k) + \frac{1}{A(k,z^{-1})}T(k) + \frac{1}{A(k,z^{-1})}e(k) \tag{5.28}$$

which shows that the output $y(k)$ is obtained by operating on the input $u(k)$ by the time variable transfer function $B(k,z^{-1})/A(k,z^{-1})$ and then adding the trend $T(k)$ and the white noise $e(k)$, both of which are operated on by the TF, $1/A(k,z^{-1})$. This form of the model may seem to be a rather unnecessary complication at this stage but the value of this TF form will become apparent below and in later chapters in this book.

It is convenient at this point to assume that $y(k)$ is stationary in the mean and remove the complication of the trend component $T(k)$. It is then easy to see that the model (5.25) can be written in the same linear regression form as considered previously but with the regression vector now denoted by $\mathbf{z}(k)$ and a parameter vector denoted by $\boldsymbol{\rho}$ in order to emphasize the differences, i.e.,

$$y(k) = \mathbf{z}^T(k)\boldsymbol{\rho} + e(k) \tag{5.29}$$

where now,

$$\begin{aligned}\mathbf{z}^T(k) = [-y(k-1) - y(k-2) \ldots -y(k-n) \\ u(k)\, u(k-1) \ldots u(k-m)]\end{aligned} \tag{5.30}$$

and $\boldsymbol{\rho}$ is defined as

$$\boldsymbol{\rho}(k) = [a_1(k)\, a_2(k) \ldots a_n(k)\, b_0(k)\, b_1(k) \ldots b_m(k)]^T$$

Note that the DAR model is simply a special case of (5.29) where,

$$\begin{aligned}\mathbf{z}^T(k) &= [-y(k-1) - y(k-2) \ldots -y(k-n)] \\ \boldsymbol{\rho}(k) &= [a_1(k)\, a_2(k) \ldots a_n(k)]^T\end{aligned} \tag{5.31}$$

and there is no deterministic excitation $u(k)$.

Using a TF formulation of this same kind, the FIR model can be written as follows:

$$\begin{aligned}y(k) &= \frac{B(z^{-1})}{A(z^{-1})}u(k) + e(k) \approx G(k,z^{-1})\, u(k) + e(k) \\ G(z^{-1}) &= g_0 + g_1 z^{-1} + g_2 z^{-2} + \cdots + g_p z^{-p} + e(k)\end{aligned} \tag{5.32}$$

Here, the coefficients provide a finite dimensional approximation to the infinite dimensional, discrete-time impulse response, as obtained when $B(z^{-1})$ is divided by $A(z^{-1})$ using long division. This is the discrete-time approximation of the well-known continuous-time convolution integral equation model, where $G(z^{-1})$ is simply a weighted linear sum of a finite number of past inputs $u(k-i), i = 1,2,\ldots,p$, with p chosen to provide a reasonable description of the impulse response and the system dynamics. The time variable parameter *Dynamic Finite Impulse Response* (DFIR) model is formulated in a similar manner to the DARX model (5.29), but now,

$$\begin{aligned}\mathbf{z}^T(k) &= [u(k)\, u(k-1)\, u(k-2) \ldots u(k-p)] \\ \boldsymbol{\rho}^T &= [g_0(k)\;\; g_1(k)\;\; g_2(k) \ldots g_p(k)]\end{aligned} \tag{5.33}$$

Clearly, since all of the above models are in a linear regression form, it is possible to recursively estimate their parameters using either the constant parameter RLS algorithm, any of the TVP estimation algorithms described in Chapter 4, or the DLR algorithm with the regressors defined accordingly. The DARX algorithm with FIS, as produced in this manner, takes the following form (*cf* the KALMSMO algorithm in Chapter 4):

Prediction:

$$\hat{\boldsymbol{\rho}}(k|k-1) = \mathbf{A}\hat{\boldsymbol{\rho}}(k-1) \tag{DARX-1}$$

$$\mathbf{P}(k|k-1) = \mathbf{A}\mathbf{P}(k-1)\mathbf{A}^T + \mathbf{D}\mathbf{Q}_{nvr}\mathbf{D}^T \tag{DARX-2}$$

Correction:

$$\begin{aligned}
\hat{\boldsymbol{\rho}}(k) &= \hat{\boldsymbol{\rho}}(k|k-1) + \mathbf{g}(k)\{y(k) - \mathbf{z}^T(k)\hat{\boldsymbol{\rho}}(k|k-1)\} && \text{(DARX-3)}\\
\mathbf{g}(k) &= \mathbf{P}(k|k-1)\mathbf{z}(k)[1 + \mathbf{z}^T(k)\mathbf{P}(k|k-1)\mathbf{z}(k)]^{-1} && \text{(DARX-4)}\\
\mathbf{P}(k) &= \mathbf{P}(k|k-1) - \mathbf{g}(k)\,\mathbf{z}^T(k)\mathbf{P}(k|k-1) && \text{(DARX-5)}\\
\mathbf{P}^*(k) &= \hat{\sigma}^2\mathbf{P}(k) && \text{(DARX-6)}
\end{aligned}$$

Smoothing:

$$\begin{aligned}
\hat{\boldsymbol{\rho}}(k|N) &= \hat{\boldsymbol{\rho}}(k) - \mathbf{P}^*(k)\mathbf{A}^T\boldsymbol{\lambda}(k) && \text{(DARX-7)}\\
\text{or } \hat{\boldsymbol{\rho}}(k|N) &= \mathbf{A}^{-1}\left[\hat{\boldsymbol{\rho}}(k+1|N) - \mathbf{D}\mathbf{Q}_{nvr}\mathbf{D}^T\boldsymbol{\lambda}(k)\right] && \text{(DARX-8)}\\
\boldsymbol{\lambda}(k-1) &= [\mathbf{I}_{np} - \mathbf{P}^*(k)\mathbf{z}(k)\mathbf{z}^T(k)]^T\\
&\qquad \{\mathbf{A}^T\boldsymbol{\lambda}(k) - \mathbf{z}(k)[y(k) - \mathbf{z}^T(k)\mathbf{A}\hat{\boldsymbol{\rho}}(k-1)]\}\\
\text{with } \boldsymbol{\lambda}(N) &= 0 && \text{(DARX-9)}\\
\mathbf{P}(k|N) &= \mathbf{P}(k) + \mathbf{P}(k)\mathbf{A}^T\mathbf{P}(k+1|k)^{-1}\\
&\qquad [\mathbf{P}(k+1|N) - \mathbf{P}(k+1|k)]\mathbf{P}(k+1|k)^{-1}\mathbf{A}\mathbf{P}(k) && \text{(DARX-10)}\\
\mathbf{P}^*(k|N) &= \hat{\sigma}^2(k)\mathbf{P}(k|N) && \text{(DARX-11)}
\end{aligned}$$

where $np = n + m + 1$ is the number of parameters in the parameter vector $\boldsymbol{\rho}$. Normally there will be little known about the stochastic evolution of the TF model parameters and $\mathbf{A}$ and $\mathbf{D}$ will be defined by a member of the GRW model set discussed in section 4.2 of Chapter 4. In the simplest vector random walk case, which is probably the most useful in many practical situations, $\mathbf{A} = \mathbf{D} = \mathbf{I}$, so that $\hat{\boldsymbol{\rho}}(k|k-1) = \hat{\boldsymbol{\rho}}(k-1)$ and equation (DARX-1) is redundant; while the $\mathbf{P}$ matrix update is reduced to simply $\mathbf{P}(k|k-1) = \mathbf{P}(k-1) + \mathbf{Q}_{nvr}$, requiring only the specification of the NVR matrix $\mathbf{Q}_{nvr}$, either by ML optimization (see section 4.5.3 of Chapter 4) or by a manual alternative.

Finally, note that both the DARX and DFIR models can easily incorporate a pure time delay effect of δ sampling intervals between the input excitation and its effect on the output $y(k)$, either implicitly, by assuming that the δ leading coefficients of the $B(k, z^{-1})$ polynomial are zero; or explicitly, by leaving the $B(k, z^{-1})$ polynomial as it is defined in (5.27b) and denoting the input variable by $u(k-\delta)$. The latter makes the presence of the time delay more obvious and will be the form most favoured in the rest of this book.

An example of DARX estimation is presented in the next section and an example of constant parameter FIR model estimation is described in section 6.5 of the next Chapter 6.

5.4.1 Example: DARX model estimation

This example is based on one originally described in Wellstead and Zarrop (1991) and considers the related results obtained by Young (2002b), using the same example and the input-output data shown in Figure 5.21. The TVP algorithms used by Wellstead and Zarrop are the standard EWP algorithm an the DF algorithm (Kulhavy, 1987) considered in section 4.1.2. Here, the results obtained using the EWP and DF algorithms are compared with those of the optimal DARX algorithm described above.

The DARX model used in the analysis is as follows:

$$y(k) = -a_1 y(k-1) - a_2 y(k-2) + b_0(k) u(k-1) + e(k)$$

where $e(k)$ is a zero mean, white noise sequence with variance 0.16 (yielding a noise/signal ratio by standard deviation (std) of 0.1: i.e. $\text{std}(e_k)/\text{std}(y_k) = 0.1$). The b_0 parameter changes from 1.0 to 2.0 at $k = 200$, and then back to 1.0 at $k = 900$; while $a_1 = -1.0$ and $a_2 = 0.25$ are time invariant. The simulated input-output data are shown in Figure 5.21: the input $u(k)$ changes from a square wave between plus and minus one to a very small amplitude square wave of plus and minus 0.002 at $k = 400$, reverting to the original large square wave at $k = 850$. This choice of input signal induces 'estimator wind-up' in the case of the standard EWP algorithm because the information content in the data during the period of low input activity is not sufficient to ensure good performance from this rather crude TVP estimation algorithm. This is illustrated in the lower panel of Figure 5.22, where the design parameter $\alpha = 0.85$ was obtained by manual tuning to yield the best overall performance: the wind-up behaviour is clearly apparent after $k = 780$.

The upper panel of Figure 5.22 shows the results obtained with the DF algorithm, which is designed specifically to limit estimator wind-up. Here, the design parameter is set to 0.85. We see that this produces a distinct improvement over the standard EWP algorithm, with the worst excesses of the wind-up no longer occurring. However, the response to the parametric change is relatively slow and there is considerable interaction between the estimates over the period of input inactivity.

By far the best results are those shown in Figure 5.23, as obtained using the optimized DARX algorithm, with the NVR hyper-parameters optimized by maximum likelihood, as discussed in section 4.5.3 of the previous Chapter 4. This is implemented by the darx and darxopt routines in the CAPTAIN Toolbox, which were used to generate the results in Figure 5.23. The diagonal elements of the NVR matrix $\mathbf{Q}_{nvr} = \text{diag}\,[\text{NVR}_{a_1}\ \text{NVR}_{a2}\ \text{NVR}_{b_0}]$ are the NVR hyper-parameters for the three TF model parameters, which are ML-optimized by darxopt, using the simplest RW option for the TVP model of the parameter variations, at the following values:

$$\text{NVR}_{a_1} = 1.5 \times 10^{-16};\ \text{NVR}_{a_2} = 4.6 \times 10^{-20};\ \text{NVR}_{b_0} = 0.0186.$$

where the NVR_{a_1} and NVR_{a_2} are both insignificantly different from zero illustrating how, quite objectively, the ML optimization has inferred from the data that the

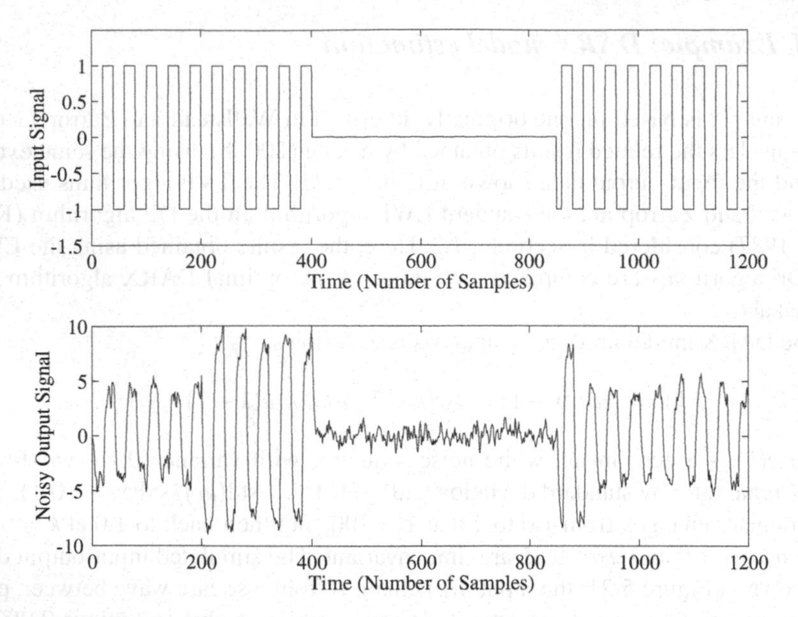

Fig. 5.21 Input $u(k)$ (upper panel) and noisy output $y(k)$ (lower panel) for the DARX simulation example.

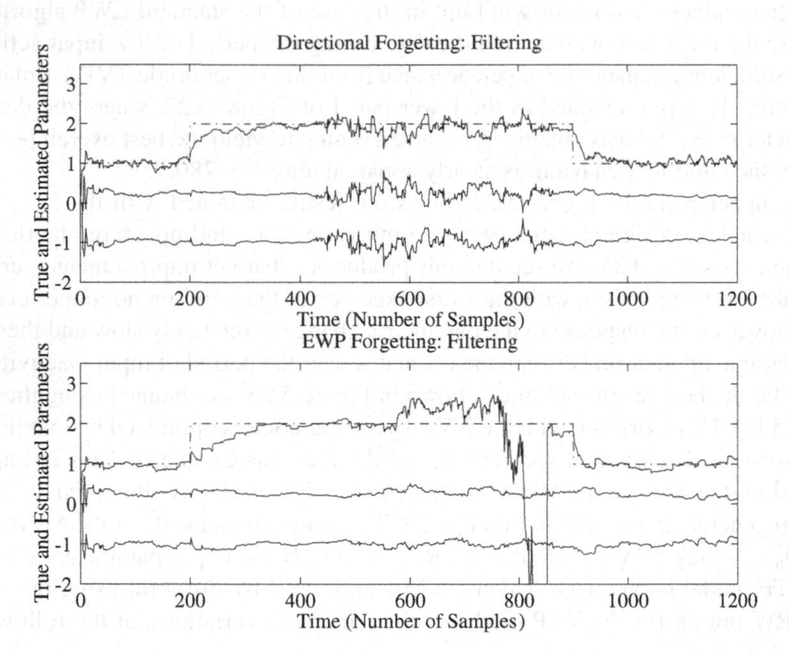

Fig. 5.22 Estimation results for the DARX simulation example using the standard EWP (lower panel) and DF (upper panel) estimation algorithms.

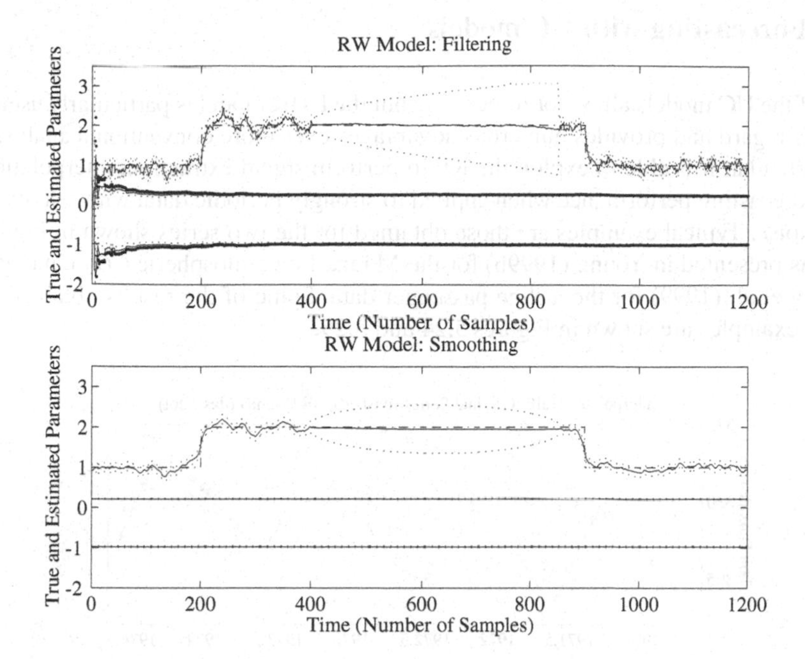

Fig. 5.23 Estimation results for the DARX simulation example using the ML optimized DARX estimation algorithm: filtering results (top panel) and smoothing results (bottom panel), with the estimated SE bounds shown dotted in both cases.

associated a_1 and a_2 parameters are time-invariant. However, NVR_{b_0} is significant and has been optimized at a value that gives good tracking of the step changes in the b_0 parameter.

The lower panel in Figure 5.23, shows the backward-pass FIS estimation results generated by the CAPTAIN darx routine. In comparison with the filtering results in the upper panel, it will be noted that the smoothed estimates are more accurate, the step changes are anticipated because the estimates are based on the whole data set; and the estimates of the constant parameters are now themselves constant. Note also that, since both the filtering and smoothing algorithms are statistical in nature, they provide information on the statistical properties of the estimates, as shown here by the standard error bounds, plotted as the dotted lines either side of the estimates. And, as expected, the SE bounds on the smoothed estimates are smaller than those of the filtered estimates, particularly over the period of input inactivity.

The superiority of the optimal TVP approach to estimation is clear from the results presented in Figures 5.21 to 5.23. Not only is the parameter tracking visibly better, despite the fact that the filtering TVP algorithm is of similar complexity to the EWP and DF algorithms, but also the data-based ML optimization of the algorithm is fully automatic, so removing the need for subjective and tiresome manual tuning.

5.5 Forecasting with UC models

All of the UC models allow for forecasting but the DHR model is particularly useful in this regard and provides numerous advantages over more conventional analysis. In particular, it is able to exploit the KF to perform signal extraction, interpolation and forecasting performance when applied to strongly periodic data, with relatively low noise. Typical examples are those obtained for the two series shown in Figure 5.1, as presented in Young (1999b) for the Mauna Loa atmospheric CO_2 data; and Young et al. (1999) for the airline passenger data. Some of the results obtained in these examples are shown in Figures 5.24 and 5.25.

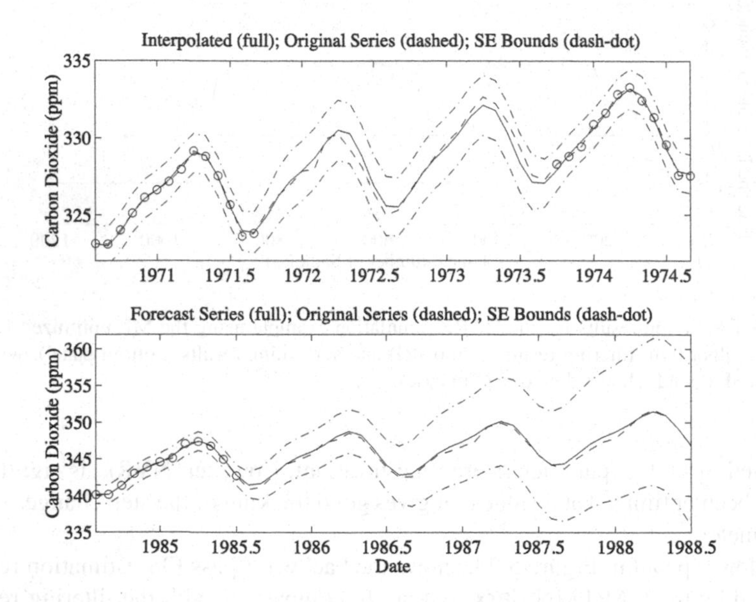

Fig. 5.24 Mauna Loa atmospheric CO_2 series: the 2 year DHR interpolations (upper plot) and 3-year-ahead forecasts (lower plot) shown as full lines, with the measured series (dashed) and standard error bounds (dash-dot). From Young (1999b).

In Figure 5.24, the upper panel shows interpolation over an artificial 2 year gap in the Mauna Loa series, demonstrating how the DHR algorithm is able to 'repair' series with missing data; while the lower panel presents the 3-year-ahead forecasting results. In both cases, the time series samples used in estimating the DHR model are shown as circular points. Figure 5.25 compares the 24-month-ahead DHR model forecasts for the airline passenger data with those obtained using the STAMP forecasting program (Koopmans et al., 1995): the excellent performance of the DHR model is due, in part, to the fact that the DHR analysis based on concatenated AR spectra (see modification 1. in section 5.2.4) identified and estimated a quasi-cyclical component in the data, in addition to the seasonal component, with spectral peaks at a fundamental period of of 51 months, with harmonics at 25.5 and 17 months, respectively. This may be due to the fluctuations in airline passengers due to the business cycle at the time.

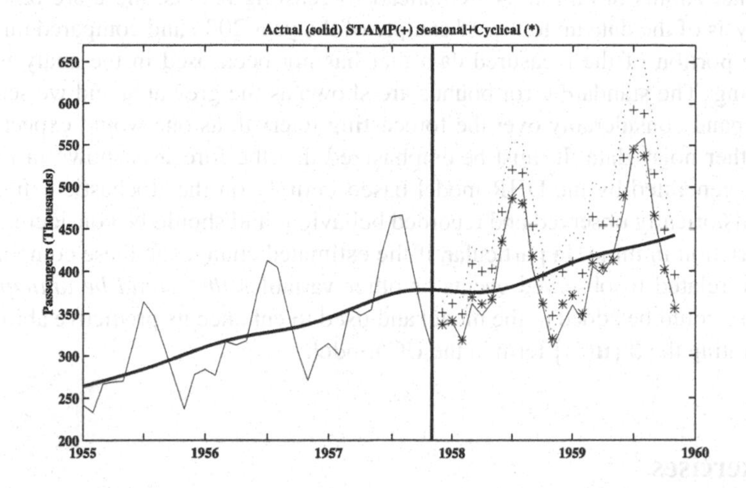

Fig. 5.25 Airline passenger series: 24 month ahead DHR forecasting performance over the latter two years of the series: actual data (solid); STAMP program results (+); and seasonal plus quasi-cyclical DHR model results (*). The estimated trend for the latter model is shown as a thick solid line. From Young et al. (1999).

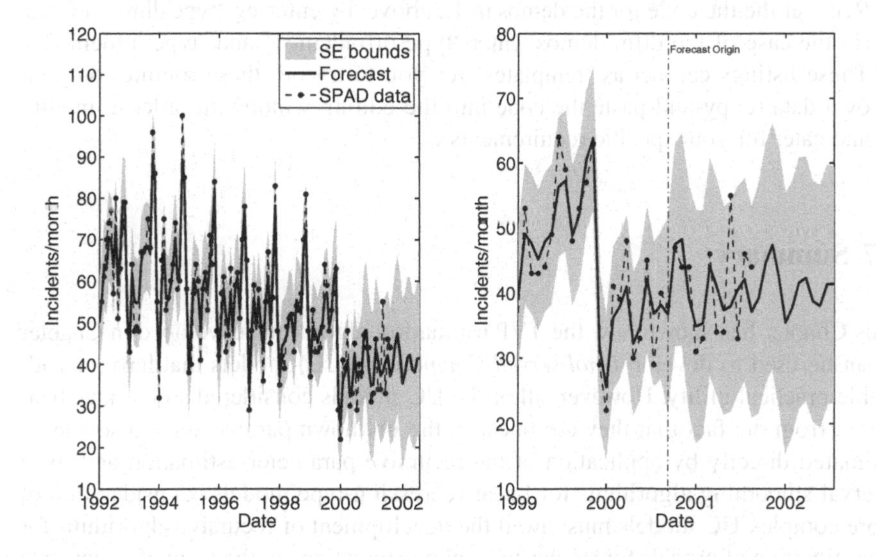

Fig. 5.26 24-step-ahead forecasting results for the SPAD data, with intervention: full data set (left panel); post-1999, showing the intervention effect more closely (right panel).

The DHR forecasting results for the SPAD data analysed in section 5.2.3 are presented in Figure 5.26. The left hand panel shows the overall results; while the right hand panel zooms in on the 24 step ahead forecasting results: these are based on the analysis of the data up to the 9th sample of the year 2000 and compared initially with the portion of the measured data that has not been used in the analysis and forecasting. The standard error bounds are shown as the grey area and we see that these expand considerably over the forecasting interval, as one would expect with these rather noisy data. It must be emphasized that the forecasts shown in Figure 5.26 are generated by the DHR model based entirely on the stochastic extrapolation of historically observed and recorded behaviour and should be considered with this restriction in mind. In particular, if the estimated changes in these components could be related to observed events or other variables *that could be anticipated*, then these could be added to the model and used to enhance its predictive ability by incorporating the $\mathcal{F}\{\mathbf{u}(k)\}$ term in the UC model.

5.6 Exercises

1. Inspect the instructions in Appendix G for running demonstration examples in the CAPTAIN Toolbox. Then, in turn, run the command line demos DLR, DHR, DAR, DARX. In each case, examine the command line instructions in order to understand how the analysis is being carried out.
2. Recover the the code for the demos in 1., above, by entering 'type dhrdemo' etc. (in the case of the dtfm demos, enter 'type dtfmdemo1' and 'type dtfdemo2'). These listings can act as 'templates' for you to try out these routines on your own data (copy-and-paste the code into the editing window in order to modify and cater for your specific requirements).

5.7 Summary

This Chapter has shown how the TVP estimation algorithms developed in Chapter 4 can be used to develop *Unobserved Component* (UC) models that have considerable practical utility. However, all of the UC models considered so far have benefitted from the fact that they are linear in the unknown parameters and so can be estimated directly by application of the recursive parameter estimation and fixed interval smoothing algorithms for linear regression-type models. Consideration of more complex UC models must await the development of recursive algorithms for the estimation of models for stochastic dynamic equations in the form of either state equations or transfer function models, as discussed in the next Chapter 6.

Part II
Recursive Estimation of Parameters in Transfer Function Models

Chapter 6
Transfer Function Models and the Limitations of Recursive Least Squares

In the first part of this book, we have considered in some detail the recursive estimation of parameters in linear regression models. Such models are, however, primarily utilized in the evaluation of static relationships between variables and, while they may well figure in some aspects of systems analysis (for example the characterization of equilibrium or steady state behaviour) they are not generally suitable for use in dynamic systems analysis. Of course, dynamic systems can be modelled in various ways and so consideration of methods for estimating the parameters of dynamic system models is dependent, to some extent at least, on the nature of the model chosen to characterize the system. This chapter considers a specific class of models for linear stochastic dynamic systems and shows how the recursive methods of estimation discussed in the previous chapter can be modified to handle the estimation of the parameters that characterize these models.

In the classical statistical literature, the models considered here have been termed 'time series' representations since they attempt to describe a time series of data, i.e. data arranged in temporal order (Box and Jenkins, 1970). While our treatment of these models will be similar to that used in statistical texts, its emphasis will be rather different: the classical approach tends to concentrate on the stochastic aspects of the time series model; while here we emphasize the relationship between the assumed deterministic input or 'exogenous' variables and that part of the measured output of the stochastic dynamic system that can be 'explained' in relation to these inputs. This emphasis arises because we are concerned with the mathematical characterization of stochastic dynamic systems for purposes such as: the better understanding of causal dynamic relationships in physical systems; the use of models for forecasting the outcome of changes in the exogenous input variables; and for designing control and management schemes.

The reader has already been introduced to one type of linear stochastic dynamic model in the last two chapters: namely the discrete time Gauss-Markov or state-space model used both as the vehicle for the development of the TVP estimation algorithms and the KFX version of the the Kalman Filter algorithm in section 4.4 of chapter 4. This seems, therefore, an appropriate starting point for our present discussion. If we consider the state space model (4.46) used in relation to the KFX algorithm, then this will serve to emphasize that we are concerned here with the modeling of a general, linear, stochastic dynamic system; rather than considering the model simply as a device for representing the variations in parameters. In this manner, we can retain the Gauss-Markov model (4.23) of chapter 4 for the latter purpose and consider its use as a model for time-variable parameter variations in later chapters.

6.1 Introduction: Direct Estimation of the State Space Model

To begin with, assume that the dynamic system is stable and stationary, so that the relevant state space description is (4.46): i.e.

$$\begin{aligned}\mathbf{x}(k) &= \mathbf{A}\mathbf{x}(k-1) + \mathbf{B}\mathbf{u}(k-1) + \mathbf{D}\boldsymbol{\zeta}(k-1)\\ \mathbf{y}(k) &= \mathbf{C}\mathbf{x}(k) + \mathbf{B}_I\mathbf{u}(k) + \boldsymbol{\xi}(k)\end{aligned} \tag{6.1}$$

where the matrices $\mathbf{A}$, $\mathbf{B}$, $\mathbf{C}$, $\mathbf{D}$ and $\mathbf{B}_I$ are now composed of time-invariant real numbers (to ensure stationarity). It is further assumed that the system is stable, with the eigenvalues of the characteristic equation

$$\det[\mathbf{I} - \mathbf{A}z^{-1}] \equiv 1 + a_1 z^{-1} + \ldots + a_n z^{-n} = 0 \tag{6.2}$$

all lying outside the unit circle in the complex z^{-1} plane (i.e. equivalent to the roots of the equation $z^n + a_1 z^{n-1} + \ldots + a_n = 0$ lying inside the unit circle in the complex z plane).

Nominally, the problem of parameter estimation for the model (6.1) is to infer, in some statistical manner, the parameters of the $\mathbf{A}$, $\mathbf{B}$, $\mathbf{C}$, $\mathbf{D}$ and $\mathbf{B}_I$ matrices, on the basis of the measurements $\mathbf{u}(k)$ and $\mathbf{y}(k)$ over some observational interval, say $k = 1, 2, \ldots, N$. In his 1960 paper, Kalman recognized this as a difficult and, at that time, unsolved problem when he noted that his approach to state variable estimation was limited by its assumption that the model parameters were known *a priori*. He concluded:

> .... {*it is*} *convenient to start with the model and regard the problem of obtaining the model itself as a separate question. To be sure, the two problems should be optimized jointly if possible; the author is not aware, however, of any study of the joint optimization problem.*

This challenge was taken up quickly and several authors suggested approaches to recursive parameter estimation that were strongly stimulated by the nature of the KF algorithm. Perhaps the best known outcome of this research effort is the *Extended Kalman Filter* (EKF), first suggested by Kopp and Orford (1963). This is a relatively straightforward approach, in which the state vector $\mathbf{x}(k)$ is augmented to include an unknown parameter vector which includes *any unknown* elements in the model matrices $\mathbf{A}$, $\mathbf{B}$, $\mathbf{C}$, $\mathbf{D}$ and $\mathbf{B}_I$; estimation then proceeds using a Kalman filter-like algorithm obtained by linearizing the now non-linear relationships (arising from the products between the unknown parameters and the unknown states) about the current estimates at each recursive step. The adjoined parametric states either constrain the parameters to be constant or allow for parameter variation by making the adjoined parametric states follow simple stochastic processes, such as the random walk models considered in chapters 4 and 5.

The EKF and other approaches to the direct estimation of the parameters in the state space model (6.1), or its continuous-time equivalent, are considered later in chapter 10. In this chapter, we consider an alternative approach which, rather than attempting to directly estimate the parameters in the state space model (6.1), considers the estimation of parameters in the 'observation space' *Transfer Function* (TF) model introduced in chapter 5. As we shall see, the TF model can be derived straightforwardly from the state space model, so establishing a link that is very important in system dynamic terms.

6.2 From State Space to Observation Space

The major difficulty with the estimation of parameters in (6.1) arises from its relative complexity as a description when viewed in parametric estimation terms. Because the model explicitly includes the unmeasured states comprising the state vector $\mathbf{x}(k)$, the problem is clearly nonlinear from an estimation standpoint with the nonlinearity, as pointed out above, arising from the product terms involving the unknown states and parameters. A more useful representation in estimation terms is one where the unknown parameters are, wherever possible, associated with measured variables in a linear model. But is such a representation possible? Or is the problem inherently nonlinear? To answer the question, it is necessary to attempt some simplification of the model (6.1) and, to do this, let us consider again the Kalman filter equations KFX in chapter 4.

The KF algorithm for the *Single Input, Single Output* (SISO) version of the model system (6.1), with the instantaneous input gain effecting the output removed, takes the following form, where the vector $\mathbf{c}^T$ replaced the matrix $\mathbf{C}$:

$$\begin{aligned}\hat{\mathbf{x}}(k) &= \mathbf{A}\hat{\mathbf{x}}(k|k-1) + \mathbf{b}u(k-1) + \mathbf{g}(k)[y(k) - \mathbf{c}^T\hat{\mathbf{x}}(k|k-1)] \\ \mathbf{g}(k) &= \mathbf{P}^*(k|k-1)\,\mathbf{c}\,[\sigma^2 + \mathbf{c}^T\mathbf{P}^*(k|k-1)\,\mathbf{c}]^{-1}\end{aligned} \tag{6.3}$$

where σ^2 the variance of the measurement noise affecting the single output $y(k)$; $u(k)$ is the single input, measured without noise; and $\mathbf{P}^*(k|k-1)$ is the covariance matrix associated with $\hat{\mathbf{x}}(k-1)$. From the discussion in previous chapters, we know that the recursive residual or innovations process $\varepsilon(k) = y(k) - \mathbf{c}^T\hat{\mathbf{x}}(k|k-1)$ has white noise properties, again with changing variance because of the variable nature of the covariance matrix $\mathbf{P}^*(k|k-1)$ and, therefore, the gain vector $\mathbf{g}(k)$. However, if the system is stable and the noise parameters are time-invariant, $\mathbf{P}^*(k)$, $\mathbf{P}^*(k|k-1)$ and $\mathbf{g}(k)$ will achieve constant asymptotic values $\mathbf{P}^*(\infty)$ and $\mathbf{g}(\infty)$ for $k \to \infty$; and will be virtually constant for quite small values of k. The KF is then operating in a 'steady state' condition and is often referred to as an asymptotic or *Steady State*

Kalman Filter. Consequently, in this condition, $\varepsilon(k)$ is a zero mean, white noise sequence with constant variance, σ^2, i.e.

$$E\{\varepsilon(j)\varepsilon(k)\} = \sigma^2\,\delta_{jk}$$

where δ_{jk} is the Kronecker delta function (see section 4.6.1 of Chapter 4 and section B.3, Appendix B).

Now, following directly from the definition of the innovations process $\varepsilon(k)$, the observation $y(k)$ can be written, in terms of the estimated state vector and $\varepsilon(k)$, as:

$$y(k) = \mathbf{c}^T\hat{\mathbf{x}}(k|k-1) + \varepsilon(k),$$

so that the KF can be written in the following alternative form:

$$\begin{aligned} \hat{\mathbf{x}}(k) &= \mathbf{A}\hat{\mathbf{x}}(k|k-1) + \mathbf{b}u(k-1) + \mathbf{g}(k)\varepsilon(k) \\ y(k) &= \mathbf{c}^T\hat{\mathbf{x}}(k|k-1) + \varepsilon(k) \end{aligned} \tag{6.4}$$

These equations are not only the steady-state KF equations, they also constitute an alternative state space representation of the stochastic, dynamic system, with the new states defined as the estimates ('conditional expectations' or 'conditional means') of the states comprising $\mathbf{x}(k)$, rather than the original state variables themselves. Here, the term 'conditional expectations' arises because $\hat{\mathbf{x}}(k)$ can be considered as the best estimate of the random process $\mathbf{x}(k)$ conditional on all the time series data up to sampling instant $k-1$, i.e. $\hat{\mathbf{x}}(k) = E\{\hat{\mathbf{x}}(k)|\mathbf{Y}(k-1)\}$, where $\mathbf{Y}(k-1)$ represents all of the time series data available up to $k-1$. The reader may find this description rather odd at first sight, with $y(k)$ being dependent of $\hat{\mathbf{x}}(k|k-1)$, but it should be recalled that the definition of the state variables is not unique and they can take on any form that is able to explain the behaviour in the observation space. And clearly the model (6.4) does this. Moreover, in terms of the input-output description, which is our main concern and which we develop in a TF form below, the definition of the state is arbitrary since it does not appear in the input-output TF.

The main advantage of the model (6.4) is its stochastic simplicity in relation to the original description, since there is only one noise term, namely the innovations process $\varepsilon(k)$. For this reason, it is usually termed the 'innovations' or 'Kalman filter' description: see e.g. Kailath (1968); Åström (1970). As a result, it is now quite straightforward to derive the following observation-space transfer function description of (6.4) by introducing the z^{-1} operator and deriving the TF from this in the usual manner (see Appendix B, equation (B.25) *et seq*). In this way, equation (6.4) can be written in the following operator form where, for notational convenience, $\varepsilon(k)$ is replaced by $e(k)$:

$$y(k) = \mathbf{c}^T[\mathbf{I} - \mathbf{A}z^{-1}]^{-1}\mathbf{b}z^{-1}u(k) + \mathbf{c}^T[\mathbf{I} - \mathbf{A}z^{-1}]^{-1}\mathbf{g}z^{-1}e(k) + e(k) \tag{6.5}$$

The reader can verify (see exercise 1 at the end of the chapter) that this yields a TF model of the form, where $A(z^{-1})$ is the determinant of $[\mathbf{I} - \mathbf{A}z^{-1}]$:

$$y(k) = \frac{B(z^{-1})}{A(z^{-1})}u(k) + \frac{D(z^{-1})}{A(z^{-1})}e(k) \tag{6.6}$$

or, multiplying throughout by $A(z^{-1})$,

$$A(z^{-1})y(k) = B(z^{-1})u(k) + D(z^{-1})e(k) \tag{6.7}$$

where

$$\begin{aligned} A(z^{-1}) &= 1 + a_1 z^{-1} + \ldots + a_n z^{-n} \\ B(z^{-1}) &= b_1 z^{-1} + \ldots + b_n z^{-n} \\ D(z^{-1}) &= 1 + d_1 z^{-1} + \ldots + d_n z^{-n} \end{aligned} \tag{6.8}$$

and

$$d_i = a_i + g_i; \quad i = 1, 2, \ldots, n \tag{6.9}$$

in which the g_i, $i = 1, 2, \ldots, n$ are the elements of the Kalman gain vector

$$\mathbf{g} = [g_1\; g_2\; \ldots\; g_n]^T.$$

Equation (6.7) can now be converted into the discrete-time model form

$$\begin{aligned} y(k) &= -a_1 y(k-1) - \ldots - a_n y(k-n) \\ &\quad + b_1 u(k-1) + \ldots + b_n u(k-n) + \eta(k) \\ \eta(k) &= e(k) + d_1 e(k-1) + \ldots + e(k-n) \end{aligned} \tag{6.10}$$

This is the required observation-space description which is linear in the system parameters and characterized by a noise disturbance $\eta(k)$ defined as a linear sum of the present and past values of the single white noise source $e(k)$.

Before proceeding, it is worth noting that, since $e(k)$ rerpresents the irreducible uncertainty in the model, the optimally filtered output $\hat{y}(k)$ can be generated by substituting $y(k) - \hat{y}(k)$ for $e(k)$ in (6.7), to give:

$$\hat{y}(k) = \frac{D(z^{-1}) - A(z^{-1})}{D(z^{-1})}y(k) + \frac{B(z^{-1})}{D(z^{-1})}u(k) \tag{6.11}$$

which provides a mechanism for generating the optimally filtered output $\hat{y}(k)$ from $y(k)$ and $u(k)$ once a model of the form (6.7) for $y(k)$ is available (Young, 1979). Indeed, because the leading term in $B(z^{-1})$ is $b_1 z^{-1}$, the model (6.7) includes an inherent one sample time delay on the input $u(k)$ so that, in this case, $\hat{y}(k+1)$ is one-step-ahead prediction $y(k+1|k)$ of $y(k+1)$, based on the measured data available at sample k. This topic is pursued further in section 7.8 of Chapter 7.

Equation (6.7) is well known in statistical and automatic control texts as the *AutoRegressive - Moving Average - eXogenous variables* (ARMAX) description (or sometimes the *Controlled AutoRegressive Moving Average* (CARMA) process): see also section B.3.3 of Appendix B. The ARMAX terminology arises because the noise effect on $y(k)$ appears in (6.6) as white noise $e(k)$ passed through a linear dynamic system or filter, with transfer function $D(z^{-1})/A(z^{-1})$. In other words, if we denote this noise term by $\xi(k)$ (note $\xi(k) \neq \eta(k)$), then

$$\xi(k) = \frac{D(z^{-1})}{A(z^{-1})} e(k) \tag{6.12}$$

or, in discrete-time (difference equation) form,

$$\begin{aligned} \xi(k) = & -a_1 \xi(k-1) - \ldots - a_n \xi(k-n) \\ & + e(k) + d_1 e(k-1) + \ldots + d_n e(k-n) \end{aligned} \tag{6.13}$$

Here, $\xi(k)$ at the latest k^{th} sampling instant is a linear function of the n previous past values of itself and a 'moving average' of the present and n previous past values of $e(k)$. Because of algebraic similarity with a linear regression relationship, the linear function of the $\xi(k)$ samples is termed an *AutoRegressive* process; while the overall relationship (6.13), or its equivalent operator notation form (6.12), are considered as *AutoRegressive-Moving Average* (ARMA) processes. The 'X' is then appended to this description to denote the presence of the input or 'exogenous' variable $u(k)$, which affects the output via the TF $B(z^{-1})/A(z^{-1})$. i.e.

$$x(k) = \frac{B(z^{-1})}{A(z^{-1})} u(k) \tag{6.14}$$

or in discrete-time (difference equation) form,

$$x(k) = -a_1 x(k-1) - \ldots - a_n x(k-n) + b_1 u(k-1) + \ldots + b_n u(k-n)$$

where $x(k)$ can be considered as a hypothetical and unmeasurable 'noise-free' output of the system. With this nomenclature (6.6) can be written as

$$y(k) = x(k) + \xi(k) \tag{6.15}$$

with the deterministic output $x(k)$ and the 'coloured' noise defined by equations (6.12) and (6.14), respectively.

The term coloured noise arises because the 'white' noise input $e(k)$, which is serially uncorrelated, is shaped by its passage through the ARMA filter and so possesses serial correlation properties, the nature of which are defined by the ARMA model coefficients, here a_i and d_i, $i = 1, 2, \ldots, n$. Sometimes the reader may also see $\xi(k)$ referred to as noise with 'rational spectral density': in simple terms, this merely means that it is derived from white noise via a 'rational' transfer function such as (6.12), which is the ratio of two polynomials in z^{-1}. Clearly the *AutoRegressive* (AR) process, i.e.

$$\xi(k) = -a_1 \xi(k-1) - \ldots - a_n \xi(k-n) + e(k) \tag{6.16}$$

and the *Moving Average* (MA) process

$$\xi(k) = e(k) + d_1 e(k-2) - \ldots - d_n e(k-n) \tag{6.17}$$

are special cases of (6.12). A detailed discussion on the statistical properties of AR, MA and joint ARMA processes is given in Box and Jenkins (1970) and Chatfield (1975); the reader is urged to consult these texts for further background on the subject, as well as related topics, such as correlation analysis (see section B.3 of Appendix B).

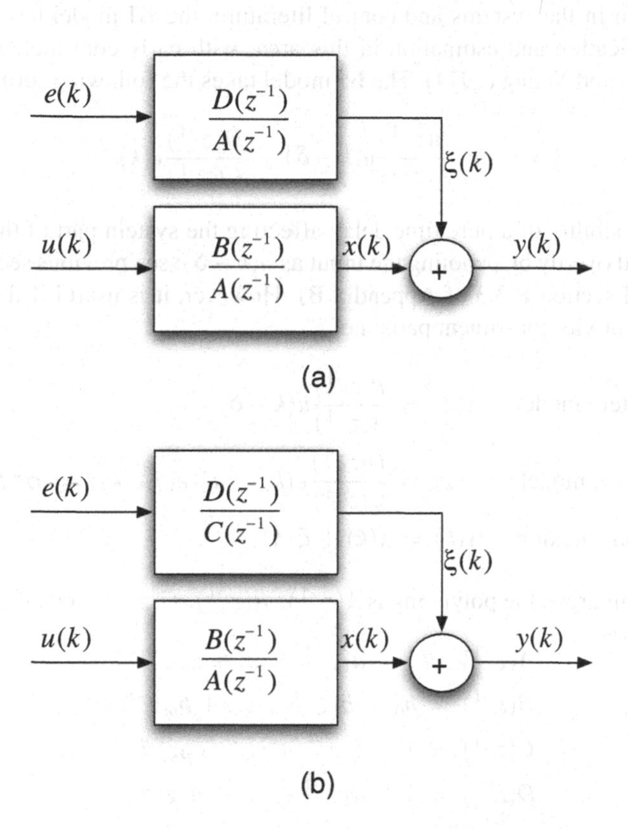

Fig. 6.1 Block diagrams of (a) ARMAX and (b) Box-Jenkins transfer function models.

Figure 6.1(a) shows a block diagram of the ARMAX model (6.6). Whilst the input signal $u(k)$ and the white noise $e(k)$ enter the system via different pathways, it is worth noting that the system transfer function $B(z^{-1})/A(z^{-1})$ and the ARMA noise transfer function $D(z^{-1})/A(z^{-1})$ are interrelated, since they share a common denominator polynomial $A(z^{-1})$. Consequently, we might consider the alternate representation in Figure 6.1(b) to be more 'general' in some sense: here, the noise transfer function is chosen as $D(z^{-1})/C(z^{-1})$ and is independent of the system transfer function $B(z^{-1})/A(z^{-1})$. Indeed, this independence is not simply a superficial visual one; as will be shown later, there is also an asymptotic statistical independence that is extremely important in estimation terms.

The time series model in Figure 6.1(b) is usually referred to as the *Box-Jenkins* (BJ) transfer function model, since it is the principal model considered by Box and Jenkins (1970) in their important and influential monograph on time series analysis. The deterministic part of this model is the classical transfer function representation of control systems analysis and, although the ARMAX model has received much more attention in the systems and control literature, the BJ model has influenced model identification and estimation in this area, with early contributions such as Bohlin (1970) and Young (1974). The BJ model takes the following form:

$$y(k) = \frac{B(z^{-1})}{A(z^{-1})}u(k-\delta) + \frac{D(z^{-1})}{C(z^{-1})}e(k) \tag{6.18}$$

where the possibility of a pure time delay affecting the system part of the model is acknowledged overtly by denoting the input as $u(k-\delta)$: see previous section 5.4 of chapter 5 and section B.3.3 of Appendix B). However, it is useful if the model is decomposed into its constituent parts: i.e.,

$$\begin{aligned}
\text{System model}: \quad x(k) &= \frac{B(z^{-1})}{A(z^{-1})}u(k-\delta) \\
\text{Noise model}: \quad \xi(k) &= \frac{D(z^{-1})}{C(z^{-1})}e(k) \qquad E\{e(j)e(k)\} = \sigma^2\delta_{jk} \\
\text{Observation equation}: \quad y(k) &= x(k) + \xi(k)
\end{aligned} \tag{6.19}$$

where, to summarize, the polynomials $A(z^{-1})$, $B(z^{-1})$, $C(z^{-1})$, and $D(z^{-1})$ are defined as follows:

$$\begin{aligned}
A(z^{-1}) &= 1 + a_1 z^{-1} + \ldots + a_n z^{-n} \\
B(z^{-1}) &= b_0 + b_1 z^{-1} + \ldots + b_m z^{-m} \\
C(z^{-1}) &= 1 + c_1 z^{-1} + \ldots + c_p z^{-p} \\
D(z^{-1}) &= 1 + d_1 z^{-1} + \ldots + d_q z^{-q}
\end{aligned} \tag{6.20}$$

where it will be noted that a b_0 term has been added to the $B(z^{-1})$ polynomial because, in general time series analysis applied to real data, such an instantaneous effect entering at the input to the system (rather than being added directly to the output measurement) is often encountered[1]. Also, in order to make the model still more flexible, the orders of the polynomials have been made different. And finally, so that the model is consistent with our assumptions of stationarity in the original state space form, it is assumed that the roots of the characteristic equation of the noise model, i.e. the roots of

$$1 + c_1 z^{-1} + \ldots + c_n z^{-n} = 0$$

[1] Although it is often considered as not 'proper' in control systems analysis because a control input computed at the kth sampling instant cannot effect the output instantaneously.

all lie completely outside the unit circle in the complex z^{-1} plane. Note, however, that Box and Jenkins also deal with non-stationary characteristics, for example trends in the data, where certain roots of the characteristic equation may lie on the unit circle ('unit roots'). In this way, they introduce summers (i.e. integrators in continuous-time) into the noise process to model these kind of phenomena. The reader is referred to Box and Jenkins (1970), or the later edition Box et al. (1994), for discussion on this topic but, as mentioned previously, should be warned that their suggested device for removing such non-stationarity, namely differencing the data, can be dangerous in practice and can lead to high noise amplification. As we have seen in the earlier chapters, there are alternatives to such differencing operations: in particular, the introduction of random walk and integrated random walk processes into the model (which have unit roots when considered in TF terms: see equation (4.2) *et seq.* in chapter 4), allows for the estimation and removal of such non-stationary effects.

In relation to the BJ model (6.18) (or (6.19)), note that, for complete generality, we might let the polynomials $A(z^{-1})$, $B(z^{-1})$, $C(z^{-1})$, and $D(z^{-1})$ have different orders, rather than specifying them all as being of nth order. In this case, if the orders of the $A(z^{-1})$ and $B(z^{-1})$ polynomials are n and m, respectively, then the system model structure would be defined by the triad [n m δ], where δ denotes any pure time delay affecting the input signal. This nomenclature will be used in the examples considered later in this chapter. However, since such modifications can be introduced simply by assuming that certain of the coefficients in the polynomials may be zero, it is convenient initially to retain the present nomenclature and so reduce the complexity of the mathematical analysis. Note also that both the $C(z^{-1})$ and $D(z^{-1})$ polynomials are chosen with a leading coefficient of unity (i.e. they are 'monic' polynomials: see section B.3.3 of Appendix B).

It is clear that the TF model (6.18) is more general than the ARMAX model in the sense that it separates out the deterministic and stochastic effects completely. But it is not really any more general as a stochastic description of the relationship between the input and output data, as we can readily see if the BJ model is multiplied throughout by $A(z^{-1})C(z^{-1})$ to yield:

$$ACy(k) = BCu(k) + DAe(k)$$

where, for convenience here, the pure time delay δ has been removed and the z^{-1} argument is omitted so that $A = A(z^{-1})$, $B = B(z^{-1})$ etc. This is, of course, an ARMAX model (6.7), with A, B and D replaced by A^*, B^* and D^*, where

$$A^* = AC; \; B^* = BC; \text{ and } D^* = DA$$

It is clear, therefore, that the ARMAX and BJ models can both be used as the description of the same stochastic dynamic system and, as such, are interchangeable descriptions. Note, however, that the immediate ARMAX description of the BJ model involves definitions of the A^*, B^* and D^* polynomials which are constrained so that there are common factors between any two polynomials (i.e. AC and BC; AC and DA). This implies that estimation of this model *based on data generated by a*

BJ *model* may need to involve some constraints on the nature of the polynomials. For example, Jakeman and Young (1981, 1983) show that this is indeed the case and that estimation problems can occur when such constraints are not imposed.

The reader will probably have realized by now that the ARMAX and BJ model forms are not the only ones that can be considered. For instance, models can be constructed that are combinations of the ARMAX and BJ forms but which, in terms of the TF model, involve special noise structures. For example the moving average noise in the ARMAX model could be replaced by ARMA noise i.e.

$$Ay(k) = Bu(k) + \frac{D}{C}e(k) \tag{6.21}$$

which, in TF model format, becomes

$$y(k) = \frac{B}{A}u(k) + \frac{D}{CA}e(k)$$

Or, alternatively, the MA noise could be replaced by AR noise, i.e.

$$Ay(k) = Bu(k) + \frac{1}{C}e(k) \tag{6.22}$$

which corresponds to a TF model

$$y(k) = \frac{B}{A}u(k) + \frac{1}{CA}e(k)$$

where the noise is also autoregressive but with a polynomial having two factors, one of which is dependent upon the system transfer function denominator A. The model (6.22) has been quite popular in the literature on recursive time series modelling because, in contrast to the MA model, the AR model is easy to estimate (see below). This model has been termed the 'dynamic adjustment' (DA) model with autogressive residuals (recursive estimation of this model was first discussed by Hastings-James and Sage, 1969). The model (6.21), which seems less popular, is simply the DA model with ARMA residuals. Other simplifications and manipulations are also possible. For example, it may be desirable in practice to simplify the BJ model (6.18) so that the noise model is AR rather than ARMA, i.e.,

$$y(k) = \frac{B(z^{-1})}{A(z^{-1})}u(k) + \frac{1}{C(z^{-1})}e(k) \tag{6.23}$$

As pointed out above and as we shall see in the next chapter 7, this model form has the advantage that the AR noise model is easier to estimate than the ARMA or MA models, being linear in the parameters of the $C(z^{-1})$ polynomial which, therefore, can be estimated by the RLS algorithm.

Some of the major contributions to TF model estimation in the systems and control literature have been made by Lennart Ljung and Torsten Soderström: see e.g. Ljung and Söderström (1983); Ljung (1999) and the prior references therein, some of which are discussed in later chapters. They prefer to consider a generalized model of the form (using their nomenclature, which is different to that used in this book):

$$Ay(k) = \frac{B}{F}u(k) + \frac{C}{D}e(k) \tag{6.24}$$

However, there is a lot of redundancy in this model form and, as they say, 'in most applications one or several of the polynomials A, B, C, D and F would be fixed to unity'. In other words, the model is useful mainly as a generalized device which includes other models as special cases.

In parameter estimation terms, linearity in the parameters is important and should be exploited wherever possible. However, the above TF models are nonlinear in the parameters, which considerably complicates their estimation in statistical terms. This has led to the idea of *Pseudo Linear Regression Models* (PLR), a concept first introduced by Young (1968c) and Panuska (1969) in the context of RLS estimation and generalized under the name PLR by Solo (1978): see also Solo (1980). As we will see in the next chapter 7, PLR models are particularly important in the context of the present book because they greatly facilitate the recursive estimation of the parameters in TF models such as those discussed above. The PLR form of the BJ model (6.18) is obtained by noting that

$$\begin{aligned}
x(k) &= -a_1 x(k-1) - \ldots - a_n x(k-n) + b_0 u(k) + b_1 u(k-1) + \ldots + b_m u(k-m) \\
\xi(k) &= -c_1 \xi(k-1) - \ldots - c_p \xi(k-p) + d_1 e(k-1) + \ldots + d_q e(k-q) + e(k)
\end{aligned}$$

so that as $y(k) = x(k) + \xi(k)$,

$$\begin{aligned}
y(k) = -a_1 x(k-1) - \ldots - a_n x(k-n) + \\
b_0 u(k) + b_1 u(k-1) + \ldots + b_m u(k-m) - \\
c_1 \xi(k-1) - \ldots - c_p \xi(k-p) + \\
d_1 e(k-1) + \ldots + d_q e(k-q) + e(k)
\end{aligned} \tag{6.25}$$

which is linear in the parameters of the polynomials $A(z^{-1})$, $B(z^{-1})$, $C(z^{-1})$ and $D(z^{-1})$. Unfortunately the model is only 'pseudo-linear' since, except for the input variables $u(k) \ldots u(k-m)$, the regression variables are not measured directly, i.e., they are 'unobserved variables'. However, the next chapter will show how such PLR relationships can be exploited in estimation terms.

The reader may well feel that, with so many different observation space model forms available, the position as regards model structure selection in any particular application is going to be rather confusing. *On the surface* this is certainly the case: there seems little to choose between the various TF model forms discussed above. All could be seen as candidate model structures in the observation space, and the selection of which format is most appropriate will depend, for example, upon the specific objectives of the analysis, as well as on the statistical properties of the time series. But we shall have much more to say on this problem of model structure 'identification' in chapter 7, where the advantages of the BJ model in statistical terms are revealed. For the moment, therefore, let us continue with the discussion in relation to the estimation of parameters in the Box-Jenkin's TF model (6.19).

6.3 Least Squares Estimation: Its Advantages and Limitations

Considering the BJ model (6.18), we see from the first equation that

$$[1 + a_1 z^{-1} + \ldots + a_n z^{-n}]x(k) = [b_0 + b_1 z^{-1} + \ldots + b_m z^{-m}]u(k)$$

so that the deterministic (noise free) output of the system is given by

$$\begin{aligned} x(k) = & -a_1 x(k-1) - \ldots - a_n x(k-n) + \\ & b_0 u(k) + b_1 u(k-1) + \ldots + b_m u(k-m) \end{aligned} \tag{6.26}$$

in other words, $x(k)$ is a function of past values of $x(k-i)$, $i = 1, 2, \ldots, n$; as well as the present and past values of $u(k-i)$, $i = 0, 1, 2, \ldots, m$. If the observation equation in the BJ model (6.18) is now substituted into (6.26), then the following relationship between the measured variables $y(k)$ and $u(k)$ is obtained

$$\begin{aligned} y(k) = & -a_1 y(k-1) - \ldots - a_n y(k-n) + \\ & b_0 u(k) + b_1 u(k-1) + \ldots + b_m u(k-m) + \eta(k) \end{aligned} \tag{6.27}$$

where the noise $\eta(k)$ is defined as

$$\eta(k) = \xi(k) + a_1 \xi(k-1) + \ldots + a_n \xi(k-n) \tag{6.28}$$

Equation (6.27) can be written in simple vector terms as

$$y(k) = \boldsymbol{\phi}^T(k)\boldsymbol{\rho} + \eta(k) \tag{6.29}$$

where

$$\begin{aligned} \boldsymbol{\phi}^T(k) &= [-y(k-1), -y(k-2), \ldots, -y(k-n), u(k), \ldots, u(k-m)] \\ \boldsymbol{\rho} &= [a_1, a_2, \ldots, a_n, b_0, \ldots, b_m]^T = [\rho_1, \rho_2, \ldots, \rho_{n+m+1}]^T \end{aligned}$$

In a similar manner, the coloured noise term $\xi(k)$ is generated by

$$\begin{aligned} \xi(k) = & -c_1 \xi(k-1) - \ldots - c_p \xi(k-p) + \\ & d_1 e(k-1) + \ldots + d_q e(k-q) + e(k) \end{aligned}$$

or in vector terms

$$\xi(k) = \boldsymbol{\psi}^T(k)\boldsymbol{\eta} + e(k) \tag{6.30}$$

where

$$\begin{aligned}\boldsymbol{\psi}^T(k) &= [-\xi(k-1),\, -\xi(k-2), \ldots, -\xi(k-p),\, e(k-1), \ldots, e(k-q)] \\ \boldsymbol{\eta} &= [c_1, c_2, \ldots, c_p, d_1, \ldots, d_q]^T\end{aligned}$$

In other words, both the system and noise model can be written in linear terms which superficially resemble the regression relationships discussed in previous chapters.

But this superficial resemblance is misleading. Equation (6.29) is obviously not a regression relationship since it is derived in a manner which demonstrates its error-in-variables or structural form (see chapter 4 and Appendix B), with each of the output variables $y(k-i)$, $i=1, 2, \ldots, n$, in the vector $\mathbf{z}(k)$ contaminated by noise $\xi(k-i)$ and with the coloured noise term $\eta(k)$, in consequence, being both correlated with the $y(k-i)$ variables and serially correlated in time. And again, while the noise expression (6.30) is apparently nearer to the regression form, with the $e(k)$ serially uncorrelated and independent (by definition) of the variables in $\boldsymbol{\psi}(k)$, it cannot be considered directly in regression terms because both $\xi(k-i)$, $i=1, 2, \ldots, p$ and $e(k-i)$,, $i=1, 2, \ldots, q$ are unavailable for measurement!

With these observations in mind, it is clear that least squares analysis is not generally applicable to the estimation of the parameters in either of the models (6.29) and (6.30). In the case of (6.29), estimation is possible but we would expect from previous discussion in chapter 4 that such analysis will normally yield estimates which are asymptotically biased and inconsistent. In order to demonstrate that this is indeed the case, let us write down the least squares 'normal' equations for the model (6.29) where, at least, we have access to all the variables in the $\boldsymbol{\phi}(k)$ vector.

After k samples, the estimate $\hat{\boldsymbol{\rho}}(k)$ of $\boldsymbol{\rho}$ satisfies the normal equations (c.f. equation (3.7) in chapter 3):

$$\left[\sum_{i=1}^{k} \boldsymbol{\phi}(i)\boldsymbol{\phi}^T(i)\right] \hat{\boldsymbol{\rho}}(k) = \sum_{i=1}^{k} \boldsymbol{\phi}(i)y(i) \tag{6.31}$$

If a vector $\mathbf{x}(k)$ of 'noise free' variables is defined as

$$\mathring{\boldsymbol{\phi}}(k) = [-x(k-1), \ldots, -x(k-n),\, u(k), \ldots, u(k-m)]^T \tag{6.32}$$

then,

$$\boldsymbol{\phi}(k) = \mathring{\boldsymbol{\phi}}(k) + \boldsymbol{\xi}(k)$$

where the noise vector $\boldsymbol{\xi}(k)$ is defined as

$$\boldsymbol{\xi}(k) = [-\xi(k-1), \ldots, -\xi(k-n), 0, \ldots, 0]^T$$

With this nomenclature and multiplying through equation (6.31) by $1/k$, we obtain:

$$\left\{\frac{1}{k}\sum_{i=1}^{k} [\mathring{\boldsymbol{\phi}}(i) + \boldsymbol{\xi}(i)]\, [\mathring{\boldsymbol{\phi}}(i) + \boldsymbol{\xi}(i)]^T\right\} \hat{\boldsymbol{\rho}}(k) = \frac{1}{k}\sum_{i=1}^{k} [\mathring{\boldsymbol{\phi}}(i) + \boldsymbol{\xi}(i)]\, y(i)$$

or, noting that $y(i) = \boldsymbol{\phi}^T(i)\boldsymbol{\rho} + \eta(i)$ from (6.29),

$$\left\{\frac{1}{k}\sum_{i=1}^{k}[\mathring{\boldsymbol{\phi}}(i)+\boldsymbol{\xi}(i)]\,[\mathring{\boldsymbol{\phi}}(i)+\boldsymbol{\xi}(i)]^T\right\}\hat{\boldsymbol{\rho}}(k)$$

$$=\frac{1}{k}\sum_{i=1}^{k}[\mathring{\boldsymbol{\phi}}(i)+\boldsymbol{\xi}(i)]\,[\mathring{\boldsymbol{\phi}}(i)+\boldsymbol{\xi}(i)]^T\boldsymbol{\rho}+\frac{1}{k}\sum_{i=1}^{k}[\mathring{\boldsymbol{\phi}}(i)+\boldsymbol{\xi}(i)]\,\eta(i)$$

As a result,

$$\left\{\frac{1}{k}\sum_{i=1}^{k}[\mathring{\boldsymbol{\phi}}(i)+\boldsymbol{\xi}(i)][\mathring{\boldsymbol{\phi}}(i)+\boldsymbol{\xi}(i)]^T\right\}[\hat{\boldsymbol{\rho}}(k)-\boldsymbol{\rho}]=\frac{1}{k}\sum_{i=1}^{k}[\mathring{\boldsymbol{\phi}}(i)+\boldsymbol{\xi}(i)]\,\eta(i)$$

or, in terms of the estimation error $\tilde{\boldsymbol{\rho}}=\hat{\boldsymbol{\rho}}(k)-\boldsymbol{\rho}$

$$\tilde{\boldsymbol{\rho}}(k)=\left[\frac{1}{k}\sum_{i=1}^{k}[\mathring{\boldsymbol{\phi}}(i)+\boldsymbol{\xi}(i)]\,[\mathring{\boldsymbol{\phi}}(i)+\boldsymbol{\xi}(i)]^T\right]^{-1}\frac{1}{k}\sum_{i=1}^{k}[\mathring{\boldsymbol{\phi}}(i)+\boldsymbol{\xi}(i)]\,\eta(i)$$

In order to examine the asymptotic situation, let us now take the probability in the limit as $k\rightarrow\infty$ (see Appendix B) i.e.,

$$\text{p.}\lim\tilde{\boldsymbol{\rho}}(k)=\text{p.}\lim\{\left[\frac{1}{k}\mathbf{C}(k)\right]^{-1}\frac{1}{k}\mathbf{b}(k)\}\tag{6.33}$$

where

$$\mathbf{C}(k)=\sum_{i=1}^{k}[\mathring{\boldsymbol{\phi}}(i)+\boldsymbol{\xi}(i)]\,[\mathring{\boldsymbol{\phi}}(i)+\boldsymbol{\xi}(i)]^T$$

$$\mathbf{b}(k)=\sum_{i=1}^{k}[\mathring{\boldsymbol{\phi}}(i)+\boldsymbol{\xi}(i)]\,\eta(i)$$

Now by Slutsky's theorem (Elgerd, 1967),

$$\text{p.}\lim\{\left[\frac{1}{k}\mathbf{C}(k)\right]^{-1}\frac{1}{k}\mathbf{b}(k)\}=\text{p.}\lim\left[\frac{1}{k}\mathbf{C}(k)\right]^{-1}\text{p.}\lim\frac{1}{k}\mathbf{b}(k)\tag{6.34}$$

provided the elements of $\frac{1}{k}\mathbf{C}(k)$ and $\frac{1}{k}\mathbf{b}(k)$ converge in probability, as they do in this case because of the stationarity and stability assumptions. Now,

$$\text{p.}\lim\frac{1}{k}\mathbf{C}(k)=\text{p.}\lim\frac{1}{k}\sum\mathring{\boldsymbol{\phi}}(i)\mathring{\boldsymbol{\phi}}(i)^T+\text{p.}\lim\frac{1}{k}\sum\boldsymbol{\xi}(i)\boldsymbol{\xi}(i)^T+$$
$$\text{p.}\lim\frac{1}{k}\sum\mathring{\boldsymbol{\phi}}(i)\boldsymbol{\xi}(i)^T+\text{p.}\lim\frac{1}{k}\sum\boldsymbol{\xi}(i)\mathring{\boldsymbol{\phi}}(i)^T$$

but, since $\mathring{\boldsymbol{\phi}}(i)$ and $\boldsymbol{\xi}(i)$ are statistically independent by definition, then the last two terms on the RHS are identically zero so that

$$\text{p.}\lim\frac{1}{k}\mathbf{C}(k)=\mathbf{R}_{\mathring{\phi},\mathring{\phi}}+\mathbf{R}_{\xi,\xi}\tag{6.35}$$

where $\mathbf{R}_{\mathring{\boldsymbol{\phi}},\mathring{\boldsymbol{\phi}}}$ and $\mathbf{R}_{\boldsymbol{\xi},\boldsymbol{\xi}}$ are the covariance matrices (Appendix B) of $\mathring{\boldsymbol{\phi}}(i)$ and $\boldsymbol{\xi}(i)$, respectively (assuming for convenience that $\mathring{\boldsymbol{\phi}}(i)$ is zero mean). Similarly, because $\mathring{\boldsymbol{\phi}}(i)$ and $\boldsymbol{\xi}(i)$ are statistically independent,

$$\text{p.lim}\,\frac{1}{k}\mathbf{b}(k) = \text{p.lim}\,\frac{1}{k}\sum[\mathring{\boldsymbol{\phi}}(i) + \boldsymbol{\xi}(i)]\,\eta(i) = \mathbf{R}_{\boldsymbol{\xi},\eta} \tag{6.36}$$

where $\mathbf{R}_{\boldsymbol{\xi},\eta}$ is a $n+m+1$ covariance vector, whose first n elements are defined as the covariances of $\xi(i-j)$ and $\eta(i)$, $j=1,2,\ldots,n$; and whose final $m+1$ elements are zero.

Substituting from (6.36) and (6.35) in (6.33) and noting (6.34),

$$\text{p.lim}\,\frac{1}{k}\tilde{\boldsymbol{\rho}}(k) = [\mathbf{R}_{\mathring{\boldsymbol{\phi}},\mathring{\boldsymbol{\phi}}} + \mathbf{R}_{\boldsymbol{\xi},\boldsymbol{\xi}}]^{-1}\mathbf{R}_{\boldsymbol{\xi},\eta} \tag{6.37}$$

Provided that the input $u(k)$ continues to perturb or 'excite' the system and, at the same time, the coefficients in the polynomial $B(z^{-1})$ are not all zero, then the output $x(i)$ will exist for all i over the observation interval. As a result, $\mathbf{R}_{\mathring{\boldsymbol{\phi}},\mathring{\boldsymbol{\phi}}}$ will be non-zero and its inverse will exist[2]. Thus the only way in which the asymptotic error can be zero is for $\mathbf{R}_{\boldsymbol{\xi},\eta}$ to be identically zero. But from the definition of $\eta(i)$ in (6.28) i.e. $\eta(k) = \xi(k) + a_1\xi(k-1) + \ldots + a_n\xi(k-n)$, we know that in, general, $\xi(i)$ and $\eta(i)$ are correlated and $\mathbf{R}_{\boldsymbol{\xi},\eta}$ will only be zero if $\xi(i)$ is zero for all i. In other words, the asymptotic error will normally be zero and the least squares estimates will be unbiased asymptotically *only if there is no noise on the data*. And if there is noise on the data, which would normally be the case with real time series data, then the estimates will be biased asymptotically to an extent determined by the relative ratio of noise to signal variance.

Nominally then, least squares estimation is not appropriate for estimating the parameters in the system TF of the BJ model, as they appear in the simple vector model (6.29) since the estimate $\hat{\boldsymbol{\rho}}$ of the parameter vector $\boldsymbol{\rho}$ will be inconsistent and the bias error will remain no matter how much data is used in the estimation. It should be remembered, however, that the importance of this bias error has to be assessed in relation to the use we make of the analytical results. For instance, if the coefficients a_i and b_i, $i = 1,2,\ldots,n$ in the TF model are, in themselves, of relevance to the problem at hand, then the bias clearly introduces some considerable difficulties. Also, if we are given *a priori* information on the covariance statistics $\mathbf{R}_{\mathring{\boldsymbol{\phi}},\mathring{\boldsymbol{\phi}}}$, $\mathbf{R}_{\boldsymbol{\xi},\boldsymbol{\xi}}$ and $\mathbf{R}_{\boldsymbol{\xi},\eta}$ (or $\mathbf{R}_{\mathring{\boldsymbol{\phi}},\mathring{\boldsymbol{\phi}}}$ and $\mathbf{R}_{\boldsymbol{\xi},\eta}$) then it would be possible to obviate these difficulties by computing the bias from (6.37) and so correcting the estimates: see e.g. Levin (1964); Vajk (2005) and section 7.4 of the next chapter 7. Even so, we might expect such information, even if available, to be rather poor in most practical situations and the corrected estimates are likely to be extremely sensitive to such errors.

---

[2] This assumes that the order of the model is identically equal to the order of the process generating the data; problems can be encountered if this is not the case.

But there are situations where the biased least squares results are useful despite the presence of bias. Johnston (1963), page 163, for example, shows in a simple error-in-variables situation that the least squares prediction based on the model is unbiased, even though the model parameter estimates are biased. Also Åström and Wittenmark (1973) showed how the recursive least squares estimates can be used directly in noisy situations to obtain 'self tuning regulator' systems for dynamic processes which exhibit optimal properties in a stochastic sense. With these examples in mind, we urge the reader to note carefully that asymptotic bias is a potential problem when using the least squares estimator, but not to assume that this negates its utility completely. It has the advantage of great simplicity and, when used carefully, can still be an important tool in time series analysis, as we see in the special case discussed in the next section. But, if abused, it can result in confusing and misleading results.

Before considering the general nonlinear estimation problem posed by the BJ model (6.18), as represented by the PLR equations (6.29) and (6.30), it is instructive to consider two special cases where recursive least squares is able to produce unbiased estimates: the ARX and FIR models.

6.4 Least Squares Estimation: ARX Model Estimation

Consider the special degenerate case of the ARMAX system shown in Figure 6.1(a) and described by equations (6.6) and (6.7), where $D(z^{-1}) = 1.0$, so that the model is in a very special ARMAX form,

$$y(k) = \frac{B(z^{-1})}{A(z^{-1})}u(k-\delta) + \frac{1}{A(z^{-1})}e(k) \tag{6.38}$$

where the noise is purely AR in form and the time delay δ has been re-introduced to allow for pure time delays on the input signal $u(k)$. An even more limited example of this situation is where the system has no 'numerator dynamics' (i.e. $B(z^{-1}) = 1.0$) and the noise enters as serially uncorrelated 'white' noise at the input to the system.

For this highly constrained model, equation (6.26) takes the reduced form

$$y(k) = \boldsymbol{\phi}^T(k)\boldsymbol{\rho} + e(k) \tag{6.39}$$

where $e(k)$ is a serially uncorrelated sequence of zero mean value, random variables with variance σ^2 and $\boldsymbol{\phi}^T(k)$ is defined as

$$\boldsymbol{\phi}^T(k) = [-y(k-1),\ \ldots, -y(k-n),\ u(k-\delta),\ \ldots,\ u(k-\delta-m)]$$

Since it is assumed that $u(k)$ is an observed, deterministic input sequence without measurements noise, it is clear (Åström and Bohlin, 1966) that this model is a simple extension of the simple autoregressive time series model with $e(k)$, by definition, statistically independent of $y(k-1)$, $i = 1, 2, \ldots, n$, (although it is not, of course completely independent of $y(k)$ over the whole observation interval, as required in true regression problems). This type of model, which is referred to as an *AutoRegressive eXogeous* (ARX) process in the systems literature[3], has received considerable attention in the statistical literature (e.g. Mann and Wald (1943); Durbin (1960) and many others), and it is well known that an asymptotically unbiased and efficient estimate of the parameter vector **a** can be obtained by simple least squares regression analysis. Referring back to the analysis in section 6.3, it is clear that the asymptotic bias in the general transfer function model case is not present here because the offending bias terms do not appear in the case of this simple model.

The least squares estimate of the parameter vector $\boldsymbol{\rho}$ in the model (6.39) is obtained from the following solution of the normal equations for this model (*cf* equation (3.8) with $\mathbf{x}(k)$ replaced by $\boldsymbol{\phi}(k)$):

$$\hat{\boldsymbol{\rho}}(k) = [\sum_{i=1}^{k} \boldsymbol{\phi}(i)\boldsymbol{\phi}^T(i)]^{-1} \sum_{i=1}^{k} \boldsymbol{\phi}(i)y(i) \tag{6.40}$$

and the Recursive ARX (RARX) algorithm in this case follows directly by reference to the SRLS algorithm: i.e.,

$$\hat{\boldsymbol{\rho}}(k) = \hat{\boldsymbol{\rho}}(k-1) + \mathbf{g}(k)[y(k) - \boldsymbol{\phi}^T(k)\hat{\boldsymbol{\rho}}(k-1)] \qquad \text{(RARX-1)}$$

$$\mathbf{g}(k) = \mathbf{P}^*(k-1)\,\boldsymbol{\phi}(k)[\hat{\sigma}^2 + \boldsymbol{\phi}^T(k)\mathbf{P}^*(k-1)\boldsymbol{\phi}(k)]^{-1} \qquad \text{(RARX-2a)}$$

$$\text{or} \quad \mathbf{g}(k) = \frac{\mathbf{P}^*(k)}{\hat{\sigma}^2}\boldsymbol{\phi}(k) \qquad \text{(RARX-2b)}$$

$$\mathbf{P}(k) = \mathbf{P}^*(k-1) - \mathbf{g}(k)\boldsymbol{\phi}^T(k)\mathbf{P}^*(k-1) \qquad \text{(RARX-3)}$$

6.4.1 Example 6.1: Estimation of a simple ARX model

Using the model structure coding triad $[n\ m\ \delta]$ introduced in section 6.2, consider the following simple first order ARX model with a single sample time delay, whose structure can be denoted by $[1\ 1\ 1]$:

$$y(k) = -a_1 y(k-1) + b_0 u(k-1) + e(k); \qquad k = 1, 2, ..., 5000 \tag{6.41}$$

with $a_1 = -0.5$ and $b_0 = 0.5$. The input $u(k)$ is a PRBS signal with switching interval of 5 samples and an amplitude 2.0; and $e(k)$ is a zero mean white noise process with variance $\sigma^2 = 0.25$, giving a noise-signal ratio based on the standard deviations of the $e(k)$ and $y(k)$ of 0.48 (48% noise), i.e., $\text{NSR}_y = std\{e(k)\}/std\{y(k)\} = 0.48$.

[3] We have already discussed the time variable parameter DARX version of this model in the previous chapter 5.

A typical section of the input-output data is shown in Figure 6.2, where the dashed line is the noise-free output obtained by generating $y(k)$ from (6.41) with $e(k)$ zero for all k (so that $y(k) = x(k)\ \forall\ k$). In relation to $x(k)$, the $\mathrm{NSR}_x = std\{e(k)\}/std\{x(k)\} = 0.58$. The noisy output shown as a dash-dot line (noisy (TF)) is referred to later.

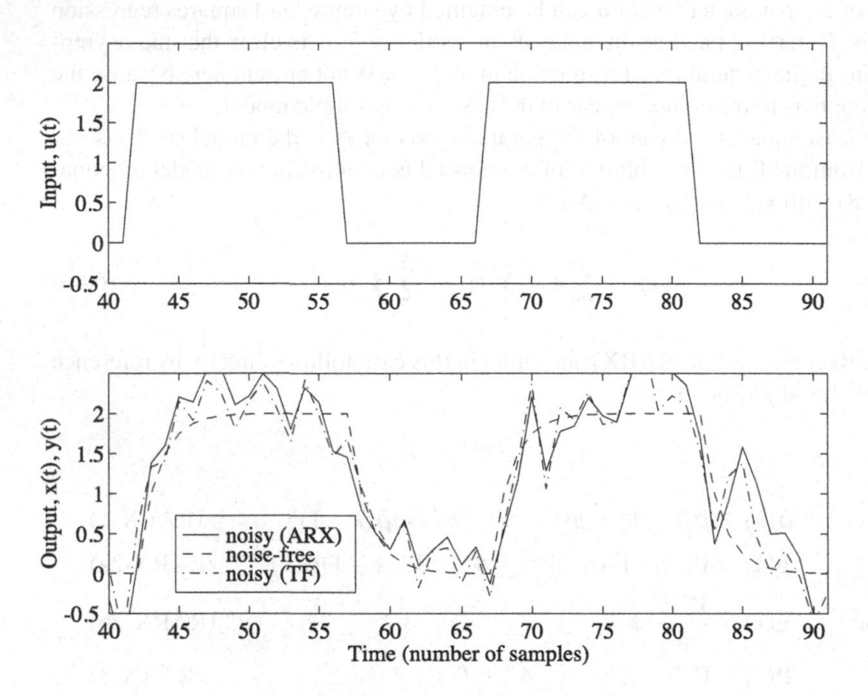

Fig. 6.2 Input-output data for simulation example 6.1.

The final RARX results for this example after 5000 samples, as obtained by least squares estimation using either the algorithm arx.m in the Matlab Identification Toolbox, or simply using Matlab to compute the RLS estimates, are as follows:

$$\hat{a}_1 = -0.492(0.008); \quad \hat{b}_0 = 0.507(0.008) \tag{6.42}$$

where the figures in parentheses are the estimated standard errors computed from the square root of the diagonal elements of the estimated covariance matrix $\mathbf{P}^*(5000)$, as supplied by the algorithm. The left panel of Figure 6.3 shows the recursive estimates and it is clear that convergence is more or less complete after about 250 samples. The grey bands show the 95% confidence bounds (twice the standard error).

The coefficient of determination R_T^2 (see Appendix B) is based on error $\hat{e}(k)$

$$\hat{e}(k) = y(k) - \hat{x}(k) \tag{6.43}$$

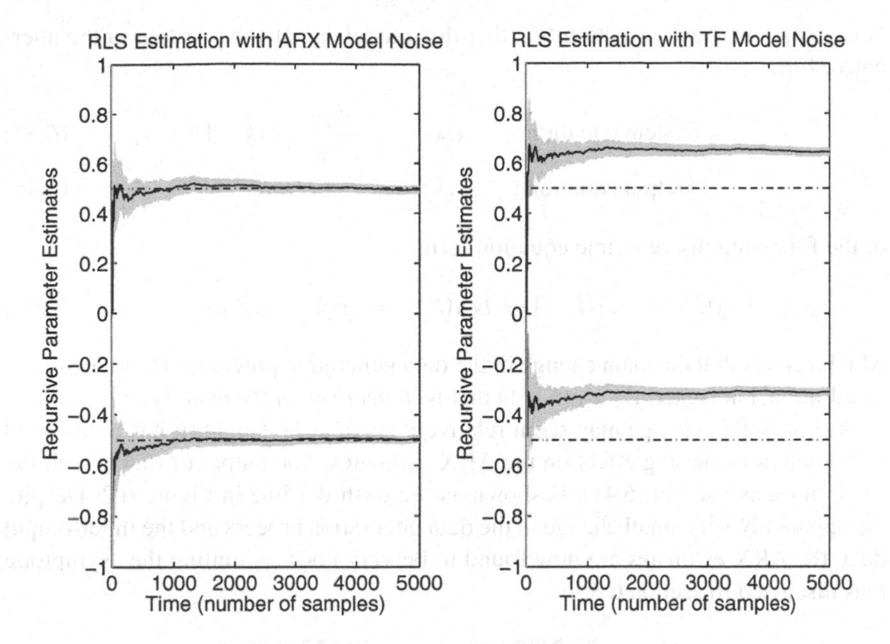

Fig. 6.3 Simulation example 6.1: recursive parameter estimates with the estimated 95% confidence bounds shown in grey.

between the measured output $y(k)$ and the noise-free output of the ARX model $\hat{x}(k)$, where,

$$\hat{x}(k) = -\hat{a}_1\hat{x}(k-1) + \hat{b}_0 u(k-1)$$

or, in TF terms,

$$\hat{x}(k) = \frac{\hat{b}_0}{1+\hat{a}_1 z^{-1}} u(k-1)$$

In this example, $R_T^2 = 0.692$, suggesting that 69.2% of the output variance is explained by the model. However, when evaluated in relation to the noise free output $x(k)$, $R_T^2 = 0.99992$; i.e. 99.992% of the noise-free output is explained by the ARX model. In other words, the ARX estimation has estimated the parameters very well and so it is able to obtain a very good estimate of the underlying dynamics and output of the system.

Despite these good results, the least squares estimation of the ARX model is quite restrictive in practice, as pointed out in earlier section 6.3. In order to illustrate these limitations, let us see what happens if the model (6.41) is modified slightly to the simplest BJ model form (6.18), with $C(z^{-1}) = D(z^{-1}) = 1.0$, i.e.,

$$y(k) = \frac{b_0}{1+a_1 z^{-1}} u(k-1) + e(k); \qquad k = 1, 2, ..., 1000 \qquad (6.44)$$

Note, as pointed out in section 6.2, that this model can also be written in the alternative form

$$\text{System equation}: \quad x(k) = \frac{b_0}{1+a_1 z^{-1}} u(k-1) \tag{6.45}$$

$$\text{Output equation}: \quad y(k) = x(k) + e(k) \tag{6.46}$$

or the following discrete-time equation form,

$$y(k) = -a_1 y(k-1) + b_0 u(k-1) + e(k) - a_1 e(k-1) \tag{6.47}$$

which reveals that the main change in the data generation process is the inclusion of an additional noise term $-a_1 e(k-1)$ that is *dependent on the model parameter* a_1.

Although this change may seem relatively small at first sight, it has a profound and deleterious biasing effect on the ARX estimates. The output of (6.45) with the same noise as used in (6.41), is shown as the dash-dot line in Figure 6.2. Despite the apparently very small change in the data generation process and the input-output data, the ARX estimates are now found to be very poor, exhibiting the asymptotic bias discussed in section 6.3:

$$\hat{a}_1 = -0.326(0.009); \quad \hat{b}_0 = 0.637(0.009)$$

The associated recursive parameter estimation results are presented in the right panel of Figure 6.3. Not only are the estimates badly biased away from the true values ($a_1 = -0.5$; $b_0 = 0.5$) but the estimated standard errors and 95% confidence bounds are much too optimistic (see below): in other words, the modeller seeing these results would be very misled about the nature of the system. The poor quality of these results is also illustrated in Figure 6.4. Here, the left panel compares the output of the ARX estimated model (full line, ARX(1) model) with the true noise-free output $x(k)$ generated by equation (6.45) (dashed line) over a small, step response portion of the data; while the right hand model shows the associated model error (full line, ARX(1) model). The other lines in these plots are discussed below.

It is possible to obtain better results with the above example using ARX model estimation, but only at the cost of estimating a higher order model. Åström and Eykhoff (1971) refer to this approach as 'repeated least squares' but a better name is high order ARX modelling. Here, ARX models of increasing order are estimated against the data, each time checking on the significance of the result in some manner. Åström and Eykhoff suggested evaluating the significance of the decrease in the sum of the squares using a Student t test (e.g. Kendall and Stuart, 1961). An alternative approach exploited below is to utilize autocorrelation (ACF) and partial autocorrelation (PACF) function tests (see section B.3, Appendix B), as well as the *Akaike Information Criterion* (AIC) (see later, section 6.9).

Given the ARX model, it is clear that the model residuals $\hat{e}(k)$ (see above, equation (6.43)) provide an estimate of the white noise input $e(k)$ and so should have no significant autocorrelation at any lag (see Appendix B). In this case, however, Figure 6.5 provides plots of the ACF and PACF functions for the $\hat{e}(k)$ obtained from the

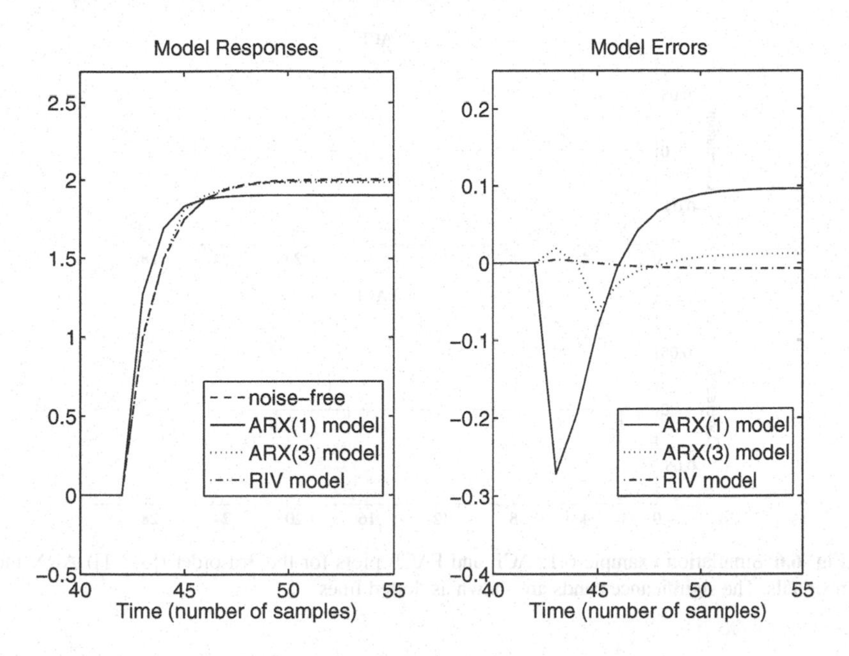

Fig. 6.4 Simulation example 6.1: comparison of model outputs $\hat{x}(k)$ with the true noise-free output $x(k)$ (left panel) and the associated errors $\hat{e}(k)$, in the right hand panel, showing the limitations of the ARX estimated model.

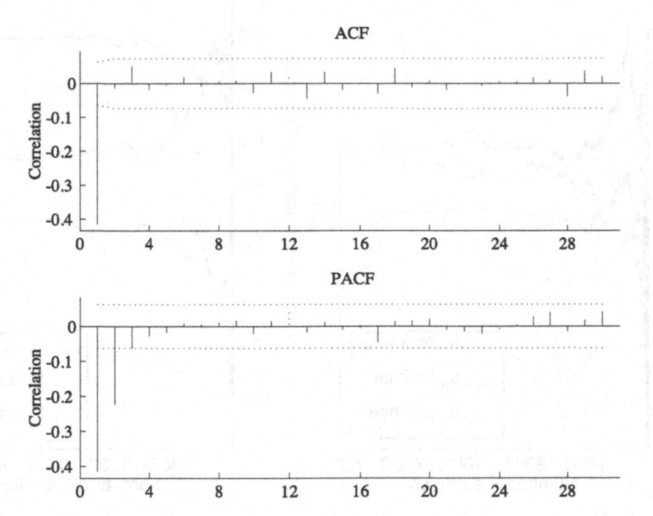

Fig. 6.5 Simulation example 6.1: ACF and PACF plots for the the 1st order ([1 1 1]) ARX model residuals. The significance bands are shown as dotted lines.

above analysis of the [1 1 1] model and shows that the series represents coloured, rather than white noise: in particular, there is significant lag 1 autocorrelation; and significant partial autocorrelation up to lag 2, or possibly 3, the latter suggesting

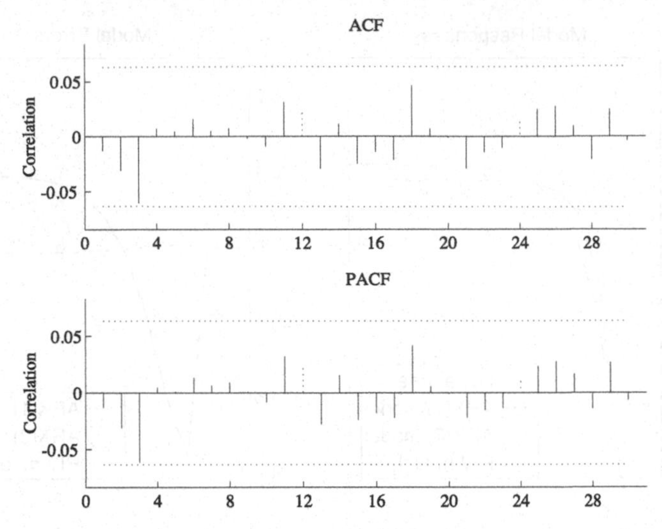

Fig. 6.6 Simulation example 6.1: ACF and PACF plots for the 3rd order ([3 3 1]) ARX model residuals. The significance bands are shown as dotted lines.

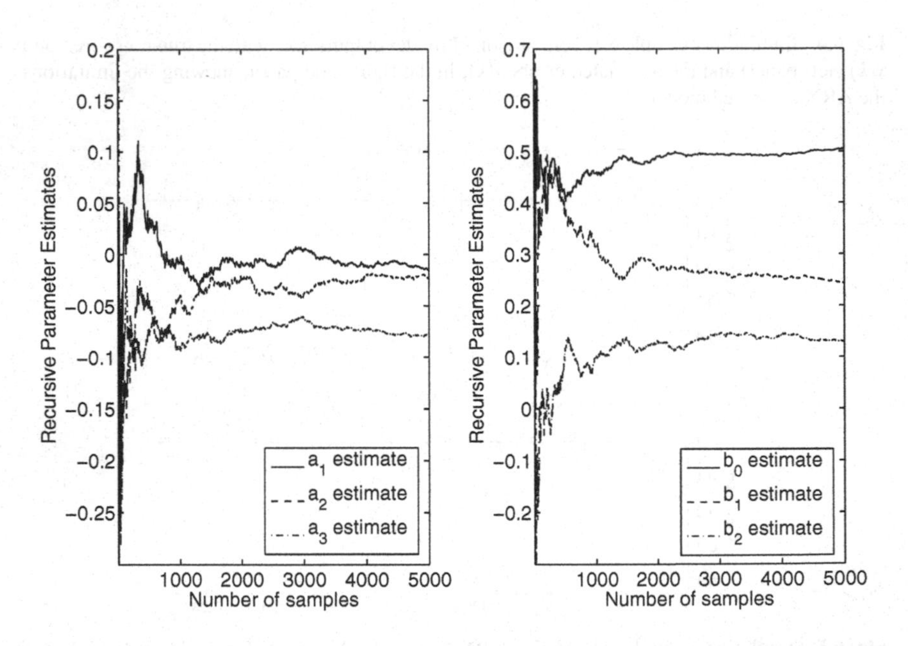

Fig. 6.7 Simulation example 6.1: recursive estimates of the parameters in the 3rd order ([3 3 1]) ARX model (95% confidence bounds not shown to avoid confusion).

an autoregressive model of order 3 is required to model $\hat{e}(k)$ (see Box and Jenkins, 1970). These results suggest that estimation of an ARX model [3 3 1] may yield white residuals and this is indeed the case, as seen in Figure 6.6, which shows that there are no significant autocorrelations or partial autocorrelations at any lags.

The ARX [3 3 1] model parameters are estimated as (rounded to 4 decimal places):

$$\hat{a}_1 = -0.0150(0.014);\ \hat{a}_2 = -0.0233(0.0133);\ \hat{a}_3 = -0.0764(0.0097);$$
$$\hat{b}_0 = 0.5052(0.0116);\ \hat{b}_1 = 0.2441(0.0175);\ \ \hat{b}_2 = 0.1290(0.0169);$$

where the $\hat{a}_i, i = 1,2$ are only just significant in relation to their standard errors, suggesting that the model is not all that well identified. This is confirmed by the plot of the recursive estimates shown in Figure 6.7, where, in contrast to the recursive estimates of the [1 1 1] model parameters in Figure 6.3, the estimates 'wander' a lot and are still changing, even after 5000 samples. Nevertheless, the model explains the data quite well, as shown by the dotted lines in Figure 6.4. Here, we see that the model response $\hat{x}(k)$ in the left panel (dotted line) matches the noise-free output $x(k)$ very well; and in the right hand panel, that the model error $\hat{e}(k)$ (again the dotted line) is quite small. Indeed this error is only a little larger than that obtained using optimal estimation, in this case the *Refined Instrumental Variable* (RIV) estimation algorithm (dash-dot line) developed in the next chapter 7, to estimate the [1 1 1] model. Of course, the difference is that, whereas the two optimal RIV parameter estimates are, as we shall see in chapter 7, statistically very well defined because the model is parametrically efficient (or 'parsimonious': see Box and Jenkins, 1970), the six ARX [3 3 1] parameter estimates, given above, are poorly defined, suggesting that the model is considerably over-parameterized.

But is there an underlying reason why the [3 3 1] model provides reasonable, albeit non-optimal results? To answer this question, let us investigate the estimation results further by considering the general ARX model in the following TF form:

$$y(k) = \frac{B(z^{-1})}{A(z^{-1})}u(k-\delta) + \frac{1}{A(z^{-1})}e(k) \tag{6.48}$$

This reveals that the additive noise is not white noise but coloured noise that is strongly influenced by the nature of the TF denominator polynomial $A(z^{-1})$. The only exception is the case where $A(z^{-1}) = 1.0$, which then forms the special *Finite Impulse Response* (FIR) model discussed in the next section.

If the ARX [3 3 1] model with the estimated parameters given above is considered in the form of (6.48), then:

$$y(k) = \frac{\hat{b}_0 + \hat{b}_1 z^{-1} + \hat{b}_2 z^{-2}}{1 + \hat{a}_1 z^{-1} + \hat{a}_2 z^{-2} + \hat{a}_3 z^{-3}}u(k-1) + \frac{1}{1 + \hat{a}_1 z^{-1} + \hat{a}_2 z^{-2} + \hat{a}_3 z^{-3}}e(k)$$

A clue to what is happening is now obtained if the poles and zeros of the system TF are computed, based on the estimated parameters. This yields:

$$\text{Roots of } \hat{A}(z^{-1}) : -0.2165 \pm 0.3519j;\ 0.4479$$
$$\text{Roots of } \hat{B}(z^{-1}) : -0.2416 \pm 0.4438j$$

So that there is a complex conjugate pair of poles in the TF denominator that are similar to a complex conjugate pair of zeros in the numerator; and these will tend to cancel each other out ('pole-zero cancellation'), resulting in a model that behaves approximately as a 1st order, rather than a 3rd order, dynamic system. As we shall see later in section 6.6, such cancellation is important as regards the 'identifiability' of the model. Moreover, if the Matlab function deconv is used to divide $\hat{A}(z^{-1})$ by $\hat{B}(z^{-1})$, the result is:

$$\frac{\hat{A}(z^{-1})}{\hat{B}(z^{-1})} = 1.9795(1-0.4982z^{-1})$$

Consequently, the model can be written in the approximate ARX form:

$$y(k) \approx 0.4982y(k-1) + 0.5052u(k-1) + e(k) \tag{6.49}$$

where the parameter estimates now compare favourably with the estimates in (6.42), as obtained by the RARX algorithm when applied to the correct ARX model form. So it is clear why the ARX [3 3 1] model is able to achieve the good response matching shown in Figure 6.4: the estimated system TF is acting dynamically as a relatively close approximation to the true 1st order system, with the higher order modal behaviour of the 3rd order model being cancelled out and playing little part in the response characteristics.

As a general approach to TF modelling, the problem with high order ARX estimation is two-fold. First, it is a rather 'messy' *ad hoc* procedure. Secondly, and more important, it yields a model with rather poorly defined estimates. For example, if we use *Monte Carlo Simulation* (MCS) analysis (see Appendix B, section B.3.9) to carry out the above model reduction, based on the covariance matrix of the ARX [3 3 1] model estimates and 5000 realizations, then the mean estimate of the parameter is -0.556, with a high standard error of 0.1. This is ten times worse than the MCS estimate of 0.4997 (0.0167) obtained by the optimal RIV algorithm. So, although least squares estimation of an ARX model seems, at first sight, to be a reasonable approach to TF model estimation, it has proven necessary to develop the other, less restrictive and more generally applicable algorithms.

6.5 Least Squares Estimation: FIR Model Estimation

The only other modelling approach that can directly exploit simple least squares is the estimation of the *Finite Impulse Response* (FIR) model, as mentioned briefly in the previous chapter. This takes the form:

$$\begin{aligned} y(k) &= \frac{B(z^{-1})}{A(z^{-1})}u(k-\delta) \approx G(z^{-1})\,u(k-\delta) \\ G(z^{-1}) &= g_{\delta}z^{-\delta} + g_{\delta+1}z^{-(\delta+1)} + g_{\delta+2}z^{-(\delta+2)} + \cdots + g_{\delta+p}z^{-(\delta+p)} \end{aligned} \tag{6.50}$$

Here, the coefficients provide a finite dimensional approximation to the infinite dimensional, discrete-time impulse response, as obtained when $B(z^{-1})$ is divided by $A(z^{-1})$. This is the discrete-time approximation of the well-known continuous-time convolution integral equation model, where $y(k)$ is a simply a weighted linear sum of a finite number of past inputs $u(k-\delta)$, $u(k-\delta-1)$,..., $u(k-\delta-p)$, with p chosen to provide a reasonable description of the impulse response and the system dynamics.

The FIR model can be formulated in a similar vector inner product manner to the ARX model, i.e.,

$$y(k) = \boldsymbol{\phi}^T(k)\boldsymbol{\rho} + e(k) \tag{6.51}$$

but now,

$$\begin{aligned} \boldsymbol{\phi}^T(k) &= [u(k-\delta)\ u(k-\delta-1)\ u(k-\delta-2)\ ...\ u(k-\delta-p)] \\ \boldsymbol{\rho}^T &= [g_\delta\ g_{\delta+1}\ g_{\delta+2} \cdots g_{\delta+p}] \end{aligned} \tag{6.52}$$

As a result, the FIR parameters can be estimated by least squares using either the recursive or *en bloc* solution. However, there are two main problems in this case. First, the model is not in a TF form and so is not particularly useful in relation to many applications. Secondly, it is not a parsimonious description, requiring as many parameters as are required to reasonably describe the impulse response at the sampling interval of the data. This will often mean that the model is severely over-parameterized, particularly when the system is characterized by long time constants, where the parameter estimates will not be estimated very well, as shown in the following Example.

6.5.1 Example 6.2: Estimation of a simple FIR model

Using the same data set as in example 6.1 (Figure 6.2 with the TF model output shown as a dash-dot line), the AIC suggests an FIR length of $p = 5$ samples. The resulting least squares estimates and standard errors for this model are:

$$\hat{g}_1 = 0.5091(0.027);\ \hat{g}_2 = 0.2729(0.037);\ \hat{g}_3 = 0.0771(0.037);$$
$$\hat{g}_4 = 0.0596(0.038);\ \hat{g}_5 = 0.0762(0.028);$$

Although this model does not explain the data as well as the ARX [3 3 1] and RIV [1 1 1] models, the result is reasonably acceptable. However, if the 1st order model parameters are changed to $a_1 = -0.95$ and $b_0 = 0.05$, the FIR model is now highly over-parameterized and performs poorly. For example, consider the results obtained with this model using a PRBS input $u(k)$ with a switching interval of 65 samples and amplitude 2.0. The noise $e(k)$ is once again a zero mean white noise process with variance $\sigma^2 = 1.0$, giving a noise-signal ratio commensurate with that used in example 6.1. Here, the AIC suggests that the FIR model requires no less that 94 parameters which, not surprisingly, are estimated with high variance. As a result, therefore, the model does not estimate the noise-free output $x(k)$ very well ($\text{var}\{x(k)-\hat{x}(k)\} = 0.049$; $R_T^2 = 0.941$) in comparison with, for example, the RIV estimated [1 1 1] model ($\text{var}\{x(k)-\hat{x}(k)\} = 0.0007$; $R_T^2 = 0.999$), whose two parameters are estimated very well. This poor FIR estimation performance arises because the high variance associated with the FIR parameter estimates means that the estimated impulse response is very noisy. On this basis, it is clear that the FIR model is unacceptable in this latter case and, indeed, in any example where the impulse response has to be defined over many sampling instants.

6.6 Identifiability

The simulation example in section 6.4.1 shows that, when the model is over-parameterized, there is an inherent redundancy in its ability to explain the input-output data that leads to a near-cancellation of poles and zeros in the estimated TF model. Clearly, any such cancellation would produce a similar effect and so, in a very real sense, the estimated model is not unique and so its 'identifiability' is called into question.

This topic of identifiability is a deep one and the reader is referred to a comprehensive text on the subject, such as Ljung (1987), page 362 *et seq*, or Ljung (1999), in order to read about the concept and its ramifications in rigorous theoretical detail. Nevertheless, it is worthwhile reviewing the problem briefly here in a non-rigorous manner in order to give the reader a feel for the topic and to see where the question of identifiability impinges on the practical aspects of stochastic model building.

The more complex transfer function models discussed previously in this chapter can be characterized by a composite parameter vector $\boldsymbol{\theta}$. For instance, in the discrete-time BJ model case, $\boldsymbol{\rho}$ can be defined as follows:

$$\begin{aligned}\boldsymbol{\theta} &= [a_1,\dots a_n, b_0,\dots b_m, c_1,\dots c_p, d_1,\dots d_q]^T \\ &= [\boldsymbol{\rho}^T \ \boldsymbol{\eta}^T]^T\end{aligned} \tag{6.53}$$

with the pure time delay δ considered separately. Model structure *Identification* is concerned with identifying an appropriate model structure, as defined in the BJ model, for instance, by the orders of the model polynomials, n,m,p,q and the pure time delay δ, with this model structure now defined by the pentad $[n\ m\ \delta\ p\ q]$. Estimation of the system part of the model is concerned with estimating the parameters that characterize $\boldsymbol{\rho}$ in (6.53), as well as the size of δ. The full process of *identification and estimation* for models such as (6.7) and (6.18) from a set of input-output data $\{y(k), u(k), k = 1,2,\dots N\}$ is discussed fully in the next chapter 7. However, as an introduction to topic of model structure identification, it is convenient to consider here the simplest approach to the estimation of a TF model: namely, least squares estimation of the following ARX model, as discussed in previous sections of this chapter:

$$y(k) = \boldsymbol{\phi}^T(k)\boldsymbol{\rho} + e(k) \tag{6.54}$$

where

$$\boldsymbol{\phi}^T(k) = [-y_{k-1} \; -y_{k-2} \cdots \; -y_{k-n} \; u_{k-\delta} \; u_{k-\delta-1} \cdots \; u_{k-\delta-m}]$$

$$\boldsymbol{\rho} = [a_1 \; a_2 \cdots a_n \; b_0 \; b_1 \cdots b_m]^T = [\rho_1 \; \rho_2 \cdots \rho_{n+m+1}]^T$$

Referring to equation (3.8) (or (B.10) in Appendix B), the least squares estimate $\hat{\boldsymbol{\rho}}$ of $\boldsymbol{\rho}$ is given by the solution of the normal equations:

$$\sum_{i=1}^{k} \boldsymbol{\phi}(i)\boldsymbol{\phi}^T(i)\hat{\boldsymbol{\rho}} = \sum_{i=1}^{k} \boldsymbol{\phi}(i)y(i)$$

so that,

$$\hat{\boldsymbol{\rho}} = \left[\sum_{i=1}^{k} \boldsymbol{\phi}(i)\boldsymbol{\phi}^T(i)\right]^{-1} \sum_{i=1}^{k} \boldsymbol{\phi}(i)y(i) \tag{6.55}$$

or

$$\hat{\boldsymbol{\rho}} = \left[\boldsymbol{\Phi}(k)\boldsymbol{\Phi}^T(k)\right]^{-1} \boldsymbol{\Phi}(k)\mathbf{y}(k) \tag{6.56}$$

where $\boldsymbol{\Phi}(k)$ is defined as $k \times (n+m+1)$ matrix with rows $\boldsymbol{\phi}^T(i)$, respectively; and $\mathbf{y}(k)$ is a $k \times 1$ vector with elements $y(i), i = 1, 2, \ldots, k$. In order for the estimate $\hat{\boldsymbol{\rho}}$ to exist, it is clearly necessary that the symmetric $n+m+1 \times n+m+1$ product matrix

$$\boldsymbol{\Phi}^T(k)\boldsymbol{\Phi}(k), \tag{6.57}$$

should be invertible. And it is the conditions under which this is possible that control the identifiability of the model.

In simple terms, identifiability is concerned with whether model structure identification and subsequent parameter estimation, based on a set of input-output data, can yield a unique model structure and set of associated parameter estimates. In the context of the transfer function model estimation problem, this depends on: (i) whether the input-output data are informative enough to allow for unique model structure identification: i.e. to distinguish between competing model structures; and (ii) whether the estimation algorithm is able to obtain a consistent and, if possible, statistically efficient, estimate $\hat{\boldsymbol{\rho}}$ of the associated model parameter vector $\boldsymbol{\rho}$.

For the purposes of the present book, a suitable definition of *stochastic identifiability* is that used by Staley and Yue (1970):

Definition 6.1. A model parameter vector $\boldsymbol{\rho}$ is said to be stochastically identifiable if every asymptotically efficient estimator $\hat{\boldsymbol{\rho}}$ of $\boldsymbol{\rho}$ converges to $\boldsymbol{\rho}$ in mean square. By this definition, if $\boldsymbol{\rho}$ is constant and stochastically identifiable, then the estimation error covariance matrix converges to zero as the number of samples approaches infinity.

From equation (3.28) in chapter 3, we know that the covariance matrix $\mathbf{P}_\rho(k)$ associated with the estimate $\hat{\boldsymbol{\rho}}(k)$ is directly related to the cross product matrix in (6.57) by the simple expression

$$\mathbf{P}_\rho = \sigma^2 \left[\boldsymbol{\Phi}^T(k)\boldsymbol{\Phi}(k)\right]^{-1} \tag{6.58}$$

and that it converges to zero as the number of samples approaches infinity provided

$$\lim_{k\to\infty} \frac{1}{k}\boldsymbol{\Phi}^T(k)\boldsymbol{\Phi}(k) \quad \text{exists and is positive definite.} \tag{6.59}$$

Now, if $\mathbf{U}(k)$ is defined as a $k \times m+1$ matrix with rows $[u(i),\ldots,u(i-m)]$ and $\mathbf{Y}(k)$ is defined as a $k \times n$ matrix with rows $[-y(i-1),\ldots,-y(i-n)]$, where $i=1,2,\ldots,k$, then $\boldsymbol{\Phi}(k)$ can be partitioned as follows

$$\boldsymbol{\Phi}(k) = \left[\mathbf{Y}(k) \,\vdots\, \mathbf{U}(k)\right]$$

As a result,

$$\mathbf{C}(k) = \boldsymbol{\Phi}^T(k)\boldsymbol{\Phi}(k) = \begin{bmatrix} \mathbf{Y}(k) \\ \cdots \\ \mathbf{U}(k) \end{bmatrix} \left[\mathbf{Y}(k) \,\vdots\, \mathbf{U}(k)\right] = \begin{bmatrix} \mathbf{Y}^T(k)\mathbf{Y}(k) & \mathbf{Y}^T(k)\mathbf{U}(k) \\ \mathbf{U}^T(k)\mathbf{Y}(k) & \mathbf{U}^T(k)\mathbf{U}(k) \end{bmatrix}$$

In order that the condition (6.59) should hold, it is necessary that for $k \to \infty$,

$$\left|\frac{1}{k}\mathbf{C}(k)\right| \neq 0 \tag{6.60}$$

After some further straightforward analysis, it can be shown that the *necessary* conditions for stochastic identifiability are that the limits

$$\begin{aligned} &\lim_{k\to\infty} \frac{1}{k}\mathbf{U}^T(k)\mathbf{U}(k) \qquad (A) \\ &\lim_{k\to\infty} \frac{1}{k}\mathbf{Y}^T(k)\mathbf{Y}(k) \qquad (B) \end{aligned} \tag{6.61}$$

exist and are positive definite.

6.6.1 Choice of input signals

It turns out that the conditions required to ensure that the input signal perturbs the system sufficiently to ensure that (6.59) holds relate to the following higher order form of the matrix $\frac{1}{k}\mathbf{U}^T(k)\mathbf{U}(k)$:

$$\frac{1}{k}\begin{bmatrix} \sum^k u^2(i) & \sum^k u(i)u(i-1) & \ldots & \sum^k u(i)u(i-n_p) \\ \sum^k u(i-1)u(i) & \sum^k u^2(i-1) & \ldots & \sum^k u(i-1)u(i-n_p) \\ \ldots & \ldots & \ldots & \ldots \\ \sum^k u(i-n_p)u(i) & \sum^k u(i-n_p)u(i-1) & \ldots & \sum^k u^2(i-n_p) \end{bmatrix}$$

where $n_p = n+m+1$ is the number of parameters in the model. Then, as Åström and Bohlin (1966) first pointed out, the requirement is that $u(i)$ are *persistently exciting of order* $np+1$, in the sense that:

$$\bar{u} = \lim \frac{1}{k} \sum^{k} u(i) \tag{6.62}$$

and

$$r_\delta = \lim \frac{1}{k} \sum^{k-\delta} (u(i) - \bar{u})(u(i+\delta) - \bar{u})$$

both exist and the covariance matrix

$$\mathbf{R} = \begin{bmatrix} r_0 & r_1 & \dots & r_n \\ r_1 & r_0 & \dots & r_{n-1} \\ \dots & \dots & \dots & \dots \\ r_n & r_{n-1} & \dots & r_0 \end{bmatrix} \tag{6.63}$$

is positive definite. In other words, the $u(i)$ should remain bounded both in mean and variance, and should continuously perturb or 'excite' the system.

These seem very reasonable, almost intuitive conditions to impose on the input signal and clearly provide a good criterion for judging their adequacy in estimation terms. For example, a pseudo-random binary (PRBS) input signal with finite variance provides persistent excitation since conditions (6.62) and (6.63) are clearly satisfied and

$$r_\delta = \begin{cases} \sigma^2; & \delta = 0 \\ 0; & \delta \neq 0 \end{cases}$$

Similarly, a square wave of an appropriate frequency provides a useful perturbing signal. On the other hand, a purely sinusoidal signal will only provide persistent excitation of order two[4] and it is well known (e.g. Young,1968c) that such a signal can only be used in the identification of a first order process.

It is interesting to note that many authors, including an early contribution from Levin (1960), through more comprehensive treatments, such as Mehra (1974) and Goodwin and Payne (1977), to recent papers, such as Gevers and Bombois (2006), have addressed the problem of optimal choice of input signals; i.e. in those situations where the input can be specified, what input signal will yield the best parameter estimation results? One problem, however is that the system characteristics need to be known to define the optimal input! This 'chicken and egg' situation is quite understandable, given the nature of the problem, but it is inconvenient as regards the application of the analysis: in effect some iterative solution must be defined in which the analysis converges on the optimal input in some manner.

[4] The reader can verify that the rank of **R** in this case is 2.

6.6.2 Restrictions on the system to be identified

Turning now to the matrix $\frac{1}{k}\mathbf{Y}^T(k)\mathbf{Y}(k)$, we see that this can be written as

$$\frac{1}{k}\begin{bmatrix} \sum^k y^2(i-1) & \sum^k y(i-1)y(i-2) & \dots & \sum^k y(i-1)y(i-n) \\ \sum^k y(i-2)y(i-1) & \sum^k y^2(i-2) & \dots & \sum^k y(i-2)y(i-n) \\ \dots & \dots & \dots & \dots \\ \sum^k y(i-n)y(i-1) & \sum^k y(i-n)y(i-2) & \dots & \sum^k y^2(i-n) \end{bmatrix}$$

As Staley and Yue (1970) pointed out, the necessary identifiabilty conditions mean that the system must not only be excited by a persistently exciting input signal $u(k)$ but it should also possess the following characteristics:

1. The system must be stable in the sense that the roots of $A(z)$ should lie inside the unit circle. This ensures that $y(k)$ remains bounded provided the noise remains bounded.
2. The coefficients $b_i, i = 0, 1, 2, \dots, m$ of the numerator polynomial $B(z)$ should not all be zero. Obviously this is necessary, because otherwise the input $u(k)$ would have no effect on the output.
3. The polynomials $A(z)$ and $B(z)$ should have no common factors. This is the observability condition: if common factors exist, then a mode of the system will be masked due to the cancellation of the factors and will be unidentifiable from the input/output measurements.

Returning to the example in section 6.4.1, while the PRBS input variable is clearly persistently exciting and conditions 2 and 3 are satisfied, the latter requirement is contravened in the case of the [3 3 1] model and this is the reason for the results obtained in this case.

6.6.3 The more general case

So far the problem of identifiability has been considered only in the case of the special ARX model with serially uncorrelated residuals. In the more general case of the ARMAX or BJ models, simple least squares analysis will not yield consistent estimates and so recourse must be made to more sophisticated method of estimation, as mentioned later and discussed in the next chapter 7. When using such algorithms, however, it seems reasonable to conjecture that the conditions outlined above must be satisfied. And this is, indeed, the case: see e.g. Ljung (1987).

6.6.4 Noise process identifiability

When considering identifiability in the general case of the BJ model, it is necessary to consider the identifiability of the ARMA model used to characterize the noise process. In addition to restrictions on $C(z^{-1})$ and $D(z^{-1})$ similar to those placed on $A(z^{-1})$ and $B(z^{-1})$, it is necessary to assume that the noise model is *minimum phase* or *invertible*, i.e. that the roots of $D(z)$ lie inside the unit circle. The term 'minimum phase' comes from the control systems literature (see e.g. Kuo and Golnaraghi, 2002), while 'invertible' is a statistical term (Priestley, 1981). The noise is non-invertible if the roots of $D(z)$ lie outside the unit circle because an attempt to generate $e(k)$ from $\xi(k)$, via the inverse of the ARMA transfer function, will fail since the resulting inverse transfer function is unstable (since $D(z)$ becomes the TF denominator and its roots are outside the unit circle) and the calculation would be unbounded. This has practical importance because, as shown in chapter 7, the estimation algorithms described there exploit prefilters that can involve the inverse of the ARMA noise model to achieve pre-whitening of the noise components in the output signal $y(k)$.

6.6.5 Some concluding comments on identifiability

The various identifiability conditions discussed briefly in this section should only be considered as a set of rules to be borne in mind when attempting to estimate the parameters of a dynamic process: it should be remembered, for instance, that the definition of identifiability on which these rules are based is somewhat restrictive, relying as it does on purely asymptotic characteristics. Before leaving the subject of identifiability, therefore, it is worth noting the following points:

1. Although certain signals may not be persistently exciting, for example a single step input, this does not necessarily mean that sensible estimates of the system parameters cannot be obtained when using such signals. It merely means that when the signal no longer continues to excite the system, no more useful information will be received and the estimates will not improve beyond that point. Nevertheless, the estimates obtained up to the point where the input signal loses its persistent excitation may well be good enough for the purposes of the experiment. In fact, as demonstrated by several examples in this book, step and impulse inputs can yield very good identification and estimation of the parameters in a dynamic system.
2. Again, while a stable system is certainly necessary to prevent the matrices becoming unbounded asymptotically, it may still be possible to obtain estimates of an unstable system provided sufficient estimation accuracy is achieved before the matrices 'blow up' (see e.g. Lee, 1964). And, as we shall see in chapter 9, unstable systems can be identified and estimated if they are enclosed in a stable closed control loop system.

3. It is true that it is not possible to fully identify a non-minimum phase (NMP) noise process, since the covariance generating function is the same for both the non-minimum and equivalent minimum phase systems (see e.g. Box and Jenkins, 1970). But this does not mean that the NMP process cannot be identified at all; it means only that the equivalent minimum phase model will be estimated from the data. Whether such information is sufficient will be very problem-dependent: in general, however, it will be sufficient for most purposes.

6.7 Recursive Estimation of Transfer Function Models

We have seen from the examples in sections 6.4.1 and 6.5.1 that linear regression analysis can yield unbiased estimates in some limited cases, but only at the cost of either assuming a restricted model form or estimating many more parameters than are necessary. In either case, therefore, it is likely that the results obtained from the application of these approaches to real, noisy data will not often prove acceptable. It is necessary, therefore, to look for estimation procedures which are directly applicable to the more general TF model (6.18) or alternatives such as the ARMAX model (6.7). The estimation problem associated with such time series models is well known and various procedures have been suggested for overcoming it. Excellent early surveys of these techniques is given by Åström and Eykhoff (1971) and Eykhoff (1974) and there have been many books on the topic published since then (e.g. Goodwin and Payne, 1977; Ljung and Söderström, 1983; Young, 1984; Norton, 1986; Söderstrom and Stoica, 1989; Wellstead and Zarrop, 1991 and many others).

Probably the best known solution is the statistically optimal Maximum Likelihood method suggested by Box and Jenkins (e.g. 1970) for the BJ model and Åström and Bohlin (1966) for the ARMAX model. Here, statistical optimality is achieved by simultaneously estimating the parameters of all the polynomials $A(z^{-1})$, $B(z^{-1})$, $C(z^{-1})$, $D(z^{-1})$, in the case of the BJ model; and $A(z^{-1})$, $B(z^{-1})$, $D(z^{-1})$ for the ARMAX model. Provided the statistical assumptions about the TF models are satisfied, the parameter estimates are asymptotically unbiased, consistent and efficient, i.e they have minimum variance (see Appendix B). However, because this is clearly a non-linear estimation problem, it requires either numerical hill-climbing procedures to determine the parameters that maximize the likelihood function, or some alternative iterative approach. In addition, the data are normally analyzed in *en bloc* form, so that conversion to a recursive solution, allowing for on-line application and time-variable parameter estimation, is not straightforward.

If a recursive solution is required, then there are a number of alternative procedures. One approach is to use the recursive form of *Generalized Least Squares* (GLS) analysis algorithm as suggested by Hastings-James and Sage (1969), which applies to the DA model with autoregressive residuals, as shown in equation (6.22). Note that the iterative GLS approach was discussed originally by Johnston (1963) and applied to time series analysis by Clarke (1967). A number of other recursive approaches have been suggested (see Eykhoff, 1974, chapter 7), which include the *Extended Matrix* method (Talmon and van den Boom, 1973) - which is related to the *Approximate Maximum Likelihood* (AML) procedure (Young, 1968c; Panuska,1969) and is also referred to as the *Recursive Maximum Likelihood 1* (RML1) by Söderström et al. (1974); the *Recursive Maximum Likelihood 2* (RML2) method of Söderström et al. (1974), which is applied to the ARMAX model; the *Prediction Error Recursion* (PER) approach of Ljung (1979b); see also Ljung and Söderström, 1983 and Ljung, 1987), which is similar to RML2 for the ARMAX model but can be applied to general time series model forms; and, finally, the various recursive algorithms suggested by Landau (1976) which are closely related to some of the other procedures mentioned above, but are derived from an alternative theoretical standpoint based on hyperstability theory. In the present context, it would serve little useful purpose to describe all of these alternative methods. It will suffice to note that all of them require simultaneous estimation of the system and noise parameters and are, therefore, relatively complex to implement. There is one procedure, however, that does not require this and is able to obtain consistent, asymptotically unbiased and relatively efficient (i.e. low, but not in general minimum variance) *en bloc* or recursive estimates of the system model parameters, without simultaneous noise model parameter estimation. This is the *Instrumental Variable* (IV) method mentioned in chapter 3 and proposed for TF model identification and estimation by the author (Young, 1965b, c; 1969b; 1970a) in connection with continuous time system estimation; and by Wong and Polak (1967) and again by the author (Young et al. 1971, 1974) in relation to discrete-time systems (see also Mayne, D.Q., 1967). Moreover, the IV method can be extended into an optimal or *Refined Instrumental Variable* (RIV) form, as mentioned previously, that allows for simultaneous noise model estimation. Indeed, as far as I am aware, the RIV algorithm is the only approach that can be applied to both discrete-time TF models, as described in the next chapter 7, and continuous-time TF models, as described in chapter 8. First, however, let us consider how the standard IV algorithm is applied to the problem of TF model estimation.

6.8 Standard Instrumental Variable (SIV) Estimation

Using the nomenclature introduced in this chapter, the *en bloc* solution (3.39) of the IV normal equations developed in chapter 3 takes the form:

$$\hat{\boldsymbol{\rho}}(k) = \left[\sum_{i=1}^{k} \hat{\boldsymbol{\phi}}(i)\boldsymbol{\phi}^T(i)\right]^{-1} \sum_{i=1}^{k} \hat{\boldsymbol{\phi}}(i)\mathbf{y}(i) \tag{6.64}$$

where $\boldsymbol{\rho}(i)$, $\boldsymbol{\phi}(i)$ and $\hat{\boldsymbol{\phi}}(i)$ replace $\mathbf{a}(i)$, $\mathbf{z}(i)$ and $\hat{\mathbf{x}}(i)$, respectively, and

$$\sum_{i=1}^{k} \hat{\boldsymbol{\phi}}(i)\boldsymbol{\phi}^T(i), \tag{6.65}$$

which has useful statistical properties and will be referred to in the next chapter 7, is called the *Instrumental Product Matrix* (IPM). The recursive IV algorithm then takes the form:

$$\hat{\boldsymbol{\rho}}(k) = \hat{\boldsymbol{\rho}}(k-1) + \mathbf{g}(k)[y(k) - \boldsymbol{\phi}^T(k)\hat{\boldsymbol{\rho}}(k-1)] \qquad \text{(SIV-1)}$$

$$\mathbf{g}(k) = \mathbf{P}(k-1)\hat{\boldsymbol{\phi}}(k)[1 + \boldsymbol{\phi}^T(k)\mathbf{P}(k-1)\hat{\boldsymbol{\phi}}(k)]^{-1} \qquad \text{(SIV-2a)}$$

$$\text{or} \quad \mathbf{g}(k) = \mathbf{P}(k)\hat{\boldsymbol{\phi}}(k) \qquad \text{(SIV-2b)}$$

$$\mathbf{P}(k) = \mathbf{P}(k-1) - \mathbf{g}(k)\boldsymbol{\phi}^T(k)\mathbf{P}(k-1) \qquad \text{(SIV-3)}$$

Note that, like its forerunner, the IV algorithm in section 3.1 of chapter 3, this is a deterministic algorithm because there is no attempt here to generate information on the uncertainty in the estimates. Unlike the situation with the RARX algorithm, therefore, the $\mathbf{P}(k)$ matrix is not an estimate of the covariance associated with the estimated parameter vector $\hat{\boldsymbol{\rho}}(k)$

As pointed out in section 3.3.2 of chapter 3, the main problem with the general application of IV algorithms is that of choosing the instrumental variables themselves. Of course, in some fortuitous circumstances, it could be that an additional variable is available that can act as an IV: for example, there could be a second output measurement contaminated by noise that is statistically independent of the nominal output noise $\xi(k)$. It is clear, however, that this does not constitute a general solution to the problem and several other, more practically meaningful solutions have been proposed.

One rather poor approach uses an IV vector composed of lagged values of $y(k)$. Provided the lag is such that the noise associated with this vector is statistically independent of the additive noise vector $\mathbf{e}_z$ defined in equation (3.36), then this satisfies the requirements of IV estimation. However, the reader might suspect that this is not a particularly good solution. Indeed, it has been used mainly for the estimation of fully stochastic AR and ARMA models (i.e. when there is no deterministic input $u(k)$ in the BJ model). In the simplest AR case, it can be shown (see e.g. Young, 1972), that the recursive IV estimates obtained in this manner are identical to those obtained from the solution of the 'Yule-Walker' equations, a set of linear equations relating the autocorrelation functions of the $y(k)$ series: see e.g. Box and Jenkins (1970); Priestley (1981). The solution of these equations has been well known for many years in the statistical literature as a simple *en bloc* method of AR model estimation. Unfortunately, the method is statistically inefficient and related approaches (see e.g. Stoica et al. (1985); Kuersteiner (1999, 2001) and the prior references therein), that are optimal *for this class of estimators*, are preferable. However, an alternative and more generally optimal IV method of ARMA model estimation is discussed in the next chapter 7.

A much more obvious source for the IVs in the case of the full BJ model is the deterministic input $u(k)$, which is clearly correlated with the noise-free part of the output, $x(k)$ and, provided there is no feedback from the output to the input (e.g. arising from feedback control), then it will be uncorrelated with the noise $\xi(k)$. But, bearing in mind that the transfer function $B(z^{-1})/A(z^{-1})$ will inject a dynamic lag between $u(k)$ and $x(k)$, in addition to any time delay δ in the model, we might expect $x(k)$ to be more highly correlated with $u(k-\delta-t_d)$, where t_d is an additional pure time delay, and that there will be some optimum choice of this time delay that maximizes this correlation. This suggests that, in general, the IV vector $\hat{\boldsymbol{\phi}}$ could be chosen of the form

$$\hat{\boldsymbol{\phi}}^T(k) = [-u(k-\delta-t_d) \ldots -u(k-\delta-t_d-n),\ u(k-\delta) \ldots u(k-\delta-n)]$$

Ideally, the time delay t_d will be chosen to maximize the correlation between the $u(k-\delta-t_d)$ and $x(k)$, recognising that this is not a simple task since $x(k)$ is not known. As a result, it is likely that several attempts may be required before a suitable value for t_d is obtained. This approach is quite similar to that first used by Joseph et al. (1961).

But why simply delay the input $u(k)$? Why not pass it through a TF model, so that the lag effects of this filter on $u(k)$ are more like those of the system TF? This will clearly be a good way to generate IVs, provided the filter has suitable dynamic properties that sufficiently resemble those of the system TF. The standard IV algorithm takes this idea to its logical conclusion and uses a procedure where the IV $\hat{x}(k)$, at the jth iteration of an iterative (or 'relaxation') estimation algorithm, is generated as the output of an iteratively updated 'auxiliary model' of the system (Levadi, 1964; Young, 1966; Wong and Polak, 1967):

$$\hat{x}(k) = \frac{\hat{B}_{j-1}(z^{-1})}{\hat{A}_{j-1}(z^{-1})} u(k-\delta) \tag{6.66}$$

where $\hat{A}_{j-1}(z^{-1})$ and $\hat{B}_{j-1}(z^{-1})$ are the TF polynomials based on the IV estimate $\hat{\boldsymbol{\rho}}_{j-1}$ of the TF model parameter vector of $\boldsymbol{\rho}$ at the $(j-1)^{th}$ iteration of the algorithm. The IV vector $\hat{\boldsymbol{\phi}}(k)$ is then defined at each iteration by:

$$\begin{aligned}\hat{\boldsymbol{\phi}}(k) = [&-\hat{x}(k-1)\ -\hat{x}(k-2) \ldots -\hat{x}(k-n)\\ &u(k-\delta)\ u(k-\delta-1) \ldots u(k-\delta-m)]^T\end{aligned} \tag{6.67}$$

and this is used in the recursive SIV algorithm or its *en bloc* equivalent. If the algorithm converges, then $\hat{x}(k)$ will converge on the noise-free output $x(k)$ and the IV estimates will have good statistical properties.

Based on this iterative strategy, the details of the standard (or 'basic') recursive-iterative algorithm are as follows:

The SIV algorithm

Step 1. Initialization: Use the RLS algorithm or its *en bloc* equivalent to obtain an estimate $\hat{\boldsymbol{\rho}}^0$ of the parameters in an ARX model. If the estimated ARX model is unstable, reflect the unstable eigenvalues of the estimated $\hat{A}(z^{-1})$ polynomial into the stable region of the complex z plane (e.g. using the Matlab routine polystab).

Step 2. Iterative IV estimation with auxiliary model updating:.

for $j = 1 : convergence$

(1) Generate the IV variable $\hat{x}(k)$ using the auxiliary model equation (6.66) with the estimates $\hat{A}(z^{-1})$ and $\hat{B}(z^{-1})$ of the TF system polynomials based on the IV estimate $\hat{\boldsymbol{\rho}}^{j-1}$ of $\boldsymbol{\rho}$ obtained at the previous $(j-1)^{th}$ iteration of the algorithm; for $j = 1$, $\hat{\boldsymbol{\rho}}^0$ is the RLS estimate obtained in Step 1. Form the IV vector $\hat{\mathbf{x}}(k)$ defined in (6.67).

(2) Use the IV algorithm or its *en bloc* equivalent to obtain the IV estimate of the TF model parameter vector $\hat{\boldsymbol{\rho}}^j$.

end

Normally, only 4 iterations are required for convergence. In general, however, it is safer to either specify more than 4 iterations or, as indicated in the algorithm above, use an automatic convergence rule based on the change in the parameter estimates.

6.8.1 Statistical properties of SIV estimates

Let us consider an analysis of the IV estimation algorithm similar to that carried out earlier in section 6.3, equation (6.31) *et seq.*

$$\frac{1}{k}\sum_{i=1}^{k}\hat{\boldsymbol{\phi}}(i)[\mathring{\boldsymbol{\phi}}(i)+\boldsymbol{\xi}(i)]^T\hat{\boldsymbol{\rho}}(k) = \frac{1}{k}\sum_{i=1}^{k}\hat{\boldsymbol{\phi}}(i)[\mathring{\boldsymbol{\phi}}(i)+\boldsymbol{\xi}(i)]^T\boldsymbol{\rho}+\eta(i) \tag{6.68}$$

But by definition, $\hat{\boldsymbol{\phi}}(i)$ is chosen to be independent of the noise inputs, so that

$$\lim_{k\to\infty}\frac{1}{k}\sum_{i=1}^{k}\hat{\boldsymbol{\phi}}(i)\boldsymbol{\xi}^T(i) = 0; \quad \lim_{k\to\infty}\frac{1}{k}\sum_{i=1}^{k}\hat{\boldsymbol{\phi}}(i)\eta^T(i) = 0$$

$$\lim_{k\to\infty}\frac{1}{k}\sum_{i=1}^{k}\hat{\boldsymbol{\phi}}(i)[\mathring{\boldsymbol{\phi}}(i) + \boldsymbol{\xi}^T(i)] = \mathbf{R}_{\hat{\boldsymbol{\phi}},\mathring{\boldsymbol{\phi}}} \neq 0$$

where $\mathbf{R}_{\hat{\boldsymbol{\phi}},\mathring{\boldsymbol{\phi}}}$ is the covariance matrix defined as

$$\mathbf{R}_{\hat{\boldsymbol{\phi}},\mathring{\boldsymbol{\phi}}} = \lim_{k\to\infty}\frac{1}{k}\sum_{i=1}^{k}\hat{\boldsymbol{\phi}}(i)\mathring{\boldsymbol{\phi}}^T(i)$$

Now (6.68) can be written

$$\frac{1}{k}\sum_{i=1}^{k}\hat{\boldsymbol{\phi}}(i)[\mathring{\boldsymbol{\phi}}(i) + \boldsymbol{\xi}(i)]^T(\hat{\boldsymbol{\rho}}(k) - \boldsymbol{\rho}) = \frac{1}{k}\sum_{i=1}^{k}\hat{\boldsymbol{\phi}}(i)\eta(i)$$

Thus, provided $\mathbf{R}_{\hat{\phi},\mathring{\phi}}$ is non-singular, the limiting value, p. lim $\tilde{\boldsymbol{\rho}}(k)$, to which $\tilde{\boldsymbol{\rho}}(k) = \hat{\boldsymbol{\rho}}(k) - \boldsymbol{\rho}$ tends in probability, is given by

$$\text{p.lim } \frac{1}{k}\tilde{\boldsymbol{\rho}} = 0 \tag{6.69}$$

and the estimates are asymptotically unbiased, as required.

It is well known that the problem of assessing the statistical properties of IV estimates is rather difficult (Durbin, 1954). In the regression situation, expressions for the covariance matrix of the estimation errors can be obtained as shown, for example, by Dhrymes (1970). The problem is, however, more difficult in the time series case (see e.g. Söderström and Stoica, 1989) but, since the iterative IV procedure seems strongly covergent, the problem is made somewhat more tractable because of the asymptotic nature of the recursive estimates.

The *en bloc* IV solution can be written [c.f. equation (6.56)]

$$\hat{\boldsymbol{\rho}}(k) = [\hat{\boldsymbol{\Phi}}^T(k)\boldsymbol{\Phi}(k)]^{-1}\hat{\boldsymbol{\Phi}}^T(k)\mathbf{y}(k) \tag{6.70}$$

if $\hat{\boldsymbol{\Phi}}(k)$ and $\boldsymbol{\Phi}(k)$ are defined as $k \times (n+m+1)$ matrices with rows $\hat{\boldsymbol{\phi}}^T(i)$ and $\boldsymbol{\phi}^T(i)$, respectively; while $\mathbf{y}(k)$ is a $k \times 1$ vector with elements $y(i)$ and where i is defined accordingly. Now, since in this case,

$$\mathbf{y}(k) = \boldsymbol{\Phi}(k)\boldsymbol{\rho} + \boldsymbol{\eta}(k)$$

where $\boldsymbol{\eta}(k)$ is a k vector with elements $\eta(i), i = 1, 2, \ldots, k$, equation (6.70) can be rewritten as

$$\hat{\boldsymbol{\rho}}(k) = [\hat{\boldsymbol{\Phi}}^T(k)\boldsymbol{\Phi}(k)]^{-1}\hat{\boldsymbol{\Phi}}^T(k)[\boldsymbol{\Phi}(k)\boldsymbol{\rho} + \boldsymbol{\eta}(k)]$$

so that the estimation error $\tilde{\boldsymbol{\rho}}(k) = \hat{\boldsymbol{\rho}}(k) - \boldsymbol{\rho}(k)$ is given by

$$\tilde{\boldsymbol{\rho}}(k) = [\hat{\boldsymbol{\Phi}}^T(k)\boldsymbol{\Phi}(k)]^{-1}\hat{\boldsymbol{\Phi}}^T(k)\boldsymbol{\eta}(k)$$

and the covariance matrix of $\tilde{\boldsymbol{\rho}}(k)$ is obtained as

$$\begin{aligned}\mathbf{P}^*(k) &= E\{\tilde{\boldsymbol{\rho}}(k)\tilde{\boldsymbol{\rho}}^T(k)\}\\ &= E\{[\hat{\boldsymbol{\Phi}}^T(k)\boldsymbol{\Phi}(k)]^{-1}\hat{\boldsymbol{\Phi}}^T(k)\boldsymbol{\eta}(k)\boldsymbol{\eta}^T(k)\hat{\boldsymbol{\Phi}}(k)[\hat{\boldsymbol{\Phi}}^T(k)\boldsymbol{\Phi}(k)]^{-1}\}\end{aligned}$$

It is not easy to progress any further with this expression since $\boldsymbol{\Phi}(k)$ contains components of $\boldsymbol{\eta}(k)$ and the expected value operator cannot be taken within the expression, as in the pure regression case.

A simpler but approximate expression for $\mathbf{P}^*(k)$ can be derived, however, by taking note that, upon convergence, the iterative IV algorithm is performing in some average manner like the linear regression solution for the equivalent regression model since $\hat{x}(k) \rightarrow x(k)$, so that, for large sample lengths, the matrix $\frac{1}{k}\mathbf{P}^*(k)$ approximates the noise-free matrix

$$\frac{1}{k}[\mathring{\boldsymbol{\Phi}}^T(k)\mathring{\boldsymbol{\Phi}}(k)]^{-1}$$

where $\mathring{\boldsymbol{\Phi}}(k)$ is a $k \times n+m+1$ matrix with rows $\mathring{\boldsymbol{\phi}}(i)$, $i = 1,2, \ldots, k$, as defined previously in equation (6.32). Thus, for the case where the $\xi(i)$ are serially independent (i.e the BJ model with $C(z^{-1}) = D(z^{-1}) = 1.0$), an estimate of $\mathbf{P}^*(k)$ can be obtained from the equivalent linear regression result, i.e.

$$\mathbf{P}^*(k) = \sigma^2[\mathring{\boldsymbol{\Phi}}^T(k)\mathring{\boldsymbol{\Phi}}(k)]^{-1} = \sigma^2\mathbf{P}(k) \tag{6.71}$$

while in the more general case where the $\xi(i)$ are serially correlated,

$$\mathbf{P}^*(k) = \mathbf{P}(k)\mathring{\boldsymbol{\Phi}}^T(k)\mathbf{W}\mathring{\boldsymbol{\Phi}}(k)\mathbf{P}(k) \tag{6.72}$$

where $\mathbf{W}$ is the covariance matrix of $\boldsymbol{\xi}(k)$ which could be estimated, for example, by reference to the estimates of the noise process. It must be emphasized, however, that these are approximate results and the optimal IV approach discussed in the next chapter 7 results in a much more satisfactory error analysis.

6.9 Model Structure Identification

So far in this book, it has been assumed that the TF model structure, as defined for example, by the triad $[n\ m\ \delta]$ or the pentad $[n\ m\ \delta\ p\ q]$, is known *a priori*. This is not normally the case, however, and the application of the RARX, SIV, or the optimal RIV algorithm developed in the next chapter 7, to any data set $Z^N = \{u(k); y(k)\}_{k=1}^N$, requires the identification of a suitable TF model structure based on these data. This topic of 'model structure identification' has received a lot of attention in various scientific disciplines, including control and systems, statistics, time series analysis and econometrics. The references cited and discussed in Ljung, 1999, for instance, provide a good review of the subject although, because he considers it from a mainly control and systems outlook, he uses the name 'model structure determination'. However, 'identification' seems a more appropriate term since model structure identification is strongly linked with model 'identifiability'; after all, the whole objective of model structure identification is the data-based inference of an identifiable model. Moreover, as the reader of this book will find when analyzing real data, the model structure is often far from being easy to 'determine'.

Rather than attempt to review all the methods of model structure identification that have been suggested over many years, attention in this section will be restricted to those methods of model structure identification that I have found to be most useful over a long period of practical evaluation. The first of these is related to the well known *Coefficient of Determination*, R^2, which derives from regression analysis, where it provides a measure of how well the regression model explains the dependent variable. In the present context, however, it is denoted by R_T^2, as defined earlier by equation (5.21) of chapter 5 but repeated below for convenience:

$$R_T^2 = 1 - \frac{\sigma^2}{\sigma_y^2}; \;\; \hat{\sigma}^2 = \frac{1}{N}\sum_{k=1}^{k=N}(\hat{e}(k) - \bar{\hat{e}}(k))^2; \;\; \sigma_y^2 = \frac{1}{N}\sum_{k=1}^{k=N}(y(k) - \bar{y}(k))^2 \quad (6.73)$$

In the present TF identification context $\hat{e}(k)$ is the error $y_k - \hat{x}_k$ between the measured output y_k and simulated TF model output $\hat{x}(k)$. In other words, R_T^2 provides a measure of how much of the variance $\sigma_{\hat{y}}^2$ associated with the measured model output $y(k)$ is explained by the deterministic part of the TF model. This deterministic output of the model is clearly very important since it explains that part of the output that is estimated as being caused by the input $u(k)$. As such, and as we shall see later in chapter 7, it plays an important role in optimal instrumental variable identification and estimation.

The R_T^2 statistic is often a more discerning measure of model adequacy than the more standard coefficient of determination R^2, which is normally defined in terms of the one-step-ahead prediction errors obtained from regression and TF models. However, R^2 provides a reasonable measure of the model's predictive ability provided there are not low frequency trends on the data, when a measure based on the differenced data seems more appropriate (see Harvey, 1989, page 268).

One problem with R_T^2, if used alone, is that it can favour over-parameterized models, which can often explain the data well: in other words, we need another measure which, while ensuring a good explanation of the data defined by a high R_T^2 value, also avoids such over-parameterization and suggests an identifiable model that is parametrically efficient or parsimonious. The most important research on such 'model order identification' criteria has been intended primarily for either purely stochastic models with no measured inputs or those, such as the BJ and ARMAX models, in which the additive noise is modelled as an ARMA process, so that the final residual error series has white noise properties.

The best known of these model order identification criteria is the *Akaike Information Criterion* (AIC) suggested in an early seminal paper by Akaike (1974) that considered the identification of AR model order and led to numerous other, related criteria that have not been limited to AR models. The AIC is defined as follows:

$$\mathrm{AIC}(np) = N\log_e \hat{\sigma}^2 + 2n_p \quad (6.74)$$

where the first term is a measure of how well the model explains the data, while the second term, with n_p equal to the number of parameters (so $n_p = n + m + 1$ for the SISO TF model) penalizes the number of parameter in the model and militates against over-parameterization.

Following from the publication of Akaike's paper, the topic of model order identification criteria has received a lot of attention and several different modifications to the AIC have been suggested. Indeed Akaike himself (Akaike, 1979) suggested a Bayesian extension of the AIC criterion: the *Bayesian Information Criterion* (BIC). Section 5.4.5 of the excellent book by Priestley (1981) provides a very good review of model order identification procedures and Shibata (1985) notes that all the AIC-like criteria have the form:

$$T + \alpha n_p$$

where T is a test statistic and the second term is the penalty term on the number of parameters in the model with $\alpha > 1$. The test statistic T can take several forms, as discussed by Shibata, but the best known is the maximum log likelihood, where the dominant term is the one step prediction error variance $\hat{\sigma}^2$, as in the AIC defined above. The most controversial point is how to choose α: which is 2 in the case of the AIC; $log_e(N)$ in the criterion suggested by Schwarz (1978) (SIC but also sometimes called the BIC since it is a simpler version of Akaike's BIC: see Priestley, 1981); and $c.log_e log_e(N)$ for some $c > 2$ in the criterion suggested by Hannan and Quinn (1979). In my experience, the AIC and BIC both work well, although the AIC is more prone to over-fitting.

The YIC (Young, 1989) is a heuristically defined identification criterion of a different type derived from the earlier EVN criterion (Young *et al.*,. 1980) and, since it exploits the special properties of the IPM, it can only be used with IV-type algorithms. Referring to equations (6.65) and (6.70), the IPM is defined as follows:

$$\text{IPM} = \sum_{i=1}^{k} \hat{\boldsymbol{\phi}}(i)\boldsymbol{\phi}^T(i) = \hat{\boldsymbol{\Phi}}^T(k)\boldsymbol{\Phi}(k)$$

and it has to be inverted, either directly for *en bloc* estimation, or indirectly via the recursive SIV algorithm. For a sample size N, the YIC criterion is defined as follows:

$$YIC = \log_e \left\{ \frac{\sigma^2}{\sigma_y^2} \right\} + \log_e EVN; \tag{6.75a}$$

$$EVN = \frac{1}{n_\rho} \sum_{i=1}^{i=n_\rho} \frac{\hat{\sigma}^2 p_{ii}}{\hat{\rho}_i^2} \tag{6.75b}$$

where n_ρ is the number of parameters in the $\boldsymbol{\rho}$ vector; p_{ii} are the diagonal elements of the inverse IPM which is, of course, the $\mathbf{P}(N)$ matrix generated by the SIV algorithm; and $\hat{\rho}_i$ are the SIV parameter estimates: see the model (6.29).

The first term in (6.75a) is simply a relative, logarithmic measure of how well the model explains the data: the smaller the model residuals the more negative the term becomes. The second term, $\log_e EVN$, on the other hand, provides a logarithmic measure of the conditioning of the IPM. If the model is over-parameterized, then it can be shown (e.g. Wellstead, 1978) that the IPM will tend to singularity and, because of its ill-conditioning, the elements of its inverse $\mathbf{P}(N)$ (on which the estimate of the parametric covariance matrix depends) will increase in value, often by several orders of magnitude: see Young et al. (1980). When this happens, the second term in the YIC tends to dominate the criterion function, indicating over-parameterization. An alternative, justification of the YIC can be obtained from statistical considerations: here, the second term in the YIC is interpreted as a suitably normalized measure of the total parameter error variance, since $\sigma^2 p_{ii}$ is the variance of the estimated error on the ith parameter: see equation (6.58).

Based on the above statistical measures, a generally useful approach to TF model structure identification *when using the* SIV *algorithm*, is as follows:

1. Use the SIV estimation algorithm to estimate a range of different models for $\min(n) \leq n \leq \max(n)$ and $\min(m) \leq m \leq \max(m)$, where the maximum and minimum values of n and m are selected by the user, and sort these by the value of R_T^2, so exposing those model structures that best explain the data in a deterministic simulation sense.
2. Amongst the best models in 1., select one that has a *relatively* low YIC value: normally this will be a large negative value, since it is a logarithmic measure, but the choice is not critical provided the associated R_T^2 is relatively high compared with that of other models. But do not select a model that has a high R_T^2 and a relatively large YIC (small negative) value, since the YIC is then indicating possible over-parameterization.
3. Use the model selected in 2. to generate an estimate of the additive noise $\xi(k)$. Then, if possible, estimate AR(p or ARMA(p,q) models for this estimated noise over a range of p and q, $\min(p) \leq p \leq \max(p)$ and $\min(q) \leq q \leq \max(q)$, again with the maximum and minimum values selected by the user, and sort these in terms of the AIC or BIC. Methods of estimating the ARMA models are described in the next Chapter 7.
4. Finally, evaluate the autocorrelation of the ARMA model residuals $\hat{e}(k)$, using the CAPTAIN routine acf, to ensure that they are reasonably white; and the cross-correlation functions between the $\hat{e}(k)$ and the input variable $u(k)$, using the CAPTAIN routine ccf, to ensure that there is no significant correlation between the residuals and the input, which would negate the assumptions on which the TF model estimation is based.
5. Sometimes it is not possible to model the noise because it does not subscribe to modelling as an ARMA process, in which case, the identification procedure stops at stage 1., and other, less formal identification strategies should be pursued. For example, the identification criteria do not always unambiguously identify the model structure and, if possible, the competing models should be evaluated in other terms, such as whether the response characteristics and the physical interpretation of the model make sense (see the later example in section 7.5.4 of the next chapter 7)

However, when the estimation algorithm involves the simultaneous estimation of a noise model, as in the case of the optimal RIV and RIVC algorithms considered in the next two chapter 7 and 8, then further stages are necessary which better exploit the AIC and BIC crtiteria. Such an extended procedure is discussed in section 7.3 of the next chapter 7.

6.9.1 Example 6.3: Recursive SIV parameter estimation for a simple TF model

Figure 6.8 shows results obtained when the standard recursive SIV algorithm is used to estimate the parameters in the first order system (6.44) considered previously in Example 6.1, section 6.4.1. The rapid convergence of the iterations is shown in the two panels. The left hand panel shows that, even at the first iteration ($j = 1$) using the biased least squares ARX parameter estimates to define the auxiliary model, the estimates have converged reasonably, with the true values lying inside the 95% confidence bounds. And, as shown in the right panel, further processing to five iterations makes only minor improvements (in fact, in this simple example, there is virtually no change after the third iteration).

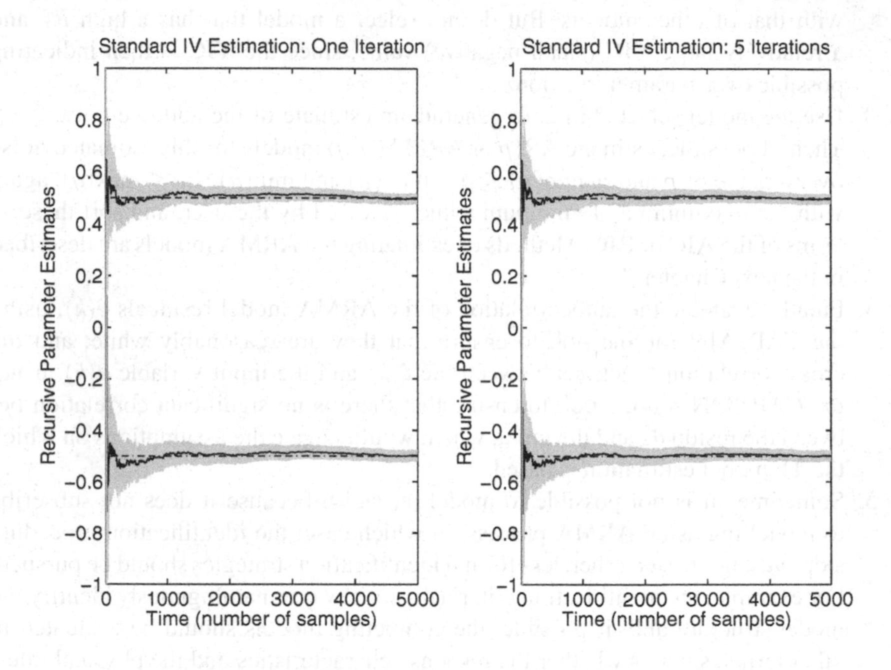

Fig. 6.8 Recursive SIV parameter estimates for example 6.3: first iteration results in the left panel and fifth iteration results in the right panel. Grey bands show the approximate 95% confidence bounds (see text).

Note that the convergence of the recursive estimates is slower than that achieved when the RARX is applied to ARX generated data (cf. Figure 6.3), with about 1000 samples required to achieve similar accuracy. This is because the IVs used in this standard algorithm, although good, are not optimal in a statistical sense (see next chapter 7). Also, because the estimates are not optimal, the computed standard errors and associated 95% confidence bounds (twice the standard errors) are only approximate (see the discussion in the previous section 6.8.1).

The nature of the error in the SIV algorithm's computed 95% confidence bounds shown in Figure 6.8 can be evaluated by *Monte Carlo Simulation* (MCS) analysis (see Appendix B, section B.3.9). Table 6.1 presents the results of such analysis and compares the 95 percentile bounds at the end of the data, after 5000 samples, revealing that the SIV estimated 95% confidence interval is a little smaller than that predicted by the SIV algorithm (but only to the third decimal place, which would be good enough for most practical purposes). In other words, the algorithm is producing estimates that are marginally better than it reports. For interest, the figures for the optimal SRIV algorithm, as discussed in the next chapter 7, are also given in Table 6.1. Here, the 95% confidence interval is smaller still, since these are the minimum variance estimates, and the SRIV estimated interval is now the same as the MCS estimated interval, as one would expect with an optimal estimation algorithm. The MCS analysis used to generate the figures in Table 6.1 involved 2000

Table 6.1 MCS Results for SIV and RIV estimation in the simulation example 6.3.

Parameter		$\hat{a}_1$	$\hat{b}_0$
True values		-0.5	0.5
SIV	mean	-0.499	0.500
(MCS)	SD	0.017	0.017
SIV	$\hat{\rho}$	-0.494	0.508
(SR)	SE	0.021	0.020
RIV	mean	-0.500	0.500
(MCS)	SD	0.015	0.014
RIV	$\hat{\rho}$	-0.490	0.508
(SR)	SE	0.015	0.014

realizations, where the SIV estimates of the TF model parameters in each realization were obtained from the analysis of input-output data in which the noise sequence $\xi(k)$, $k = 1,2, \ldots, 5000$, was selected using the Matlab routine randn, with zero mean value and the same variance, 0.25, as that used in the single simulation exercises described previously.

6.9.2 Example 6.4: SIV estimation of tracer experiment data

Research on the transport and dispersion in rivers and wetlands is aided by the ability to conduct simple planned experiments using conservative tracer materials, such as the fluorescent red dye, Rhodamine WT. Small quantities of such tracers can be injected into the environment, for example into a river system, and then the subsequent low concentrations can be measured using special equipment, such as a fluorometer in the case of Rhodamine WT (see e.g. Wallis et al, 1989). Typical results of a tracer experiment in a wetland area are shown in Figure 6.9, in this case using conservative potassium bromide (KBr) as the tracer material.

The wetland area is located in Florida, USA, and it receives treated domestic wastewater which travels slowly through the wetland to allow for further nutrient removal. The tracer experiment was part of a study carried out by Chris Martinez

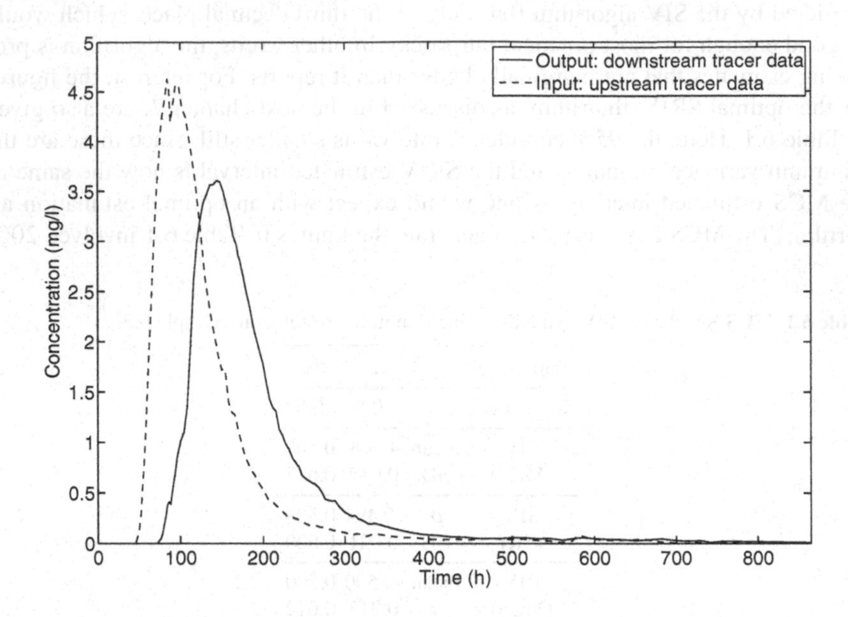

Fig. 6.9 Tracer example: experimental tracer data showing upstream and downstream measurements.

and William R. Wise of the Environmental Engineering Sciences Department, University of Florida for the City of Orlando (Martinez and Wise, 2003).

BEST 12 models (Rt2)

na	nb	td	nc	nd	YIC (TF)	Rt2	BIC (ARMA)	AIC
1	1	16	0	0	-10.148	0.9946	-2251.97	-16.584
1	2	16	0	0	-2.612	0.9942	-2226.75	-16.391
1	3	16	0	0	-2.365	0.9942	-2230.88	-16.367
1	4	16	0	0	0.249	0.9941	-2232.67	-16.333
1	5	16	0	0	3.872	0.9941	-2237.98	-16.320
1	6	16	0	0	-0.055	0.9940	-2239.82	-16.291
2	6	16	0	0	0.361	0.9939	-2225.40	-16.251
2	3	16	0	0	2.015	0.9938	-2198.88	-16.234
2	2	16	0	0	-1.001	0.9938	-2190.09	-16.225
2	4	16	0	0	0.176	0.9937	-2200.50	-16.202
2	5	16	0	0	3.068	0.9937	-2204.87	-16.187
2	1	16	0	0	-4.478	0.9934	-2158.62	-16.092

The model order identification results are show above, as produced (verbatim) by the rivbjid routine in CAPTAIN using the SIV option setting and ordered in terms of the R_T^2 values (labelled Rt2), as defined in (6.73). According to this and the other YIC, BIC and AIC meaures, the first order [1 1 16] model is the best identified model. The output of this model is compared with the measured output in Figure 6.10. However, as we shall see subsequently in chapters 7 and 8, optimal refinements of the SIV algorithm, for the estimation of both discrete and continuous-time models, favour [2 3 16] and [2 2 16] structures, respectively, and so the output of such a higher order model is also shown, for comparison, as a dash-dot line (SRIVC TF model). It is clear that this model provides a visibly better explanation of the data and no SIV estimated models are able to match this. There is no doubt, however, that, *as far as* SIV *identification alone is concerned*, the [1 1 16] has to be accepted.

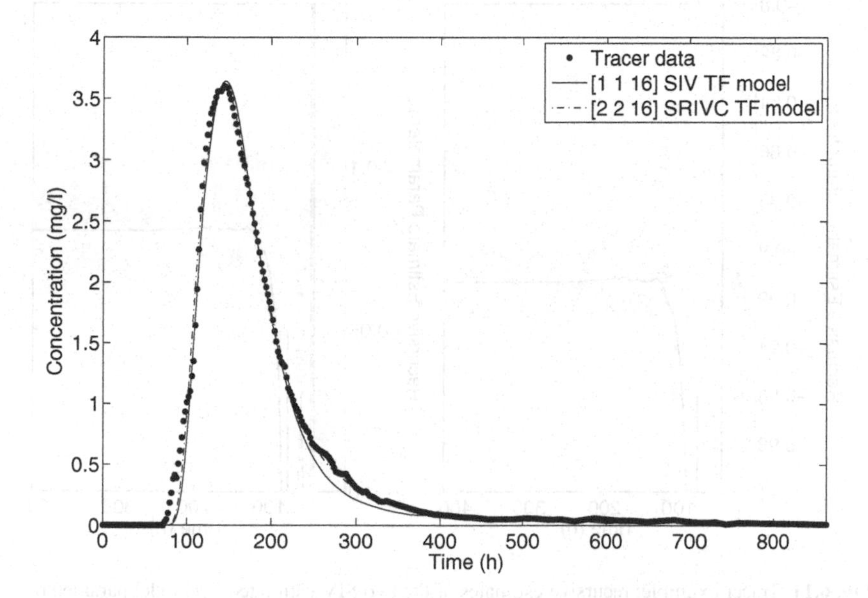

Fig. 6.10 Tracer example: comparison of SIV estimated model output (full line) and downstream tracer data (dots). Also shown, output of optimally estimated continuous-time model (dash-dot).

This estimated [1 1 16] model takes the form

$$y(k) = \frac{\hat{b}_0}{1+\hat{a}_1} u(k-16) + \xi(k) \tag{6.76}$$

and the estimated parameters are as follows

$$\hat{a}_1 = -0.914(0.008) \;\; \hat{b}_0 = 0.0805(0.007)$$

where the figures in parentheses are the SIV estimated standard errors. The recursive estimates are plotted in Figure 6.11. Note that the dye first arrives at the upstream site after 48 hours and at the downstream site after 76 hours, so there is very little information in the data before this. As a result, the grey 95% uncertainty bounds are very large at the start of the estimation because the dye concentration is zero and the recursive algorithm is started with a diffuse prior (see chapter 3, section 3.2). After about 90 hours, however, there is rapid convergence of the estimates, with very small change after about 130 hours. In this same connection, note that the input signal here is certainly not persistently exciting (see section 6.6.1): the measured output has a predominantly low frequency content resulting from the original impulsive input[5] of the dye being dispersed considerably before it reaches the downstreamstream site.

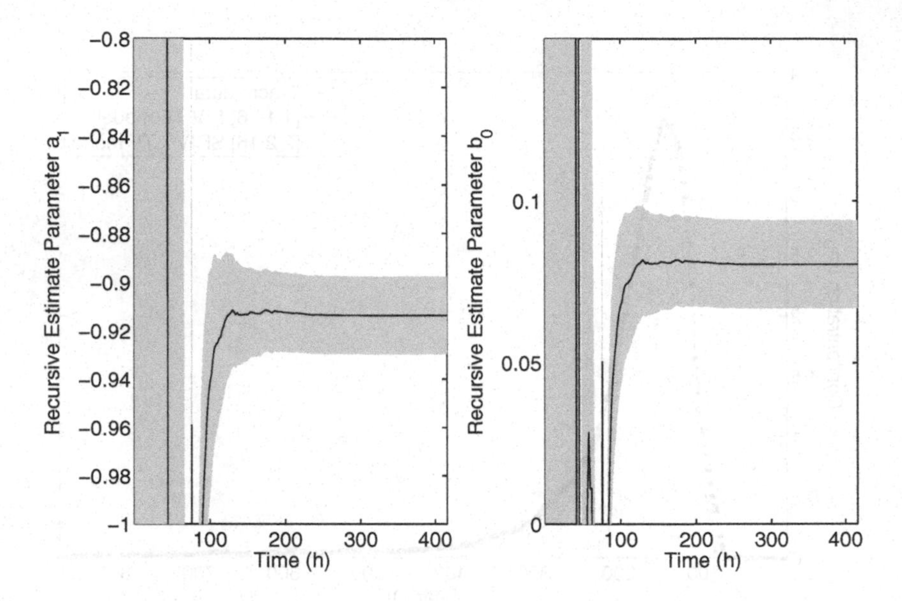

Fig. 6.11 Tracer example: recursive estimates of the two SIV estimated TF model parameters.

The discrete-time model (6.76) can be transformed either manually (see Appendix B, section B.3.7.1) or by using the d2c routine in Matlab, into the following continuous-time model:

$$y(t) = \frac{0.0421}{s+0.0451}u(t-16) + \xi(t); \tag{6.77a}$$

or in differential equation form:

$$\frac{dy(t)}{dt} = -0.0451y(t) + 0.0421u(t-16) + \eta(t) \tag{6.77b}$$

that reveals more obviously the nature of the physical system.

[5] Termed a 'gulp' input in hydrology.

The above analysis can be considered as an exercise in a procedure that I have called *Data-Based Mechanistic* (DBM) modelling because it attempts to infer the physical nature of the system from the estimated model. DBM modelling is considered fully in chapter 12 but it is clear in this example that the model (6.77) makes sense in physical terms: it defines a first order system with a steady state gain of 0.933 (i.e. 0.421/0.451) and a time constant of 22.17 (i.e. $1/0.0451$) hours; and this conforms with the *Aggregated Dead Zone* (ADZ) model for solute transport and dispersion in a river system (Beer and Young, 1983; Wallis et al., 1989).

In purely deterministic terms, the estimated steady state gain of 0.933 is less than unity and, at first, this suggests some loss of dye between the input and output. One check of the steady state gain is to compute the area under each of the measured curves in Figure 6.9, which provides a measure of the dye mass passing each site, and then divide the area under the downstream curve by that under the upstream curve. This yields a estimated steady state gain of 0.977, which is quite a bit larger than the model estimated 0.933, but still indicates a loss of dye when considered in this deterministic manner.

Such an apparent loss of dye is quite possible in this experiment since the wetland area is not a normal, physically constrained, river channel and so not all the dye leaving the input measurement site will necessarily pass through the output site. However, this is a stochastic model and so we can check on the significance of the difference between the model estimated value and unity. In this case, the difference does not appear to be significant if the computation is based on the estimated uncertainty derived from the SIV estimation. However, we know that this is approximate, see section 6.8.1, equation (6.71), and because subsequent optimal discrete and continuous-time analysis of these same tracer data is considered in chapters 7 and 8, it is best to delay such uncertainty considerations until then.

6.10 Dynamic Transfer Function Models

The recursive SIV estimation equations developed in the last section are, of course, closely related to the RLS equations in section 3.1 of chapter 3. As a result, they can be extended in any of the ways discussed in the chapters 4 and 5 to allow for the estimation of time variable parameters. Using the terminology introduced in chapter 5 for unobserved component models, the TVP version of the SIV algorithm is the *Dynamic Transfer Function* (DTF) model, which takes the form:

$$y(k) = \frac{B(k,z^{-1})}{A(k,z^{-1})}u(k) + \xi(k) \tag{6.78}$$

where $\xi(k)$ is general noise sequence: it is simply that part of $y(k)$ that cannot be explained by the effects of $u(k)$. This model can be written in the following vector equation form:

$$y(k) = \boldsymbol{\phi}^T(k)\boldsymbol{\rho}(k) + \mu(k) \tag{6.79}$$

where,

$$\begin{aligned}\boldsymbol{\phi}^T(k) &= [-y(k-1)\ -y(k-2)\ \ldots\ -y(k-n)\ u(k)\ldots\ u(k-m)] \\ \boldsymbol{\rho}(k) &= [a_1(k)\ a_2(k)\ \ldots\ a_n(k)\ b_0(k)\ \ldots\ b_m(k)]^T\end{aligned} \tag{6.80}$$

and $\mu(k) = A(k,z^{-1})\xi(k)$. As a result, even if $\xi(k)$ is a white noise sequence, $\mu(k)$ will be autocorrelated except in the unlikely situation that $\xi(k) = \{1/A(k,z^{-1})\}e(k)$, where $e(k)$ is a white noise sequence (i.e. the simple ARX model). It is obvious from the previous discussion of TF model estimation that the recursive least squares estimates obtained by applying the DARX algorithm of chapter 5 to data generated by this DTF model will be asymptotically biased away from their 'true' values. For this reason, it is necessary to modify the DARX algorithm to avoid these biasing problems. This can be achieved by attempting to model the noise $\mu(k)$ in some manner, such as that used in the *Approximate Maximum Likelihood* or *Extended Least Squares* algorithms (e.g. Young, 1968c; Norton, 1975, 1986). However, since $\mu(k)$ is a complicated, non-stationary, noise process, its complete estimation is not straightforward.

An alternative approach, which does not require modelling $\mu(k)$, provided it is independent of the input $u(k)$, is the recursive *Dynamic Transfer Function* (DTF) algorithm (Young, 2000), which has the following form (cf. the IV algorithm in section 3.3.2 of chapter 3, the KALMSMO algorithm in section 4.5.2 of chapter 4 and the DARX algorithm of chapter 5):

1. Forward-Pass Symmetric IV Equations (iterative)

Iterate the following recursive filtering equations DTF-1 to DTF-7 for $j = 1,2,\ldots,I_T$, with $\hat{\boldsymbol{\phi}}(k) = \boldsymbol{\phi}(k)$ for $j=1$; and $\boldsymbol{\phi}(k) = \hat{\boldsymbol{\phi}}(k)$ for $j = I_T$:

Prediction:

$$\hat{\boldsymbol{\rho}}(k|k-1) = \mathbf{A}\hat{\boldsymbol{\rho}}(k-1) \tag{DTF-1}$$

$$\mathbf{P}(k|k-1) = \mathbf{A}\mathbf{P}(k-1)\mathbf{A}^T + \mathbf{D}\mathbf{Q}_{nvr}\mathbf{D}^T. \tag{DTF-2}$$

Correction:

$$\hat{\boldsymbol{\rho}}(k) = \hat{\boldsymbol{\rho}}(k|k-1) + \mathbf{g}(k)\{y(k) - \boldsymbol{\phi}(k)\hat{\boldsymbol{\rho}}(k|k-1)\} \tag{DTF-3}$$

$$\mathbf{g}(k) = \mathbf{P}(k|k-1)\hat{\boldsymbol{\phi}}^T(k)\left[1 + \boldsymbol{\phi}(k)\mathbf{P}(k|k-1)\hat{\boldsymbol{\phi}}^T(k)\right]^{-1} \tag{DTF-4}$$

$$\mathbf{P}(k) = \mathbf{P}(k|k-1) + \mathbf{g}(k)\boldsymbol{\phi}(k)\mathbf{P}(k|k-1) \tag{DTF-5}$$

$$\mathbf{P}^*(k) = \hat{\sigma}^2\mathbf{P}(k) \tag{DTF-6}$$

where,

$$\hat{\boldsymbol{\phi}}^T(k) = [-\hat{x}(k-1) - \hat{x}(k-2), \ldots, -\hat{x}(k-n), u(k), \ldots, u(k-m)]$$

$$\text{Auxiliary Model:} \quad \hat{x}(k) = \frac{\hat{B}_{j-1}(k,z^{-1})}{\hat{A}_{j-1}(k,z^{-1})} u(k). \qquad \text{(DTF-7)}$$

Now apply the following fixed interval smoothing equations for a single pass:

2. Backward-Pass Fixed Interval Smoothing IV (FISIV) equations (single pass)

$$\hat{\boldsymbol{\rho}}(k|N) = \boldsymbol{\rho}(k) - \mathbf{P}(k)\mathbf{A}^T\boldsymbol{\lambda}(k) \qquad \text{(DTF-8)}$$

$$\text{or } \hat{\boldsymbol{\rho}}(k|N) = \mathbf{A}^{-1}\left[\hat{\boldsymbol{\rho}}(k+1|N) - \mathbf{D}\mathbf{Q}_{nvr}\mathbf{D}^T\boldsymbol{\lambda}(k)\right] \qquad \text{(DTF-9)}$$

$$\boldsymbol{\lambda}(k-1) = [\mathbf{I}_p - \mathbf{P}(k)\hat{\boldsymbol{\phi}}(k)\hat{\boldsymbol{\phi}}^T(k)]^T$$
$$\{\mathbf{A}^T\boldsymbol{\lambda}(k) - \hat{\boldsymbol{\phi}}(k)[y(k) - \hat{\boldsymbol{\phi}}^T(k)\mathbf{A}\hat{\boldsymbol{\rho}}(k-1)]\}$$

$$\text{with } \boldsymbol{\lambda}(N) = 0 \qquad \text{(DTF-10)}$$

$$\mathbf{P}(k|N) = \mathbf{P}(k) + \mathbf{P}(k)\mathbf{A}\mathbf{P}(k+1|k)^{-1}$$
$$[\mathbf{P}(k+1|N) - \mathbf{P}(k+1|k)]\mathbf{P}(k+1|k)^{-1}\mathbf{A}\mathbf{P}(k)$$

$$\mathbf{P}^*(k|N) = \hat{\sigma}^2(k)\mathbf{P}(k|N) \qquad \text{(DTF-11)}$$

The main difference between the DTF algorithm and the standard filtering and smoothing algorithms in chapter 4 is the introduction of 'hats' on the $\hat{\boldsymbol{\phi}}(k)$ vector, as well as the use of an iterative IV solution in the forward-pass algorithm. In (DTF-4) and subsequently, $\hat{\boldsymbol{\phi}}(k)$ is the *IV vector*, which is used by the algorithm in the generation of all the $\mathbf{P}(k)$ terms and is the main vehicle in removing the bias from the TVP estimates. The subscript $j-1$ on $\hat{A}_{j-1}(k,z^{-1})$ and $\hat{B}_{j-1}(k,z^{-1})$ indicates that the estimated DTF polynomials in the 'auxiliary model', which generates the instrumental variables $\hat{x}(k-i)$, $i = 1, 2, \ldots, n$, that appear in the definition of $\hat{\boldsymbol{\phi}}(k)$, are updated in an iterative manner, starting with the least squares estimates of these polynomials (note $\hat{\boldsymbol{\phi}}(k) = \boldsymbol{\phi}(k)$ for $j = 1$). Iteration is continued for I_T iterations, until the forward pass (filtered) IV estimates of the TVPs are no longer changing significantly: normally only 3 or 4 iterations are required; or I_T can be decided by an automatic convergence criterion. Then, in the final I_T^{th} iteration, the symmetric version of the IV algorithm is used with $\boldsymbol{\phi}(k) = \hat{\boldsymbol{\phi}}(k)$, in order to generate the symmetric $\mathbf{P}(k)$ matrix that is required for the backwards FIS estimation pass through the data. A further justification for using this symmetric version of the IV algorithm is provided by the subsequent development of the optimal RIV algorithm in the next chapter 7.

This DTF algorithm is the one currently implemented in the CAPTAIN Toolbox as the dtfm routine. However, it is a rather heuristically designed algorithm and alternative implementations are possible. The heuristic nature of the DTF algorithm also means that, while it can yield good results, it is not particularly robust in practical application. For instance, it does not always converge, particularly when applied to noisy data sets where the system has fairly complex dynamics: for example, when the system is dynamically 'stiff' and characterized by widely spaced eigenvlaues (see some of the constant parameter examples considered in the next chapter 7).

6.10.1 Example 6.5: DTF model estimation

As a simple example of DTF modelling, consider the estimation of the parameters in the following first order TVP model:

$$x(k) = \frac{b_0}{1 + a_1(k)z^{-1}} u(k-2) \qquad u(k) = \mathcal{N}(0, 6.25)$$
$$y(k) = x(k) + \xi(k) \qquad \xi(k) = \mathcal{N}(0, 2.56),$$

where the $b_0 = 0.5$ is constant and $a_1(k)$ varies sinusoidally, as $0.9\sin(0.02k)$. Estimation is based on the measurements of $y(k)$ and $u(k)$, $t = 1,2,\ldots,2000$. The overall noise/signal ratio on the output measurement $y(k)$ is high (0.84 by standard deviation).

It is assumed that no information is available on the variation of the parameters and so RW models are chosen for both of the two unknown parameters. ML optimization of the NVRs, using the dtfmopt routine in the CAPTAIN Toolbox, then yields $\mathbf{Q}_{nvr} = \text{diag}[0.0011 \;\; 1.91 \times 10^{-20}]$, where the NVR for the b_0 parameter is insignificantly different from zero, indicating that the parameter is identified as being time invariant. The DTF estimated TVP $a_1(k|N)$, obtained using the dtfm routine in CAPTAIN, is shown in Figure 6.12 (upper panel), where it is compared with the DARX estimates (lower panel).

The superiority of the DTF estimates is particularly clear: not only are the DTF estimates much better than the equivalent, biased DARX estimates, but the estimated standard errors (shown dotted) are more realistic. As in the case of similar situations with constant parameter models, the least squares DARX standard errors are too optimistic and, in contrast to the DTF standard errors, they do not encompass the true variation of the parameters. The DTF model with these estimated parameters explains the data well: the coefficient of determination based on the simulated model output compared with the noise-free output is $R_T^2 = 0.93$ (93% of the output variance explained by the TVP model); whilst for the DARX model, this is reduced to $R_T^2 = 0.85$. The model residuals (normalized innovations) for the DTF model are also superior: they have an approximately normal amplitude distribution; and, as required, the ACF of the normalized recursive residuals and the cross correlation function (CCF) between the residuals and the input $u(k)$, are insignificant at all lags. In contrast, the CCF for the DARX model residuals shows significant correlation with $u(k)$ at some lags.

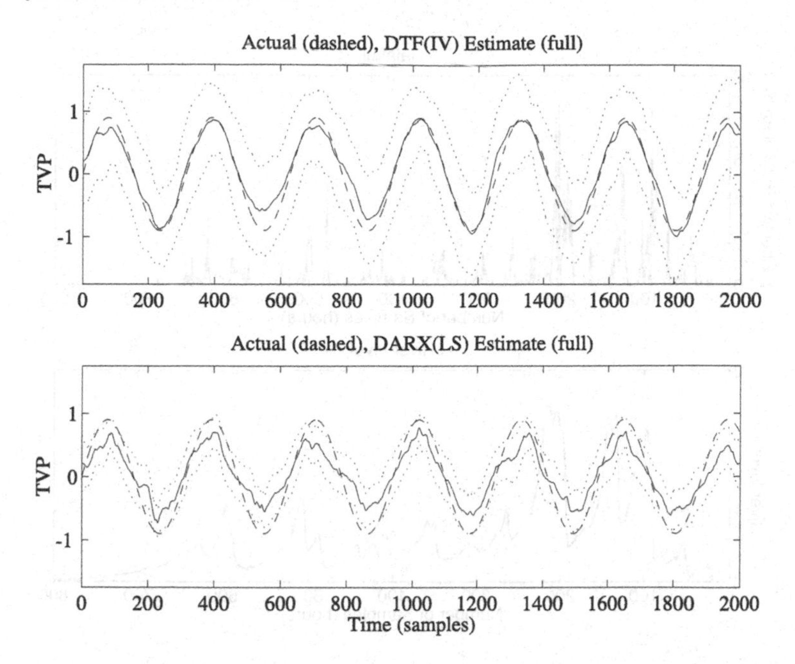

Fig. 6.12 Simulated DTF model. Top panel: DTF estimate of $a_1(k)$ (full line) compared with actual variation (dashed), with standard error bound shown dotted. Lower panel: biased DARX estimate of the TVP under the assumption that the model is DARX (lines as for upper panel). From Young (2000).

Note that, if both parameters are allowed to vary in a similar manner (e.g. $a_1(k)$ sinusoidal, as here, and $b_0(k)$ with a similar frequency, cosine variation between 0.9 and -0.9), then the associated NVRs are optimized as $\mathbf{Q}_{nvr}$= diag [0.0010 0.0023], showing that the ML optimization has found strong evidence of temporal changes in both parameters. As a result, they are both estimated well (although, as might be expected, there is some deterioration in the $\hat{a}_1(k)$ estimate, when compared with the above results).

6.10.2 Example: Using DTF estimation for model diagnosis

The ability to estimate time variable parameters using the DTF algorithm can be useful in various ways for diagnosing limitations in estimated constant parameter models. For instance, if the estimated time variable parameters change in a statistically significant manner, then this suggests that, depending on the application, the model may require improvement: e.g. by the inclusion of additional inputs; or allowing the parameters to be time variable or state-dependent (see Chapter 11). Also, the estimated uncertainties in the TVP estimates will indicate which parameters are estimated well and where they are less well-defined.

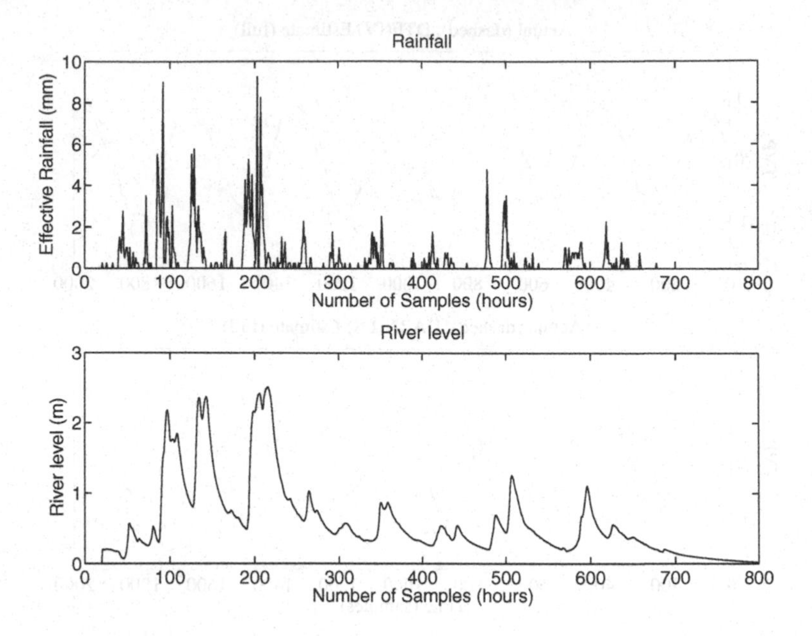

Fig. 6.13 Rainfall-level data for Meifod on the River Severn in the UK.

An example is the estimation of a model for the rainfall-flow dynamics at the village of Meifod on the River Severn in the UK, based on the hourly 'effective rainfall' and river level data plotted in Figure 6.13. The effective rainfall is a non-linear transformation of the measured rainfall, where the transformation accounts for changes in the catchment soil moisture: see section 10.5.1 of Chapter 10. In the present case, the rainfall plotted in Figure 6.13 is computed as an average of the rainfall gauge measurements at Vyrnwy and Llanfyllin in the river catchment. The only other processing of the data, prior to model identification and estimation, is the removal of the minimum value from the level data, so the level measurements are changes above this minimum level (0.929 *m*). The constant parameter [2 2 4] TF model identified and estimated using the rivbjid routine in the CAPTAIN Toolbox (see the next Chapter 7) explains the river level data well, with a coefficient of determination $R_T^2 = 0.95$. However the error between the measured flow and the simulated output of the model has significant 'spikes' at times following rainfall events and this is difficult to model adequately as an AR or ARMA process (see the further discussion on this model in the Epilogue to this book). It makes sense, therefore, to investigate the model further using the dtfmopt and dtfm routines in CAPTAIN.

In relation to the TF parameter vector, $\boldsymbol{\rho} = [a_1\ a_2\ b_0\ b_1]^T$ in this case, the dtfmopt routine, based on an RW assumption for the parameter variations, yields NVR values of $[6.8719 \times 10^{-7}\ 0.015449\ 0.50104\ 5.4228]$, suggesting that there is significant variation in the numerator parameter b_1. As expected, the DTF model estimated by the dtfm routine, with a diagonal $\mathbf{Q}_{nvr}$ matrix defined by these NVR values, explains the data very well, with $R_T^2 = 0.992$. The associated TVP estimates of the parameters are plotted in Figure 6.14 over a short section of the data, with the standard error bounds shown in grey. Although there is very little estimated variations in the denominator parameters, which can be assumed sensibly constant, both numerator parameters show variations, although they are much more notable in the case of the $\hat{b}_1$ variations, where they are statistically significant when the standard error bounds become narrow. Also shown on the plot is the scaled rainfall series and it is clear that this sharp narrowing of the SE bounds is particularly apparent immediately after a rainfall event has occurred and the rainfall is having its *initial* effect on the subsequent level change (noting that the model has a four hour pure time delay). This confirms that it is at these times where the information in the rainfall-flow data is at its highest in relation to the estimates of the numerator parameters. Thereafter, the SE bounds 'balloon' because the information content is low in this regard.

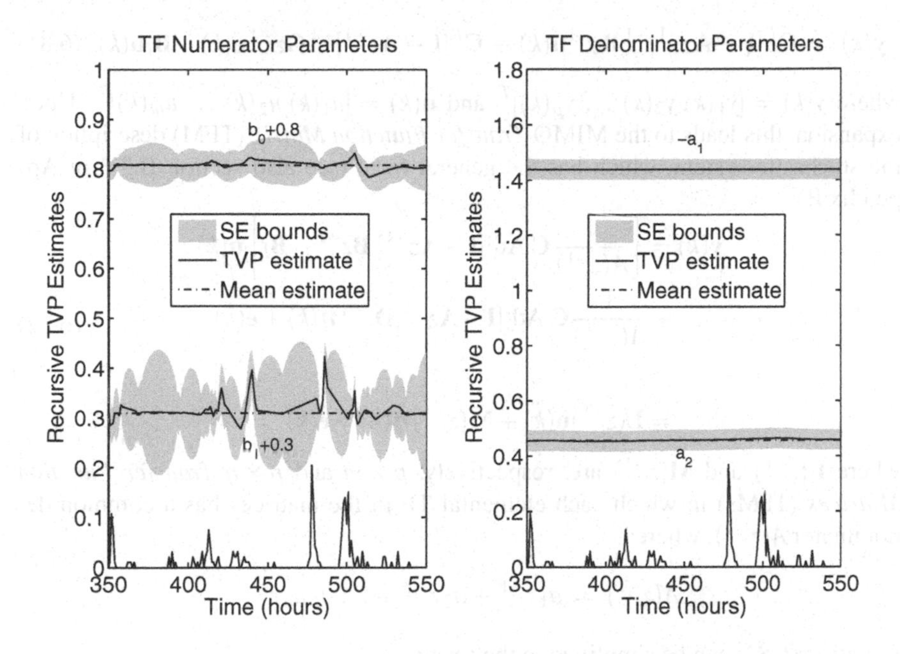

Fig. 6.14 Recursive DTF parameter estimates for the Meifod effective rainfall-river level data.

The results in Figure 6.14 suggest that most errors in the model are related to the effects of input rainfall. This could be due to errors in the rainfall series, so that the estimation is suffering from the 'errors-in-variables' problem (see section 7.4 of the next Chapter 7); or it could mean that the model can be improved in some manner. For instance, the errors may arise from limitations in the effective rainfall transformation, which is a notoriously difficult problem in hydrology. The DTF results also suggest that, if the constant parameter model is retained and used in forecasting applications, then it would be worthwhile making the model adaptive by incorporating real-time recursive estimation of the model (see section 10.5.1 of Chapter 10). Or more simply, since the DTF estimated parameter variations only affect the TF numerator parameters, by adding a single adaptive gain element (see e.g. Lees et al. (1993); Young (2002a); Romanowicz et al. (2006); Young (2010c)); indeed it is TVP results such as those above that led to such innovations in flood forecasting.

6.11 Multivariable Transfer Function Models

In general, dynamic systems are multivariable, with the possibility of several inputs and outputs. For example, the *Multi-Input, Multi-Output* (MIMO) version of equation (6.5), with m inputs and p outputs, is as follows

$$\mathbf{y}(k) = \mathbf{C}^T[\mathbf{I}-\mathbf{A}z^{-1}]^{-1}\mathbf{B}z^{-1}\mathbf{u}(k) + \mathbf{C}^T[\mathbf{I}-\mathbf{A}z^{-1}]^{-1}\mathbf{G}z^{-1}\mathbf{e}(k) + \mathbf{B}_I\mathbf{u}(k) \quad (6.81)$$

where $\mathbf{y}(k) = [y_1(k)\ y_2(k)\ \ldots\ y_p(k)]^T$ and $\mathbf{u}(k) = [u_1(k)\ u_2(k)\ \ldots\ u_m(k)]^T$. Upon expansion, this leads to the MIMO *Transfer Function Matrix* (TFM) description of the stochastic system, which has the general form (see also section B.3.3 of Appendix B):

$$\begin{aligned}\mathbf{y}(k) &= \left[\frac{1}{A(z^{-1})}\mathbf{C}\,\mathrm{Adj}[\mathbf{I}-\mathbf{A}z^{-1}]\,\mathbf{B}z^{-1} + \mathbf{B}_I\right]\mathbf{u}(k) \\ &\quad + \frac{1}{A(z^{-1})}\mathbf{C}\,\mathrm{Adj}[\mathbf{I}-\mathbf{A}z^{-1}]\mathbf{D}z^{-1}\boldsymbol{\eta}(k) + \mathbf{e}(k) \qquad (6.82)\\ &= \mathbf{L}(z^{-1})\mathbf{u}(k) + \mathbf{M}(z^{-1})\boldsymbol{\eta}(k) + \mathbf{e}(k)\end{aligned}$$

where $\mathbf{L}(z^{-1})$ and $\mathbf{M}(z^{-1})$ are, respectively, $p \times m$ and $p \times q$ *Transfer Function Matrices* (TFMs) in which each elemental TF in the matrices has a common denominator $A(z^{-1})$, where

$$A(z^{-1}) = a_1 z^{-1} + a_2 z^{-2} + \ldots + a_n z^{-n}$$

Equation (6.82) can be simplified to the form:

$$\mathbf{y}(k) = \mathbf{L}(z^{-1})\mathbf{u}(k) + \boldsymbol{\xi}(k) \quad (6.83)$$

where $\boldsymbol{\xi}(k) = \mathbf{M}(z^{-1})\boldsymbol{\eta}(k) + \mathbf{e}(k)$ is a $p \times 1$ stochastic input vector and

$$\mathbf{L}(z^{-1}) = \begin{bmatrix} L_{11}(z^{-1}) & L_{12}(z^{-1}) & \cdots & L_{1m}(z^{-1}) \\ L_{21}(z^{-1}) & L_{22}(z^{-1}) & \cdots & L_{2m}(z^{-1}) \\ \cdots & \cdots & \cdots & \cdots \\ \cdots & \cdots & \cdots & \cdots \\ L_{p1}(z^{-1}) & L_{p2}(z^{-1}) & \cdots & L_{pm}(z^{-1}) \end{bmatrix}$$

with

$$L_{ij} = \frac{B_{ij}(z^{-1})}{A(z^{-1})},\ i = 1,2,\ \ldots,\ p;\ j = 1,2,\ \ldots,\ m$$

The TFM model (6.83) clearly represents a formidable identification and estimation problem and, although recursive algorithms have been developed for this (e.g. Jakeman and Young, 1979), they are not easy to use in practice. An alternative, much simpler and practical, approach is to consider the set of *p Multi-Input, Single Output* (MISO) models formed by considering each row of (6.83) in turn: e.g. in the case of the rth row,

$$y_r(k) = \frac{B_{r1}(z^{-1})}{A(z^{-1})}u_1(k) + \ldots + \frac{B_{rm}(z^{-1})}{A(z^{-1})}u_m(k) + \xi_r(k)$$

$$\text{i.e.} \quad y_r(k) = \sum_{j=1}^{m} \frac{b_{0j}^r + \ldots + b_{m_j j}^r z^{-m_j}}{1 + a_1 z^{-1} + \ldots + a_n z^{-n}} u_j(k) + \xi_r(k) \tag{6.84}$$

for $r = 1,2,\ \ldots,\ p$. Clearly, the first equation here can also be written as,

$$A(z^{-1})y_r(k) = B_{r1}(z^{-1})u_1(k) + B_{r2}(z^{-1})u_2(k) + \ldots + B_{rm}(z^{-1})u_m(k) + \eta_r(k)$$

where $\eta_r(k) = A(z^{-1})\xi_r(k)$. It is now clear that, because the $A(z^{-1})$ polynomial is common to all the elemental TFs, this is an obvious extension of the SISO model and can be handled in exactly the same manner, requiring only the insertion of the additional input terms. For example, in the case of the ARX model (6.29) where pure time delays are assumed to affect the input, the data and parameter vectors in the MISO case are as follows:

$$\begin{aligned}\boldsymbol{\phi}^T(k) = [&-y(k-1)\ \cdots\ -y(k-n)\ u_1(k-\delta_1)\ \cdots\ u_1(k-\delta_1-m_1)\\ &u_2(k-\delta_2)\cdots\ u_2(k-\delta_2-m_2)\ \cdots\ u_m(k-\delta_m)\ \cdots\ u_m(k-\delta_m-m_m)]\\ \boldsymbol{\rho} = [&a_1 \cdots a_n\ b_{10}\ b_{11} \cdots b_{1m_1}\ b_{20}\ b_{21} \cdots b_{2m_2} \cdots b_{m0}\ b_{m1} \cdots b_{mm_m}]^T\end{aligned} \tag{6.85}$$

where it will be noted that each input $u_1(k),\ u_2(k),\ \ldots,\ u_m(k)$, can have a different order polynomial $(m_1,\ m_2,\ \ldots,\ m_m)$ and time delay $(\delta_1,\ \delta_2,\ \ldots,\delta_m)$ associated with it. Note also that the r superscript has been dropped for convenience in these definitions. This implies that there is only one output in the multivariable system (6.83) (see the example 7.3 of the next chapter 7). The MISO ARX model is then seen to be the dynamic TF equivalent of the multivariate regression model, with the lagged outputs $y(k-1),\ y(k-2),\ \ldots,\ y(k-n)$ introducing the main dynamic behaviour and the b_{ij} parameters being analogous to the standard regression coefficients multiplying the input terms.

The MIMO version of the Box-Jenkins model (6.18) is very complicated if the noise processes for each output are mutually correlated. However, if we assume they are independent, the MISO model for the rth output $y_r(k)$ takes the form:

$$x_r(k) = \sum_{j=1}^{m} \frac{b^r_{0j} + \ldots + b^r_{m_j j} z^{-m_j}}{1 + a^r_{1j} z^{-1} + \ldots + a^r_{n_j j} z^{-n_j}} u_j(k - \delta^r_j)$$

$$\xi_r(k) = \frac{1 + d^r_1 z^{-1} + \ldots + d^r_{q_r} z^{-q_r}}{1 + c^r_1 z^{-1} + \ldots + c^r_{p_r} z^{-p_r}} e_r(k) \tag{6.86}$$

$$y_r(k) = x_r(k) + \xi_r(k)$$

The separate recursive estimation of the parameters in each such MISO equation, or its common denominator equivalent, can be accomplished using the SIV algorithm or the optimal RIV estimation algorithm described in the next Chapter 7.

Finally, following from the nomenclature used to define the model structure in the SISO case, the general structure of the MISO model will be denoted by $[n_1 \ldots n_m \; m_1 \ldots m_m \; \delta_1 \ldots \delta_m \; p \; q]$ or, simply $[n \; m_1 \ldots m_m \; \delta_1 \ldots \delta_m \; p \; q]$, in the constrained common denominator case.

6.12 Exercises

1. Evaluate the ARMAX model (6.7) from the state space model (6.4) by reference to equation (6.5) if:

$$\mathbf{A} = \begin{bmatrix} -a_1 & 1 \\ -a_2 & 0 \end{bmatrix}; \quad \mathbf{b} = \begin{bmatrix} b_1 \\ b_2 \end{bmatrix}; \quad \mathbf{c}^T = \begin{bmatrix} 1 & 0 \end{bmatrix}; \quad \mathbf{g} = \begin{bmatrix} g_1 \\ g_2 \end{bmatrix}$$

2. Write a Matlab m-file script to simulate the simple first order transfer function model (6.44) *et seq* and generate data such as that shown in Figure 6.2.
3. Write a Matlab m-file script to implement the recursive SIV algorithm and apply this to the data generated in exercise 2. Hint: use the CAPTAIN del routine to generated delayed vectors and the Matlab filter routine to simulate the auxiliary model (6.66).
4. Consider how the script produced in excercise 2 would be extended to handle higher order models and implement this for a [2 2 δ] model, with user-specified δ.
5. Apply the m-file script in 4. to some suitable data of your own, or suitable data available in the CAPTAIN command-line demos (see the instructions in Appendix G for running demonstration examples in CAPTAIN). Subsequently, for comparison, analyze the same data using the optimal rivbj algorithm in CAPTAIN, *after you have read the next chapter* 7.

6.13 Summary

The purpose of this chapter has been to introduce the reader to the problem of time series model estimation; in particular, the estimation of parameters in simple transfer function representations of linear stochastic, dynamic systems. It has shown how the ubiquitous RLS algorithm fails to yield satisfactory estimates of the parameters in such systems, except in the special cases of the ARX and FIR models; and it has explored various ways of circumventing these limitations, most importantly by the use of instrumental variable methods. Only the sub-optimal, standard (SIV) method has been considered in this chapter but the next two chapters 7 and 8 consider in detail how this simple recursive-iterative SIV algorithm can be converted into the more complex, optimal RIV form by the introduction of iteratively adaptive prefilters.

Chapter 7
Optimal Identification and Estimation of Discrete-Time Transfer Function Models

The previous chapter 6 has set the scene for the identification and estimation of transfer function models of stochastic, dynamic systems by revealing the limitations of the linear least squares estimation in this context and showing how instrumental variable modifications to the recursive least squares estimation equations can overcome these limitations in a sub-optimal manner. The reason for these limitations is, of course, that general transfer function models of the kind considered in chapter 6 represent inherently nonlinear and stochastic parameter estimation problems. For reasons that will become obvious as we proceed, the present chapter considers the most interesting of these models, the Box-Jenkins (BJ) model, and shows how the standard, recursive-iterative instrumental variable (SIV) algorithm, as developed and evaluated in chapter 6, can be modified to estimate this model in a statistically optimal manner.

For convenience, let us first write down again the details of the Box-Jenkins model:

$$y(k) = \frac{B(z^{-1})}{A(z^{-1})}u(k-\delta) + \frac{D(z^{-1})}{C(z^{-1})}e(k) \quad e(k) = \mathcal{N}(0,\sigma^2) \tag{7.1}$$

or, in decomposed form:

$$\text{System model}: \quad x(k) = \frac{B(z^{-1})}{A(z^{-1})}u(k-\delta) \tag{7.2a}$$

$$\text{ARMA noise model}: \quad \xi(k) = \frac{D(z^{-1})}{C(z^{-1})}e(k) \quad e(k) = \mathcal{N}(0,\sigma^2) \tag{7.2b}$$

$$\text{Observation equation}: \quad y(k) = x(k) + \xi(k) \tag{7.2c}$$

where, in general, the polynomials $A(z^{-1})$, $B(z^{-1})$, $C(z^{-1})$ and $D(z^{-1})$ are defined as follows:

$$\begin{aligned} A(z^{-1}) &= 1 + a_1 z^{-1} + a_2 z^{-2} + \ldots + a_n z^{-n} \\ B(z^{-1}) &= b_0 + b_1 z^{-1} + b_2 z^{-2} + \ldots + b_m z^{-m} \\ C(z^{-1}) &= 1 + c_1 z^{-1} + c_2 z^{-2} + \ldots + c_p z^{-p} \\ D(z^{-1}) &= 1 + d_1 z^{-1} + d_2 z^{-2} + \ldots + d_q z^{-q} \end{aligned} \tag{7.3}$$

This can be considered as a Stochastic Transfer Function (STF) model, where the addition of 'stochastic' into the TF name is in order to emphasize that the model contains a stochastic element, in the form of the ARMA noise model. This is important because it is the presence of the ARMA noise model that constitutes the primary reason for the nonlinearity in estimation terms and, at the same time, provides a mechanism for considering statistically optimal solutions to this nonlinear estimation problem. However, the reader new to this subject might ask why one adds this stochastic element, if it causes so much added complexity and, in any case, why it is assumed to take the ARMA model form.

The reason for the addition of the stochastic model for $\xi(k)$ is simply to allow for the kind of real problems that are encountered in practice: the noise is that part of the data that cannot be accounted for by the effects of the input signal: it represents the total uncertainty in the model, including not only the noise on the measured data but also the uncertainty *in our knowledge of the system*. In other words, it is a 'catch-all' for all that we cannot explain about the system from the data. On this basis, it is not surprising that the noise component of $y(k)$ will not be a simple additive 'white' noise to the model when written in the regression equation form, as in the ARX model, but 'coloured': i.e. $\xi(k)$ will most often exhibit serial correlation, with a frequency spectrum that is not flat for all frequencies but is accentuated for some frequency bands and attenuated for others. And, as we have seen in the previous chapter 6, this can deleteriously affect the estimation process in various ways.

In order to account for this colour in the noise $\xi(k)$, it is necessary to model it in some manner. The reason for the selecting the ARMA model (7.2b) is convenience and analytical tractability. In particular, it allows for the presence of coloured noise with rational spectral density of the kind that is commonly encountered in practice. Moreover, it is in a linear TF model form that provides a rather obvious addition to the TF component (7.2a) which represents the important causal, input-output dynamics of the system. In this manner, it produces a rather convenient, unified description of the stochastic dynamic system in which the system dynamics and the noise dynamics are represented as similar but separate mathematical objects: one deterministic, excited by the measured input $u(k)$; the other stochastic, with an unobservable and unpredictable white noise input $e(k)$. Of course, this is not the only description that could be used but it is one that has stood the test of time since it was introduced by Box and Jenkins forty years ago: it is, in other words, an analytically elegant and practical useful model form.

Models such as (7.1) can only be considered in a linear estimation context by formulating them in *Pseudo-Linear Regression* (PLR) terms, as in equation (6.25) of the previous chapter 6. But, as its name prescribes, the PLR model is not truly linear in the parameters because it contains unobservable variables and so it cannot be estimated directly using linear least squares. The major attraction of the PLR formulation, in relation to the other nonlinear formulations mentioned in chapter 6 is, as we shall see in the present chapter, that it allows for the retention of a 'linear-like' solution to the parameter estimation problem, albeit one that requires an iterative or 'relaxation' type of solution. In recursive estimation terms, this is very valuable because it permits the immediate application of recursive algorithms, such as recursive least squares or recursive instrumental variables, that yield exactly the same estimates as those of the associated *en bloc* algorithms. Other methods based on nonlinear least squares optimization, such as the gradient optimization methods used in standard maximum likelihood estimation of TF models, are not so convenient in this regard. While recursive algorithms that are linked with such methods, such as the *Recursive Prediction Error Minimization* (RPEM) algorithm (see Ljung, 1999, and chapter 10), have been developed, they do not necessarily yield exactly the same estimates as their *en bloc* progenitors (see later, chapter 10).

This PLR formulation has one other, important advantage. It allows for the use of an instrumental variable (IV) solution to the estimation equations. This is important because the assumptions built into a model such as (7.1) may well not be satisfied in practical situations, even if the system has linear dynamics, as in the BJ model case. For example, the noise $\xi(k)$ may not have a normal amplitude distribution; the variance of the white noise may change over time (heteroscedastic noise: see the discussion of this topic in chapter 5); or, more fundamentally, it may be 'odd' noise that does not conform to a rational spectral density assumption at all. In this situation, the practitioner will not be impressed by the elegance of the mathematical description alone and will demand a solution, such as that provided by IV methods, which is robust to the violation of the theoretical assumptions.

This chapter and the next chapter 8 outline the main aspects of the statistically optimal *Refined Instrumental Variable* approach to the identification and estimation of both discrete-time (RIV) and continuous-time (RIVC) STF models (Young, 1976b; Young and Jakeman, 1979b; Jakeman and Young, 1979b; Young and Jakeman, 1980a; Young, 1984). They also demonstrate how, provided the statistical assumptions are satisfied, this approach yields parameter estimates with optimal statistical properties. As pointed out in chapter 6, these RIV/RIVC algorithms together appear to constitute the only unified time domain family of recursive algorithms that provide such statistically optimal solutions for both discrete and continuous-time system models. Moreover, as we shall see, the instrumental variable formulation means that the RIV and RIVC algorithms are able to exploit the properties of the *Instrumental Product Matrix* (IPM) for model order identification Wellstead (1978); Young et al. (1980).

The present chapter also points out how the discrete-time Box-Jenkins model form exploited by the RIV algorithm has certain advantages over the ARMAX alternative (although the RIV algorithm can be used to estimate the ARMAX model, if this is required). In particular, a theorem due to Pierce (1972) is presented that shows how the *Maximum Likelihood* (ML) estimates of the TF system model parameters are asymptotically independent of the ARMA noise model parameters. The chapter shows how this property proves very useful for both the theoretical justification and the practical implementation of the RIV and RIVC algorithms.

7.1 Refined Instrumental Variable Estimation

As we shall see, it is convenient to consider the unknown parameters in the BJ model as the two parameter vectors $\boldsymbol{\rho}$ and $\boldsymbol{\eta}$ that are defined as follows in terms of the parameters in the TF system polynomials $\{A(z^{-1})\ B(z^{-1})\}$ and the ARMA noise model polynomials $\{C(z^{-1})\ D(z^{-1})\}$:

$$\begin{aligned} \boldsymbol{\rho} &= [a_1 \cdots a_n\ b_0 \cdots\ b_m]^T \\ \boldsymbol{\eta} &= [1\ c_1 \cdots\ c_p\ d_1 \cdots d_q\,]^T \end{aligned} \tag{7.4}$$

where the need for the unity entry in $\boldsymbol{\eta}$ will become clear as we proceed. In these terms, the estimation problem posed by the BJ model (7.1) is to estimate $\boldsymbol{\rho}$ and $\boldsymbol{\eta}$ based on the uniformly sampled input-output data $Z^N = \{u(k); y(k)\}_{k=1}^N$ at a sampling interval of Δt time units, where N is the number of samples (the observation interval). The recursive RIV solution to this problem is based on forming the estimation equations for the system and noise model parameters into special, PLR relationships, as mentioned earlier. These estimation equations then provide the basis for an iterative (relaxation) algorithm that estimates the system and noise model parameters in separate but linked iterative sub-algorithms, where the linear estimation problems formed in this manner at each iteration are solved using recursive or *en bloc* IV algorithms.

I originally devised the RIV algorithm (Young, 1976b) from the manipulation of the ML optimization expressions related to the BJ model (7.1), as described in Appendix F. The aim of the following sub-sections is to provide the basic motivation behind this formulation of the algorithm in a simpler and more transparent form than that presented in Appendix F. They also demonstrate how such a formulation allows for the inclusion of recursive parameter estimation equations, something which is not possible with alternative methods of ML estimation that utilize gradient-based numerical optimization methods, such as the PEM algorithm. Following this motivational discussion and analysis, the recursive and iterative forms of the full RIV and SRIV algorithms are presented in section 7.2.

7.1.1 Recursive-iterative instrumental variable estimation

A practical estimation methodology should be robust to violation of the theoretical assumptions on which the associated estimation procedure is based. And, as we have seen previously in chapter 3 and 6, a major technique for engendering such robustness is the exploitation of IV estimation. But the question raised in chapter 3 concerning the optimality of an IV solution remains unanswered. Before proceeding to develop the optimal RIV algorithm, therefore, let us consider briefly the more general problem of optimal IV estimation. Within the context of linear and pseudo-linear models, a typical estimation model has the following *general* form:

$$\text{General Estimation Equation}: \; v(k) = \boldsymbol{\varphi}^T(k)\boldsymbol{\vartheta} + e(k) \tag{7.5}$$

where $\boldsymbol{\vartheta}$ is the vector of all the parameters in both the system and noise TF models. One rather obvious thought is that the IV estimate of the parameter vector $\boldsymbol{\vartheta}$ in this model optimizes the sum of the squares of the residuals associated with the IV normal equations. In more formal terms, it is easy to see that this is indeed the case if one considers the IV normal equations in the case of the estimation model (7.5) with an IV vector $\hat{\boldsymbol{\varphi}}$: i.e.

$$\left[\sum_{k=1}^{N} \hat{\boldsymbol{\varphi}}(k)\boldsymbol{\varphi}^T(k)\right]\boldsymbol{\vartheta} - \left[\sum_{k=1}^{N} \hat{\boldsymbol{\varphi}}(k)v(k)\right] = 0 \tag{7.6}$$

then the optimization of the quadratic form:

$$\hat{\boldsymbol{\vartheta}} = \arg\min_{\boldsymbol{\vartheta}} \left\| \left[\sum_{k=1}^{N} \hat{\boldsymbol{\varphi}}(k)\boldsymbol{\varphi}^T(k)\right]\boldsymbol{\vartheta} - \left[\sum_{k=1}^{N} \hat{\boldsymbol{\varphi}}(k)v(k)\right] \right\|^2 \tag{7.7}$$

with respect to $\boldsymbol{\vartheta}$ does, in fact, yield the IV solution:

$$\hat{\boldsymbol{\vartheta}}_N = \left[\sum_{k=1}^{N} \hat{\boldsymbol{\varphi}}(k)\boldsymbol{\varphi}^T(k)\right]^{-1} \sum_{k=1}^{N} \hat{\boldsymbol{\varphi}}(k)v(k) \tag{7.8}$$

In chapter 8 of Söderström and Stoica (1989), the authors develop this same argument in a more general *Extended Instrumental Variable* (EIV) context that includes the use of both extended IV vectors and prefilters. Such prefilters depend of the form of STF model being considered and, in the case of the BJ model, they are the same as those proposed in Young (1976b); Young and Jakeman (1980b). These were considered in the first edition of the present book and their importance is explained in subsequent sections of this present chapter.

Returning to the specific context of the BJ model, $\boldsymbol{\vartheta}(k)$, $\boldsymbol{\varphi}(k)$ and $v(k)$ in the above general IV equations are replaced by $\boldsymbol{\theta}(k)$, $\boldsymbol{\phi}(k)$ and $y(k)$, where,

$$\boldsymbol{\theta} = \begin{pmatrix} \boldsymbol{\rho} \\ \boldsymbol{\eta} \end{pmatrix} \tag{7.9}$$

and $\boldsymbol{\phi}(k)$ is defined below in equation (7.15a). As in the standard IV algorithm described in section 6.8 of chapter 6, the optimal RIV algorithm is iterative, with either recursive or *en bloc* estimation of the model parameters at each iteration. In the sub-sections that follow, therefore, let us consider the general j^{th} iteration of such a procedure, where we have access to the estimated model parameter vector

$$\hat{\boldsymbol{\theta}}^{j-1} = \begin{pmatrix} \hat{\boldsymbol{\rho}}^{j-1} \\ \hat{\boldsymbol{\eta}}^{j-1} \end{pmatrix} \tag{7.10}$$

obtained previously at iteration $j-1$, where, $\hat{\boldsymbol{\rho}}$ and $\hat{\boldsymbol{\eta}}$ are, respectively, the parameter vectors associated with the system and noise models in the BJ model, as defined in (7.4). These are estimated in a *separate but linked* manner and the next two subsections consider the formulation of their estimation equations in the form (7.5) and demonstrate how they are updated at each iteration.

7.1.2 The system TF estimation model and RIV estimation

In the present situation, the Gaussian normality assumptions on $e(k)$ suggest that a suitable error function $\varepsilon(k)$ for estimating the parameters in the system model is defined as follows:

$$\varepsilon(k) = \frac{C(z^{-1})}{D(z^{-1})}\left[y(k) - \frac{B(z^{-1})}{A(z^{-1})}u(k-\delta)\right]$$

which can be written as,

$$\varepsilon(k) = \frac{C(z^{-1})}{D(z^{-1})A(z^{-1})}\left[A(z^{-1})y(k) - B(z^{-1})u(k-\delta)\right] \tag{7.11}$$

Minimization of a least squares criterion function in $\varepsilon(k)$, measured at the sampling instants, provides the basis for optimal stochastic estimation. However, since the polynomial operators commute in this linear case, the prefilter:

$$f_1(z^{-1}) \triangleq \frac{C(z^{-1})}{D(z^{-1})A(z^{-1})} \tag{7.12}$$

can be taken inside the square brackets to yield:

$$\varepsilon(k) = A(z^{-1})y_{f_1}(k) - B(z^{-1})u_{f_1}(k-\delta) \tag{7.13}$$

or, on expansion,

$$\begin{aligned} \varepsilon(k) = {} & y_{f_1}(k) + a_1 y_{f_1}(k-1) + \ldots + a_n y_{f_1}(k-n) \\ & - b_0 u_{f_1}(k-\delta) - \ldots - b_m u_{f_1}(k-\delta-m) \end{aligned} \tag{7.14}$$

where the subscripts f_1 denote that the associated variables have been prefiltered by $f_1(z^{-1})$. As a result, it is now possible to formulate the following *estimation equation* in the pseudo-linear form (cf. (7.5)):

$$\text{System model estimation equation: } \; y_{f_1}(k) = \boldsymbol{\phi}^T(k)\boldsymbol{\rho} + e(k) \tag{7.15}$$

where,

$$\boldsymbol{\phi}^T(k) = [-y_{f_1}(k-1) \cdots -y_{f_1}(k-n) \;\; -u_{f_1}(k-\delta) \cdots \;\; -u_{f_1}(k-\delta-m)] \quad (7.15a)$$

$$\boldsymbol{\rho} = [a_1 \cdots a_n \; b_0 \cdots \; b_m]^T \quad (7.15b)$$

This estimation equation is simply a way of presenting the BJ model (7.1) in a pseudo-linear form, so that it can provide a suitable basis for the recursive or *en bloc* estimation of the TF system parameter vector $\boldsymbol{\rho}$. Thus, provided we assume that $A(z^{-1})$, $C(z^{-1})$ and $D(z^{-1})$ are known *a priori*, it can be used for the definition of a likelihood function and ML optimization, so ensuring statistical efficiency with the parametric error covariance matrix achieving the Cramer-Rao lower bound. Note that the prefilter $f_1(z^{-1})$ makes obvious physical sense when considered as the serial connection of the filter $1/A(z^{-1})$ and the 'inverse noise model' filter $C(z^{-1})/D(z^{-1})$: the former filter attenuates all signals outside the passband of system, including high frequency noise; while the latter pre-whitens the noise that remains within the pass-band of the system. The averaging operation that is inherent in the estimation procedure then 'averages out' the effects of this pre-whitened noise on the estimated parameters.

There are two problems with this formulation. The most obvious one is, of course, that the $A(z^{-1})$, $C(z^{-1})$ and $D(z^{-1})$ polynomials are *not* known *a priori*: they are, of course, the object of the estimation procedure! The less obvious one is that, in practical applications, the theoretical assumptions on which it is based may not be satisfied and, in particular, the noise may not have rational spectral density and cannot, therefore, be described very well by an ARMA model. As we shall see later, both of these problems are solved by the RIV algorithm. This employs an IV optimization procedure that iteratively adjusts the parameters that characterize the unknown polynomials in the BJ model (7.1), applying IV estimation at each iteration, until they converge on an optimal solution.

Given the pseudo-linear estimation equation (7.15), the IV estimation normal equations are obtained by reference (7.6) as:

$$\left[\sum_{k=1}^{N} \hat{\boldsymbol{\phi}}(k)\boldsymbol{\phi}^T(k)\right]\boldsymbol{\rho} - \left[\sum_{k=1}^{N} \hat{\boldsymbol{\phi}}(k)y_{f_1}(k)\right] = 0 \quad (7.16)$$

and the *en bloc* solution is then:

$$\hat{\boldsymbol{\rho}}_N = \left[\sum_{k=1}^{N} \hat{\boldsymbol{\phi}}(k)\boldsymbol{\phi}^T(k)\right]^{-1} \sum_{k=1}^{N} \hat{\boldsymbol{\phi}}(k)y_{f_1}(k) \quad (7.17)$$

The recursive estimation of the system model parameters follows from a *stochastic version* of the recursive SIV algorithm in chapter 6 and takes the form:

$$\hat{\boldsymbol{\rho}}(k) = \hat{\boldsymbol{\rho}}(k-1) + \mathbf{g}(k)[y_{f_1}(k) - \boldsymbol{\phi}^T(k)\hat{\boldsymbol{\rho}}(k-1)] \quad \text{(RIV-1)}$$
$$\mathbf{g}(k) = \hat{\mathbf{P}}_\rho(k-1)\hat{\boldsymbol{\phi}}(k)[\sigma^2 + \boldsymbol{\phi}^T(k)\hat{\mathbf{P}}_\rho(k-1)\hat{\boldsymbol{\phi}}(k)]^{-1} \quad \text{(RIV-2)}$$
$$\hat{\mathbf{P}}_\rho(k) = \hat{\mathbf{P}}_\rho(k-1) - \mathbf{g}(k)\boldsymbol{\phi}^T(k)\hat{\mathbf{P}}_\rho(k-1) \quad \text{(RIV-3)}$$

where, because of the stochastic nature of pseudo-linear model formulation, the σ^2 term can be introduced in equation (RIV-2), as in the RARX algorithm of section 6.4 of chapter 6, and then $\hat{\mathbf{P}}_\rho(k)$ is recognized as the estimated error covariance matrix. In these recursive equations, $\hat{\boldsymbol{\phi}}(k)$ is the system model IV vector defined as follows,

$$\hat{\boldsymbol{\phi}}(k) = [-\hat{x}_{f_1}(k-1) \cdots -\hat{x}_{f_1}(k-n) \; -u_{f_1}(k-\delta) \cdots \; -u_{f_1}(k-\delta-m)]^T \quad (7.18)$$

in which $\hat{x}_{f1}(k)$ is the prefiltered output of the system 'auxiliary model',

$$\hat{x}(k) = \frac{\hat{B}(z^{-1})}{\hat{A}(z^{-1})} u(k-\delta) \quad (7.19)$$

where $\hat{A}(z^{-1})$ and $\hat{B}(z^{-1})$ are the iteratively updated estimates of the system TF model polynomials. In this manner, $\hat{x}(k)$ provides an iteratively updated estimate of the noise-free output $x(k)$ (see later) and so qualifies as an instrumental variable with optimal connotations. Note that, in the ideal but entirely hypothetical situation where the noise-free output $x(k)$ is available for measurement, the IV vector in (7.18) would be replaced by,

$$\mathring{\boldsymbol{\phi}}(k) = [-x_{f_1}(k-1) \cdots -x_{f_1}(k-n) \; u_{f_1}(k-\delta) \cdots \; u_{f_1}(k-\delta-m)]^T \quad (7.20)$$

This vector and $\hat{\mathbf{P}}_\rho(k)$ are referred to later in section 7.2.3; while the recursive RIV equations RIV-1 to RIV-3, or their *en bloc* equivalent, form the basis for the system model parameter estimation steps of the iterative RIV algorithm described later in the section 7.2.

7.1.3 The ARMA noise estimation model and IVARMA estimation

In the simplest situation where the noise $\xi(k)$ in (7.1) is purely white, so that $C(z^{-1}) = D(z^{-1}) = 1$, the prefilter $f_1(z^{-1})$ in (7.12) reduces to $1/A(z^{-1})$ and no noise model estimation is required. Estimation is particularly simple in this case and the iterative optimization approach has been termed the *Simplified Refined Instrumental Variable* (SRIV) algorithm (Young, 1985a). In the more general situation where the noise can be represented by an ARMA model, the full RIV algorithm is naturally more complex and involves the separate but linked estimation of the ARMA noise model parameters.

Any method of recursive ARMA model estimation can be used; or, alternatively, the ARMA model can be approximated by a high order AR model that can be estimated straightforwardly by recursive, or *en bloc*, linear least squares estimation. Indeed, this is the method used until recently for the implementation of the RIV algorithm in the CAPTAIN Toolbox. Here, however, we will consider another approach (Young, 2006b, 2008b) that has the advantage of being able to generate recursive estimates of the full ARMA noise model parameters and has has been incorporated into CAPTAIN as an improvement on an earlier, related algorithm (Young, 1985b).

This *Instrumental Variable* ARMA (IVARMA) estimation algorithm is motivated in a similar manner to that used above for TF system model parameter estimation. In particular, a suitable error function $\varepsilon_n(k)$ in this case is defined as follows:

$$\varepsilon_n(k) = \frac{C(z^{-1})}{D(z^{-1})}\xi(k) - e(k) \tag{7.21}$$

Now, by introducing the prefilter $f_2(z^{-1})$ defined as,

$$f_2(z^{-1}) \triangleq \frac{1}{D(z^{-1})} \tag{7.22}$$

the error function can be written in the pseudo-linear form,

$$\varepsilon_n(k) = C(z^{-1})\xi_{f_2}(k) \; - \; D(z^{-1})e_{f_2}(k) \tag{7.23}$$

where the subscript f_2 indicates that the associated variables have been prefiltered by the filter $f_2(z^{-1})$ in (7.22). Consequently, in an ideal but entirely hypothetical situation, the prefiltered noise variables can be related by the equation,

$$e_{f_2}(k) = \mathring{\boldsymbol{\psi}}^T(k)\boldsymbol{\eta} \tag{7.24}$$

where,

$$\mathring{\boldsymbol{\psi}}^T(k) = [\xi_{f_2}(k) \cdots \xi_{f_2}(k-p) \; -e_{f_2}(k-1) \cdots -e_{f_2}(k-q)] \tag{7.24a}$$

$$\boldsymbol{\eta} = [1 \; c_1 \cdots c_p \; d_1 \cdots d_q]^T \tag{7.24b}$$

It may seem strange here that $e_{f_2}(k)$ in (7.24) is selected as the dependent variable. However, this is required so that the estimation problem can be formulated in a manner that allows for optimal IV estimation, as shown below.

Similar to the situation encountered in the TF system model case, there are clear problems with the utilization of the equation (7.24) for estimation purposes: first, we do not have access to $e(k)$ nor $\xi(k)$; and second, the $D(z^{-1})$ polynomial is not known *a priori*. But again, there are solutions to these problems. First, if it is assumed that $A(z^{-1})$ and $B(z^{-1})$ are available, then the noise-free output $x(k)$ can be obtained from (7.2a) and $\xi(k)$ can be obtained by reference to equation (7.2c), i.e.,

$$\xi(k) = y(k) - x(k) \tag{7.25}$$

Accessing $e(k)$ is more difficult but it is well known that a high order AR model of $\xi(k)$ yields residuals that provide a good estimate of $e(k)$ (Hannan and Rissanen, 1982; Young, 1985b). This reasoning is the basis of the IVARMA algorithm.

First, an estimate $\hat{\xi}(k)$ of $\xi(k)$ is obtained using the latest, iteratively updated estimates of the $A(z^{-1})$ and $B(z^{-1})$ polynomials obtained from the above RIV algorithm. Then, high order AR estimation is used to obtain a high order AR model for $\hat{\xi}(k)$ and to generate an estimate of $e(k)$ from the high order AR model residuals, denoted here by $\tilde{e}(k)$. These estimates have to be prefiltered by $f_2(z^{-1})$, which is updated in the iterative IVARMA algorithm (see later). The prefiltered variables are then used to construct the estimate of the vector $\mathring{\psi}(k)$ given by,

$$\boldsymbol{\psi}(k) = [\hat{\xi}_{f_2}(k) \cdots \hat{\xi}_{f_2}(k-p) \ -\tilde{e}_{f_2}(k-1) \cdots -\tilde{e}_{f_2}(k-q)]^T \tag{7.26}$$

This can now replace the unobservable vector $\mathring{\psi}(k)$ and be used to formulate the following pseudo-linear noise model estimation equation (cf. (7.24) and (7.5)):

$$\text{Noise Model Estimation Equation}: \ \tilde{e}_{f_2}(k) = \boldsymbol{\psi}^T(k)\boldsymbol{\eta} + \varepsilon(k) \tag{7.27}$$

where $\varepsilon(k)$ is the approximation error[1] arising from the need to replace the unobserved variables in $\mathring{\psi}(k)$ by their estimates.

As in the system model case, given the pseudo-linear estimation equation (7.27), the recursive estimation of the noise model parameters again follows from the recursive IV equations and takes the form:

$$\begin{aligned}
\hat{\boldsymbol{\eta}}(k) &= \hat{\boldsymbol{\eta}}(k-1) + \mathbf{g}_n(k)[\tilde{e}_{f_2}(k) - \boldsymbol{\psi}^T(k)\hat{\boldsymbol{\eta}}(k-1)] && \text{(IVARMA-1)}\\
\mathbf{g}_n(k) &= \hat{\mathbf{P}}_\eta(k-1)\hat{\boldsymbol{\psi}}(k)[\sigma^2 + \boldsymbol{\psi}^T(k)\hat{\mathbf{P}}_\eta(k-1)\hat{\boldsymbol{\psi}}(k)]^{-1} && \text{(IVARMA-2)}\\
\hat{\mathbf{P}}_\eta(k) &= \hat{\mathbf{P}}_\eta(k-1) - \mathbf{g}_n(k)\boldsymbol{\psi}^T(k)\hat{\mathbf{P}}_\eta(k-1) && \text{(IVARMA-3)}
\end{aligned}$$

where $\hat{\boldsymbol{\psi}}(k)$ is the noise model IV vector defined as follows,

$$\hat{\boldsymbol{\psi}}(k) = [\hat{\xi}_{f_2}(k) \cdots \hat{\xi}_{f_2}(k-p) \ -\hat{e}_{f_2}(k-1) \cdots -\hat{e}_{f_2}(k-q)]^T \tag{7.28}$$

in which $\hat{e}_{f_2}(k)$ is the prefiltered output of the noise auxiliary model:

$$\hat{e}(k) = \frac{\hat{C}(z^{-1})}{\hat{D}(z^{-1})}\hat{\xi}(k). \tag{7.29}$$

[1] Note that $\varepsilon(k)$ is not the same as $\varepsilon_n(k)$

where $\hat{C}(z^{-1})$ and $\hat{D}(z^{-1})$ are the iteratively updated estimates of the noise model polynomials $C(z^{-1})$ and $D(z^{-1})$. Here, it is important to note the difference between $\hat{e}(k)$, which is the source of the noise model instrumental variables, and the estimate $\tilde{e}(k)$ of the white noise, which is obtained *via* high order AR estimation. Finally, $\hat{\mathbf{P}}_\eta(k)$ is the estimated error covariance matrix discussed later in section 7.2.3. The recursive equations IVARMA-1 to IVARMA-3, or their *en bloc* equivalent, form the basis for the recursive noise model parameter estimation steps of the iterative RIV algorithm described in the next section 7.2.

7.2 The Recursive-Iterative Algorithms

Based on the motivational discussion and analysis given in the previous sections, the recursive and iterative forms of the full RIV and SRIV algorithms are summarized below, together with their theoretical justification. The RIV algorithm is implemented in the rivbj routine of the CAPTAIN Toolbox, where the SRIV algorithm is the simplest option and is obtained by specifying the noise model polynomial orders as zero: i.e. $[n\ m\ \delta\ 0\ 0]$. The rivbjid routine in CAPTAIN allows for model structure identification, of the kind outlined in the later section 7.3, by repeatedly calling the rivbj over a range of user-specified system and noise model polynomial orders and providing the associated R_T^2, YIC, AIC and BIC identification criteria.

A block diagram showing the main aspects of the full RIV algorithm organization is shown in Figure 7.1.

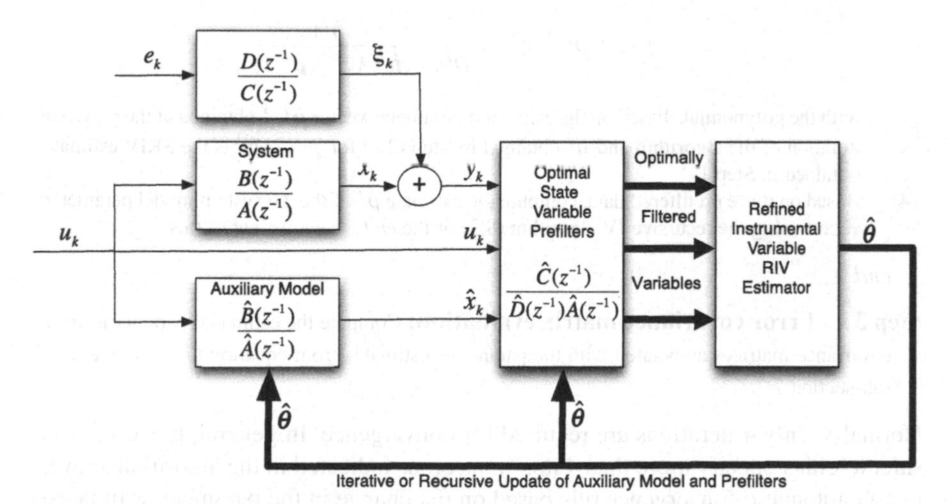

Fig. 7.1 Block diagram of the full discrete-time RIV estimation algorithm.

7.2.1 Implementation of the RIV and SRIV algorithms

The full RIV algorithm (optimal for additive ARMA noise)

Step 1. Initialization: Apply the SRIV algorithm (see later) in the standard manner and compute an initial estimate of the TF system model parameter vector $\hat{\boldsymbol{\rho}}^0$.

Step 2. Iterative or recursive-iterative IV estimation with prefilters:

for $j = 1 : convergence$

(1) If the estimated TF model is unstable, reflect the unstable eigenvalues of the estimated $\hat{A}(z^{-1}, \hat{\boldsymbol{\rho}}^{j-1})$ polynomial into the stable region of the complex z plane (e.g. using the Matlab routine polystab)[2]. Generate the IV series $\hat{x}(k)$ from the system auxiliary model:

$$\hat{x}(k) = \frac{\hat{B}(z^{-1}, \hat{\boldsymbol{\rho}}^{j-1})}{\hat{A}(z^{-1}, \hat{\boldsymbol{\rho}}^{j-1})} u(k - \delta)$$

with the polynomials based on the estimated parameter vector $\hat{\boldsymbol{\rho}}^{j-1}$ obtained at the previous iteration of the algorithm; for $j = 1$, $\hat{\boldsymbol{\rho}}^0$ is the SRIV estimate obtained in Step 1.

(2) Obtain the latest estimate $\hat{\boldsymbol{\eta}}^j$ of the noise model parameter vector based on the estimated noise sequence $\hat{\xi}(k)$ from the equation:

$$\hat{\xi}(k) = y(k) - \hat{x}(k) \tag{7.30}$$

and the procedures described in section 7.1.3, using the recursive IVARMA estimation algorithm or the *en bloc* equivalent of this.

(3) Prefilter the input $u(k)$, output $y(k)$ and instrumental variable $\hat{x}(k)$ signals by the filter

$$f_1(z^{-1}, \hat{\boldsymbol{\rho}}^{j-1}, \hat{\boldsymbol{\eta}}^j) = \frac{\hat{C}(z^{-1}, \hat{\boldsymbol{\eta}}^j)}{\hat{D}(z^{-1}, \hat{\boldsymbol{\eta}}^j)\hat{A}(z^{-1}, \hat{\boldsymbol{\rho}}^{j-1})}$$

with the polynomials based on the estimated parameter vector $\hat{\boldsymbol{\rho}}^{j-1}$ obtained at the previous iteration of the algorithm and $\hat{\boldsymbol{\eta}}^j$ obtained in Step (2). ; for $j = 1$, $\hat{\boldsymbol{\rho}}^0$ is the SRIV estimate obtained in Step 1.

(4) Based on these prefiltered data, compute the estimate $\hat{\boldsymbol{\rho}}^j$ of the TF system model parameter vector using the recursive IV algorithm (8.3) or the *en bloc* equivalent of this.

end

Step 3. Error covariance matrix evaluation: Compute the estimated parametric error covariance matrices associated with the parameter estimates from equation (7.31) in the later sub-section 7.2.3.

Normally, only 4 iterations are required for convergence. In general, however, it is safer to either specify more than 4 iterations or, as indicated in the algorithm above, use an automatic convergence rule based on the change in the parameter estimates. This algorithm is implemented as the rivbj routine in the CAPTAIN Toolbox and both of these options are available, although the latter is recommended

[2] This is not essential to the functioning of the algorithm: it allows for rarely occurring situations, normally with very poor data, when the initially estimated model is found to be unstable.

The SRIV Algorithm (optimal for additive white noise)

As noted previously, the RIV estimation problem is considerably simplified if it is assumed that the additive noise is white, *i.e.* $C(z^{-1}) = D(z^{-1}) = 1$. In this case, SRIV estimation involves only the parameters in the $A(z^{-1})$ and $B(z^{-1})$ polynomials and the prefiltering only involves the prefilter $f_1(z^{-1}, \hat{\boldsymbol{\rho}}^{j-1}) = 1/A(z^{-1}, \hat{\boldsymbol{\rho}}^{j-1})$. Consequently, the main steps in the SRIV algorithm are the same as those in the RIV algorithm, except that the noise model estimation in stage (2) of the iteration is omitted, with only this simpler prefiltering operation required, and the estimation is initiated by least squares (ARX model) estimation of the system model parameters. The SRIV algorithm is implemented as an option of the rivbj routine in CAPTAIN when the model structure is specified with the noise model order parameters p and q set to zero (i.e. $[n\ m\ \delta\ 0\ 0]$)

7.2.2 *Convergence of the iterative algorithms*

As pointed out above, the SRIV and RIV algorithms are rapidly convergent, normally converging to a stationary solution in only a few iterations (typically 3-4). Moreover, they have been available for many years in the CAPTAIN Toolbox, where they have been very robust in simulation and practical applications, converging in all cases where the model is identifiable from the data. In this regard, it is interesting note that the SRIV algorithm is quite similar to the iterative least squares algorithm of Steiglitz and McBride (1965), the convergence of which has been established in the case of white additive noise (Stoica and Söderström, 1981). Moreover, the inherent optimal instrumental variable nature of the SRIV algorithm removes the limitations of the Steiglitz and McBride-type algorithm in the coloured noise situation.

Until recently, however, the convergence of the full refined IV algorithms has not been considered fully in theoretical terms. Fortunately, during the writing of this book, a paper has appeared (Liu et al., 2011) that proves the local convergence of the iterative procedure used in the RIVC algorithm, as described in the next chapter 8, which clearly applies also to the RIV algorithm. This is, however, a proof of *local* convergence and so the authors conclude that:

> When the RIVC method is compared with the well-known prediction error method (PEM), there does not appear to be any evidence of the multiple minima problem which results in non-convergence of the gradient optimization used in PEM. So an interesting question is whether the global convergence can be established for the RIVC method. For a finite number of data samples, the actual convergence behaviour of the RIVC method is another open problem.

Certainly practical experience, coupled with these theoretical results, would suggest that RIV and RIVC algorithms do not have any major convergence problems, except in poorly identifiable situations or when the sample size is very small (i.e. situations where most estimation algorithms experience difficulties).

7.2.3 Theoretical justification of the RIV method

The original maximum likelihood development of the RIV method (see Appendix F and Young (1976b); Young and Jakeman (1979b); Young (1984)), as well as the related motivation outlined in the previous section, are based on the decomposition of the estimation problem into two separate but inter-linked sub-problems: first, the estimation of the system transfer function model parameters under the assumption that the noise model parameters are known; and second, the estimation of the ARMA noise model parameters under the assumption that the TF system model parameters are known. This process, which is sometimes termed 'bootsrapping' is justified by the following theorem due to Pierce (1972) and formulated in the present control theoretic form by Young and Jakeman (1979b, 1980a).

Theorem (Pierce (1972)) If, in the Box-Jenkins TF model (7.1):

(i) the $e(k)$ are independent and identically distributed with zero mean, variance σ^2 and skewness and kurtosis κ_1 and κ_2 ;
(ii) the parameter values are admissible (that is the model is stable and identifiable), and
(iii) the $u(k)$ are persistently exciting;

then the ML parameter estimates, obtained from a data set of N samples, possess a limiting normal distribution, such that the following results hold:

1. the asymptotic covariance matrix of the estimation errors associated with the estimate of the system parameters $\{a_i; b_j\}$ is of the form:

$$\mathbf{P}_\rho = \frac{\sigma^2}{N}\left[\text{p.lim}\,\frac{1}{N}\sum_{k=1}^{N}\mathring{\boldsymbol{\phi}}(k)\mathring{\boldsymbol{\phi}}^T(k)\right]^{-1}$$

   where $\mathring{\boldsymbol{\phi}}(k)$ is the underlying 'ideal' IV vector, defined by equation (7.20) for the system model estimation model.
2. the estimates of the noise model parameters $\{c_i; d_j\}$ are *asymptotically independent* of the $\{a_i; b_j\}$ estimates and have an error covariance matrix of the form:

$$\mathbf{P}_\eta = \frac{\sigma^2}{N}\left[E\left\{\sum_{k=1}^{N}\mathring{\boldsymbol{\psi}}(k)\mathring{\boldsymbol{\psi}}^T(k)\right\}\right]^{-1}$$

   where $\mathring{\boldsymbol{\psi}}(k)$ is the underlying 'ideal' IV vector, defined by equation (7.24a) for the noise model estimation
3. the estimate $\hat{\sigma}^2$ has asymptotic variance $(2\sigma^4/N)(1+0.5\kappa_2)$ and, if $\kappa_1 = 0$, is independent of the above estimates.

Proof See Pierce (1972).

Comments:

1. If it is assumed that, in all identifiable situations, the RIV algorithm converges in the sense that $\hat{\boldsymbol{\rho}} \Rightarrow \boldsymbol{\rho}$ and $\hat{\boldsymbol{\eta}} \Rightarrow \boldsymbol{\eta}$, then the elements of the vectors $\hat{\boldsymbol{\phi}}(k)$ and $\hat{\boldsymbol{\psi}}(k)$ will converge in probability to the equivalent elements of the ideal vectors $\mathring{\boldsymbol{\phi}}(k)$ and $\mathring{\boldsymbol{\psi}}(k)$. Consequently, $\hat{\boldsymbol{\phi}}(k)$ and $\hat{\boldsymbol{\psi}}(k)$ can be used to compute estimates $\hat{\mathbf{P}}_\rho$ and $\hat{\mathbf{P}}_\eta$ of the parametric covariance matrices $\mathbf{P}_\rho$ and $\mathbf{P}_\eta$, respectively, as defined in the Theorem, *i.e.*,

$$\hat{\mathbf{P}}_\rho = \hat{\sigma}^2 \left[\sum_{k=1}^{N} \hat{\boldsymbol{\phi}}(k) \hat{\boldsymbol{\phi}}^T(k) \right]^{-1} \quad ; \quad \hat{\mathbf{P}}_\eta = \hat{\sigma}^2 \left[\sum_{k=1}^{N} \hat{\boldsymbol{\psi}}(k) \hat{\boldsymbol{\psi}}^T(k) \right]^{-1} \tag{7.31}$$

   In the implementation of the RIVC algorithm, therefore, $\hat{\mathbf{P}}_\rho$ in (7.31) is provided as the estimate of the parametric error covariance matrix and will provide good empirical estimates of the uncertainty in the parameter estimates in $\hat{\boldsymbol{\rho}}$. And, noting the Monte Carlo results of both Pierce (1972) and Young and Jakeman (1979b), we might also assume that they will provide a good indication of the error covariance properties, even for small sample size N. This is confirmed in the later simulation examples.
2. In the control and systems literature, optimal IV estimation is sometimes considered by application of the "extended-IV" approach to estimation (Söderström and Stoica, 1983). In the open-loop situation, this shows that the IV optimization in (7.7) achieves the Cramer-Rao lower bound on $\mathbf{P}_\rho$ when $\boldsymbol{\varphi}(k) = \boldsymbol{\phi}(k)$ and $\hat{\boldsymbol{\varphi}}(k) = \hat{\boldsymbol{\phi}}(k)$, thus confirming the optimality of the RIV prefilters and the Pierce Theorem.
3. The selection of the TF model form is often decided by the predilection of the analyst. However, the above theorem suggests that the ML estimation of the BJ model has one particularly attractive statistical advantage: namely, the asymptotic independence of the system and noise model parameter estimates (that is the covariance matrix is block diagonal), thus justifying the system-noise model decomposition that is an essential element of the iterative RIV/RIVC algorithms. In addition, Jakeman and Young (1981, 1983) have investigated this aspect of the RIV algorithm further and showed that, while a BJ model estimation algorithm, such as RIV, is able to estimate an ARMAX model, without any approximation, from data generated by a stochastic ARMAX system (the ARMAX model is simply a constrained BJ model), the reverse is not true. If the polynomials are not to share common factors, then the ARMAX form of the BJ model is :

$$A(z^{-1})y(k) = B(z^{-1})u(k-\delta) + F(z^{-1})e(k)$$
$$F(z^{-1}) = \frac{D(z^{-1})A(z^{-1})}{C(z^{-1})}$$

   so that the noise polynomial is not normally of finite dimension and estimation of a finite size polynomial will imply an approximation.

4. Finally, note that the Pierce results suggest a symmetric gain version of the recursive RIV algorithm, with $\boldsymbol{\phi}(k) = \hat{\boldsymbol{\phi}}(k)\ \forall\ k$ in (7.17) and the RIV algorithm. I proposed and evaluated this many years ago (Young, 1970a, 1984) but, not surprisingly, it lacks the practical robustness of the standard 'asymmetric gain' RIV algorithm discussed in previous sections of the chapter.

7.3 Model Structure Identification

The model structure identification procedure suggested in section 6.9 of chapter 6 for the SIV algorithm needs to be extended when using the RIV algorithm. As shown below, the first few steps are similar, except that they now involve the SRIV algorithm, rather than the SIV algorithm. However, steps 4. and 5. are added so that the optimal properties of the RIV algorithm can be exploited.

1. Use the SRIV estimation algorithm (i.e. the SRIV option in the CAPTAIN rivbjid identification routine) to estimate a range of different models for $\min(n) \leq n \leq \max(n)$ and $\min(m) \leq m \leq \max(m)$, where the maximum and minimum values of n and m are selected by the user, and sort these by the value of R_T^2, so exposing those model structures that best explain the data in a deterministic simulation sense.
2. Amongst the best models in 1., select one that has a *relatively* small YIC value: normally this will be a large negative value, since it is a logarithmic measure. The choice is not critical provided the associated R_T^2 is relatively high compared with that of other models, but do not select a model that has a high R_T^2 and a relatively large YIC (small negative) value, since the YIC is then probably indicating over-parameterization. Use this selected model to generate an estimate of the additive noise $\xi(k)$. Then, if possible, estimate ARMA(p,q) models for this estimated noise over a range of p and q, $\min(p) \leq p \leq \max(p)$ and $\min(q) \leq q \leq \max(q)$, again with the maximum and minimum values selected by the user, and sort these in terms of the AIC or BIC.
3. Sometimes it is not possible to model the noise because it does not subscribe to modelling as an ARMA process, in which case, the identification procedure stops at this stage and other, less formal identification strategies should be pursued. For example, the identification criteria do not always unambiguously identify the model structure and, if possible, the competing models should be evaluated in other terms, such as whether the response characteristics and the physical interpretation of the model make sense (see the later example in section 7.5.4)
4. If the noise can be modelled as an ARMA process, re-estimate the full model using the CAPTAIN rivbjid routine, with the ARMA noise model included, over a smaller range of n, m, p and q, based on the results obtained in steps 1. to 3., using the full RIV algorithm and sorting on the basis of R_T^2, with evaluation taking into account the YIC, AIC or BIC values: the model should have a relatively high R_T^2 and relatively large negative values for the YIC, AIC and BIC. As pointed out

in step 3. above, however, such measures do not always unambiguously identify the model structure and a similar, less formal and physically more meaningful, approach should be used whenever this is possible.

5. Check the recursive estimates of the parameters generated by the CAPTAIN rivbj routine to confirm that these converge satisfactorily, without any signs of unusual movement that could arise from over-parameterization (see the later example in section 7.5.3).
6. Finally, evaluate the autocorrelation of the ARMA model residuals $\hat{e}(k)$, using the CAPTAIN routine acf, to ensure that they are reasonably white; and the cross-correlation functions between the $\hat{e}(k)$ and the input variable $u(k)$, using the CAPTAIN routine ccf, to ensure that there is no significant correlation between the residuals and the input, which would negate the assumptions on which the TF model estimation is based.

Note that, in some cases, particularly if the system is 'stiff' and characterized by widely spaced eigenvalues (see the later examples in this chapter), it may be that the model estimated by RIV (or alternative optimal algorithms such as PEM) in step 4. provides a poor estimate of the deterministic system output compared with the SRIV estimated model output. In this case, it is better to use the SRIV estimate, together with the ARMA noise model estimated in 2, to define the full stochastic model. In this case, the covariance matrix provided by the SRIV algorithm will not necessarily provide a good estimate of the parametric uncertainty and will need to be corrected. This nominally sub-optimal approach is called SRIV-ARMA estimation: see Young (2010d). Its good performance in practical applications raises questions about the need for a wider interpretation of optimality; questions that are explored in the Epilogue of the present book, which includes an interesting rainfall-river level modelling and forecasting example.

7.4 Input Noise and Errors-in-Variables

The RIV approach to TF model estimation partly solves the problem of *Errors-in-Variables* (EIV), or 'measurement error models' caused by noise on the output variable. However, it does not directly solve the EIV problem when there is measurement noise on the input variable (or variables): i.e., where the measurement $v(k)$ of $u(k)$ is contaminated by measurement noise $\eta(k)$ that does not pass through the system (otherwise it could be considered as output noise):

$$v(k) = u(k) + \eta(k) \tag{7.32}$$

A later example illustrates the nature of this problem and the biasing effect that input noise has on the TF parameter estimates.

It is well known that, in general, EIV systems are not uniquely identifiable from second order statistical properties (Anderson and Deistler, 1984; Deistler, 1986; Solo, 1986; Agüero and Goodwin, 2008). However, specific prior knowledge can be used to restrict the class of models consistent with the given data. Indeed, the IV approach can still be used in this input noise situation but, analogous to the case of the additive output noise, it requires the existence of an IV that is highly correlated with the input variable but not correlated with the noise. The existence of such a variable is likely to be problem dependent: for instance, it is possible that, because multiple rainfall measurements are normally made at various locations in a river catchment, one of these measurements could act as an IV for another, since the measurement errors could be assumed to be uncorrelated. In this case, the RIV vector would be replaced by:

$$\hat{\boldsymbol{\phi}}(k) = [-\hat{x}_{f_1}(k-1) \cdots -\hat{x}_{f_1}(k-n)\ \hat{u}_{f_1}(k-\delta) \cdots \hat{u}_{f_1}(k-\delta-m)]^T \tag{7.33}$$

where $\hat{u}_{f_1}(k)$ would be the prefiltered second measurement of the input variable. Another possibility in this hydrological context occurs in the case of a 'routing' model between flow at two locations B and C down a river. In this situation, an additional TF model could be estimated between a measured flow further *upstream* at a location A and the input flow at location B. The deterministic output of this additional model could then be used as an IV for the input and the IV vector would be defined again as in (7.33), this time with $\hat{u}_{f_1}(k)$ defined as the prefiltered output of the additional TF model.

Of course IV solutions such as these are not optimal because the RIV approach is only optimal if all the assumptions are satisfied and these require a noise-free input variable. Moreover, these solutions are problem dependent and it is unlikely if conditions such as these will be possible in most practical situations. One exception is in the context of automatic control systems, where an external input is normally activating a feedback control system and so can act as a source of IVs for the noisy internal control signal. However, this special situation is discussed in chapter 9.

This general EIV problem caused by measurement noise affecting both the output and input variables has received a lot of attention in the control and systems literature: see for example Van Huffel et al. (2002); Söderström (2007) and the prior references therein; and some of these approaches ('bias-eliminating least squares' (BELS), the 'Frisch scheme' and 'extended compensated least squares' (ECLS) methods) are related (Hong and Söderström, 2009). Although these various approaches will not be considered here, because EIV is a major subject in its own right, it is worth sketching out the general idea behind the methods. Hopefully, this brief analysis should act as a useful introduction for the reader who is interested in finding out more about this topic. In this regard, the paper by Söderström is a particularly valuable review that provides deep insight into the nature of the problem and the different estimation procedures that have been suggested.

Let us consider the least squares estimate of the ARX model discussed in section 6.4 of chapter 6, i.e.,

$$y(k) = \boldsymbol{\phi}^T(k)\boldsymbol{\rho} + e(k) \tag{7.34}$$

and partition the data vector $\boldsymbol{\phi}(k)$ as follows, so that the contributions of the input and output variables are clear:

$$\boldsymbol{\phi}^T(k) = [\boldsymbol{\phi}_y^T(k)\ \boldsymbol{\phi}_u^T(k)]^T \tag{7.35}$$

where,

$$\boldsymbol{\phi}_y^T(k) = [-y(k-1)\ \cdots\ -y(k-n)];\ \boldsymbol{\phi}_u^T(k) = [u(k)\ \cdots\ u(k-n)]$$

Then, using the nomenclature for covariance matrices and vectors used in section 6.3 of chapter 6, the asymptotic least squares estimate $\hat{\boldsymbol{\rho}}$ of $\boldsymbol{\rho}$ satisfies the normal equations

$$\begin{bmatrix} \mathbf{R}_{\boldsymbol{\phi}_y,\boldsymbol{\phi}_y} & \mathbf{R}_{\boldsymbol{\phi}_y,\boldsymbol{\phi}_u} \\ \mathbf{R}_{\boldsymbol{\phi}_y,\boldsymbol{\phi}_u} & \mathbf{R}_{\boldsymbol{\phi}_u,\boldsymbol{\phi}_u} \end{bmatrix} \hat{\boldsymbol{\rho}} = \begin{bmatrix} \mathbf{R}_{\boldsymbol{\phi}_y,y} \\ \mathbf{R}_{\boldsymbol{\phi}_y,u} \end{bmatrix} \tag{7.36}$$

Now, suppose that, contrary to the assumptions of the ARX model, the input variable $u(k)$ is contaminated by noise and the output noise is coloured, rather than white. For similar reasons to those discussed in section 6.3 of chapter 6 within the more limited context of just output noise, the estimate of $\boldsymbol{\rho}$ obtained in this manner will be asymptotically biased. But, in this input noise situation, the bias depends not only on the statistical properties of the output noise but also on statistical properties of the input noise; for instance, the variance σ_η^2, in the white noise case.

If these latter statistical properties of the noise are known, it is possible to compensate for the bias by finding an estimate $\hat{\boldsymbol{\rho}}$ that satisfies the following normal equations, which follow directly from the analysis in section 6.3 of chapter 6 but accounting for the additional effects of the input noise:

$$\left(\begin{bmatrix} \mathbf{R}_{\boldsymbol{\phi}_y,\boldsymbol{\phi}_y} & \mathbf{R}_{\boldsymbol{\phi}_y,\boldsymbol{\phi}_u} \\ \mathbf{R}_{\boldsymbol{\phi}_y,\boldsymbol{\phi}_u} & \mathbf{R}_{\boldsymbol{\phi}i_u,\boldsymbol{\phi}_u} \end{bmatrix} - \begin{bmatrix} \mathbf{R}_{\boldsymbol{\xi}_y,\boldsymbol{\xi}_y} & \mathbf{0} \\ \mathbf{0} & \sigma_\eta^2 \mathbf{I}_{m+1} \end{bmatrix}\right) \hat{\boldsymbol{\rho}} = \begin{bmatrix} \mathbf{R}_{\boldsymbol{\phi}_{y,y}} \\ \mathbf{R}_{\boldsymbol{\phi}_{y,u}} \end{bmatrix} - \begin{bmatrix} \mathbf{R}_{\boldsymbol{\xi}_{y,\eta}} \\ \mathbf{0} \end{bmatrix} \tag{7.37}$$

where $\mathbf{R}_{\boldsymbol{\xi}_y,\boldsymbol{\xi}_y}$ is the covariance matrix of the noise on the output elements of $\boldsymbol{\phi}(k)$; $\mathbf{R}_{\boldsymbol{\xi}_{y,\eta}}$ is the covariance vector of the noise on the output elements of $\boldsymbol{\phi}(k)$ and $\eta(k)$ (see section 6.3). Hence, the principle of the bias compensating least squares approach is to estimate the noise parameters and then to compute the estimate $\hat{\boldsymbol{\rho}}$ as the solution of the compensated normal equations (7.37). As pointed out in section 6.3, this solution will be quite sensitive to any errors in the estimates of the noise parameters and is the reason why, in the situation where it can be assumed that there is only output noise, solutions such as RIV are preferred.

The current EIV solution that best fits within the context of the present book is *the recursive Extended Bias Compensating Instrumental Variables* (EBCIV), which avoids the need for ouput noise compensation since this is accommodated by the IV solution. The IV sequence can be defined in various ways but the approach used by Linden et al. (2009), for example, is particularly straightforward because it uses the delayed inputs as instrumental variables. Of course, these instruments are not optimal in any way, so the statistical efficiency will not be high and very much dependent on the nature of $u(k)$.

There is no doubt that EIV is an important topic in theoretical terms and has important implications in some practical contexts. From my own personal experience, however, I do not consider that input noise is nearly as important a problem in practice as output noise. The main reasons for this view are listed below.

1. First, asymptotic bias on the TF model parameters caused by the presence of input noise is of most importance if the model parameter *values* are relevant in themselves: for instance because they have a physical interpretation that is important to the modeller and the associated scientific study. If this is not the case then, provided the noise is not too large and the resulting bias relatively small, the model will often suffice for most practical purposes, despite the presence of bias. This is particularly the case if the spectral characteristics of the noise are likely to remain the same when the model is used in practice, for then the bias will be consistent in these future uses of the model.
2. Second, as we see in the example 7.1 below, the bias is often not excessive, even if the noise is higher than might be expected in most practical situations. Moreover, if the input noise is white, as assumed often in EIV analysis, it affects only the coefficients of the estimated TF numerator polynomial $\hat{B}(z^{-1})$ and so does not lead to any bias on the estimates of the system eigenvalues, which are defined entirely by the estimated TF denominator polynomial $\hat{A}(z^{-1})$. As a result, parameters derived from the estimation results, such as time constants, natural frequencies and damping coefficients, are not affected by the bias. In the case of coloured input noise, the $\hat{A}(z^{-1})$ coefficients are generally biased, but this bias can be quite small.
3. Finally, the bias is affected by the presence of the prefilters, which attenuate the effect of noise on all of the measured signals, including the input, as discussed in previous sections of this chapter. Of course, the effect of this attenuation is a function of the input noise characteristics and to what extent they are different from those on the output variable (since, in the RIV algorithm, the prefilters are a function of the latter).

These comments do not negate the utility of EIV estimation which, at the very least, can be used to diagnose whether there may be a input noise-induced bias that is serious enough to affect the use for which the model is intended. However, given the debate on what is the best solution to the general EIV problem and bearing on the comments above, Matlab toolboxes such as CAPTAIN and SID do not currently provide routines that allow for input noise.

7.4.1 Example 7.1: Errors-in-Variables: the effects of input noise

This example gives some idea of how the presence of reasonable levels of input noise can effect the estimates of TF model parameters. The simulation model takes the form:

$$
\begin{aligned}
x(k) &= \frac{1+0.5z^{-1}}{1-1.5z^{-1}+0.7z^{-2}}u(k) \qquad & u(k) &= \mathcal{N}(0,1.0) \\
\xi(k) &= \frac{1.0}{1-0.85z^{-1}}e(k) & e(k) &= \mathcal{N}(0,2.1) \\
y(k) &= x(k)+0.5\xi(k) & & \\
v(k) &= u(k)+\eta(k) & \eta(k) &= \mathcal{N}(0,0.1)
\end{aligned}
\tag{7.38}
$$

This yields a noise/signal ratio on both the output and input signals of 0.31 by standard deviation, which represents the kind of maximum noise levels that might be encountered in practical applications.

In order to the establish a reference against which to evaluate the effects of the input noise, $\eta(k)$ is first set to zero $\forall k$ and the model is used to generate 5000 samples of $u(k)$, $x(k)$ and $y(k)$. RIV identification using the rivbjid routine in CAPTAIN correctly identifies the correct [2 2 0 1 0] model structure with the following parameter estimates after 5000 samples, where the figures in parentheses are the estimated standard errors:

$$
\hat{a}_1 = -1.501(0.003);\ \hat{a}_2 = 0.700(0.003);\ \hat{b}_0 = 1.004(0.01);\ \hat{b}_1 = 0.499(0.013)
$$
$$
\hat{c}_1 = -0.850(0.00005)
$$

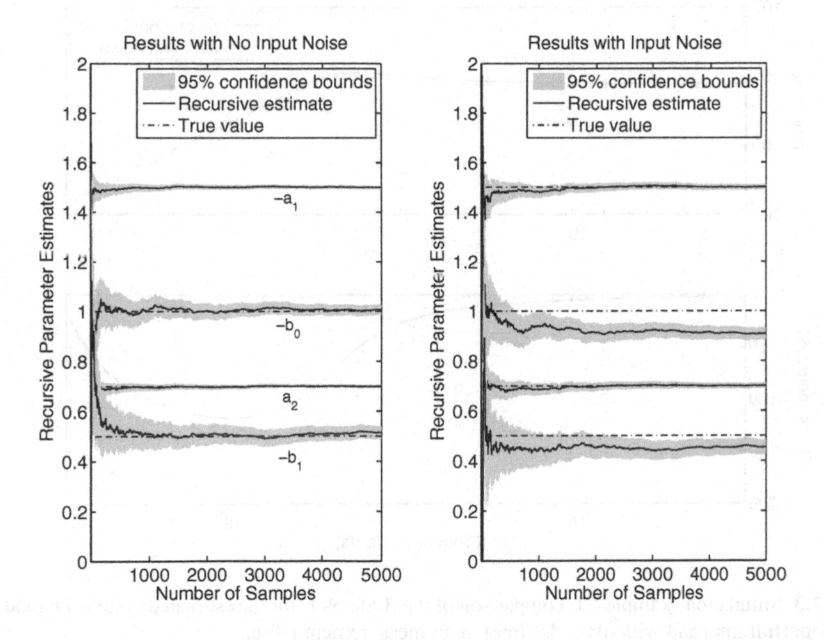

Fig. 7.2 Simulation example 7.1: recursive RIV estimates of the system TF model parameters without (left-hand panel) and with (right-hand panel) input measurement noise.

The associated recursive estimation results are shown in the left hand panel of Figure 7.2; while the right hand panel shows the comparative results obtained from a data set in which the noise $\eta(k)$ is added to the input signal. For clarity, the recursive estimates of the noise model parameters are not plotted here. In the noisy input case, the noise-free input signal and the output signal are exactly the same as those used to obtain the results in the left hand panel but, because of the noise on the input signal, the model structure identified by rivbjid is [2 2 0 3 0] and the parameter estimates after 5000 samples are changed to:

$$\hat{a}_1 = -1.500(0.006);\ \hat{a}_2 = 0.700(0.005);\ \hat{b}_0 = 0.905(0.012);\ \hat{b}_1 = 0.448(0.014)$$
$$\hat{c}_1 = -1.107(0.140);\ \hat{c}_2 = 0.162(0.021);\ \hat{c}_3 = 0.141(0.014)$$

The recursive estimates in the right hand panel of Figure 7.2 demonstrate that the input white noise biases the estimates of the $B(z^{-1})$ polynomial parameters but that the $A(z^{-1})$ polynomial parameters are not affected in this white noise case. Moreover, even with this reasonable level of 31% input noise, the biases are not too large; indeed, the bias on $\hat{b}_1$ is only just significant in relation to the 95% confidence interval. As a result, the estimated model properties are not very different from the true properties, as shown in Figures 7.3 and 7.4.

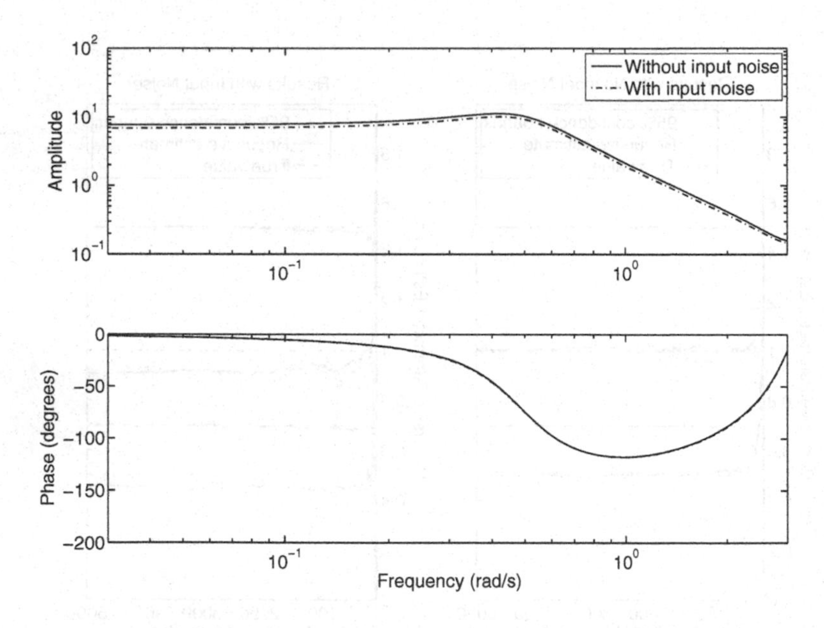

Fig. 7.3 Simulation example 7.1: comparison of the Bode plots for the estimated system TF models without (full line) and with (dash-dot line) input measurement noise.

Figure 7.3 compares the Bode plots of the the models estimated without (full line) and with (dash-dot line) input noise, showing that the gain plots are marginally different because the biased model has a smaller steady gain: 6.94 compared with 7.56; but the phase is not affected. These results are reflected in the estimated natural frequency $\omega_n = 0.49$ rad/sec and damping ratio $\zeta = 0.36$, which are the same for both models and agree with the true values. Figure 7.4 shows that these small differences also characterize the noise-free outputs of the the two estimated models, which are compared with the true noise-free output (dots). The model estimated without input noise is virtually the same as the true output; while the model estimated with input noise is just a little different, but the errors are well within the 95% confidence bounds associated with the input noise-free model output shown in grey.

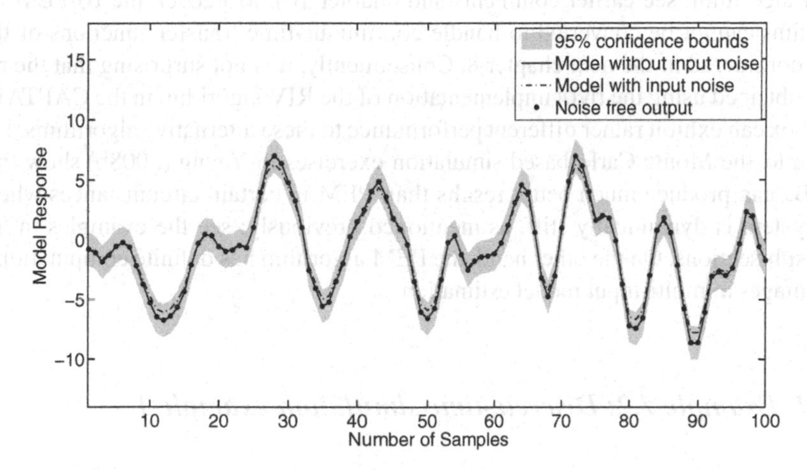

Fig. 7.4 Simulation example 7.1: the deterministic outputs of the estimated TF system models compared with the noise free output (dots).

If the input noise is coloured, all the estimates are biased to some extent, as we see from the following estimates which are obtained when the input noise is generated by an AR noise model with $C(z^{-1}) = 1 - 0.7z^{-1}$, but with the noise/signal ratio by standard deviation maintained at 0.31:

$$\hat{a}_1 = -1.472(0.006);\ \hat{a}_2 = 0.683(0.005);\ \hat{b}_0 = 0.908(0.012);\ \hat{b}_1 = 0.458(0.014)$$
$$\hat{c}_1 = -1.235(0.140);\ \hat{c}_2 = 0.193(0.022);\ \hat{c}_3 = 0.200(0.014)$$

However, the biasing effect is still not too large and the estimated model characteristics are still quite close to those of the true model, with $SSG = 6.47$, $\omega_n = 0.51$ rad/sec and $\zeta = 0.37$. As a result, the errors in the noise-free output response are still well within the 95% confidence bounds associated with the input noise-free model: i.e. only a little worse than those shown in Figure 7.4.

7.5 Examples of RIV Identification and Estimation: SISO Models

Although the RIV algorithm yields optimal ML estimates of the parameters in the discrete-time BJ model, its iterative implementation is considerably different from other ML algorithms, such as the original BJ algorithm and the PEM algorithm, that utilize gradient-type numerical optimization. For example, as we have seen, the RIV formulation allows for the direct recovery of recursive estimates whose final estimates after N samples exactly match the estimates obtained by the *en bloc* solution. This is not possible with these other algorithms (although a *separate* recursive PEM (RPEM) algorithm is available that is not the exact equivalent of the PEM algorithm: see earlier comments and chapter 10). Moreover, the BJ/PEM algorithms cannot be converted to handle continuous-time transfer functions of the kind considered in the next chapter 8. Consequently, it is not surprising that the results obtained using the rivbj implementation of the RIV algorithm in the CAPTAIN Toolbox can exhibit rather different performance to these alternative algorithms. For example, the Monte Carlo based simulation exercises in Young (2008b) show that RIVBJ can produce much better results than PEM in certain circumstances where the system is dynamically stiff, as mentioned previously: see the examples in the next sub-sections. On the other hand, the PEM algorithm has definite computational advantages in multi-input model estimation.

7.5.1 Example 7.2: Discrete-time simulation example 1

Here, the RIV algorithm is applied to $N = 1700$ samples of simulated data generated by the following Box-Jenkins TF model:

$$y(k) = \frac{0.016 + 0.026z^{-1} - 0.0375z^{-2}}{1 - 1.6252z^{-1} + 0.642z^{-2}} u(k) + \frac{1 + 0.5z^{-1}}{1 - 0.85z^{-1}} e(k)$$
$$u(k) = \mathcal{N}(0, 8.8) \qquad e(k) = \mathcal{N}(0, 0.0009)$$

The simulation is based on a real data example (Young et al., 2004; Young, 2008a) concerned with the modelling of daily rainfall-flow data from the River Canning, an 'ephemeral' river in Western Australia that stops flowing in Summer. The input variable to the above model $u(k)$ is the 'effective rainfall', i.e. the rainfall that is effective in causing flow changes in the river. This is a stiff dynamic system, with widely spaced eigenvalues, and it has a reasonable noise level (noise/signal ratio of 0.33 by standard deviation). The TF can be decomposed into a parallel connection of three TFs: one a simple gain, 0.015964; and two with first order TFs having time constants of 2.6 and 18.7 samples (days in this case). This is a common model form in the environmental sciences but it appears to pose some problems for the PEM algorithm. Table 7.1 compares the results of single run and MCS analysis (see Appendix B, section B.3.9), based on 100 realizations, for the RIV, SRIV, PEM and IV4

Table 7.1 Monte Carlo simulation results for simulation example 7.2: $\hat{\boldsymbol{\theta}} = [\hat{\boldsymbol{\rho}}^T \hat{\boldsymbol{\eta}}^T]^T$ is the estimated full parameter vector (see (7.10)); SE denotes the standard error on the estimates; SD the standard deviation of the MCS realizations; SR the single run results; and MCS the results from MCS analysis based on 100 random realizations.

Parameter		$\hat{a}_1$	$\hat{a}_2$	$\hat{b}_0$	$\hat{b}_1$	$\hat{b}_2$	$\hat{c}_1$	$\hat{d}_1$	Failures
True Values		-1.6252	0.642	0.016	0.026	-0.0375	-0.85	0.5	
RIV	$\hat{\boldsymbol{\theta}}$	-1.6112	0.6331	0.0162	0.0264	-0.0375	-0.845	0.494	
(SR)	SE	0.0449	0.0334	0.0003	0.0007	0.0018	0.01	0.02	
RIV	mean	-1.6255	0.642	0.0160	0.0260	-0.0375	-0.847	0.501	
(MCS)	SD	0.0254	0.0200	0.0002	0.0004	0.0010	0.01	0.02	
SRIV	$\hat{\boldsymbol{\theta}}$	1.6312	0.649	0.0165	0.0256	-0.0379	-0.848	0.485	
(SR)	SE	0.0199	0.0159	0.0007	0.0014	0.0009	0.010	0.02	
SRIV	mean	-1.6164	0.635	0.0159	0.0262	-0.0372	-	-	
(MCS)	SD	0.0535	0.0428	0.0005	0.0013	0.0014	-	-	
PEM	$\hat{\boldsymbol{\theta}}$	-1.5939	0.6204	0.0161	0.0266	-0.0367	-0.849	0.502	
(SR)	SE	0.0650	0.0480	0.0003	0.0010	0.0026	0.01	0.02	
PEM	mean	-1.6166	0.6354	0.0160	0.0261	-0.0371	-0.849	0.502	9
(MCS)	SD	0.0400	0.030	0.0003	0.0007	0.00164	0.01	0.02	
RIV-AR	$\hat{\boldsymbol{\theta}}$	-1.611	0.6328	0.0162	0.0263	-0.0375	-	-	
(SR)	SE	0.0440	0.0328	0.0003	0.0007	0.0017	-	-	
RIV-AR	mean	-1.619	0.6371	0.016	0.0261	-0.0373	-	-	
(MCS)	SD	0.0356	0.0270	0.0002	0.0006	0.0014	-	-	
IV4	$\hat{\boldsymbol{\theta}}$	-1.568	0.6019	0.0162	0.0270	-0.0357	-	-	
(SR)	SE	0.0991	0.0716	0.0003	0.0016	0.0040	-	-	
IV4	mean	-1.6090	0.6298	0.0160	0.0262	-0.0369	-	-	
(MCS)	SD	0.0513	0.0384	0.0002	0.0008	0.0021	-	-	

algorithms (the latter two from the Matlab R2010b SID Toolbox). It also presents the results obtained with the RIV algorithm when the ARMA model is approximated by an AIC identified AR(5) model (indicated by the acronym RIV-AR). As would be expected because of their common basis in maximum likelihood estimation, both of the RIV and PEM algorithms perform similarly when convergence occurs, with the single run predicted standard errors on the parameter estimates matching reasonably the standard deviations computed from the MCS analysis. However, in this example, the PEM algorithm fails to converge correctly in 9 of the 100 realizations (i.e. the estimates at convergence were far removed from the actual parameter values), but these realizations were removed in computing the statistics shown in Table 7.1; while the RIV algorithm does not fail at all.

The SRIV algorithm also performs well : in fact, in terms of the single run and MCS estimated mean parameter estimates, it performs better than PEM in the sense that it has no failures. However, we see that its estimated standard errors are too optimistic, as might be expected, and no attempt has been made to correct them. Also, note that an estimate of an ARMA model for the noise can be obtained by applying the IVARMA algorithm separately to the residual noise estimate obtained from the SRIV estimation results. For illustration, this is shown in Table 7.1 only for the single run case. Of course, it had no influence of the TF system model parameter estimates and it was not computed at all in the MCS simulation results presented below this. Note that this SRIV algorithm is computationally a quite efficient algorithm (a little faster than PEM in this case and only marginally slower than IV4) and so, as suggested in section 7.3, it provides the best algorithm for initial model structure/order identification analysis.

The results in Table 7.1 are typical of the performance comparison in the case of examples such as that considered here. The poorer performance of the PEM algorithm appears to be due to the stiff nature of the TF model in this example and a consequent failure to converge from the initial conditions specified for the parameter estimates in the PEM gradient optimization algorithm. It is clear from the results that the RIV algorithm does not suffer from this problem and is always providing statistically consistent and efficient estimates. However, these results are not representative: normally both the RIV and PEM algorithms perform well and produce similar estimation results.

It will be noted that the RIV-AR results are comparable with those of RIV (for convenience, the AR(5) noise model parameter estimates are not shown). They demonstrate how this 'approximate' implementation of the RIV algorithm is appealing because it is computationally much more efficient than RIV and yet performs similarly in most cases. The IV4 algorithm, which also uses an AR noise model, produces reasonable results but they are noticeably poorer than those of RIV and SRIV. Like them, however, it has no failures amongst the MCS realizations. This is not always the case, as can be seen in the next simulation example .

7.5.2 Example 7.3: Discrete-time simulation example 2

This example is concerned with a simulation model based on a $[n\ m\ \delta\ p\ q] = [2\ 2\ 4\ 0\ 0]$ TF model identified and estimated from the real effective rainfall-flow data. It is a re-appraisal of an example used in a previous comparative study Young (2003b) but now using the latest available versions of the RIV, PEM and IV4 algorithms[3]. The hourly flow $y(k)$ is measured in a river over most of a year (7500 hours or 312.5 days); while the associated "effective rainfall" $u(k)$ is processed from the actual rainfall measurements. The simulation data are generated by passing this effective rainfall input through the model, with its parameters set to those estimated from the real data. The output is then contaminated by white noise with variance $\sigma^2 = 5$, giving a noise/signal ratio (by standard deviation) of 0.62.

[3] The previous results obtained with the PEM algorithm were considerably worse than those shown here but it would appear that some improvements have been introduced into the PEM algorithm since the original analysis was carried out.

Table 7.2 Monte Carlo simulation results for simulation example 7.3: SE denotes the standard error on the estimates; SD the standard deviation of the MCS realizations; SR the single run results; and MCS the results from MCS analysis based on 100 random realizations (out of 124 total realizations).

Parameter		$\hat{a}_1$	$\hat{a}_2$	$\hat{b}_0$	$\hat{b}_1$	Failures
True Values		-1.8563	0.8565	0.0545	-0.0542	
RIV	$\hat{\rho}$	-1.8575	0.8578	0.0543	-0.0541	
(SR)	SE	0.0028	0.0027	0.0009	0.0009	
RIV	mean	-1.8560	0.8563	0.0545	-0.0543	
(MCS)	SD	0.0027	0.0026	0.0008	0.0008	
PEM	$\hat{\rho}$	-1.8561	0.8563	0.0546	-0.0543	
(SR)	SE	0.0028	0.0028	0.0008	0.0008	
PEM	mean	-1.8585	0.8587	0.0541	-0.0539	24
(MCS)	SD	0.0027	0.0027	0.0008	0.0009	
IV4	mean	0.0954	-0.8806	0.0585	0.0481	114
(MCS)	SD	13.0	11.6	0.0224	0.694	

Table 7.2 compares the results of the RIV estimation (here effectively SRIV because the additive noise is white) with those obtained using the PEM and IV4 algorithms. Again, as would be expected, both of the RIV and PEM algorithms perform similarly *when convergence occurs*, with the single run predicted standard errors on the parameter estimates matching the standard deviations computed from the MCS analysis. However, the PEM algorithm has a quite high failure rate of 19.4%: it fails to converge satisfactorily in 24 of the 124 realizations (these realizations were removed in computing the statistics shown in Table 7.2) while the RIV algorithm does not fail at all. The performance of IV4 is much worse: it fails to converge satisfactorily in 114 of the 124 realizations, an unacceptable failure rate of 91.9%, so no single run results are shown.

This model is a very stiff dynamic system, with time constants of 6.5 and 605 hours[4] and, again, PEM's problems seem to be connected with this property. When the algorithm fails to converge on the correct system, it most often converges on a false optimum with one root of the denominator polynomial $A(z^{-1})$ negative and very close to the unit circle; while the other is positive and just greater than 0.9 (a typical example is $\{-0.99943, 0.91657\}$).

[4] Note that these time constants are sensitive to the estimated model parameter values and were computed from estimates with more decimal places than those shown in Table 7.2.

7.5.3 Example 7.4: Evaluating over-parameterization

One useful application of recursive estimation is to help diagnose identifiability problems when identifying model structure. In this regard, let us look briefly at a real example, where the parameters of a daily rainfall-flow model are being updated each day, in real-time, using the recursive option of the rivbj estimation algorithm in the CAPTAIN Toolbox. The exact nature of these data, the associated model and the method of recursive estimation are not important at this time, since we are concerned only with the general consequences of over-parameterization and poor identifiability on real-time updating. However, in order to place the results in context, they relate to the analysis of daily effective rainfall-flow series from the Leaf River in Mississippi, USA, where the effective rainfall is the rainfall adjusted nonlinearly to account for catchment storage (soil moisture) effects. These series are part of a data set that is used later in chapter 10, which presents a case study of forecasting and real-time state/parameter updating. Consequently, the identifiability results presented here are relevant to the selection of the rainfall-flow model used in this subsequent example.

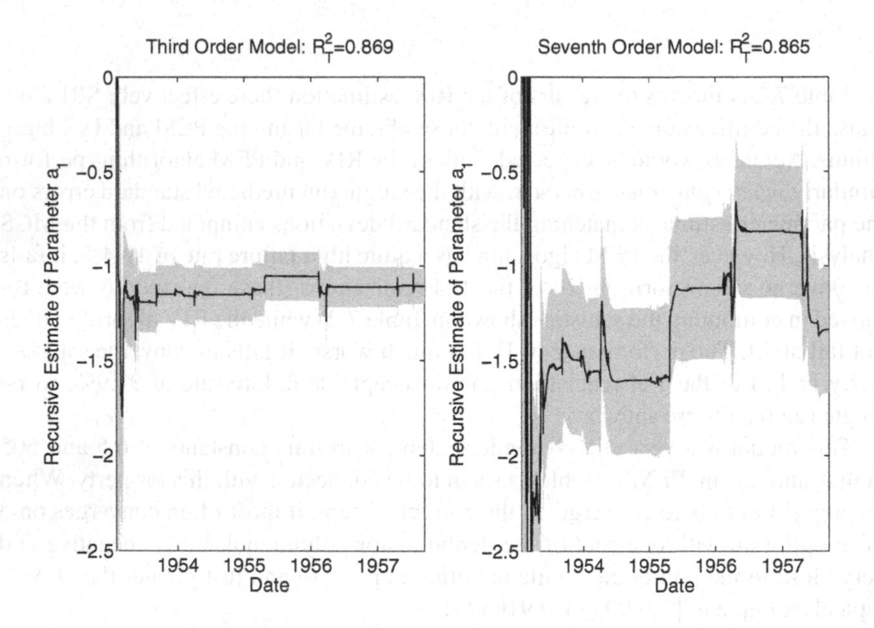

Fig. 7.5 Rainfall-flow modelling example: comparison of recursive estimates of the parameter a_1 obtained for an identifiable 3rd order model (left panel) and a poorly identifiable 7th order model (right panel). The rivbj estimated standard error bounds are shown in grey.

Figure 7.5 shows the recursive estimates for the assumed time-invariant a_1 parameter, as obtained over several years (a total of 1681 days) of data. The results for a reasonably identifiable, third order model, are shown in the left hand panel; and those for an over-parameterized and poorly identifiable, seventh order model, are plotted in the right hand panel. We see that the recursive estimates in the left hand panel converge rapidly and, although they fluctuate to some degree, as one would expect with estimation from noisy data, this is far less than that encountered in the right hand panel. Here, the estimated parameter not only fluctuates much more widely, but it also has much wider estimated uncertainty bounds, shown in grey. This is typical of all the parameters estimated in the higher order model.

In spite of the extremely volatile behaviour seen in the right hand panel of Figure 7.5, the 7th order model defined by the estimates of its parameters at any point in the data set explains the flow data to virtually the same extent as the well identified model: both have a coefficient of determination based on the simulated (deterministic) model output response of around $R_T^2 = 0.86$, suggesting that the model is much better defined parametrically than it really is. Moreover, it might be mistakenly thought that the significant changes in the parameter estimates are redolent of non-stationarity in the model, with the parameters needing to change to reflect changing effective rainfall-flow dynamics. But this is clearly not the case, since the 3rd order model has an equivalent ability to explain the observed flow variation and its parameter estimates change very little. In other words, the estimation results (recursive or *en bloc*) in this over-parameterized situation are highly misleading.

7.5.4 Example 7.5: RIV estimation of tracer experiment data

Here, we consider again the analysis of the tracer experiment data shown previously in Figure 6.9 of chapter 6, this time using the RIV algorithm. The model order identification results, as produced by the rivbjid routine in CAPTAIN using the SRIV option setting and ordered in terms of the R_T^2 values are given below. The SRIV option is selected because of the non-standard, heteroscedastic noise in this case, as mentioned in the previous chapter and discussed later.

Although the BIC and AIC measures favour the [3 3 16 0 0] model with the best $R_T^2 = 0.9959$ value, the YIC prefers the [1 1 16 0 0] and [2 2 16 0 0] models, both of which have fewer parameters and comparable R_T^2 values. Note also that the [3 2 16 0 0] and [2 3 16 0 0] explain the output variance almost as well as the [3 3 16 0 0] and have one less parameter. And of these two, the [2 3 16 0 0] model is the most appealing because it is only a second order system; but the YIC strongly favours the [2 2 16 0 0] model.

BEST 9 models (Rt2)

na	nb	td	nc	nd	YIC (TF)	Rt2	BIC (ARMA)	AIC
3	3	16	0	0	-8.436	0.9959	-2369.33	-31.439
3	2	16	0	0	-8.518	0.9956	-2337.88	-29.238
2	3	16	0	0	-4.986	0.9956	-2342.56	-28.375
2	2	16	0	0	-13.351	0.9955	-2332.29	-28.617
1	3	16	0	0	-7.725	0.9952	-2314.79	-21.499
1	2	16	0	0	-7.872	0.9948	-2278.10	-21.539
1	1	16	0	0	-14.625	0.9948	-2269.89	-21.390

In this kind of confusing situation, it is necessary to look closer at the competing models. First, the six parameter, [3 3 16 0 0] model has complex poles (eigenvalues) and, since there is no physical mechanism for this, it would appear that the model is probably over-fitting the data and so over-parameterized. It is possible to estimate the model with poles constrained to be real but there is no real justification in this case since the other, five parameter, models are explaining the data nearly as well as this six parameter one. Moreover, the [3 2 16 0 0] has real roots and so appears at first sight to be more physically meaningful. However, if this TF is decomposed by partial fraction expansion, either manually or using the residue routine in Matlab (see section B.3.7 of Appendix B), then one of the residues is negative, implying a physically meaningless negative flow. So this model can also be eliminated on physical grounds, leaving the [2 3 16 0 0] and [2 2 16 0 0] models, the outputs of which are compared with the measured output in Figure 7.6. This also shows the output of the [2 3 16 0 0] model estimated by the pem algorithm in the Matlab SID Toolbox: here, the output in the recession part of the response is very similar to that of the [2 2 16 0 0].

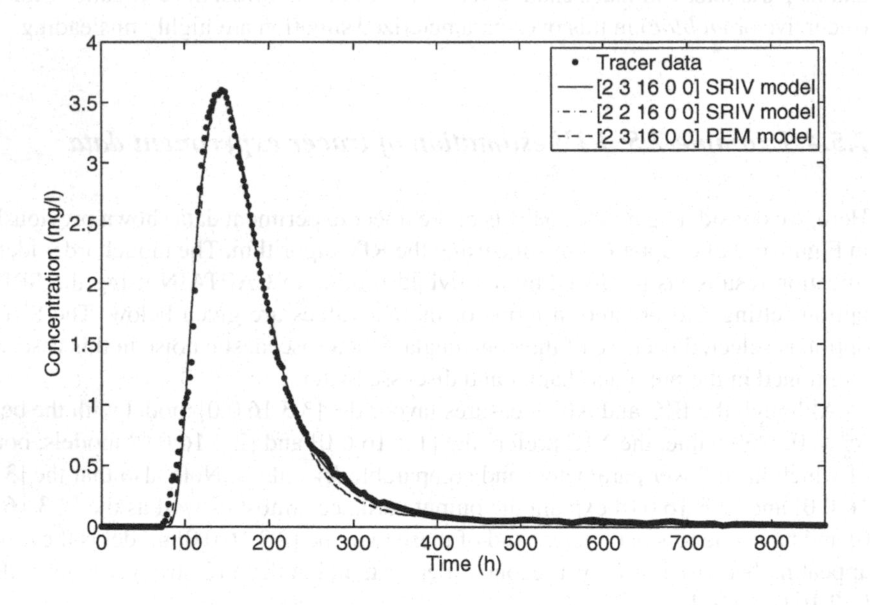

Fig. 7.6 Tracer example: comparison of RIV and PEM estimated model responses.

In hydrology, the recession part of the tracer curve is important because it best defines the dispersion characteristics. In this case, it is clear that the SRIV estimated [2 3 16 0 0] model is a little better than both the SRIV estimated [2 2 16 0 0] and the PEM estimated [2 3 16 0 0] models in explaining this recession and so, on these grounds, it is selected as the best model. The $R_T^2 = 0.996$ and the SRIV estimates of the model parameters obtained using the SRIV option in rivbj are as follows, with the estimated standard errors shown in parentheses:

$$\hat{a}_1 = -1.8680(0.009);\ \hat{a}_2 = 0.8706(0.009);$$
$$\hat{b}_0 = 0.0641(0.016);\quad \hat{b}_1 = -0.0333(0.034);\ \hat{b}_2 = -0.0284(0.018)$$

Converting this model to continuous-time, the steady state gain is 0.967 and the time constants associated with its real poles are 17.4 and 83.7 hours, respectively. The decomposed continuous-time model obtained by the residue routine in Matlab and interpreted in parallel flow terms, has an instantaneous effect after 32 hours (recall that the sampling interval Δt is two hours) and two parallel pathways, each characterized by a first order TF:

$$y(t) = \left\{0.064 + \frac{0.134}{1+83.7s} + \frac{0.769}{1+17.4s}\right\} u(t-32) + \xi(t) \tag{7.39}$$

These results can be compared with those given in example 6.4 (see section 6.9.2 of the previous chapter 6), where the SSG is estimated as 0.933 and the single time constant of the SIV identified first order system is 22.17. It is clear that the SRIV estimation has indeed 'refined' the SIV results and is suggesting that there is evidence of the much longer 83.7 hours time constant that helps to explain better the recession limb of the tracer curve.

The recursive estimates for the [2 3 16 0 0] model are not shown here but the most notable aspect is that the estimates are very volatile for the first 170 hours, until the tracer curve has started its recession. This shows that the parameters are most poorly defined statistically during this first phase in the response, where the tracer concentration is rising rapidly after its first arrival downstream and where the model fit is worst. The same [2 3 16 0 0] model structure is identified as the best continuous-time, differential equation model using the RIVCBJ algorithm described in the next chapter 8. However, since a related multi-input example is presented in 8.6.3 of the next chapter 8, further evaluation of the model results is delayed until then.

7.6 MISO Model Estimation

As discussed in section 6.11 of the previous chapter 6, the extension of the RIV algorithm to the MISO model case with common denominator transfer functions is straightforward, since it simply involves extending the data and parameter vectors to include the additional input terms, so that the definitions of $\boldsymbol{\phi}$ and $\boldsymbol{\rho}$ in pseudo-linear system estimation equation (7.15) are changed to those given in equation (6.85). However, the situation is not nearly so simple in the case where the TF denominators are different because it is not possible to manipulate the MISO equations directly into a suitable pseudo-linear form except by increasing the order of the model substantially (i.e. converting it to the high order ARMAX form: e.g. in the examples of section 7.7, below, the RIV estimated model structure would need to be 6th order with 4th order numerator polynomials). This is possible but it will normally lead to much higher variance estimates and possible identifiability problems. A more parsimonious solution is the SRIV approach suggested by Young and Jakeman (1980a), evaluated fully by Jakeman et al. (1980) and used recently in a continuous-time context by Garnier et al. (2007). In order to explain this SRIV MISO algorithm, it is worth considering a 3-input simulated system that is related to the real system considered in the next section 7.7.1. Here, the simulated system is:

$$\begin{aligned} x(k) &= x_1(k) + x_2(k) + x_3(k) \\ &= \frac{B_1(z^{-1})}{A_1(z^{-1})}u_1(k) + \frac{B_2(z^{-1})}{A_2(z^{-1})}u_2(k) + \frac{B_3(z^{-1})}{A_3(z^{-1})}u_3(k) \\ y(k) &= x(k) + \xi(k) \end{aligned} \tag{7.40}$$

where.

$$\begin{aligned} A_1(z^{-1}) &= 1 - 1.9868z^{-1} + 0.9875z^{-2}; \; B_1(z^{-1}) = 8.8108 \times 10^{-5} \\ A_2(z^{-1}) &= 1 - 1.7641z^{-1} + 0.7702z^{-2}; \; B_2(z^{-1}) = 7.689 \times 10^{-3} \\ A_3(z^{-1}) &= 1 - 1.9229z^{-1} + 0.9267z^{-2}; \; B_3(z^{-1}) = 8.396 \times 10^{-4} \end{aligned}$$

and $\xi(k)$ is an ARMA(14,1) noise model with a zero mean, white noise input with variance $\sigma^2 = 6.07 \times 10^{-6}$ and parameters:

$$\begin{array}{lllll} c_1 = -0.22; & c_2 = -0.34; & c_3 = -0.13; & c_4 = -0.19; & c_5 = 0.0062 \\ c_6 = -0.12; & c_7 = 0.0057; & c_8 = -0.040; & c_9 = 0.021; & c_{10} = -0.049 \\ c_{11} = 0.033; & c_{12} = -0.017; & c_{13} = 0.041; & c_{14} = 0.034; & d_1 = 0.068 \end{array}$$

This results in a highly coloured noise signal with an associated noise/signal level of about 0.07 (7%) by standard deviation. This level and degree of colour are realistic since they are defined by the estimation results obtained from the analysis of the real data. This example is also selected because it presents a reasonable estimation challenge. From the above parameter values, it will be seen that the steady state gains of the TFs associated with the three inputs $u_1(k)$, $u_2(k)$ and $u_3(k)$ are 0.132, 1.252 and 0.219, respectively. Consequently, the effects of the $u_1(k)$ and $u_3(k)$ on $y(k)$ are much less that $u_2(k)$: indeed, a SRIV estimated 2nd order model between $u_2(k)$ and $y(k)$ explains 97.7% of the variance in $y(k)$. As a result, the effective noise/signal level on the first and third TF components of the MISO model will be relatively high and their parameters will be more difficult to estimate than those of the second TF component.

From (7.40), we see that

$$y(k) = x_1(k) + x_2(k) + x_3(k) + \xi(k)$$

so that, if we define 'modified outputs' as follows,

$$\begin{aligned} y_1(k) &= y(k) - x_2(k) - x_3(k) \\ y_2(k) &= y(k) - x_1(k) - x_3(k) \\ y_3(k) &= y(k) - x_1(k) - x_2(k) \end{aligned} \tag{7.41}$$

then,

$$\begin{aligned} y_1(k) &= \frac{B_1(z^{-1})}{A_1(z^{-1})}u_1(k) + \xi(k) \\ y_2(k) &= \frac{B_2(z^{-1})}{A_2(z^{-1})}u_2(k) + \xi(k) \\ y_3(k) &= \frac{B_3(z^{-1})}{A_3(z^{-1})}u_3(k) + \xi(k) \end{aligned} \tag{7.42}$$

This shows that, if we were able to carry out the operations in (7.41), the MISO problem would be converted into the three SISO estimation problems in (7.42) that could be solved using the SRIV algorithm described in previous sections of this chapter. And, bearing in mind the nature of the iterative algorithms considered previously in this book, it suggests that we might be able to construct an iterative 'back-fitting' algorithm (see e.g. Young, 2000) in which estimates $\hat{y}_1(k)$, $\hat{y}_2(k)$ and $\hat{y}_3(k)$ of $y_1(k)$, $y_2(k)$ and $y_3(k)$, respectively, are generated iteratively, starting with some initial estimates $\hat{x}_1^0(k)$, $\hat{x}_2^0(k)$ and $\hat{x}_3^0(k)$ of the noise-free outputs $x_1(k)$, $x_2(k)$ and $x_3(k)$ based, for example, on either the SRIV common denominator MISO estimates; or, alternatively, on the SRIV SISO model estimates obtained from the relationships:

$$\begin{aligned} y(k) &= \frac{B_1(z^{-1})}{A_1(z^{-1})}u_1(k) + \zeta_1(k) \\ y(k) &= \frac{B_2(z^{-1})}{A_2(z^{-1})}u_2(k) + \zeta_2(k) \\ y(k) &= \frac{B_3(z^{-1})}{A_3(z^{-1})}u_3(k) + \zeta_3(k) \end{aligned} \tag{7.43}$$

where,

$$\begin{aligned} \zeta_1(k) &= x_2(k) + x_3(k) + \xi(k) \\ \zeta_2(k) &= x_1(k) + x_3(k) + \xi(k) \\ \zeta_3(k) &= x_1(k) + x_2(k) + \xi(k) \end{aligned} \tag{7.44}$$

are additive 'noise' variables composed from the effects of the actual noise $\xi(k)$ and components arising from the effects of the other input variables that do not appear in each SISO relationship. The latter approach works because of the inherent IV form of the SRIV algorithm. However, the results may not be all that good if the noise/signal level is large, as it can be in this MISO situation. Consequently, the former, common denominator estimation is preferable.

Let us now consider a SRIV solution to the the general MISO model estimation problem, based on the back-fitting ideas above. The general MISO model follows from (6.85) of chapter 6 with the noise model parameter vector $\boldsymbol{\eta}$ as defined in the earlier section 7.1.3 and the system model now defined by the following parameter vector:

$$\boldsymbol{\rho} = [a_{11} \cdots a_{1n_1}\ a_{21} \cdots a_{2n_2} \cdots\cdots a_{mn_m}\ b_{10} \cdots b_{1m_1}\ b_{20} \cdots b_{2m_2}\ \cdots\cdots\ b_{m_m}]^T$$

where a_{ij}, $i = 1,2, \ldots, m$; $j = 1,2, \ldots, n_j$ are the parameters of the m different denominator polynomials in the MISO model. The above reasoning for the 3-input example results in the following general *Simplified Refined Instrumental Variable for Different Denominator* (SRIVDD) algorithm:

The SRIVDD algorithm for MISO models with different denominators

Step 1. Standard SRIV MISO model estimation: apply the SRIV algorithm to the MISO data in the standard manner to obtain a common denominator MISO model. Use this MISO model to generate initial estimates $\hat{x}_j^0(k)$ of the m unobserved outputs $x_j(k), j = 1,2, \ldots, m$ in the MISO model.

Step 2. Back-fitting estimation using the SISO SRIV algorithm:. Iterate the following loop for $p = 1,2, \ldots, p_c$ until the parameter estimates in the vector $\hat{\boldsymbol{\rho}}$ have not changed significantly according to some chosen criterion:

for $i = 1 : m$

Form the modified output variable $y_i(k) = y(k) - \Sigma_{j \neq i} \hat{x}_j^p$, where $\hat{x}_j^p, j = 1,2, \ldots, m$ are the outputs estimated at the previous iteration and, each time, apply the SRIV algorithm to estimate the SISO model between $y_i(k)$ and $u_i(k)$.

end

Step 3. Noise model estimation: form the final MISO system model parameter estimate $\hat{\boldsymbol{\rho}}$ from the m TF models estimated in Step 3 and re-estimate a single ARMA noise model from final error series $y(k) - \hat{x}(k)$ in order to provide the estimates of the parameters in the noise model parameter vector $\hat{\boldsymbol{\eta}}$.

Access to the standard RIV algorithm rivbj in CAPTAIN makes this a quite simple to implement in Matlab (see later, exercise 3) since it only requires calls to the rivbj routine, which provides the required SRIV algorithm. However, although convergence is normally reasonable, it can be computationally rather slow compared with rivbj and should be used only where the MISO model with different denominator TF models is essential[5].

Note that the back-fitting procedure used in SRIVDD means that the parameters of the m constituent TF sub-models in an m-input MISO model are each estimated, in turn, as SISO models by application of the SRIV algorithm. As a result, m separate covariance matrices are produced that provide the estimated uncertainties associated with the system model parameters. In addition, a separate covariance matrix is produced for the estimated noise model parameters. Thus, for the winding process data considered later in section 7.7.2, three 3×3 covariance matrices are produced for the system model, plus one 15×15 noise model covariance matrix. This means that the covariance matrix for all the MISO model parameters in the

[5] A computationally more efficient implementation is planned for CAPTAIN.

$\boldsymbol{\theta} = [\boldsymbol{\rho}^T \boldsymbol{\eta}^T]^T$ parameter vector is constrained to be block diagonal with no cross-covariance estimates provided for parameters between the sub-models or with the noise model parameters. This contrasts with the equivalent covariance matrix produced by standard maximum likelihood algorithms, such as PEM, which provide a full, unconstrained covariance matrix: e.g. in the case of the winder process example this is of dimension 24×24.

The constrained structure of the SRIVDD covariance matrix and the use of the SRIV algorithm within a back-fitting context means that SRIV-MISO estimation is clearly not optimal in maximum likelihood terms. However, the SRIV algorithm ensures that the estimates are consistent, asymptotically unbiased, and they should relatively efficient, particularly in cases where the input variables are not correlated. Indeed, as we shall seen in the examples below, the SRIVDD algorithm can even produce better results than PEM in some cases. On the other hand, the SRIVDD algorithm is not as computationally efficient as PEM and questions remain about its implementation: for example, whether the ordering of the back-fitting operations can affect the estimation results; or whether starting the algorithm with common denominator MISO estimation always leads to convergence. So there is a clear need further research in this regard. For this reason, while a prototype rivbjdd routine has been developed for CAPTAIN and is being evaluated, it has not yet been introduced into the Toolbox (see Appendix G).

7.7 Examples of SRIVDD Identification and Estimation for MISO Models

This section first describes two, related examples of SRIVDD estimation applied to a three input winding process similar to that used in a number of industrial processes. The first simulation example allows us to evaluate the performance of the algorithm with this kind of process; and the second is an application to a set of real data from a laboratory scale pilot process. A third, simulation example evaluates the SRIVDD algorithm using Monte Carlo simulation analysis.

7.7.1 *Example 7.6: Simulation results for a 3-input MISO model*

Here, the SRIVDD algorithm is applied to 2942 input-output samples of simulated data, as generated by the 3-input model (7.40) considered above, with three independent PRBS signals used as inputs. This model is a simulated version of the winding system considered in the next example. The simulated data are shown in Figure 7.7. For comparison, two other estimation algorithms are applied to the same data set: the PEM algorithm from the Matlab SID Toolbox; and the standard, common denominator, RIV algorithm, in order to estimate a common denominator model.

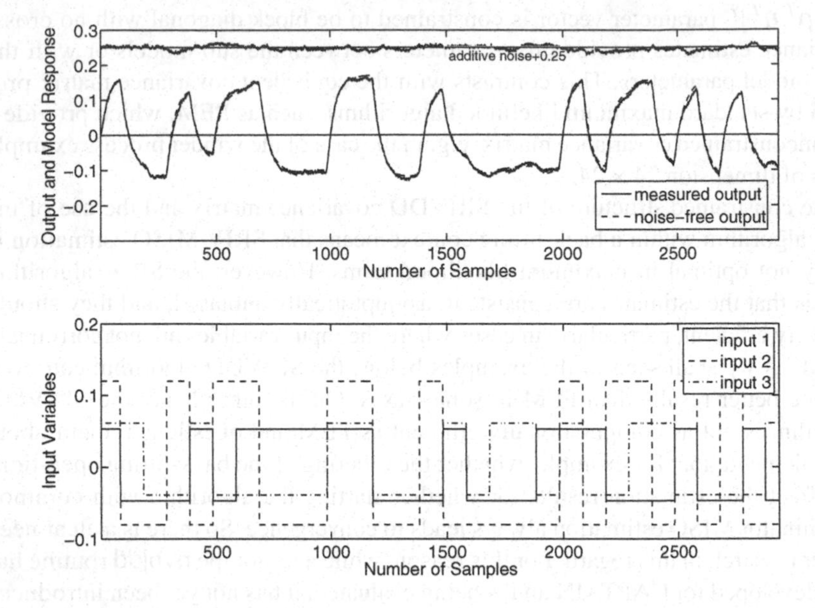

Fig. 7.7 MISO simulation example 7.6: inputs (lower panel) and output (upper panel) signals used for estimation. The additive noise(+0.25) is shown above to indicate more clearly the noise level.

Table 7.3 Estimation results for simulation example 7.6: $\hat{\rho}$ is the estimated parameter vector for the system model; The ARMA(14,1) noise model parameter estimates are not shown to conserve space; SE denotes the estimated standard error on the estimates.

Parameter		$\hat{a}_{11}$	$\hat{a}_{12}$	$\hat{a}_{21}$	$\hat{a}_{22}$	$\hat{a}_{31}$	$\hat{a}_{32}$	$\hat{b}_{10}$	$\hat{b}_{20}$	$\hat{b}_{30}$
True Values		-1.9868	0.9875	-1.7641	0.7702	-1.9229	0.9267	8.811e-05	0.007689	0.0008396
SRIVDD	$\hat{\rho}$	-1.9904	0.99108	-1.7576	0.7639	-1.8998	0.9043	8.176e-05	0.007873	0.001073
$R_T^2 = 0.9968$	SE	0.0058	0.0058	0.0079	0.0077	0.0179	0.0174	3.132e-05	0.00025	0.00017
PEM	$\hat{\rho}$	0.0139	-0.9846	-1.7612	0.7674	-1.8350	0.9837	0.003659	0.007777	-0.003799
$R_T^2 = 0.9803$	SE	0.019	0.019	0.011	0.011	0.018	0.019	0.0021	0.00033	0.00032
RIV	$\hat{\rho}$	-1.7641	0.77025	-1.7641	0.77025	-1.7641	0.77025	0.0006631	0.007682	0.001915
$R_T^2 = 0.9927$	SE	0.0099	0.0097	0.0099	0.0097	0.0099	0.0097	0.00017	0.00031	0.00018

The estimation results are presented in Table 7.3 and Figure 7.8 compares the deterministic outputs from the three estimated models with the noise-free measured output. It is clear from these results that the SRIVDD estimated model matches the simulated data extremely closely with $R_T^2 = 0.9998$ and the parameter estimates are defined well in relation to the true values. Rather surprisingly, given that these are simulated data, the PEM results ($R_T^2 = 0.9863$ and quite large errors) are not nearly as good as those obtained from the standard RIV with common denominator TFs, which does quite well with $R_T^2 = 0.9968$. As expected and shown in Figure 7.9, the estimation of the first (top panel) and third (bottom panel) component TF models is not as good as the estimation of the most important, second TF (middle panel) due to their much lower steady state gains.This also seems to be the reason for the poorer PEM estimation results, where the first and third TF models are not estimated well.

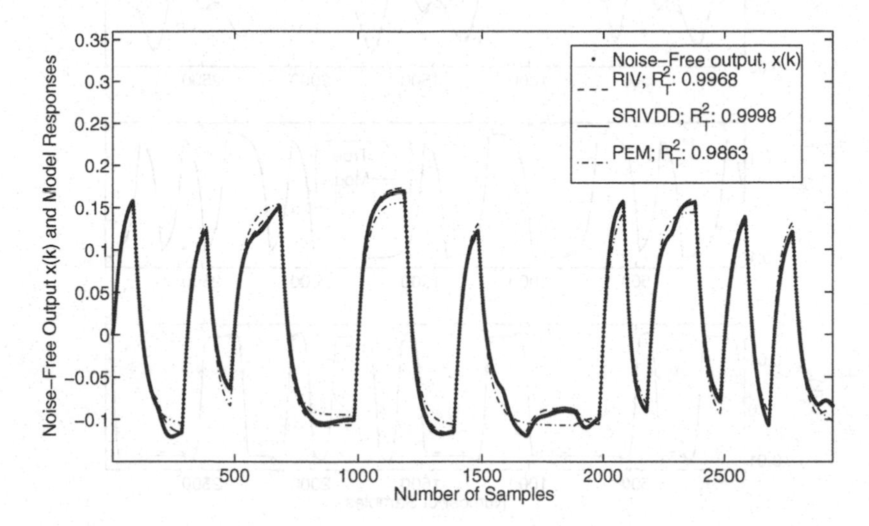

Fig. 7.8 MISO simulation example 7.6: comparison of RIV, SRIVDD and PEM estimated model responses.

7.7.2 Example 7.7: SRIVDD modelling of a MISO winding process

This example utilizes real data collected from the pilot scale 3-input winding process. Winding systems are encountered in a wide variety of industrial plants such as rolling mills in the steel industry, plants involving web conveyance including coating, paper making and polymer film extrusion processes. The main role of a winding process is to control the web conveyance in order to avoid the effects of friction and sliding, as well as the problems of material distortion which can also damage the quality of the final product. A diagram of the process is given in Figure 6 of the Garnier et al. (2007) paper. The main part of the pilot plant is a winding process composed of a plastic web and three reels. Each reel is coupled with a direct-current motor via gear reduction. The angular speed of each reel is measured by a tachometer, while the tensions between the reels are measured by tension meters. At a second level, each motor is driven by a local controller. Two Proportional-Integral (PI) control loops adjust the motor currents 1 and 3; and a double PI control loop drives the angular speed. Driving a winding process essentially comes down to controlling the web linear velocity and the web tensions around a given operating point. The data set used in this example is obtained from the authors of the paper by Bastogne et al. (1998), which contains full information on the experimental design and implementation.

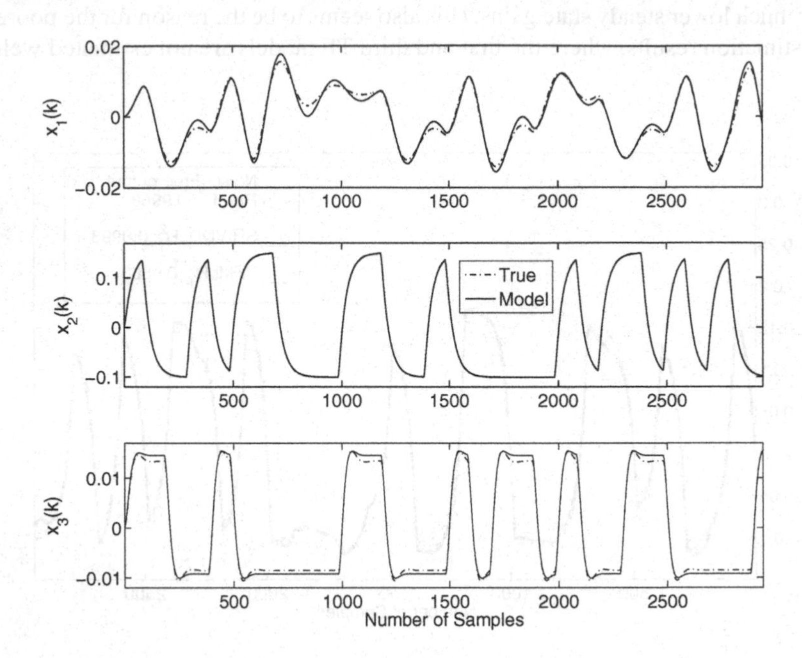

Fig. 7.9 MISO simulation example 7.6: SRIVDD estimated component model outputs compared with true, noise-free, simulated outputs.

The MISO simulation model discussed in the previous sub-section is based on the SRIVDD estimation of the winder data considered in this sub-section, so the three inputs are those plotted in the lower panel of Figure 7.7. These are the input to adjust motor current 1, the input controlling the motor angular speed and the input to adjust the motor current 3, respectively. The measured output is the motor angular speed and this is plotted in Figure 7.10. This pilot process has also been considered previously by Garnier et al. (2007), who identify and estimate a continuous-time MISO model based on the same sampled data employed in the present example, but using an estimation algorithm of the SRIVCDD type discussed in the next chapter.

Figure 7.10, compares the measured output of the winding process with the deterministic outputs of the SRIVDD, standard RIV and PEM estimated MISO models. Once again, the PEM algorithm does rather badly on these real data. The SRIVDD parameter estimates are shown below, with the estimated standard errors shown in parentheses. The system model parameter estimates are well defined but some of the noise model parameters have large relative standard errors, suggesting that they are statistically insignificant and could be set to zero (a subset ARMA process).

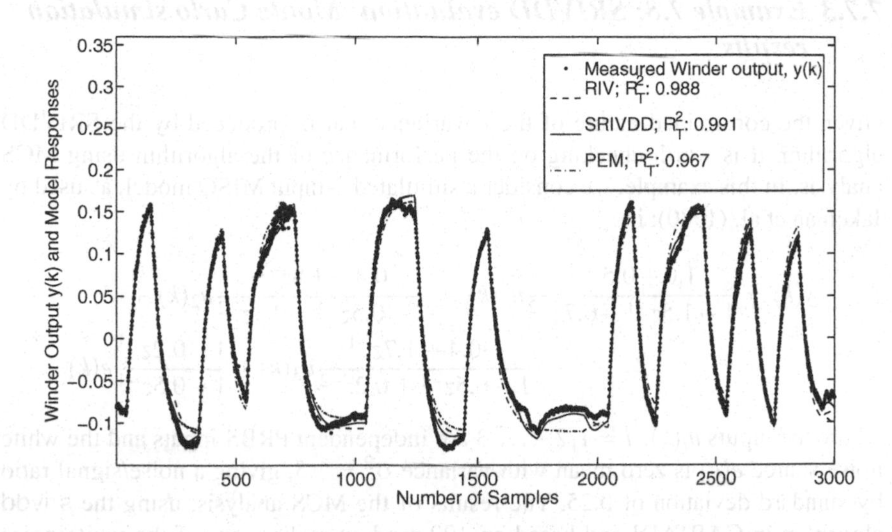

Fig. 7.10 Winding process example 7.7: estimated MISO model deterministic responses compared with the measured output.

$$
\begin{aligned}
&\hat{a}_{11} = -1.9868(0.0072);\ \hat{a}_{12} = 0.9875(0.0072);\ \hat{b}_{10} = 8.8108 \times 10^{-5}(3.3 \times 10^{-5}) \\
&\hat{a}_{21} = -1.7641(0.0094);\ \hat{a}_{22} = 0.7702(0.0092);\ \hat{b}_{20} = 7.689 \times 10^{-3}(3.0 \times 10^{-3}) \\
&\hat{a}_{31} = -1.9229(0.0178);\ \hat{a}_{32} = 0.9267(0.0174);\ \hat{b}_{30} = 8.396 \times 10^{-4}(1.6 \times 10^{-4}) \\
&\hat{c}_1 = -0.2227(0.60); \quad \hat{c}_2 = -0.3410(0.18); \quad \hat{c}_3 = -0.1303(0.19) \\
&\hat{c}_4 = -0.1888(0.07); \quad \hat{c}_5 = 0.006204(0.11); \quad \hat{c}_6 = -0.1226(0.02) \\
&\hat{c}_7 = 0.005717)0.08); \quad \hat{c}_8 = -0.03962(0.02); \quad \hat{c}_9 = 0.02108(0.03) \\
&\hat{c}_{10} = -0.04876(0.02); \quad \hat{c}_{11} = 0.03321(0.03); \quad \hat{c}_{12} = -0.01747(.03) \\
&\hat{c}_{13} = 0.04066(0.02); \quad \hat{c}_{14} = 0.03412(0.03); \quad \hat{d}_1 = 0.06798(0.6) \\
&\hat{\sigma}^2 = 6.07 \times 10^{-6}
\end{aligned}
$$

However, the variance of the estimated noise $\hat{\xi}(k)$ is 14 times the variance of the final residuals $\hat{\sigma}^2$ and most of the colour has been purged from $\hat{\xi}(k)$, so that the model seems satisfactory in this regard. Finally, the whole MISO model performs very well on the validation data (see Garnier et al., 2007) with an $R_T^2 = 0.989$ that is only marginally less than that achieved on the estimation data.

7.7.3 Example 7.8: SRIVDD evaluation: Monte Carlo simulation results

Given the constrained nature of the covariance matrix produced by the SRIVDD algorithm, it is worth checking on the performance of the algorithm using MCS analysis. In this example, we consider a simulated 3-input MISO model, as used by Jakeman et al. (1980): i.e.,

$$y(k) = \frac{1.0+0.5z^{-1}}{1-1.5z^{-1}+0.7z^{-2}}u_1(k) + \frac{0.4-1.8z^{-1}}{1-0.5z^{-1}-0.3z^{-2}}u_2(k) + \frac{-0.4+1.7z^{-1}}{1-0.5z^{-1}+0.2z^{-2}}u_3(k) + \frac{1-0.2z^{-1}}{1-0.5z^{-1}}e(k)$$

where the inputs $u_i(k),\ i=1,2,\ \ldots,\ 3$ are independent PRBS inputs and the white noise source $e(k)$ is zero mean with variance $\sigma^2 = 1.5$, giving a noise/signal ratio by standard deviation of 0.25. The results of the MCS analysis, using the srivdd algorithm in CAPTAIN and based on 100 random realizations of the white noise $e(k)$, are shown in Tables 7.4 and 7.5. These show that the parameters are estimated well and that the standard errors estimated by the SRIVDD algorithm are close to those evaluated empirically by the MCS analysis. A similar analysis using the pem algorithm in the Matlab SID Toolbox produced virtually the same results; so that, *in this example*, the constrained nature of the SRIVDD covariance matrix makes no difference and the estimation performance is close to optimal. The generality of this result is not clear at this time and further research is required to evaluate this aspect of the algorithmic performance.

Table 7.4 System model estimation results for simulation Example 7.8: $\hat{\mathbf{p}}$ is the estimated parameter vector for the system model; SE denotes the SRIVDD estimated standard error on the estimates and SD denotes the standard deviation computed by the MCS analysis.

Parameter	$\hat{a}_{11}$	$\hat{a}_{12}$	$\hat{b}_{10}$	$\hat{b}_{12}$	$\hat{a}_{21}$	$\hat{a}_{22}$	$\hat{b}_{20}$	$\hat{b}_{21}$	$\hat{a}_{31}$	$\hat{a}_{32}$	$\hat{b}_{30}$	$\hat{b}_{31}$
True Values	-1.5	0.7	1	0.5	-0.5	-0.3	0.4	-1.8	-0.5	0.2	-0.4	1.7
$\hat{\mathbf{p}}$	-1.4982	0.6986	0.9946	0.5109	-0.5184	-0.2841	0.3489	-1.7343	-0.5078	0.2083	-0.4066	1.7103
SE	0.011	0.007	0.075	0.105	0.090	0.078	0.110	0.148	0.089	0.061	0.129	0.162
SD	0.010	0.007	0.073	0.101	0.092	0.080	0.112	0.149	0.098	0.072	0.118	0.164

Table 7.5 Noise model estimation results for simulation example 7.8: $\hat{\boldsymbol{\eta}}$ is the estimated parameter vector for the ARMA(1,1) noise model; SE denotes the SRIVDD estimated standard error on the estimates and SD denotes the standard deviation computed by the MCS analysis.

Parameter	$\hat{c}_1$	$\hat{d}_1$
True Values	-0.5	-0.2
$\hat{\boldsymbol{\eta}}$	-0.4976	-0.1999
SE	0.037	0.044
SD	0.038	0.042

7.8 Optimal Prefilters and an Adaptive Kalman Filter

Given the optimality of the prefilters used in RIV estimation, we might suspect that they can be linked with the optimal Kalman Filter. This is indeed the case: Young (1979) has shown how the estimates $\hat{\mathbf{x}}(k)$ of the state $\mathbf{x}(k)$ of a stochastic, discrete-time, SISO system can be obtained as a linear function of the outputs of the adaptive prefilters used in the RIV algorithm when it is applied to the ARMAX model (6.7) in chapter 6.

The analysis starts with equation (6.11) of chapter 6, which is repeated below for convenience:

$$\begin{aligned} \hat{y}(k) &= \frac{D(z^{-1})-A(z^{-1})}{D(z^{-1})}y(k) + \frac{B(z^{-1})}{D(z^{-1})}u(k) \\ \hat{y}(k) &= (d_1-a_1)y_{f_2}(k-1) + \ldots + (d_n-a_n)y_{f_2}(k-n) \\ &\quad b_1 u_{f_2}(k-1) + \ldots + b_n u_{f_2}(k-n) \end{aligned} \tag{7.45}$$

where, as in sub-section 7.1.3 above, the subscript f_2 indicates that the associated variable is prefiltered by $f_2(z^{-1})$ defined in equation (7.22). In other words, the optimally filtered output $\hat{y}(k)$ at the kth sampling interval can be generated by a linear sum of the prefiltered variables $y_{f_2}(k-i)$ and $u_{f_2}(k-i)$, $i=1,2,\ldots,n$, over previous sampling intervals.

We know that the state vector is not unique so, in order to generate an estimate of a whole state vector, it is necessary to select a specific state vector. If we consider the state space model defined by **A, b** and **c** in the canonical form (B.26) of section B.3.3 in Appendix B, then the following expression for the optimal estimate $\hat{\mathbf{x}}(k)$ of the whole state vector $\mathbf{x}(k)$ in this state space model can be obtained in a similar but less straightforward manner (Young, 1979):

$$\hat{\mathbf{x}}(k) = \mathbf{Z}(k)\mathbf{p} \tag{7.46}$$

where

$$\begin{aligned} \mathbf{Z}(k) &= [\mathbf{N}_1\boldsymbol{\zeta}_1(k) \ldots \mathbf{N}_n\boldsymbol{\zeta}_1(k)\mathbf{N}_1\boldsymbol{\zeta}_2(k) \ldots \mathbf{N}_n\boldsymbol{\zeta}_2(k)] \\ \mathbf{p}^T &= [\mathbf{d}^T - \mathbf{a}^T \vdots \mathbf{b}^T] \end{aligned} \tag{7.47}$$

Here $\boldsymbol{\zeta}_2(k)$ and $\boldsymbol{\zeta}_2(k)$ are the following vectors of prefiltered variables,

$$\boldsymbol{\zeta}_1(k) = [y_{f_2}(k-1),\ldots,y_{f_2}(k-n)]\,; \quad \boldsymbol{\zeta}_2(k) = [u_{f_2}(k-1),\ldots,u_{f_2}(k-n)]$$

while $\mathbf{N}_i$ are $n \times n$ matrices, $i=1,2,\ldots,n$, composed of the numerator coefficients of $[\mathbf{I}_n - \mathbf{F}z^{-1}]^{-1}\delta_i$, where δ_i is the ith unit vector and $\mathbf{F}$ is the following matrix

$$\mathbf{F} = \begin{bmatrix} -d_1 & 1 & 0 & \ldots & 0 \\ -d_2 & 0 & 1 & \ldots & 0 \\ -d_3 & 0 & 0 & \ldots & 0 \\ \ldots & \ldots & \ldots & \ldots & 1 \\ -d_n & 0 & 0 & \ldots & 0 \end{bmatrix} \tag{7.48}$$

Finally, the vectors **a**, **b** and **d** in (7.47) are defined as

$$\mathbf{a} = [a_1, \ldots, a_n]^T; \ \mathbf{b} = [b_1, \ldots, b_n]^T; \ \mathbf{d} = [d_1, \ldots, d_n]^T \tag{7.49}$$

The state estimation filter (7.46) can be implemented in practice by replacing $\mathbf{p}$ by its estimate $\hat{\mathbf{p}}$ obtained from a recursive RIV algorithm and using the outputs of the prefilters in the same algorithm to define $\mathbf{Z}(k)$. A typical example of the performance of an adaptive state estimator of this kind, for a discrete-time second order system, is given in Young and Jakeman (1980a).

This approach to state estimation can also be applied in the purely stochastic situation where no input variable $u(k)$ is present: here the IVARMA algorithm would provide the source of parameter estimates and prefiltered variables. Also, it is possible to extend (7.46) to multivariable systems (Jakeman and Young, 1979a) but this suffers from the disadvantages of complexity associated with all such multivariable 'black box' methods and has not yet been implemented in practice.

7.9 Exercises

1. Implement the Matlab m-file script below (requires the Control System Toolbox) for the given values of the parameters in the first line.

```
zeta1=0.1;wn1=1.0;zeta2=0.2;wn2=10;G1=1.0;G2=1.0;
A=[1 2*zeta1*wn1 wn1^2];B=G1*wn1^2;%system
C=[1 2*zeta2*wn2 wn2^2];D=1;%noise
DA=conv(D,A);%prefilter denominator
M1=idpoly(1,B,1,1,A,0,0,0);%system model object
M2=idpoly(1,D,1,1,C,0,0,0);%noise model object
M3=idpoly(1,C,1,1,DA,0,0,0);%prefilter model object
M1d=c2d(M1,0.1,'zoh');M2d=c2d(M2,0.1,'zoh');
M3d=c2d(M3,0.1,'zoh');%conversion to discrete-time
figure(1);clf;bode(M1d,M2d,M3d);%discrete-time Bode
figure(2);clf;bode(M1,M2,M3)%continuous-time Bode
```

   Repeat for different values of the parameters and consider what the results mean in relation to the function of the optimal prefiltering operations in the RIV algorithm.

2. Compare the least squares solution of (7.36) with the solution (6.40) in section 6.4 of chapter 6 and show that they are the same (hint: refer to (4.6) in chapter 4 in the limit as $k \to \infty$).

3. Using the CAPTAIN rivbj and ivarma routines, write a Matlab m-file script to implement the SRIVDD algorithm for a 2-input system and test this out on the simulation example in section 7.7.3, with the third input removed.
4. Write a Matlab m-file script to simulate the transfer function model used in example 7.1 and check that you obtain similar estimation results to those reported in section 7.7.3.
5. Starting from equation (7.45), obtain the full state estimation equation (7.46).

7.10 Summary

This chapter has described the formulation and implementation of the *Refined Instrumental Variable* (RIV) algorithm for the identification and estimation of discrete-time TF models. While this algorithm has been implemented in full multivariable (MIMO) form (see Jakeman and Young, 1979), the SISO and MISO versions are of much more practical utility and the emphasis here has been on these simpler implementations. Under the normal statistical assumptions associated with the Box-Jenkins TF model, the RIV algorithm yields statistically optimal estimates of the system and noise model parameters and, because of the inherent instrumental variable nature of the algorithm, the estimates remain consistent even if the noise model assumptions are violated.

RIV routines for discrete-time δ operator models (Middleton and Goodwin, 1990) have been developed (Young et al., 1998; Chotai et al., 1998)) and such models, which provide an interesting bridge between discrete and continuous-time models, are very useful in automatic control applications. However, these δ operator routines lead on naturally to fully continuous-time models, as discussed in the next chapter 8 and so they are not discussed in the present book.

Finally, in case there is any confusion, it must be emphasized that the discrete-time RIV algorithm described in this paper is quite different to the IV4 algorithm available in the Matlab SID Toolbox, which is formulated on the basis of an ARARX model form, under the assumption of an AR$(n+m)$ noise model. For instance, Söderström and Stoica (1989) are quite misleading when they state that: "*The optimal IV method … has been analyzed by Stoica and Söderström … For an approximate implementation see also the schemes suggested by Young and Jakeman …*". Although they have some conceptual elements in common, the IV4 is a four *step* algorithm, while the essential element of the RIV algorithm is its exploitation of multiple *iterations*. And, as we have seen in this paper, it is through this iterative updating of the instrumental variables and the prefilters that these algorithms are not "approximate implementations" but, upon convergence, possess the same optimal statistical properties (consistency and asymptotic efficiency) as the more conventional ML and PEM gradient algorithms. Most importantly in practical terms, simulation and practical experience has shown that the RIV algorithm is more reliable than IV4.

Chapter 8
Optimal Identification and Estimation of Continuous-Time Transfer Function Models

8.1 Introduction

Continuous-time (CT) models have advantages in many practical situations. First, most scientific laws, such as the conservation laws (mass, energy, momentum etc.) are posed in terms of CT equations so that, normally, the estimated parameters in the CT model can be interpreted directly in physically meaningful terms. Second, provided the sampling interval used in the high order model simulation is small enough, the parameters of the CT model are well defined and unique: unlike the DT model, they are not a function of the sampling interval Δt. Finally, if required (e.g. for control or forecasting system design), the DT model at any selected Δt can be generated straightforwardly by transformation from the CT model (see next section 8.2).

The main disadvantage of CT models is that the direct estimation of the parameters in a continuous-time transfer function (TF) model identification raises several technical issues. In particular, unlike the difference equation model, the differential equation model contains time-derivative terms that are not normally available for direct measurement and cannot be obtained by direct differentiation because of the noise on the data. However, early research (e.g. Valstar, 1963; Young, 1964, 1965a, 1966) established that judiciously selected prefilters on the measured input and output signals yielded filtered derivatives that could be used to replace the direct derivatives in parameter estimation algorithms.

Various prefiltering techniques have been devised to deal with the need to reconstruct these time-derivatives (see e.g. Young, 1981; Unbehauen and Rao, 1987, 1990, 1998; Sinha and Rao, 1991; Young et al., 1991; Garnier et al., 2003). Each of these methods is characterized by specific advantages, such as mathematical convenience, simplicity in numerical implementation and computation, handling of initial conditions, physical insight, and accuracy. Initially, most of these methods were largely deterministic, in the sense that they did not explicitly model the additive noise process nor attempt to quantify the statistical properties of the parameter estimates. Instead, consistent estimates were obtained using basic, sub-optimal

Instrumental Variable (IV) methods, such as the standard IV method described in chapter 6. Indeed, one deterministic approach of this type, known as the 'state-variable filter' (SVF) method (Young, 1964, 1965a, 1966), dates from the days of analog and hybrid computers.

Although the sub-optimal methods are still utilized to some extent because they involve the use of fixed, rather than iteratively adaptive, prefilters and so are simpler to implement, they will not be considered here since the optimal *Refined Instrumental Variable for Continuous-time Systems* (RIVC) algorithm, as described in the present chapter, is easily available in both the CAPTAIN and CONTSID[1] Toolboxes for Matlab and seems robust in general use, as we shall see in this chapter. Also, some of the sub-optimal implementations are options in the CAPTAIN Toolbox implementation of the RIVC algorithm, rivcbj; and most of the sub-optimal algorithms are routines in the CONTSID Toolbox.

During the original development of the RIV algorithm, as discussed in the previous chapter, it was realized (Young and Jakeman, 1980a) that the optimal prefiltering also provided an approach to the optimal estimation of continuous-time TF models that was a logical extension of sub-optimal methods developed previously (Young, 1970a,b). Originally this optimal RIVC algorithm was developed in a *Simplified* (SRIVC) form, analogous to the SRIV algorithm described in chapter 7, which is optimal in the case of white additive noise. This had the advantage that it did not require the specific inclusion of a continuous-time ARMA noise model, which introduces some theoretical and practical problems. In particular, the concept of a continuous-time white noise process presents a major difficulty, since the noise does not exist in a rigorous sense because it implies infinite variance (see Appendix B, section B.3.5 and Priestley (1981), page 156 *et seq*). In order to avoid this difficulty, it was suggested that a hybrid model should be considered, with the additive noise modelled as a discrete-time ARMA process. This suggestion was taken up much later by Young, Garnier and Gilson (2008) and the resulting optimal RIVC algorithm is described in the next section 8.3.

It should be mentioned that the estimation of continuous-time ARMA processes from sampled data has received considerable attention. Information on this area of research is available in Larsson et al. (2008), Gillberg and Ljung (2009) and the prior references therein. It is possible, therefore, that such methods could be utilized in the context of SRIVC estimation, but this has not been considered because the methods proposed are not useful in the present recursive estimation context.

However, before proceeding to consider the optimal RIVC algorithm, it is important that we consider the relationship between continuous-time and discrete-time models since this provides the necessary background to the development of the RIVC algorithm.

[1] The CONTSID Toolbox can be downloaded from http://www.cran.uhp-nancy.fr/contsid/.

8.2 TF Conversion Between Continuous and Discrete-Time

This section deals with the relationship between discrete and continuous-time models, as well as how one converts from continuous to discrete-time models and vice versa. This is best considered in general terms from the standpoint of the deterministic state space model:

$$\begin{aligned} \frac{d\mathbf{x}(t)}{dt} &= \mathbf{F}\mathbf{x}(t) + \mathbf{G}\mathbf{u}(t) \qquad (i) \\ y(t) &= \mathbf{H}\mathbf{x}(t) + \mathbf{G}_I\mathbf{u}(t) \qquad (ii) \end{aligned} \tag{8.1}$$

Section B.3.7.1 of Appendix B shows how this can be converted to the following discrete-time form provided it is assumed that the input vector $\mathbf{u}(k)$ is constant over the sampling interval Δt, i.e. the zero-order hold (ZOH) assumption:

$$\begin{aligned} \mathbf{x}(k) &= \mathbf{A}\mathbf{x}(k-1) + \mathbf{B}\mathbf{u}(k-1) \qquad \text{(i)} \\ y(k) &= \mathbf{C}\mathbf{x}(k) + \mathbf{B}_I\mathbf{u}(k) \qquad \text{(ii)} \end{aligned} \tag{8.2}$$

where,

$$\begin{aligned} \mathbf{A}(k) &= e^{\mathbf{F}\Delta t} \\ \mathbf{B}(k) &= \int_0^{\Delta t} e^{\mathbf{F}\tau} d\tau\, \mathbf{G} \end{aligned} \tag{8.3}$$

For this reason, it is often called the ZOH equivalent of (8.1). The reason why this assumption (or the alternative first order hold, FOH) is usually made relates to the normal situation, when confronted with sampled data, that there is no information about how the input variables are changing over the sampling interval. As a result, some form of interpolation over the sampling interval is necessary and the ZOH interpolation (the signal remains constant) is the simplest. Given the stochastic nature of the model estimation problem, it would be possible to use some form of optimal interpolation (e.g. based upon the fixed interval smoothing algorithms discussed in chapters 4 and 5), but this would add considerable complexity and, as we shall see later in this chapter, it does not seem to be justified in the present context.

A simple but useful example of the above conversion is for a scalar, first order discrete-time model of the following form, in which the nomenclature and sign conventions are those used in the later sections of this chapter[2]

$$\frac{dx(t)}{dt} = -\alpha x(t) + \beta u(t) \tag{8.4}$$

where the discrete-time equivalent, using the above results with their ZOH assumption, is of the form

$$x(k) = -ax(k-1) + bu(k-1) \tag{8.5}$$

where the conversion relationships are

[2] TF and state space models have the same form in this scalar situation.

$$
\begin{aligned}
a &= -e^{-\alpha \Delta t} && \text{for conversion from continuous to discrete-time } (i) \\
\alpha &= -\frac{log_e(-a)}{\Delta t} && \text{for conversion from discrete to continuous-time } (ii) \\
b &= \beta(1+a) && \text{the relationship between numerator coefficients } (iii)
\end{aligned}
\tag{8.6}
$$

These relationships are useful because they allow for a quick evaluation of the decomposed TF models (see section B.3.7 of Appendix B) where any decomposed first order TFs associated with real eigenvalues have this form. In particular, the time constant T of the first order system (a continuous-time concept) associated with the estimate of a in (8.5) is given by $1/\alpha$, i.e.

$$
T = -\frac{\Delta t}{log_e(-a)} \tag{8.7}
$$

a relationship that is often used when interpreting the physical nature of decomposed RIV and RIVC estimated TF models. Other aspects of such physical interpretations are discussed in section B.3.7 of Appendix B.

Note from (8.6)(i) that as Δt gets smaller, so the magnitude of a approaches unity (i.e. the eigenvalue approaches the unit circle in the complex z domain) and the discrete-time equation (8.5) approaches the equation for an integrator. This has important implications for discrete-time TF model identification and estimation since the estimate of a, or in general the parameters in the TF denominator $A(z^{-1})$, become increasingly poorly defined and difficult to estimate. This is discussed further in section 8.7 of the present chapter.

The above theoretical background to the relationship between continuous and discrete-time models clarifies the nature of the relationships but in practice, of course, such conversion is nearly always carried out in a computer. In Matlab, for instance the conversions are accomplished by the obviously named c2d and d2c routines, which work for both state space and TF models. In both routines, the user can select a variety of interpolation methods, in addition to ZOH ('zoh'), for the required assumption of how the input variable changes over the sampling interval: linear interpolation or first order hold FOH ('foh'); Impulse-invariant discretization ('impulse'); bilinear approximation or Tustin ('tustin'); and the matched pole-zero method ('matched': for SISO systems only). However, the ZOH assumption is probably the most useful option in many practical situations. More information of the background to these methods is available at the Web site `http://www.mathworks.com/help/toolbox/control/ref/c2d.html` and in Moler and Loan (2003); Higham (2005).

8.3 Hybrid Continuous-Time RIVC Estimation

The stochastic version of the deterministic state space model (8.1) takes the form of equation (B.36) in Appendix B: i.e.,

$$\begin{aligned}\frac{d\mathbf{x}(t)}{dt} &= \mathbf{F}\mathbf{x}(t) + \mathbf{G}\mathbf{u}(t) + \mathbf{J}\boldsymbol{\eta}(t) \qquad (i)\\ \mathbf{y}(t) &= \mathbf{H}\mathbf{x}(t) + \mathbf{G}_I\mathbf{u}(t) + \mathbf{e}(t) \qquad (ii)\end{aligned} \tag{8.8}$$

The hybrid continuous-time RIVC algorithm is concerned with the estimation of the parameters in the TF model derived from this state space model and its associated differential equation model form. In particular, in the SISO situation, it involves the identification and estimation of parameters in the following multi-order, continuous-time (CT), differential equation model based on discrete-time, sampled data measurements of the input and output variables:

$$\frac{d^n x(t)}{dt^n} + \alpha_1 \frac{d^{n-1}x(t)}{dt^{n-1}} + \cdots + \alpha_n x(t) = \beta_0 \frac{d^m u(t-\tau)}{dt^m} + \cdots + \beta_m u(t-\tau) \tag{8.9}$$

This equation can be written in the following alternative, simpler form, where the bracketed superscript denotes the differentiation order,

$$x^{(n)}(t) + \alpha_1 x^{(n-1)}(t) + \cdots + \alpha_n x^{(0)}(t) = \beta_0 u^{(m)}(t-\tau) + \cdots + \beta_m u^{(0)}(t-\tau) \tag{8.10}$$

Here, the pure time delay τ, in appropriate time units, is often assumed to be an integer number related to the sampling time, as in the DT case: that is $\tau = \delta\Delta t$ but this is not essential: in this CT environment, "fractional" time delays can be introduced if required (for example, see Young, 2006). In transfer function (TF) terms, the above differential equation takes the form:

$$x(t) = \frac{B(s)}{A(s)} u(t-\tau), \tag{8.11}$$

with

$$\begin{aligned}B(s) &= \beta_0 s^m + \beta_1 s^{m-1} + \cdots + \beta_m,\\ A(s) &= s^n + \alpha_1 s^{n-1} + \cdots + \alpha_n,\end{aligned}$$

where s is the differential operator, *i.e.* $s^p x(t) = \frac{d^p x(t)}{dt^p}$. It is assumed that the input signal $\{u(t), t_1 < t < t_N\}$ is applied to the system and that this input and the output $x(t)$ are sampled at discrete times $t_1, \cdots, t_N$, not necessarily uniformly spaced.

In the case of uniformly sampled data at a sampling interval Δt, the measured output $y(t_k)$, where $t_k = k\Delta t$, is assumed to be corrupted by an additive measurement noise $\xi(t_k)$,

$$y(t_k) = x(t_k) + \xi(t_k) \tag{8.12}$$

where $x(t_k)$ is the deterministic, noise-free output of the system and, as in the discrete-time case, $\xi(t_k)$ is modelled as a discrete-time ARMA process, *i.e.*,

$$\xi(t_k) = \frac{D(z^{-1})}{C(z^{-1})} e(t_k) \qquad e(t_k) = \mathcal{N}(0, \sigma^2) \tag{8.13}$$

Note that the nomenclature for sampled signals used here, i.e. $x(t_k)$ etc., is changed from that used for purely discrete-time models in chapter 7, i.e. $x(k)$ etc., in order to emphasize that we are dealing here with the sampling of signals from an underlying continuous-time system.

Taken together, equations (8.11), (8.12) and (8.13) constitute the *Hybrid Box-Jenkins* model. The estimation problem posed by this model is to estimate the parameters of the continuous-time TF model (8.11) from N sampled measurements of the input and output $Z^N = \{u(t_k); y(t_k)\}_{k=1}^{N}$. In the case of non-uniform sampling, the sampling interval Δt will itself be a function of the sampling integer k, that is $\Delta t = \Delta t(k)$. In this hybrid, continuous-discrete time situation, the ARMA noise estimation model remains in the same form (7.27) as in the fully discrete-time model case, since the noise model is not changed: i.e.

$$\tilde{e}_{f_2}(k) = \boldsymbol{\psi}^T(k)\boldsymbol{\eta} + \varepsilon(k) \tag{8.14}$$

with the nomenclature defined in chapter 7. However, the TF system estimation model at the kth sampling instant is written in the following pseudo-linear regression form:

$$y_f^{(n)}(t_k) = \boldsymbol{\phi}^T(t_k)\boldsymbol{\rho}_c + e(t_k) \tag{8.15}$$

$$\boldsymbol{\phi}^T(t_k) = [-y_f^{(n-1)}(t_k) \cdots - y_f^{(0)}(t_k)\ u_f^{(m)}(t_k - \tau) \cdots u_f^{(0)}(t_k - \tau)] \tag{8.16}$$

$$\boldsymbol{\rho}_c = [\alpha_1 \ \ldots \ \alpha_n \ \beta_0 \ \ldots \ \beta_m]^T \tag{8.17}$$

where now the subscript f denotes *hybrid prefiltering* which is completely analogous in function to that used in the discrete-time modelling situation, but involves a combination of continuous and discrete-time filters, as shown in Figures 8.1 and 8.2.

First, the prefiltered derivatives are obtained as the inputs to the integrators in the continuous-time implementation of the initial prefilter $1/A(s)$, as shown in Figure 8.1. Note that the inputs to these continuous-time prefilters are interpolated, normally by a FOH operation. Optimal interpolation is possible but these simpler options seem to be sufficient unless the sampling rate is very slow. These prefiltered derivatives are then sampled at the sampling interval Δt, prior to discrete-time prefiltering using the inverse noise filter $C(z^{-1})/D(z^{-1})$, as shown in Figure 8.2. Following the same approach used in the discrete-time situation, these prefilters are based on estimates of the model polynomials and both they, and the polynomials of the continuous-time auxiliary model, are updated iteratively based on the iterative estimates $\hat{\boldsymbol{\rho}}_c$ and $\hat{\boldsymbol{\eta}}$ of parameter vectors $\boldsymbol{\rho}_c$ and $\boldsymbol{\eta}$.

Except for the obvious changes in nomenclature and the nature of the prefiltering, the recursive algorithms for the noise and system model parameters follow straightforwardly because of the unified nature of the RIV/RIVC algorithms. The recursive algorithm for the former has exactly the same equations as the IVARMA algorithm in chapter 7, except that the k argument is replaced by t_k; while the RIVC algorithm for the latter takes the following form, which is almost identical to the RIV algorithm in chapter 7, except for the definition of the vectors and matrices.

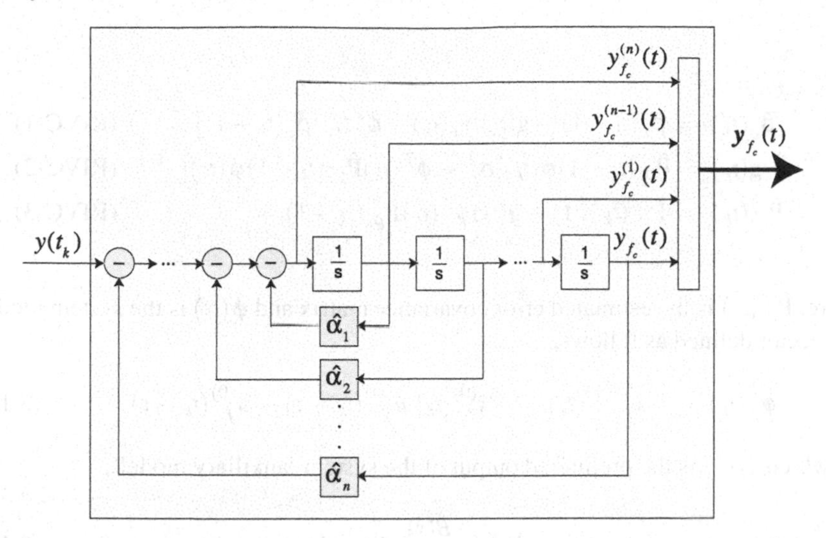

Fig. 8.1 Generation of prefiltered derivatives for the output $y(t)$ by the prefilter $f_c(s) = 1/A(s, \boldsymbol{\rho}_c^j)$, where $\hat{\alpha}_i, i = 1, 2, \ldots, n$ are the iteratively updated parameters of the model polynomial $A(s, \boldsymbol{\rho}_c^j)$ at the jth iteration. This is the inside of the block marked **A** in Figure 8.2 (based on Figure 2(b) in Young et al., 1980a.)

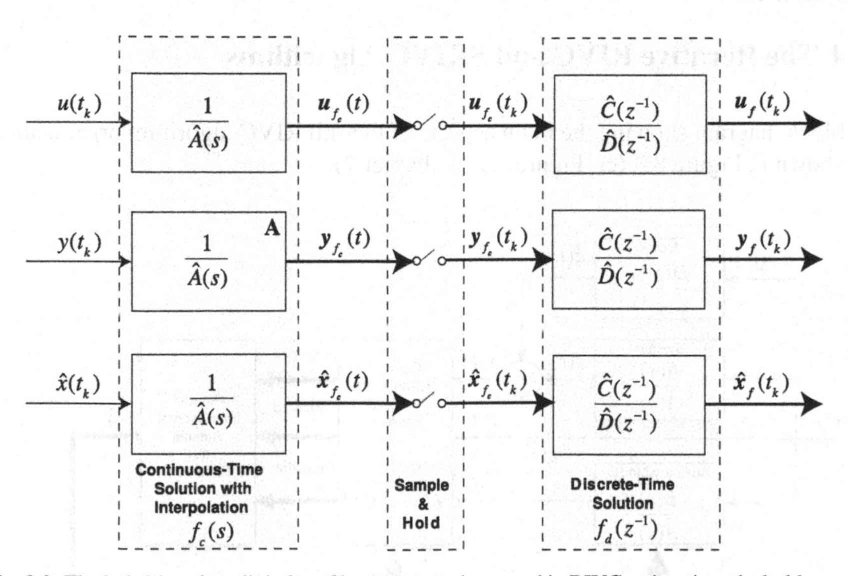

Fig. 8.2 The hybrid analog-digital prefiltering operations used in RIVC estimation: the bold arrows and associated bold-face variables denote vector quantities with elements defined as the appropriate prefiltered derivatives of $u(t_k)$, $y(t_k)$ and $\hat{x}(t_k)$, while $\hat{A}(s) = A(s, \hat{\boldsymbol{\rho}}_c^{j-1})$, $\hat{C}(z^{-1}) = C(z^{-1}, \hat{\boldsymbol{\eta}}^j)$ and $\hat{D}(z^{-1}) = D(z^{-1}, \hat{\boldsymbol{\eta}}^j)$ are the iteratively updated model polynomials that define the prefilters (based on Figure 2 in in Young et al., 1980a). The detail of the block marked **A** is shown in Figure 8.1.

$$\hat{\boldsymbol{\rho}}_c(t_k) = \hat{\boldsymbol{\rho}}_c(t_k-1) + \mathbf{g}(t_k)[y_f(t_k) - \boldsymbol{\phi}^T(t_k)\hat{\boldsymbol{\rho}}_c(t_k-1)] \qquad \text{(RIVC-1)}$$
$$\mathbf{g}(t_k) = \hat{\mathbf{P}}_{\rho_c}(t_k-1)\hat{\boldsymbol{\phi}}(t_k)[\sigma^2 + \boldsymbol{\phi}^T(t_k)\hat{\mathbf{P}}_{\rho_c}(t_k-1)\hat{\boldsymbol{\phi}}(t_k)]^{-1} \qquad \text{(RIVC-2)}$$
$$\hat{\mathbf{P}}_{\rho_c}(t_k) = \hat{\mathbf{P}}_{\rho_c}(t_k-1) - \mathbf{g}(t_k)\boldsymbol{\phi}^T(t_k)\hat{\mathbf{P}}_{\rho_c}(t_k-1) \qquad \text{(RIVC-3)}$$

Here, $\hat{\mathbf{P}}_{\rho_c}(t_k)$ is the estimated error covariance matrix and $\hat{\boldsymbol{\phi}}(t_k)$ is the system model IV vector defined as follows,

$$\hat{\boldsymbol{\phi}}^T(t_k) = [-\hat{x}_f^{(n-1)}(t_k) \cdots - \hat{x}_f^{(0)}(t_k)\; u_f^{(m)}(t_k-\tau) \cdots u_f^{(0)}(t_k-\tau)] \qquad (8.18)$$

in which $\hat{x}_f(k)$ is the prefiltered output of the system 'auxiliary model',

$$\hat{x}(t) = \frac{\hat{B}(s)}{\hat{A}(s)}u(t-\tau). \qquad (8.19)$$

where $\hat{A}(s)$ and $\hat{B}(s)$ are the iteratively updated estimates of the system TF model polynomials.

8.4 The Iterative RIVC and SRIVC Algorithms

A block diagram showing the main aspects of the full RIVC algorithm organization in shown in Figure 8.3 (cf. Figure 7.1 of chapter 7).

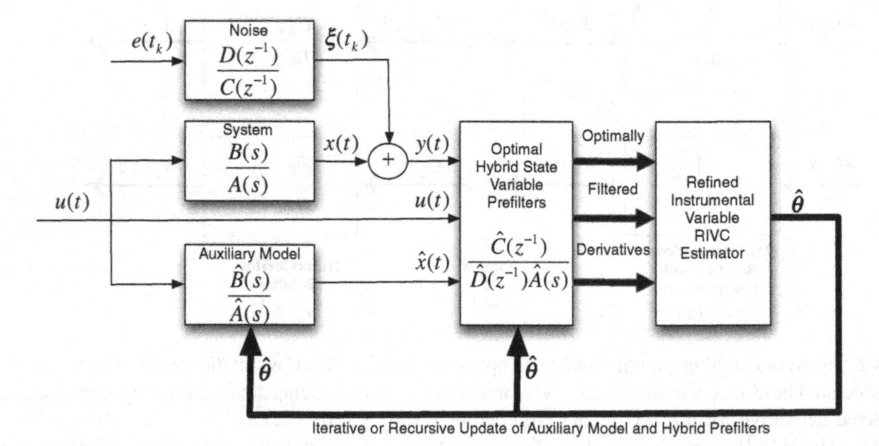

Fig. 8.3 Block diagram of the full hybrid continuous-time RIVC estimation algorithm. Note that the mixed z^{-1} and s operators are used in the optimal hybrid state variable prefilter block to emphasize that this is a hybrid filtering operation.

The SRIVC algorithm was first suggested and implemented by Young and Jakeman (1980a), while the full RIVC algorithm has been implemented recently by Young et al. (2006, 2008). Since the algorithms are so similar to their discrete-time counterparts, they will not be discussed in detail here but simply summarized below.

The full RIVC algorithm (optimal for additive ARMA noise)

Step 1. Initialization: apply the SRIVC algorithm based on an initial $A(s,\hat{\boldsymbol{\rho}}_c^0)$ polynomial (see below) and generate an initial estimate of the TF model parameter vector $\hat{\boldsymbol{\rho}}_c^0$.

Step 2. Iterative or recursive-iterative IV estimation with hybrid prefilters:

for $j = 1 : convergence$

(1) If the estimated TF model is unstable, reflect the unstable eigenvalues of the estimated $A(s,\hat{\boldsymbol{\rho}}_c^{j-1})$ polynomial into the stable region of the complex s plane (e.g. using the Matlab routine polystc)[3]. Generate the IV variable $\hat{x}(t)$ from the auxiliary model:

$$\hat{x}(t) = \frac{B(s,\hat{\boldsymbol{\rho}}_c^{j-1})}{A(s,\hat{\boldsymbol{\rho}}_c^{j-1})}u(t-\tau)$$

with the polynomials based on the estimated parameter vector $\hat{\boldsymbol{\rho}}_c^{j-1}$ obtained at the previous iteration of the algorithm.

(2) Generate the discrete-time series $u(t_k)$, $y(t_k)$ and $\hat{x}(t_k)$ by sampling $u(t)$, $y(t)$ and $\hat{x}(t)$ at the discrete-time sampling interval Δt. Then, as in the discrete-time RIV case, obtain an estimate $\hat{\boldsymbol{\eta}}^j$ of the noise model parameter vector $\boldsymbol{\eta}$ based on the estimated noise sequence $\hat{\xi}(t_k) = y(t_k) - \hat{x}(t_k)$.

(3) Prefilter the input $u(t)$, output $y(t)$ and instrumental variable $\hat{x}(t)$ signals by the continuous-time filter,

$$f_c(s,\hat{\boldsymbol{\rho}}_c^{j-1}) = \frac{1}{A(s,\hat{\boldsymbol{\rho}}_c^{j-1})}, \tag{8.20}$$

in order to generate the filtered derivatives that are available at the inputs to the integrators in the prefilter, as shown in Figure 8.1.

(4) Sample the continuous-time prefiltered input and output derivative signals from (3) and filter them by the discrete-time inverse noise filter,

$$f_d(z^{-1},\hat{\boldsymbol{\eta}}^j) = \frac{C(z^{-1},\hat{\boldsymbol{\eta}}^j)}{D(z^{-1},\hat{\boldsymbol{\eta}}^j)}, \tag{8.21}$$

as shown in Figure 8.2.

(5) Based on these sampled, prefiltered derivatives and the estimation model (8.15), generate the estimate $\hat{\boldsymbol{\rho}}_c^j$ of the parameter vector $\boldsymbol{\rho}_c$ using the *en bloc* RIVC algorithm or its recursive equivalent.

end

Step 3. Compute the estimated parametric error covariance matrices $\hat{\mathbf{P}}_{\rho_c}$ and $\hat{\mathbf{P}}_{\eta}$ associated with the parameter estimates from equivalent expressions to those in (7.31), as used in the discrete-time algorithms.

---

[3] This is not essential to the functioning of the algorithm: it allows for rarely occurring situations, normally with very poor data, when the initially estimated model is found to be unstable.

This full iterative algorithm requires access to the SRIVC algorithm, which is described below.

The SRIVC algorithm (optimal for additive white noise)

As in the case of the discrete-time RIV algorithm, the RIVC estimation problem is considerably simplified if it is assumed that the additive noise is white, i.e., $C(z^{-1}) = D(z^{-1}) = 1$. In this case, SRIVC estimation involves only the parameters in the $A(s)$ and $B(s)$ polynomials and the prefiltering only involves the prefilter $1/A(s, \hat{\boldsymbol{\rho}}_c^{j-1})$. Consequently, the main steps in the SRIVC algorithm are the same as those in the RIVC algorithm, except that the noise model estimation in stage (2) of the iteration and the discrete-time prefiltering in stage (4) are both omitted. However, the initiation of the SRIVC algorithm requires the specification of $A(s, \hat{\boldsymbol{\rho}}_c^0)$. This is discussed fully in Young et al. (2008) but the two simplest approaches are: (i) estimate the model in discrete-time and convert this to continuous-time, so providing an initial estimate of the $A(s)$ polynomial; and (ii) select a single breakpoint parameter λ (breakpoint frequency in radians/time unit) of the *state variable filter*,

$$f(s, \boldsymbol{\rho}_c^0) = \frac{1}{A(s, \boldsymbol{\rho}_c^0)} = \frac{1}{(s+\lambda)^n} \tag{8.22}$$

which is chosen so that it is equal to, or larger than, the bandwidth of the system to be identified. This latter filter form was suggested a long while ago (Young, 1964, 1965a) but has proven popular ever since. Both of these approaches are available as options in the rivcbj routine in the CAPTAIN Toolbox. When using the latter approach in practice, it is necessary to find a λ value that induces satisfactory convergence. However, the choice is not critical since convergence occurs over a fairly wide range of values.

8.5 Model Structure Identification and MISO Model Estimation

The approach to both model structure identification for hydrid continuous-time models and the RIVC estimation of MISO TF model parameters is virtually the same as that used in the RIV case and the reader is referred to sections 7.3 and 7.6 of chapter 7 for the details. The same general approach to structure identification is followed, since the various criteria apply equally well to the hybrid continuous-time model. The rivcbjid routine in CAPTAIN performs in exactly the same way as the rivbjid routine. And the RIVC estimation of the common denominator MISO model is once again a simple extension of the SISO case and is implemented in the CAPTAIN Toolbox routine rivcbj. In the different denominator case, the SRIVCDD algorithm uses the same iterative, back-fitting approach as the SRIVDD algorithm, normally initiated with the common denominator RIVC results. Finally, the comments in section 7.3 of chapter 7 on the identification of 'stiff' systems also apply in this continuous-time situation and it is better to use the SRIVC-ARMA algorithm, the hybrid continuous-time equivalent of SRIV-ARMA.

8.6 Examples of RIVC Identification and Estimation

The next section 8.7 considers the important topic of how the choice of sampling interval Δt can affect the estimates of parameters in both continuous and discrete-time TF models. Before this, however, the present section presents a number of simulation and practical examples that demonstrate the utility of RIVC estimation. As in the case of discrete-time RIV estimation discussed in previous chapter 7, MCS results confirm the utility and statistical efficiency of the RIVC algorithm, but the most imporatant of these are reported fully in Young et al. (2008) and will not be repeated here.

This first example is concerned with the analysis of data from a 'benchmark' simulation model prepared for the IFAC SYSID'06 Symposium in Newcastle, NSW, Australia (for data, see http://sysid2006benchmark.cran.uhp-nancy.fr/). Unfortunately, the associated Benchmark Session at SYSID was cancelled because referees felt that insufficient submitted papers were acceptable (only one of the papers submitted to the proposed benchmark session got even close to the correct model, demonstrating the difficulty of the benchmark exercise). The present example is a modified version of the original benchmark example, in which the simple white additive noise of the original is replaced by ARMA noise, making it still more difficult. The example is the following fourth order system with widely separated modal frequencies and an ARMA(2,1) noise model:

$$x(t) = \frac{-120s^2 - 1560s + 3600}{s^4 + 30.2s^3 + 3607s^2 + 750s + 3600} u(t - 0.035)$$

$$\begin{aligned} y(t_k) &= x(t_k) + \frac{1 + 0.5z^{-1}}{1 - 1.4z^{-1} + 0.7z^{-2}} e(t_k) \\ e(t_k) &= \mathcal{N}(0, 0.0025) \end{aligned}$$

The input signal is a pseudo-random binary sequence (± 1.0) and the complete data set consists of 6138 input-output samples with a sampling interval of $\Delta t = 0.005$ time units (that is total time 30.69 time units). A section of the input-output data used in the example is plotted in Figure 8.4: the noise-signal-ratio (by standard deviation) is 0.8[4].

8.6.1 Example 7.1: Continuous-time simulation example

In the results obtained using the SRIVC option of the rivcbjid routine in CAPTAIN, the YIC function (section 6.9 of chapter 7) clearly identifies the [4 3 7 0 0] order model. The subsequent single run SRIVC and RIVC estimation results obtained

[4] A more comprehensive version of this example, which includes plots of the recursive parameter estimates, is given in the 'rivbjdemo2' demonstration example in the CAPTAIN Toolbox.

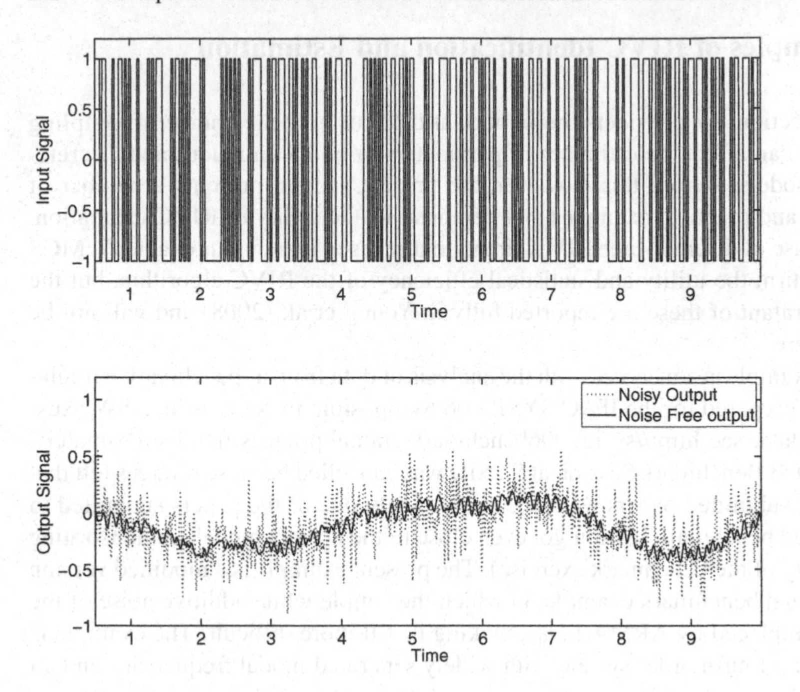

Fig. 8.4 A section of the modified IFAC SYSID'06 Benchmark data.

from the rivcbj routine are shown in Table 8.1, where we see that both algorithms provide good estimates of the parameters. The main difference is that the SRIVC provides rather optimistic estimates of the standard error (SE) on the parameters; while RIVC provides more realistic estimates of this uncertainty. Figure 8.5 shows that the Bode plots of the single run RIVC estimated model (full line) are hardly distinguishable from those of the true model (dash-dot line). These Bode plots were generated by the bode routine in the control systems toolbox of Matlab, with the system model object defined by the tf routine. The SRIVC estimated model produced a very similar Bode plot. Indirect estimation using the discrete-time RIV, PEM and IV4 algorithms, followed by conversion to continuous-time using the Matlab d2c (or the earlier but still useful d2cm) functions, completely failed at this fast sampling rate because the algorithms did not converge on acceptable discrete-time models.

It is worth noting that, in this example, the computation time of the RIVC algorithm is about three times as long as that of the SRIVC algorithm. As a result, it is advantageous to use the SRIVC algorithm for initial model order identification and only employ the full RIVC algorithm in those situations where the theoretical assumptions are satisfied and it is essential to have the most efficient parameter estimates, as well as better estimates of the uncertainty on the parameters. For day-to-day usage, the SRIVC algorithm provides a quick and reliable approach to

Table 8.1 RIVC and SRIVC-ARMA estimation results for the benchmark example: $\hat{\boldsymbol{\theta}}_c = [\hat{\boldsymbol{\rho}}_c^T \ \hat{\boldsymbol{\eta}}^T]^T$ (other nomenclature as for earlier Tables).

Parameter		$\hat{\alpha}_1$	$\hat{\alpha}_2$	$\hat{\alpha}_3$	$\hat{\alpha}_4$	$\hat{\beta}_0$	$\hat{\beta}_1$	$\hat{\beta}_2$	$\hat{c}_1$	$\hat{c}_2$	$\hat{d}_1$
True		30.2	3607	750	3600	-120	-1560	3600	-1.4	0.7	0.5
RIVC	$\hat{\boldsymbol{\theta}}_c$	31.2	3622	799	3645	-131.9	-1608	3650	-1.399	0.698	0.483
(SR)	SE	5.9	224	59	229	19	132	241	0.01	0.01	0.013
SRIVC-	$\hat{\boldsymbol{\theta}}_c$	34.4	3710	820	3732	-145.5	-1644	3729	-1.399	0.698	0.482
ARMA	SE	3.82	141	39	145	12.7	89	155	0.01	0.01	0.013

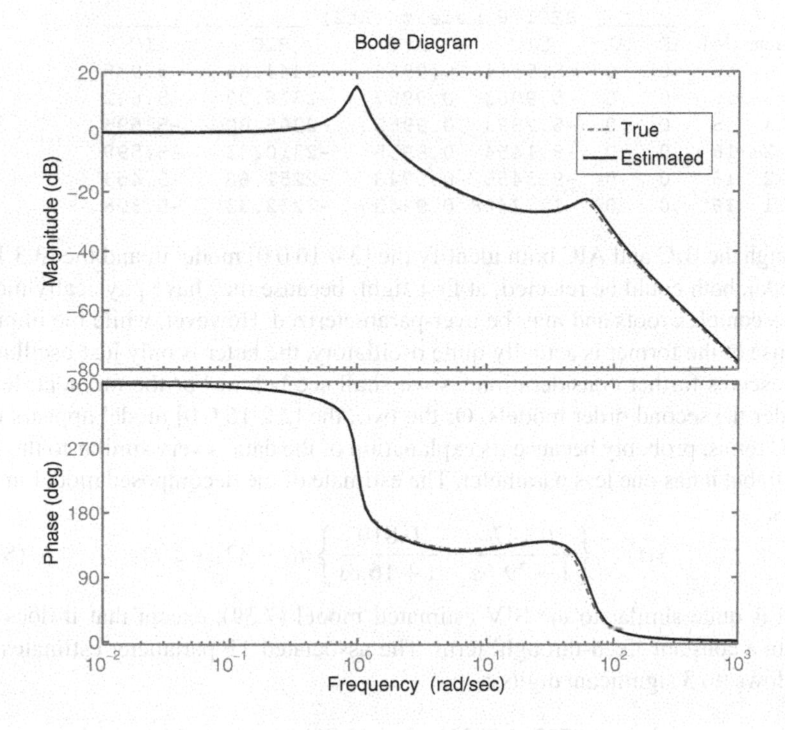

Fig. 8.5 Comparison of true and RIVC estimated model Bode diagrams.

continuous-time model identification and estimation and has been used for many years as the algorithm of choice for this in the CAPTAIN Toolbox.

8.6.2 Example 7.2: RIVC SISO estimation of tracer experiment data

Here, we consider again the tracer data analysed previously in chapters 6 and 7. The model order identification results, as produced by the rivcbjid routine in CAPTAIN using the SRIVC option setting and ordered in terms of the R_T^2 values are given below. Once again, as in the previous chapters, the simplified algorithm is selected in this example because of the non-standard, heteroscedastic noise, which is discussed later.

BEST 6 models (Rt2)

den	num	del	AR	MA	YIC	Rt2	BIC	AIC
3	4	16	0	0	-5.5314	0.9968	-2444.05	-5.935
3	3	16	0	0	-5.9063	0.9957	-2316.98	-5.632
2	3	16	0	0	-5.2893	0.9955	-2305.00	-5.595
2	2	16	0	0	-8.1454	0.9955	-2310.43	-5.598
1	2	16	0	0	-9.3455	0.9948	-2251.68	-5.453
1	1	16	0	0	-13.7408	0.9945	-2232.33	-5.398

Although the BIC and AIC both identify the [3 4 16 0 0] model it, and the [3 3 16 0 0] model, both could be rejected, at first sight, because they have physically meaningless complex roots and may be over-parameterized. However, while the impulse response of the former is actually quite oscillatory, the latter is only just oscillatory and deserves further consideration, as we shall see below. For the moment, let us consider the second order models. Of the two, the [2 2 16 0 0] model appears best in YIC terms, probably because its explanation of the data is very similar to the [2 3 16 0 0], but it has one less parameter. The estimate of the decomposed model in this case is:

$$y(t) = \left\{ \frac{0.147}{1+79.6s} + \frac{0.819}{1+16.8s} \right\} u(t-32) + \xi(t) \tag{8.23}$$

which is quite similar to the SIV estimated model (7.39), except that it does not contain a constant 'feed-through' term. The associated TF parameter estimates are as follows (to 3 significant digits):

$$\hat{\alpha}_1 = 0.0722(0.0035);\ \hat{\alpha}_2 = 0.000749(0.00013);$$
$$\hat{\beta}_0 = 0.0507(0.0008);\ \hat{\beta}_1 = 0.000724(0.00013);$$

This would appear to be a well identified model that explains the data very adequately, with $R_T^2 = 0.9955$.

However, so far we have only considered unconstrained estimation. What if we constrain the poles to be real and re-consider the [3 3 16 0 0] model in these constrained terms? It is fairly easy to answer this question when using the CAPTAIN rivbj and rivcbj routines because it is straightforward to embed them within an optimization procedure that constrains the denominator to have real poles. In the case of the present example, this is achieved by using the lsqnonlin optimization routine in Matlab, which is called by the statement:

```
[Xir,resnorm,F,exitf,out,lam,J]=...
lsqnonlin('CTconstrained_opt',e0,lb,ub,OPTS,y,u,m,disc);
```

where lb and ub can be specified to constrain the range of the optimized parameters (here the denominator poles). The associated optimization routine CTconstrainedopt is as follows:

```
function Q=CTconstrainedopt(X,y,u,m,disc)
A=conv([1 X(1)],[1 X(2)]);A=conv(A,[1 X(3)]);
if disc
    [th,M,D,stats,e,eef]=...
        rivbj([y u],m,[0.00001 -1 -1 1],...
        [A(2:end) 0 0 0],diag([0 0 0 10^6 10^6 10^6]));
else
    [th,M,Mn,D,stats,e,eef]=...
        rivcbj([y u],m,[0.00001 2 1 0],...
    -0.1,[A(2:end) 0 0 0],diag([0 0 0 10^6 10^6 10^6]));
end
  xh=y-e;plot([y xh]);drawnow
  Q=e;
  if ~isfinite(Q);Q=1e+23;end
```

In this listing, m (here [3 3 16 0 0]) is the model structure pentad; 'disc' is a flag to select discrete or continuous-time estimation; and X is the 1×3 vector of unknown real poles that have to be estimated (optimized). First, the Matlab routine conv is called to form the denominator polynomial $A(s)$ or $A(z^{-1})$ from the three real poles specified by X; and this then forms an input to the rivcbj routine, with the option to estimate the parameters recursively, so allowing for the specification of the initial parameter vector $\boldsymbol{\rho}(0)$ and the associated covariance matrix $\mathbf{P}(0)$. These are chosen by the vectors:

$$\boldsymbol{\rho}(0) = [\mathrm{A(2{:}end)}\ 0\ 0\ 0]; \quad \mathbf{P}(0) = \mathrm{diag}([0\ 0\ 0\ 10^6\ 10^6\ 10^6])$$

The first three zero diagonal elements in $\mathbf{P}(0)$ are set to zero, informing the algorithm that these are known exactly (i.e. no uncertainty); while the next three diagonal elements are set to 10^6, informing the algorithm that these are not known at all (the 'diffuse prior' setting: see chapter 3, section 3.2). In this way, the rivcbj algorithm simply estimates the numerator polynomial coefficients, with the denominator polynomial polynomial coefficients set at the latest values defined by the optimization algorithm. The optimization continues until the sum of the squares of the model error Q (equal to the error 'e' returned from rivbj) is minimized. The plot statement is not essential: it simply allows the user to monitor how well the estimated model output $\hat{x}(k)$, as denoted by 'xh' and computed by subtracting the estimated error 'e' from the measured output 'y' (=$y(k)$), is explaining the measured output as the optimization proceeds. Finally, the estimates of the standard errors on the estimates of the denominator polynomial coefficients are provided by reference to the confint routine in the CAPTAIN Toolbox:

```
[conf,var,covar] = confint(Xir,F,J);
```

and those for the estimates of the numerator polynomial coefficients are provided by reference to the covariance matrix $\mathbf{P}^*(N)$ returned by the rivcbj routine.

In this case, optimization convergence occurs satisfactorily but with all three of the estimated poles virtually the same. As a result, the confidence intervals on the parameters returned by the conf routine are very large and so the model is not satisfactory. However, a new run of the routine is carried out, this time assuming that all three poles of $A(s)$ are the same, i.e. $A(s) = (s+\alpha)^3$, so that there is only one parameter to optimize. This optimization converges well, with $R_T^2 = 0.9957$ but now the estimate of the single pole α is very well-defined with low standard error. The resulting constrained TF model takes the form:

$$y(t) = \frac{\beta_0 s^2 + \beta_1 s + \beta_2}{(s+\alpha)^3} u(t-32) + \xi(t) \tag{8.24}$$

with the following parameter estimates and standard errors:

$$\hat{\alpha} = 0.0328(0.0012); \qquad \hat{\beta}_0 = 0.0487(0.0004);$$
$$\hat{\beta}_1 = 0.00224(9.3\times 10^{-6}); \qquad \hat{\beta}_2 = 3.388\times 10^5(1.305\times 10^{-7});$$

In fact, this model explains the data marginally better than the second order, unconstrained model, despite the fact that both models have the same number of parameters. The SSG of this model is 0.963 and the three identical poles correspond to three identical residence times of $T_i = 30.6$ hours, $i = 1,\ 2,\ 3$ hours. These estimates can be compared with the second order model (8.23), where the SSG $= 0.966$ and the time constants are $T_1 = 79.6$; $T_2 = 16.4$ hours. Since both the SSG's and the sum of the time constants are so similar, it is clear that the overall dynamics of the two models are also virtually the same when the uncertainty on the estimates is taken into account.

8.6.3 Example 7.3: SRIVCDD MISO modelling of the tracer experiment data

In this example, we continue with the analysis of the tracer data. For interest and illustration here, however, the output tracer concentration is modelled as a MISO system in relation to two input tracer concentration series. All of these data are plotted in Figure 8.6, which also shows the output of the SRIVCDD estimated model, as discussed below.

The best identified model has the following structure:

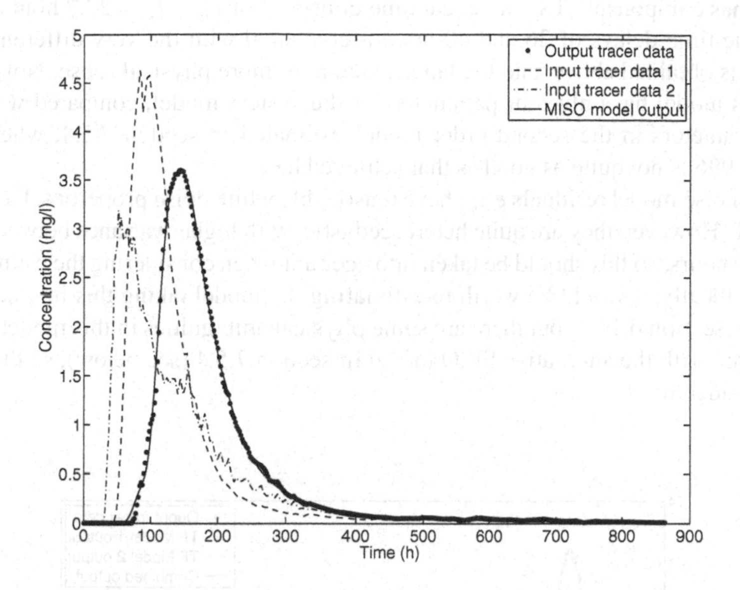

Fig. 8.6 Tracer example: plots of the output and the two input tracer concentration changes, together with the deterministic output of the SRIVCDD estimated model.

$$\begin{aligned} x(t) &= \frac{b_{10}}{s+a_{11}} u_1(t-26) + \frac{b_{20}}{s+a_{21}} u_2(t-40) \\ \text{or} \quad x(t) &= \frac{G_1}{1+T_1 s} u_1(t-26) + \frac{G_2}{1+T_2 s} u_2(t-40) \\ y(t_k) &= x(t_k) + \xi(t_k) \end{aligned} \tag{8.25}$$

where the additive noise is modelled by an AIC identified AR(12) process. The associated parameter estimates for the system model obtained by the prototype srivcbjdd routine in CAPTAIN (see earlier comments about the srivbjdd routine in chapter 7 and Appendix G), are as follows:

$$\hat{a}_{11} = 0.09544(0.0015);\ \hat{a}_{21} = 0.02582(0.00085);\ T_1 = 10.5 \text{ h};\ T_2 = 38.7 \text{ h}$$
$$\hat{b}_{10} = 0.06703(0.0010);\ \hat{b}_{20} = 0.008270(0.00025);\ G_1 = 0.70;\ G_2 = 0.32$$
$$\sigma^2_{\xi(t_k)} = 0.0016;\ \sigma^2_{e(t_k)} = 0.00017$$

The coefficient of determination based on the simulated output error is $R_T^2 = 0.998$ and, since the noise is modelled in this case, it is possble to compute the coefficient of determination based on the residual white noise (one step-ahead prediction errors), $e(t_k)$, which is evaluated as $R^2 = 0.9998$. By contrast, the common denominator MISO model has $R_T^2 = 0.994$ and $R^2 = 0.9997$, so there is some advantage here in terms of explaining the data. More importantly, the common denominator model has component TFs with equal time constants of $T_1 = T_2 = 27.7$ hours (with the same time delays of 26 and 40 hours), compared with the very different time constants obtained above, and the latter make a lot more physical sense. Note also, that this model has only four parameters in the system model, compared with the five parameters in the second order model estimated in section 7.5.4, where the $R_T^2 = 0.996$ is not quite as good as that achieved here.

The noise model residuals $e(t_k)$ have reasonable white noise properties, based on the ACF. However, they are quite heteroscedastic, with higher variance between 200 and 300 hours, so this should be taken into account when considering the estimation results. Ideally, it would be worth re-estimating the model taking this into account (see e.g. section 6.10.2) but there are some physical ambiguities in this model when compared with the alternative SISO model in section 7.5.4 (see below), so this was not considered.

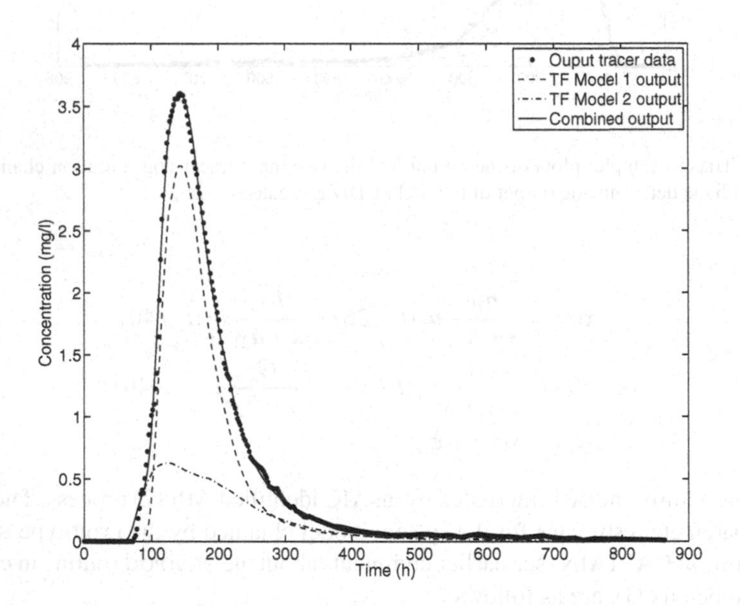

Fig. 8.7 Tracer example: SRIVCDD estimation results showing the contribution that the two additive component TF model outputs make to the total output.

Figure 8.7 compares the outputs of the two, first order, transfer functions, as well as their combined effect (i.e. the deterministic output of the full model, which is the same as the model output plotted in Figure 8.6), with the output tracer series. Figure 8.8 shows the recursive estimation results, indicating very good convergence after the initial uncertainty, with well defined estimates of all the parameters. Figure 8.9 shows the results of MCS analysis, based on 10000 realizations, which is carried out to evaluate the uncertainty on the derived steady state gain and time constant parameters that are so important in this case since they embody the physical meaning of the model. All of these suggest quite well defined estimates, with the most uncertainty, as would be expected, in the time constant of the second TF.

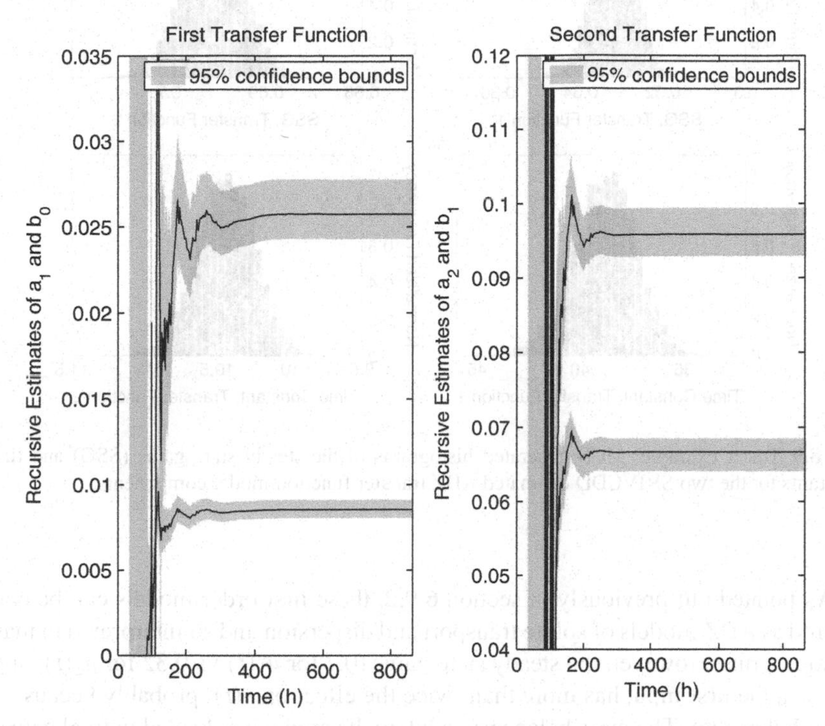

Fig. 8.8 Tracer example: Recursive SRIVCDD estimation results.

Finally, we need to compare the MISO model (8.25) with the SISO models estimated above in section 8.6.2 and in the previous chapter 7. All the models explain the 'output' tracer concentration variations well but in the SISO cases this is due to a single input $u_1(t) = u(t)$; while in the MISO case, an additional input $u_2(t)$ is added. But we have not referred at all to the physical location of the measurement sites in the wetland and the prevailing flow. In fact, the locations are probably across the prevailing flow to some extent, although this is difficult to gauge from the information in the paper (Martinez and Wise, 2003), with the output location in the middle of the other two. But the flows at the three locations are very different: 0.02 for $y(t)$), 0.224 for $u_1(t)$ and 0.038 for $u_2(t)$, all in units of cubic metres/sec. So much the greatest flow and mass transfer is occurring through the first input location. In other words, the mass transfer between these 'inputs' and the 'output' is probably not the main mass transfer in this system. In this situation, it makes sense to have two inputs, since the three concentration series are likely to be interacting and it is unlikely if all of the concentration changes at the 'output' location are just due to the changes measured at u_1, as in the SISO model.

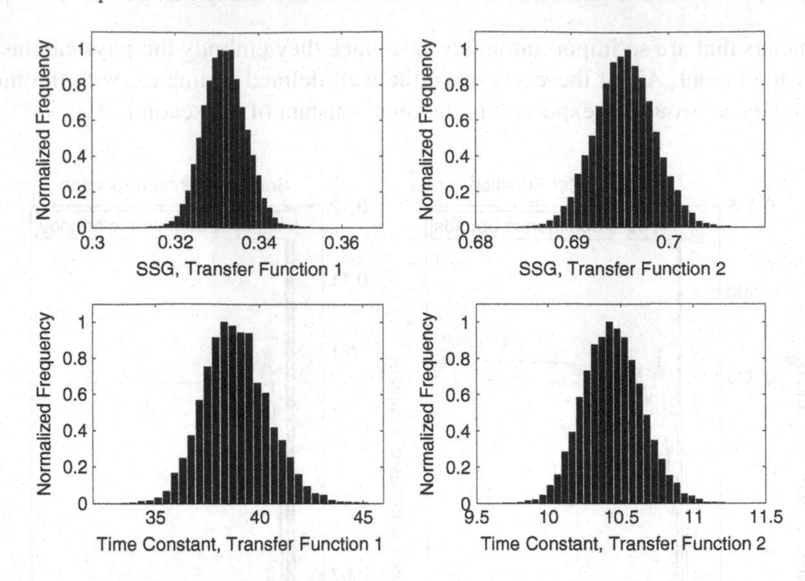

Fig. 8.9 Tracer example: MCS generated histograms of the steady state gains (SSG) and time constants for the two SRIVCDD estimated ADZ transfer function model components.

As pointed out previously in section 6.9.2, these first order models can be considered as ADZ models of solute transport and dispersion and so interpreted in mass balance terms. However, the steady state gains (0.7 for $u_1(t)$ vs 0.32 for $u_2(t)$) suggest that the first input has more than twice the effect on $y(t)$, probably because of its high flow rate. The mass balance created by the model is reflected in total gain of 1.02, which is insignificantly different from unity and suggests mass conservation (no tracer lost).

We can see, therefore, that the physical nature of the interactions between the 'inputs' and 'outputs' is not clear from the data used in the above analysis: there is some underlying ambiguity that can only be resolved by much greater knowledge of the system and analysis that utilizes all of the available data collected in the experiment, such as that described by Martinez and Wise (2003) using the deterministic distributed parameter OTIS model[5]. Indeed, the very nature of the experiment and the difficulty of collecting data on a wetland area such as this probably means that some ambiguity will remain even if all of the experimental data are available for modelling. With this in mind, it must be emphasized that this example is intended only to illustrate various aspects of the SISO and MISO model identification and estimation and is in no sense whatsoever a complete modelling study.

For instance, the limited data availability has also prevented any predictive validation of the tracer data models considered above and in the previous chapters.

[5] See `http://csdms.colorado.edu/wiki/Model:OTIS`

It could, however, constitute the first step in a much more comprehensive research project, such as that carried out into the ADZ model by Wallis et al. (1989) on single channel systems (so removing some of the difficulty and ambiguity of the present wetland example). This established that, for many riverine and man-made channels, the 'dispersive fraction', defined by the ratio $T/(T+\tau)$, is relatively invariant over most of the flow régime so that, once this is established, the ADZ model is able to describe pollution transport and dispersion for any defined flow conditions (which is not possible in the case of the OTIS model: see Martinez and Wise, page 217). It would be very interesting, therefore, to see whether the limited results presented in this chapter are confirmed in this wider context and with a much larger data base.

The limited stochastic analysis in this section and the related sections in the previous chapters is useful, however, because it suggests the most likely interaction between the sites considered, so that similar analysis using all the data from the tracer experiment could assist in the total modelling effort. And, of course, the analysis of these data has been primarily useful, in this and the previous chapters, for illustrating how the various estimation algorithms can be applied to real data.

8.6.4 Example 7.4: Evaluation of a Global Circulation Model

The *Global Circulation Models* (GCMs) used to study the possibility of global warming are amongst the largest computer simulation models ever constructed. In this example, we use the RIVC algorithm to identify and estimate a continuous-time model between emissions forcing $u(t)$ and the global mean temperature perturbation $y(t)$ obtained from a standard forcing experiment performed on the U.K. HadCM3 A-OGCM GCM model, as shown in Figure 8.10. The results reveal the inadequacy of the data set and demonstrate the power of the recursive RIVC parameter estimation in highlighting this problem.

The identification statistics from the rivcbjid routine in CAPTAIN suggest a second order [2 2 24] system model and an associated AR(3) or MA(5) noise model: using the more parsimonious AR(3) model, the RIVC estimated [2 2 24 3 0] model then takes the following form:

$$\begin{aligned} x(t) &= \frac{0.1653s+0.0006454}{s^2+0.07509s+0.0001692}u(t-24) \\ y(t_k) &= x(t_k)+\frac{1}{1-0.491z^{-1}+0.133z^{-2}-0.137z^{-3}}e(t_k) \\ e(t_k) &= \mathcal{N}(0,0.017) \qquad \Delta t = 1\ year \end{aligned} \tag{8.26}$$

The explanation of the data is good, as we see in Figure 8.10, with $R_T^2 = 0.993$ and low residual variance. The ACF shows that the noise is a little coloured and the identified AR(3) noise model yields white residuals, as required.

The recursive estimates of the system and noise model parameters are shown in Figures 8.11 and 8.12, respectively. It is clear that, while the noise model parameters

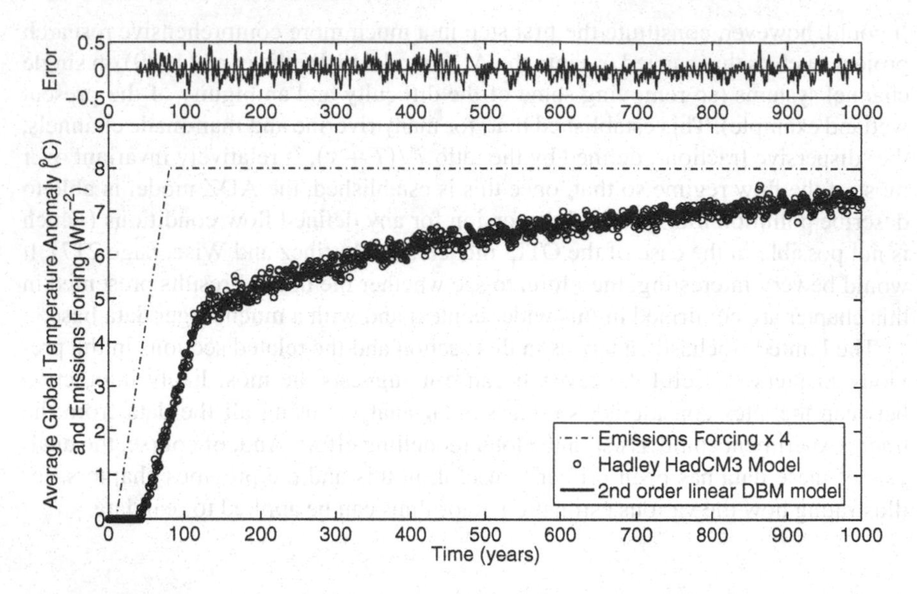

Fig. 8.10 Hadley GCM data and RIVC estimated model output, with the estimated noise $\xi(t_k)$ shown above. The emissions forcing is multiplied by 4 for clarity.

have converged reasonably after about 800 years, the system parameters are still changing at the end of the data and the 95% confidence bounds are quite large. The reason for this is clear from the nature of the data set in Figure 8.10, where the temperature response is not complete and does not reach its steady state value: in other words, the information content in the data is not sufficient to obtain well defined parameter estimates. Presumably the experiment to obtain these data was computationally intensive and so it was curtailed before it should have been. The value of the recursive estimates is clear: without them, it could be assumed that the model parameters are well defined, with all the *en bloc* statistics suggesting a reasonably estimated model.

The above data-based model (8.26) is an example of a *Dynamic Emulation Model* (DEM). This represents a stage in the *Data-Based Mechanistic* (DBM) approach to modelling, which is discussed fully in chapter 12. An important aspect of DBM modelling is that the model, although obtained using essentially black-box methods of identification and estimation, should have a sensible physical interpretation. In this regard, the model (8.26) can be decomposed straightforwardly into a positive feedback connection of two first order processes (see Appendix B). This is easily interpreted in energy balance terms as a feedback process with quick forward path (SSG=2.32; time constant 14 years) and slow positive feedback path (SSG=0.169 and time constant 256 years) modes, an interpretation that is immediately useful in scientific terms and can be used to compare the nature of the reduced order model with that of the very high order GCM from which it has been derived. However,

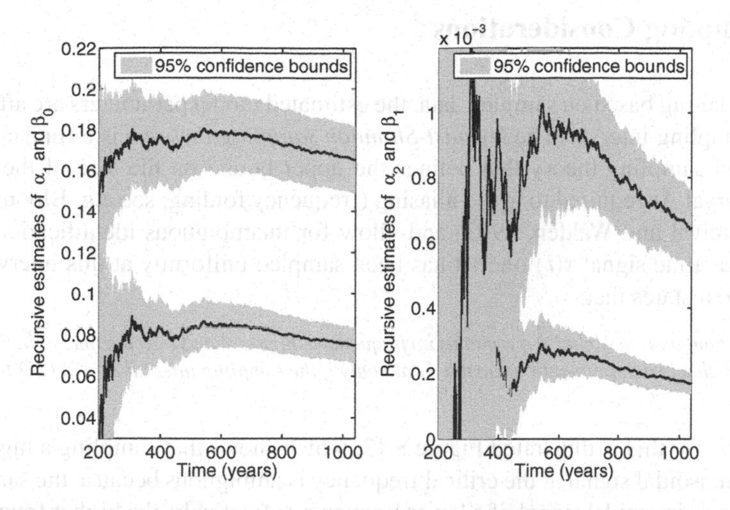

Fig. 8.11 Hadley Model: recursive RIVC estimates of system TF model parameters.

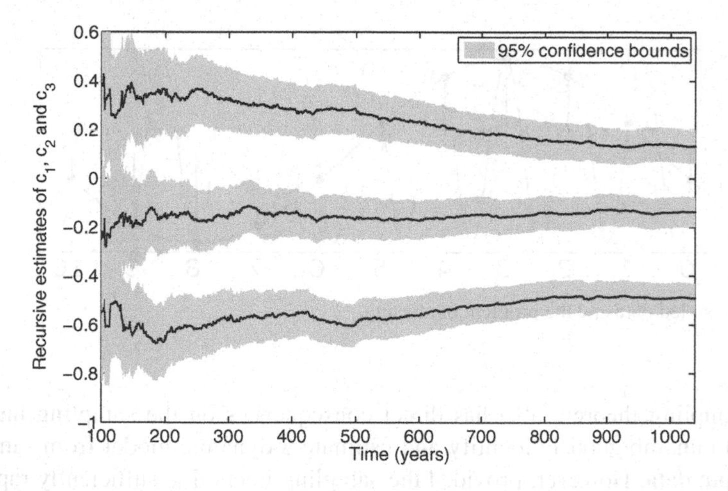

Fig. 8.12 Hadley Model: recursive estimates of AR(3) noise model parameters.

given the inadequacy of the data set in this case and the associated uncertainty in the model parameters, it is necessary to be cautious about these inferences. Certainly another experiment with a 'sufficiently exciting' input and longer duration is advisable (but would the climate change community understand such advice?).

8.7 Sampling Considerations

In all modelling based on sampled data, the estimated model parameters are affected by the sampling interval. The *Nyquist-Shannon sampling theorem* is a fundamental concept in sampling theory that defines the upper bound on the size of the sampling interval Δt required to avoid aliasing (frequency folding: see e.g. Bloomfield, 1967; Percival and Walden, 1993) and allow for unambiguous identification of a continuous-time signal $x(t)$ once it has been sampled uniformly at this interval. In particular, it states that:

> *If a continuous-time signal $x(t)$ contains no frequencies higher than Ω cycles/time unit, then it is completely determined from sampled data only if the sampling interval $\Delta t \leq 1/2\Omega$ time units.*

The reason for this is illustrated Figure 8.13, which shows that sampling a high frequency sinusoidal signal at the critical frequency is ambiguous because the sampled data lie on a sinusoidal signal of a lower frequency: effectively, the higher frequency signal is hidden by the crude sampling and cannot be identified uniquely from the sampled data.

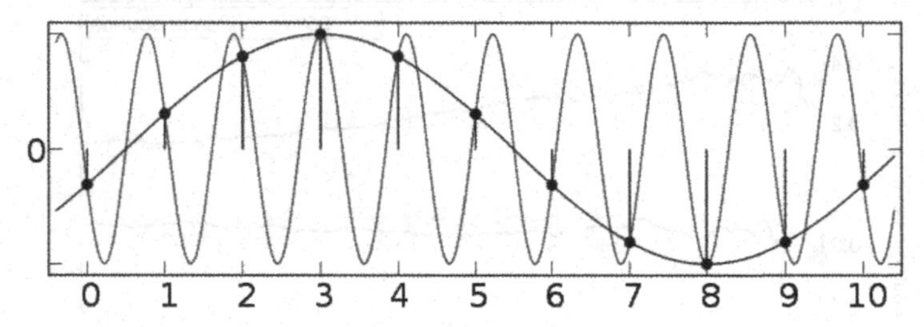

Fig. 8.13 Sampled data and the problem of aliasing.

The sampling theorem also has direct consequences on the sampling interval required to unambiguously identify and estimate a dynamic model from sampled input-output data. However, provided the sampling interval is sufficiently rapid to avoid these problems, it is possible to estimate a continuous-time model that is independent of the sampling interval and then derive discrete-time models from this continuous-time model at any specified sampling interval. In sub-section 8.7.1 below, simulation results are discussed that reveal the consequences of estimating a CT model from data at widely different sampling intervals. The model used for these simulations is based originally on hourly rainfall and flow data from the River Wye at Cefn Brwyn (see Littlewood et al, 2010). The RIVCBJ estimated CT model in the decomposed form is as follows:

$$y(t_k) = \left[\frac{0.473}{1+2.919s} + \frac{0.476}{1+180.8s} \right] u(t_k) + \xi(t_k) \tag{8.27}$$

The advantage of CT model estimation in terms of model interpretation is clear here, since this model shows that the system, as in numerous other examples considered in this book, is 'stiff' in dynamic terms, with widely different time constants of 2.92 and 180.8 hours. This model can be written in the alternative form,

$$y(t_k) = \left[\frac{0.162}{s+0.3426} + \frac{0.00263}{s+0.00553}\right] u(t_k) + \xi(t_k) \tag{8.28}$$

showing that the bandwidth of the quick-flow mode, which dictates the sampling strategy in the present context, is 0.3426 rad/hour or $0.3426/2\pi$ cycles/hour. Consequently, for this mode, $1/\Omega = 2\pi/0.3426$ hours and the critical sampling interval associated with this Nyquist frequency is $\Delta t = \pi/0.3426 = 9.17$ hours. At first sight, therefore, hourly data should be sufficient to avoid any ambiguity and yield a well-identified continuous-time, differential equation model.

There are, however, several rules-of thumb that assist in the more detailed selection of the sampling interval for model estimation and reduce it to values that are well removed from the critical Nyquist-Shannon upper limit. One suggestion (see Ljung, 1999, page 386) is that the sampling frequency should be ten times the bandwidth: in relation to the bandwidth of the highest frequency, quick-flow mode in the present example, this results in a sampling interval of $\Delta t = 2\pi/3.426 = 1.83$ hours. Another, that I have used for many years, is that there should be 6 samples or more over the residence time (time constant) of the highest frequency mode: in this example, $T_q = 2.92$ hours, which suggests a smaller sampling interval of about 0.5 hours. Another interesting discussion on the selection of sampling intervals *for discrete-time models* is given in Åström (1969), which shows that there is a safe band of sampling intervals that will ensure low variance parameter estimates.

So, taking all these factors into consideration, it should be relatively safe to use hourly data, although half-hourly data would be preferable if it were available. But what kind of estimation results does one obtain at other finer and coarser sampling intervals than these? And is there a band of safe sampling intervals that can be used with some confidence? These questions are addressed by the simulation example in the next section.

8.7.1 The effects of sampling interval on estimation accuracy

In order to evaluate the effects of the sampling interval on the estimation of model parameters, the model (8.28) is simulated in Matlab, using the `lsim` routine at a very small and, therefore, accurate sampling interval of one second, over a total simulation time of 6500 hours (23,400,000 seconds!). In order to ensure good parametric identifiability, the input is selected as a persistently exciting (see section 6.6.1, chapter 6), repeated unit step signal with six steps over the 6500 hour total period so that, at each step (which excites all frequencies), the full dynamic behaviour is simulated, with the modelled flow output reaching near steady state each time and so providing good estimation of the steady state gain. This requires a substantial simulation time since T_s is very large.

In order to simulate uncertainty in the measurements, the additive noise signal, $\xi(t_k)$ is selected as a zero mean, serially uncorrelated random variable (discrete-time white noise) with a noise to signal ratio by standard deviation of 0.1. In this way, the complicating effects of colour in the noise are avoided and the estimation results reflect more clearly the effects of the sampling interval. This means that the results obtained here are probably the best that could be obtained and that a less persistently exciting signal and a more coloured noise sequence would probably lead to less estimation accuracy than is seen in the results presented below.

The one second sampling interval data generated by the above simulation experiment is sub-sampled with sub-sampling intervals ranging from 5 seconds to 10 hours. Three types of sub-sampling are compared: simple sub-sampling based on the samples occurring at the sub-sampling interval; 'decimation', i.e. sampling of the same kind but carried out after the data have been prefiltered by an anti-aliasing filter (e.g. the Matlab decimate routine; or the irwsm routine in CAPTAIN); and 'accumulation', which is common in hydrology, where accumulated totals of rainfall and flow over the selected interval are used (equivalent to averaging over the sampling interval).

For each sub-sampled data set obtained in the above manner, the continuous-time model parameters are estimated in three ways. First, directly, using the CT model estimation algorithm rivcbj routine in the CAPTAIN Toolbox. Second, indirectly, using the DT estimation algorithm rivbj in CAPTAIN to obtain estimates of the parameters in the DT model (with structure [2 2 1]) and then using the d2c routine in Matlab to convert this model to continuous-time with the zero-order hold (ZOH) assumption. Third, using the DT estimation algorithm pem in the Matlab SID Toolbox, with the same conversion to CT, this acting as a check on the results obtained by the rivbj algorithm. However, the results were quite similar and so are not shown in the results below (but see Exercise 2 at the end of the chapter).

Figure 8.14 shows the estimates of the the two time constants T_q and T_s plotted against the sampling interval when the first sub-sampling approach is used. The true values of the time constants are shown as the dash-dot lines and, as expected, the rivcbj direct estimate is very much superior for the small sampling intervals, with rivbj (and pem) estimation producing very erratic results at these very small sampling intervals where the estimated DT model eigenvalues (roots of the denominator polynomial) are very close to the unit circle in the complex z domain (so that the model increasingly appears like an integrator).

At the other extreme, all the estimation algorithms appear to perform similarly, with increasing variance as the critical Nyquist-Shannon frequency is approached. Of course, this increased variance is partly due to the fact that the sample size is reducing as the sub-sampling interval is increased: the original sample size of 23,400,000 is reduced to only 650 samples when the sub-sampling interval is ten hours. The conclusion from these results is that the direct CT model estimation is more reliable, over a wider range of sub-sampling intervals, than the indirect estimation.

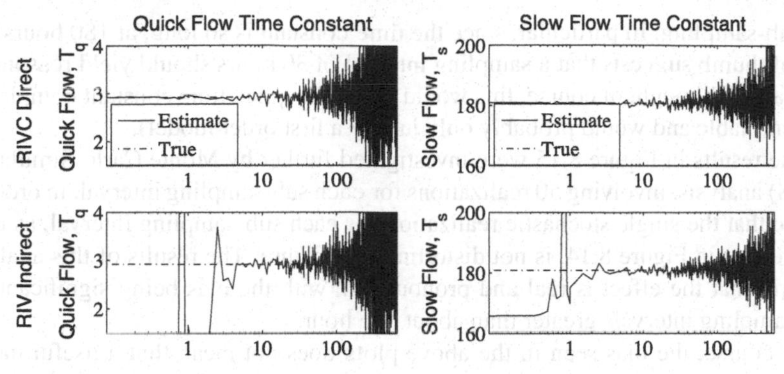

Fig. 8.14 The effects of the sampling interval on the estimation of the CT model parameters T_q and T_s when using simple sub-sampling.

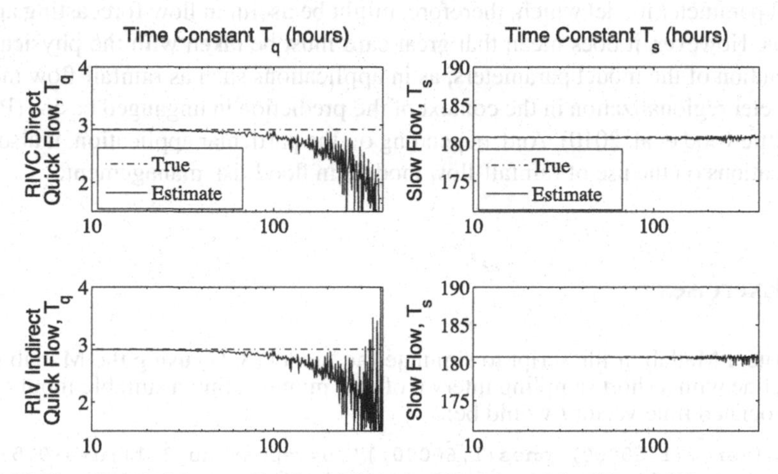

Fig. 8.15 The effects of the sampling interval on the estimation of the CT model time constants T_q and T_s when using accumulation sub-sampling.

The same experiments were repeated for the two other methods of sub-sampling. The main difference is that the estimates of the quick-flow time constant obtained with the accumulation method of sub-sampling, of the kind often favoured in hydrology, becomes progressively biased when the sub-sampling interval is increased, as shown in Figure 8.15. In particular, the introduction of the pre-processing (effectively simple average-type filtering) into the sub-sampling operations has the effect of introducing a pronounced and clearly visible reduction of the estimated value obtained for the quick time constant by all the estimation algorithms, including rivcbj, for coarser sampling intervals greater than about one hour. However, as might be expected, the estimates of the slow time constant are not biased. This is because the information in the data on this time constant is still quite considerable, despite the sub-sampling. In particular, since the time constant is so long, at 180 hours, the rule of thumb suggests that a sampling interval of 30 hours should yield reasonable estimates (although, of course, this would make the quick time constant completely unidentifiable and would probably only justify a first order model).

The results in Figure 8.15 were investigated further by Monte Carlo Simulation (MCS) analysis, involving 50 realizations for each sub-sampling interval, in order to ensure that the single stochastic realization for each sub-sampling interval, as used for the plot in Figure 8.14, is not distorting the picture. The results of this analysis confirm that the effect is real and pronounced, with the bias being significant for sub-sampling intervals greater than about one hour.

Of course, the bias seen in the above plots does not mean that a useful model cannot be obtained from data derived at larger sampling intervals: this will depend upon the application for which the model is intended. In the case of rainfall-flow data, for instance, the flow data can still be explained reasonably well by such a biased parameter model which, therefore, might be useful in flow forecasting applications. However, it does mean that great care must be taken with the physical interpretation of the model parameters, as in applications such as rainfall-flow model parameter regionalization in the context of the prediction in ungauged basins (PUB: see Littlewood et al, 2010). And, depending on the particular application, it also has implications on the use of rainfall-flow models in flood risk management.

8.8 Exercises

1. Write a Matlab m-file script to simulate the model (8.27) using the Matlab lsim routine with a short sampling interval of one minute. Hint: a suitable input u and associated time vector t would be:

```
uu=[zeros(1,60000) ones(1,60000)]';u=repmat(uu,3,1);u=u-0.5;
u=[zeros(1,15000) u']'; t=(0:1/60:(length(u)-1)*1/60)';
```

2. Investigate sub-sampling performance at small sub-sampling intervals using the following m-file script, where x is the deterministic output of lsim in 1.:

```
for i=1:5
    e=0.1*std(x)*randn(size(x));%additive noise
    y=x+e;%add noise to simulated data
    dt=i;disp(dt);% sub-sampling interval
    ym=y(1:dt:length(y));%sub-sample y
    um=u(1:dt:length(y));%sub-sample u
    xm=x(1:dt:length(x));%sub-sample x
    %RIVC estimation
    [th,M,Mn,Dc,stats,e,eef,var,P] = ...
        rivcbj([ym um],[2 2 0 0 0],[0.00001 dt/60 1 0],-0.3);
    %RIV DT estimation
    [thd,Md,D,statsd,ed,eefd,vard,Psd]=...
        rivbj([ym um],[2 2 1 0 0],0.00001);
    %PEM DT estimation
    Mp = pem(D,[0 2 0 0 2 1]);
    %insert sampling interval into object description
```

```
    Md.Ts=dt/60;Mp.Ts=dt/60;
    %convert to continuous-time
    Mddc = d2c(Md,'zoh');
    Mpdc = d2c(Mp,'zoh');
    %Compare responses
    compare(Dc,Mpdc,Mddc,M)
    pause
end
```

It is worth running this several times to see the volatility of the DT/conversion estimation.

3. Add the following code to the m-file script above in order to investigate how the estimates of the time constants are affected:

```
%Decompose TFs
[Rc,Pc,Kc] = residue(M.b,M.f);
[Rd1,Pd1,Kd1] = residue(Mddc.b,Mddc.f);
[Rd2,Pd2,Kd2] = residue(Mpdc.b,Mpdc.f);
%Compute and display time constants
TCc=-1./Pc';TCd1=-1./Pd1';TCd2=-1./Pd2';
disp([TCc;TCd1;TCd2])
```

4. Select data of your own and use the rivbjid and rivcbjid routines in the CAPTAIN toolbox to identify and estimate a suitable TF model. Compare the results obtained by the two routines based on your reading of the present chapter. Which model do you think would be best in your application?
5. Run the 'RIVCBJ-Winding pilot' demonstration example in the CAPTAIN Toolbox (selected from the 'Upgraded (BJ) TF demos' window: see Appendix G) and compare the results with those obtained using discrete-time estimation in section 7.7.2 of chapter 7. Note how the recursive estimation results help to identify the better parameterized model.

8.9 Summary and Conclusions

As far as I am is aware, the RIV and RIVC algorithms described in this and the previous chapter 7 constitute the only unified, time domain family of algorithms that not only provide statistically optimal solutions to the estimation of both discrete-time and "hybrid" continuous-time TF Models of the Box-Jenkins type, but also have an inherent ability to provide recursive estimates. Moreover, because of their IV heritage, the simpler SRIV/SRIVC options in these algorithms can be applied in circumstances where the noise process does not conform to the rational spectral density assumptions. Consequently, they have proven to be very robust, providing excellent, computationally efficient, day-to-day approaches to TF model identification and estimation that have been applied successfully to a wide range of practical applications, in areas ranging from ecology, through engineering, to economics. Also, the RIV/RIVC algorithms can be extended easily for use in a feedback situation provided an "external" input is being applied, as we shall see in the next chapter.

The main advantage of the full RIV/RIVC estimation algorithms is that they allow for improved statistical efficiency, with better estimation of the standard error bounds on the parameter estimates. The main disadvantages are : (i) they depend on concurrent noise model estimation and so have a longer computation time; and (ii), in the case of RIVC based on rapidly sampled data, estimation may be somewhat less robust when the roots of the noise model denominator polynomial $C(z^{-1})$ approach the unit circle (although this has not yet been evaluated comprehensively and no problems have been reported).

At high sampling frequencies, direct SRIVC/RIVC identification is much superior to the alternative of indirect discrete-time estimation (with associated continuous-time model conversion), as we have seen in section 8.7. Also, physical interpretation of the TF models is facilitated by their differential equation form and the uniqueness of the continuous-time model parameters, which are not a function of the sampling interval Δt. However, the SRIVC/RIVC estimation computation time is longer and the algorithms are not quite so easy to apply as the SRIV/RIV duo.

The asymptotic independence of the system and noise model parameter estimates is a most attractive feature of the Box-Jenkins family of models that forms the basis for RIV/RIVC estimation and is exploited by these algorithms for computational advantage. This, together with the iterative form of the RIV/RIVC algorithms, seems to engender advantages in some practical situations. For example, as we have seen in the case of stiff dynamic systems with widely spaced eigenvalues, the alternative gradient optimization algorithms, such as PEM and standard ML, can experience convergence problems. However, the latter, non-recursive, algorithms are normally more computationally efficient and always produce *en bloc* estimates even in poorly identifiable situations (although the estimates will normally be poor in these situations). In such cases, the poor identifiability is detected in the RIV/RIVC algorithms by the near singularity of the IV cross-product matrix and convergence may not occur (in the CAPTAIN rivbj and rivcbj routines, this is reported as 'poorly identifiable').

Chapter 9
Identification of Transfer Function Models in Closed-Loop

9.1 Introduction

The identification and estimation of transfer function models in a closed-loop situation has received a lot of attention in the control systems literature: see e.g. Ljung (1999); Söderström and Stoica (1989); Verhaegen (1993); Van den Hof (1998); Gilson and Van den Hof (2005). Provided there is an external command input signal, simple, sub-optimal transfer function estimation within a closed automatic control loop has always been straightforward when using *Instrumental Variable* (IV) estimation methodology (Young, 1970b). However, the more recent RIV and RIVC algorithms discussed in previous chapters of this book provide a new stimulus to the development of statistically optimal methods of closed-loop estimation and a number of possible solutions are discussed by Gilson et al (2008, 2009), within a continuous-time setting.

These latest optimal RIV procedures are fairly complicated and so, in the spirit of the present book, we will consider here a new and particularly simple generalized refined IV method for estimating discrete *and* continuous-time transfer function models enclosed within a feedback control system. This 'three-stage' method derives from a simple two-stage algorithm (Young et al., 2009) that yields consistent, but statistically inefficient, parameter estimates. The additional third stage allows for statistically efficient estimation of the enclosed TF model parameters when the system within the closed-loop is stable. The attraction of this new approach is its relative simplicity: in particular, the resulting *Closed-Loop* RIV (CLRIV) and *Closed-Loop* RIV *for Continuous-time models* (CLRIVC) algorithms are straightforward to implement since they use the existing RIV and RIVC estimation algorithms described in previous chapters of this book and implemented by the rivbj and rivcbj routines in the CAPTAIN Toolbox. As a result, the coding of the algorithms in CAPTAIN is straightforward, requiring only three calls to these existing algorithms.

In order to illustrate the unique, unified aspect of the CLRIV and CLRIVC algorithms, they are considered in general operational terms, with the identification and estimation performance evaluated by simulation examples for both discrete and continuous-time simulated systems enclosed in *Proportional-Integral-Plus* (PIP) optimal control loops (see e.g. Young et al. (1987); Taylor et al. (2000) and the prior references therein), the design of which is accomplished using the PIP control system design tools in the CAPTAIN Toolbox (see Appendix G). Linear systems that are unstable or marginally stable in open-loop are also considered, but only the sub-optimal two-stage algorithms work satisfactorily in this more difficult environment.

9.2 The Generalized Box-Jenkins Model in a Closed-Loop Context

The *Generalized Box-Jenkins* (GBJ) model takes the following form, which relates to the feedback control system shown in Figure 9.1:

$$\begin{aligned} v_x(t_k) &= G(\rho)v_u(t_k) \\ y(t_k) &= v_x(t_k) + \xi(t_k) \\ \xi(t_k) &= H(z^{-1})e(t_k) \qquad e(t_k) = \mathcal{N}(0,\sigma^2) \end{aligned} \tag{9.1}$$

where $G(\rho) = B(\rho)/A(\rho)$ is the System model and $H(z^{-1}) = D(z^{-1})/C(z^{-1})$ is the ARMA Noise model. The generalized operator ρ should not be confused with the bold $\boldsymbol{\rho}$ used previously to denote a parameter vector. In this chapter, ρ represents the backward shift operator z^{-1} in the discrete-time case and the inverse of the derivative operator s^{-1}, where $s = d/dt$, in the hybrid continuous-time case[1].

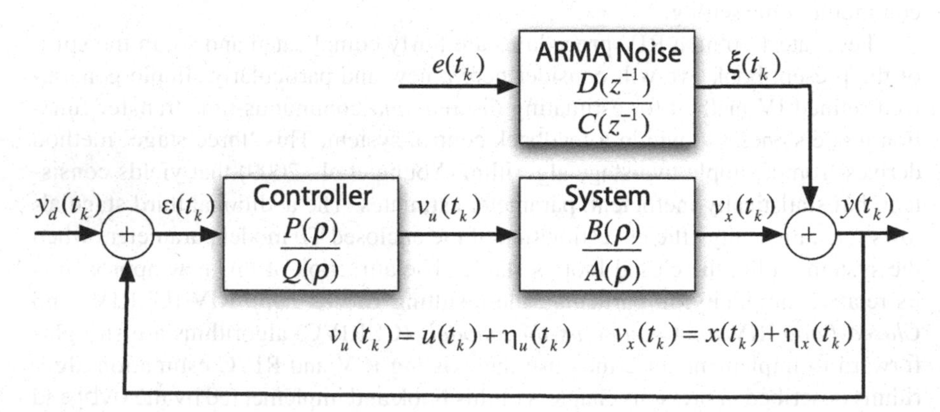

Fig. 9.1 Closed-loop system block diagram.

[1] ρ is used because it is employed as a generalized operator, although not as here, by Middleton and Goodwin (1990).

Referring to Figure 9.1, $y_d(t_k)$ is the command input to the closed-loop system and $\varepsilon(t_k) = y_d(t_k) - y(t_k)$ is the control system error that the Controller $P(\rho)/Q(\rho)$ is designed to minimize in some manner (see later discussion). The control input to the system, $v_u(t_k)$, is generated by the Controller and is affected by additive circulatory noise $\eta_u(t_k)$, with $u(t_k)$ denoting the underlying 'noise-free' control input to the system that would be measured at this point if $\xi(t_k) = \eta_u(t_k) = 0 \; \forall \; t_k$. The output from the system $v_x(t_k)$ is also affected by additive circulatory noise $\eta_x(t_k)$, with the noise-free output $x(t_k)$ defined in a similar manner to the noise-free input. Finally, the noisy measured output $y(t_k)$ is the sum of $v_x(t_k)$ and the additive *AutoRegressive, Moving Average* (ARMA) noise $\xi(t_k)$, while $e(t_k)$ is the zero mean, normally distributed white noise source to the ARMA noise model. All of these signals are sampled at a uniform sampling interval Δt. The system and noise transfer functions, $G(\rho)$ and $H(z^{-1})$, are defined as the following ratios of rational polynomials in the ρ and z^{-1} operators:

$$G(\rho) = \frac{B(\rho)}{A(\rho)}; \quad H(z^{-1}) = \frac{D(z^{-1})}{C(z^{-1})} \tag{9.2}$$

More specifically,

$$\begin{aligned} A(\rho) &= 1 + a_1\rho + a_2\rho^2 + \ldots + a_n\rho^n \\ B(\rho) &= b_0 + b_1\rho + b_2\rho^2 + \ldots + b_m\rho^m \\ C(z^{-1}) &= 1 + c_1z^{-1} + c_2z^{-2} + \ldots + c_pz^{-p} \\ D(z^{-1}) &= 1 + d_1z^{-1} + d_2z^{-2} + \ldots + d_qz^{-q} \end{aligned} \tag{9.3}$$

Note that the ARMA noise model is always considered in discrete-time terms and that ρ could also denote δ^{-1}, where in this context, δ is the delta operator (Middleton and Goodwin, 1990), where

$$\delta = \frac{z-1}{\Delta t}; \quad \text{i.e. } \delta y(t_k) = \frac{y(k+1) - y(k)}{\Delta t}$$

which is the discrete-time equivalent of the derivative operator s. However, while RIV routines for δ operator models have been developed (Young et al., 1998; Chotai et al., 1998)), they are not currently available in the CAPTAIN Toolbox.

The model (9.1) is assumed to be contained within a feedback system, such as that shown in Figure 9.1, which represents the specific case of a unity feedback digital controller with the forward path control element represented as an assumed known ratio of rational polynomials $P(\rho)$ and $Q(\rho)$ in the generalized ρ operator. The command input to the control system is denoted by $y_d(t_k)$ and this is assumed to be statistically independent of the additive noise $\xi(t_k)$. It must be stressed that the identification and estimation procedures presented in this and subsequent sections of the chapter are not limited to this particular feedback control structure, but this is the forward path PIP control structure used in the illustrative examples considered later, where the controller is implemented in discrete-time (i.e. $\rho = z^{-1}$ in the controller transfer function) for both discrete and continuous-time systems.

The primary aim of this chapter is to show how the various refined IV algorithms described in previous chapters of the book can be exploited for closed-loop TF model identification and estimation. In order to simplify the presentation, however, the acronyms GRIV and GSRIV are used in unified situations where both discrete and continuous-time estimation is being considered, so that all of the RIV/RIVC and SRIV/SRIVC acronyms apply simultaneously in the context of the model (9.1).

9.3 Closed-Loop Identification and Estimation

In this section, the first sub-section 9.3.1 outlines two very simple methods of closed-loop model estimation that exploit the GSRIV algorithms and yield consistent, asymptotically unbiased estimates of the parameters in both discrete or hybrid continuous-time transfer function models of the enclosed system. These simple, two-stage, algorithms can be applied reasonably successfully even if the enclosed system is inherently unstable, although convergence cannot be guaranteed in this unstable situation. Sub-section 9.3.2 then goes on to describe how one of these simple algorithms can be enhanced by the addition of a third stage to induce asymptotic efficiency and ensure that the estimates have desirable minimum variance characteristics, but only *when the enclosed system is stable.*

Note that an important aspect of all these methods, not referred to specifically in the following descriptions of the estimation procedures, is the identification of appropriate orders for the TF model polynomials in (9.3). In all of the examples presented later in section 9.4, however, these are identified using standard RIV structure identification statistics available in the rivbjid and rivcbjid routines in the CAPTAIN Toolbox (see section 6.9 of chapter 6).

9.3.1 Simple CLSRIV and CLSRIVC two-stage closed-loop estimation

Provided the command input $y_d(t_k)$ is free of noise and persistently exciting (see section 6.6.1 of chapter 6) it can be used as a source on instrumental variables. Note that neither $\varepsilon(t_k) = y_d(t_k) - y(t_k)$ nor $v_u(t_k)$ could be used as the source of the instrumental variables because both are contaminated by the circulatory noise in the closed-loop system. Now, because both $v_u(t_k)$ and $y(t_k)$ are available for measurement, in addition to $y_d(t_k)$, two rather obvious GSRIV approaches to closed-loop estimation are possible:

Method 1. Estimate the parameters of the TF model between between $y_d(t_k)$ and the control input $v_u(t_k)$ using the appropriate GSRIV algorithm. Note that $v_u(t_k)$

is a function of $y(t_k)$ and so is affected by the component of the noise circulating around the closed-loop at this location, denoted by $\eta_u(t_k)$, i.e.

$$v_u(t_k) = u(t_k) + \eta_u(t_k) \tag{9.4}$$

where $u(t_k)$ is the underlying noise-free control input to the system (see Figure 9.1 and the associated definition of the variables). As a result, the deterministic output of this estimated TF model provides a good but sub-optimal estimate $\hat{u}(t_k)$ of the noise-free input $u(t_k)$ to the enclosed system. As a result, the GSRIV algorithm can be used again to estimate the required transfer function between $\hat{u}(t_k)$ and the noisy $y(t_k)$. This is the two-stage approach suggested by Young (2008b) and evaluated fairly comprehensively by Young et al. (2009)[2].

Method 2. Estimate the parameters of the TF model for the whole closed-loop system between $y_d(t_k)$ and the measured, noisy output $y(t_k)$. The deterministic output of this model then provides a good estimate $\hat{x}(t_k)$ of the noise-free output from the system and the appropriate GSRIV algorithm can be used again, this time to estimate the transfer function between the two estimated variables $\hat{u}(t_k)$ and $\hat{x}(t_k)$. This approach is less satisfying than the first method in statistical terms because the final estimation involves two estimated noise-free variables, without direct reference to the measured output $y(t_k)$.

As pointed out previously and confirmed in the later simulation examples of section 9.4, these algorithms can be used even if the enclosed system is unstable. However, it is necessary to use the ordinary IV implementation of the GSRIV algorithms at the second stage of the estimation, where only the first iteration of the algorithm is implemented. This is to avoid the use of an unstable auxiliary model and prefilters (see chapters 7 and 8).

9.3.2 Three-stage CLRIV and CLRIVC closed-loop estimation

Here, the three-stages of the estimation algorithm are as follows, where it will be noted that the first two stages are very similar to the simple two-stage Method 1, outlined in the previous sub-section, except that full GRIV, rather than GSRIV algorithms are utilized throughout:

Stage 1. Estimate the TF between the command input $y_d(t_k)$ and the noisy control input $v_u(t_k)$ using the appropriate GRIV algorithm, and generate an estimate $\hat{u}(t_k)$ of the underlying noise-free control input $u(t_k)$ using this model.

Stage 2. Use the appropriate GRIV algorithm to obtain initial, two stage estimates $\hat{A}(\rho)$ and $\hat{B}(\rho)$ of the system TF model polynomials $A(\rho)$ and $B(\rho)$, respectively, based on the estimated noise-free control input signal $\hat{u}(t_k)$ obtained in Stage 1 and the noisy measured output signal $y(t_k)$. Note that (see Figure 9.1):

[2] This is conceptaully similar to the two-stage algorithm suggested by Van den Hof and Schrama (1993) but they used output error estimation of an FIR model, rather than SRIV.

$$y(t_k) = v_x(t_k) + \xi(t_k) = x(t_k) + \eta_x(t_k) + \xi(t_k) \tag{9.5}$$

where $x(t_k)$ is the underlying 'noise-free' output of the system that would be measured at this point if $\xi(t_k) = \eta_u(t_k) = 0 \; \forall \; t_k$. Note also that, because of this,

$$y(t_k) - \eta_x(t_k) = x(t_k) + \xi(t_k) \tag{9.6}$$

which is referred to below.

Stage 3. Compute the estimate $\hat{\eta}_u(t_k) = u(t_k) - \hat{u}(t_k)$ of the circulatory noise component of the control input signal, $\eta_u(t_k)$, and transfer this to the output of the system using the system model obtained in Stage 2, i.e.

$$\hat{\eta}_x(t_k) = \frac{\hat{B}(\rho)}{\hat{A}(\rho)} \hat{\eta}_u(t_k) \tag{9.7}$$

where it provides an estimate of the component of the circulatory noise at the output of the system, $\eta_x(t_k)$, that derives from $\eta_u(t_k)$. Consequently, if this estimate is subtracted from the measured output it yields the following estimate of the output signal *that does not include the circulatory noise component from the closed-loop*,

$$\hat{y}(t_k) = y(t_k) - \hat{\eta}_x(t_k) \tag{9.8}$$

It is, therefore, an estimate of the noise-free output $x(t_k)$ plus only the additive noise $\xi(t_k)$ (see equation (9.6)). As a result, the data set $\{\hat{u}(t_k); \hat{y}(t_k)\}$ provides an estimate of the data set that would have been obtained if the system was being estimated in the open-loop situation. Finally, therefore, use the appropriate open-loop GRIV algorithm for a second time to re-estimate the system model based on this constructed data set.

In computational terms, this three-stage procedure is straightforward to implement because it only makes use of estimation routines already available in the CAPTAIN Toolbox (see later, section 9.4).

9.3.3 Unstable systems

One major reason why one would wish to identify and estimate a system within a closed-loop is when the system is open-loop unstable, so that open-loop estimation is difficult, if not impossible. However, the three-stage estimation procedure described in the previous sub-section 9.3.2 includes explicit simulation of the estimated model in the form of the iteratively updated 'auxiliary model' and utilization of the estimated system denominator polynomial $\hat{A}(\rho)$ in the iteratively updated prefilters. In the normal CAPTAIN Toolbox implementation of the GRIV-type algorithms for open-loop stable systems, therefore, the auxiliary model and prefilter denominator polynomials are stabilized at each iteration in order to avoid any (rare) transient violations of stability during the iterative updating process (using the

Matlab routines polystab for discrete-time models and polystc for continuous-time models). Despite these precautions, the full three-stage algorithm fails to produce satisfactory estimation results when applied to open-loop unstable systems, even when they are contained within a stable closed-loop.

Fortunately, however, the simple two-stage methods outlined in section 9.3.1 function satisfactorily in this closed-loop environment, as we shall see in the next section 9.4. Comprehensive Monte Carlo analysis suggests that they normally yield consistent, asymptotically unbiased parameter estimates with acceptable confidence bounds provided sufficient data are available from the closed-loop experiments when the data are noisy.

9.4 Simulation Examples

In this section, we consider a number of simulation examples, all of which concern the estimation of a TF model for a system contained within a digital *Proportional-Integral-Plus* (PIP) closed-loop control system (see e.g. Taylor *et al*, 2000, and the prior references therein), with the control system designed using the PIP design routines in the CAPTAIN Toolbox (see Appendix G). The three-stage CLRIV and CLRIVC algorithms (section 9.3.2) are implemented using the CAPTAIN *rivbj* and *rivcbj* routines, respectively. The simple two-stage methods (See section 9.3.1) use the SRIV and SRIVC options of these same routines.

In all of the examples, the TF system is based on the following second order, non-minimum phase, continuous-time model:

$$\frac{B(s)}{A(s)} = \frac{-0.5s+1}{s^2+2\zeta\omega_n s+\omega_n^2} \tag{9.9}$$

where ζ and ω_n take on various values; while the discrete-time ARMA noise process always takes the form

$$\frac{D(z^{-1})}{C(z^{-1})} = \frac{1+0.5z^{-1}}{1-0.85z^{-1}} \tag{9.10}$$

and σ^2 in (9.1) is adjusted to provide different levels of additive noise. In the case of continuous-time estimation, this system is simulated in continuous-time within Simulink for 300 secs., using the variable step length ODE45 (Dormer and Prince) solver and enclosed, as shown in Figure 9.1, within a discrete-time, PIP control loop based on optimal *Linear-Quadratic* (LQ) design (i.e. optimization of a linear system control performance based on a quadratic performance index). For discrete-time model estimation, the Simulink simulation uses the discrete-time equivalent of the continuous-time model (9.9), as obtained by conversion using the c2d routine in Matlab, with the zero-order-hold option at the selected sampling interval. The reader not acquainted with the PIP control system design methods should consult the previously cited references and run the demonstration example pipdemo1 in the CAPTAIN Toolbox.

The first example evaluates the discrete-time estimation performance with $\zeta = 0.05$ and $\omega_n = 2$ rad/sec., so the sampling interval for both estimation and control system design is set at $\Delta t = 0.2$ seconds (1500 samples), which ensures that the eigenvalues of the discrete-time model are not too close to the unit circle in the complex z-plane. In this case, the forward path PIP control transfer function obtained by application of the pipopt design routine in CAPTAIN is:

$$\frac{P(z^{-1})}{Q(z^{-1})} = \frac{7.8553z^{-1} - 15.682z^{-2} + 7.8331z^{-3}}{1 - 2.4443z^{-1} + 2.0597z^{-2} - 0.61532z^{-3}} \tag{9.11}$$

In the second example, a continuous time model is estimated, again with $\zeta = 0.05$ and $\omega_n = 2$ rad/sec., but the sampling interval is set ten times smaller at $\Delta t = 0.02$ seconds (15000 samples), which is more appropriate for good continuous-time estimation. The pipopt designed forward path PIP control transfer function in this case is:

$$\frac{P(z^{-1})}{Q(z^{-1})} = \frac{4.2091z^{-1} - 7.971z^{-2} + 4.0917z^{-3}}{1 - 1.0985z^{-1} + 0.27071z^{-2} - 0.17222z^{-3}} \tag{9.12}$$

Note that no attempt is made here to control the closed-loop non-minimum phase behaviour, although this is possible in PIP design using command input anticipation (Taylor et al., 1994), as implemented in the CAPTAIN pipcom routine. Also, the system is not designed to maximize disturbance rejection, so the circulatory noise effects can be quite high.

9.4.1 Example 9.1: Closed-loop, stable, discrete-time system estimation

In this example, the command input signal $r(t_k)$ is a ± 1.0 PRBS signal of length 1500 samples and the white noise input signal $e(t_k)$ has variance $\sigma^2 = 0.1$. The resulting noise-to-signal ratios, by standard deviation, are 0.35 at the output and 0.30 at the control input. A typical segment of the data is shown in Figure 9.2, where we see that the circulatory noise is quite highly coloured.

At the first stage in a typical single run of the three-stage algorithm, using the SRIV option of the rivbjid routine in CAPTAIN, the order identification criteria (see section 6.9 of chapter 6) clearly identify a third order, [3 3 1 0 0] model between $r(t_k)$ and $u(t_k)$; and, despite the quite high noise level on the control input signal, the SRIV option of the *rivbj* routine produces an estimated model with a coefficient of determination between the model output (the estimated control signal $\hat{u}(t_k)$) and the actual noise-free output (the noise-free control signal $u(t_k)$) of $R_T^2 = 0.999$.

The results of the *Monte Carlo Simulation* (MCS) analysis for the complete three-stage algorithm are presented in Table 9.1, together with associated results for the simple two stage algorithms. Also shown for reference and comparison are the results obtained using the PEM algorithm in the Matlab SID Toolbox, which is a (non-recursive) alternative to the RIV algorithm in this discrete-time example. These

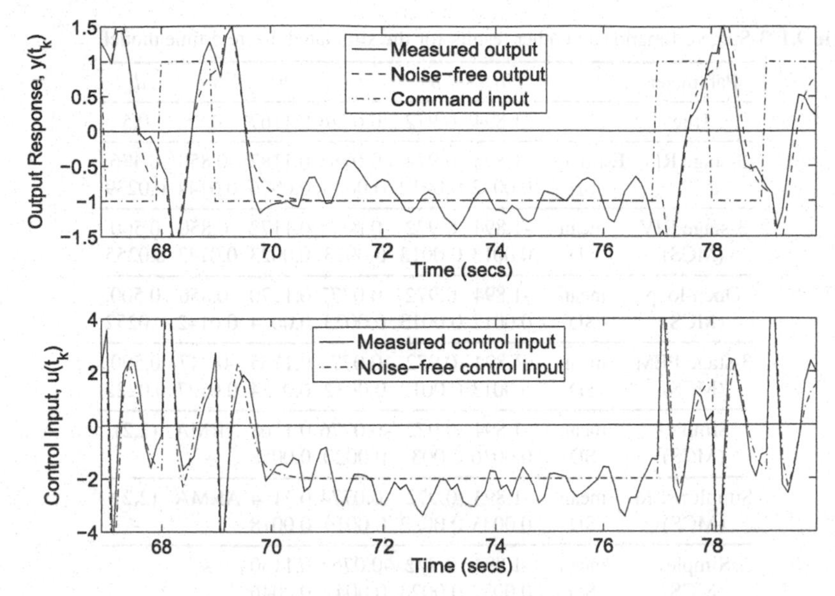

Fig. 9.2 Segment of the data used in the closed-loop estimation.

were obtained from 100 Monte Carlo realizations based on random selections of the white noise sequence $e(t_k), k = 1,2,\ldots,1500$. Figures 9.3 and 9.4 show the resultant ensembles of the Bode plots and impulse responses. It is clear that the three-stage algorithm is working as expected, with the single run and associated standard error estimates matching those obtained from the MCS analysis. Furthermore, the estimation results also match the optimal results obtained in the equivalent open-loop situation using the rivbj routine in CAPTAIN, demonstrating the optimality of the CLRIV estimation.

The simple two stage methods described in section 9.3.1 also perform quite well, yielding consistent estimates. The standard deviations are higher than those obtained using the three-stage algorithm but the standard deviations of the Method 1 match those obtained when the PEM algorithm from the SID Toolbox in Matlab is used in place of RIVBJ at the second stage. In both cases, the BIC identifies an ARMA(2,2) model for the additive noise but, in order to simplify the presentation, the parameter estimates for these ARMA models are not shown.

Finally, the quality of the estimation is demonstrated graphically in Figures 9.3 and 9.4. Here, the uncertainty bounds in the estimated Bode diagram are very small, as we see in Figure 9.3, and even in the case of the estimated impulse response, which is normally very sensitive to parametric uncertainty, the 95% confidence bounds are still quite small.

Table 9.1 3-Stage estimation and other results for the simulated discrete-time model.

Parameter		a_1	a_2	b_0	b_1	c_1	d_1
True		-1.894	0.972	-0.0776	0.1167	-0.85	0.5
3-stage RIV	Estimates	-1.896	0.974	-0.0788	0.1187	-0.850	0.486
(SR)	SE	0.0012	0.0012	0.0026	0.0026	0.0144	0.0239
3-stage RIV	mean	-1.894	0.972	-0.0779	0.1172	-0.850	0.500
(MCS)	SD	0.0013	0.0013	0.0023	0.0023	0.0143	0.0255
Open-loop	mean	-1.894	0.972	-0.0777	0.1170	-0.850	0.500
(MCS)	SD	0.0012	0.0013	0.0023	0.0024	0.0142	0.0252
3-stage PEM	mean	-1.894	0.972	-0.0774	0.1165	-0.847	0.500
(MCS)	SD	0.0013	0.0012	0.0032	0.0034	0.0127	0.0222
Simple†	mean	-1.894	0.972	-0.0776	0.1168	ARMA	(2,2)
(MCS)	SD	0.0036	0.0033	0.0020	0.0018	-	-
Simple†PEM	mean	-1.893	0.972	-0.0774	0.1164	ARMA	(2,2)
(MCS)	SD	0.0035	0.0032	0.0019	0.0018	-	-
Simple‡	mean	-1.894	0.972	-0.0765	0.1150	-	-
(MCS)	SD	0.0032	0.0028	0.0041	0.0046	-	-

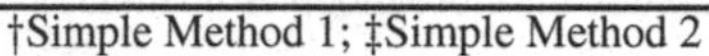
†Simple Method 1; ‡Simple Method 2

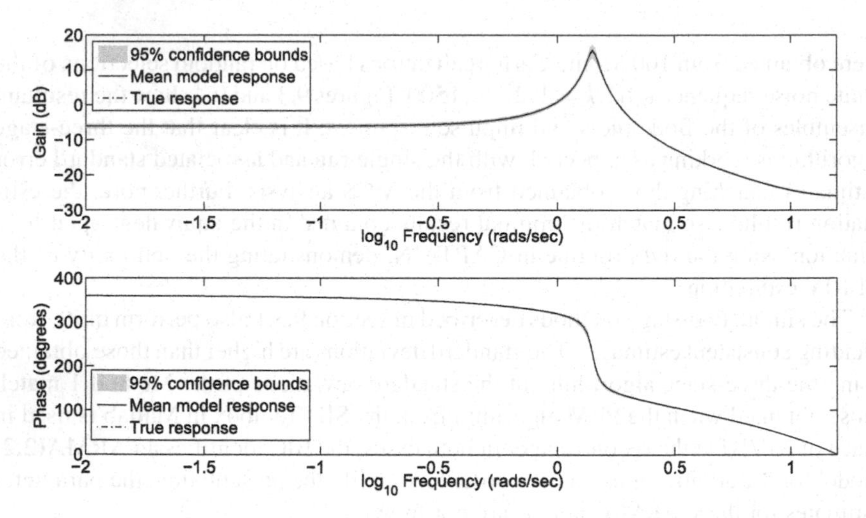

Fig. 9.3 MCS-based comparison of the true and CLRIV estimated model Bode plots.

9.4.2 Example 9.2: Closed-loop, stable, continuous-time system estimation

In this example, the command input signal $r(t_k)$ is a ± 1.0 PRBS signal of length 15000 samples. The MCS analysis is the same as in the discrete-time example

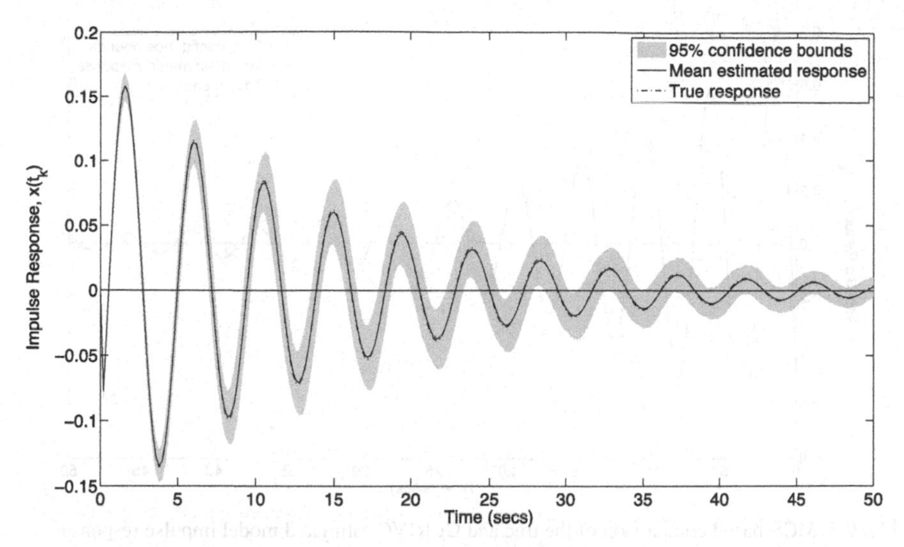

Fig. 9.4 MCS-based comparison of the true and CLRIV estimated model impulse responses.

above, based on 100 realizations, but the white noise variance σ^2 is selected as 0.05 in order to yield a noise-to-signal ratio on the control input of 0.30 by standard deviation, which is the same that used in the discrete-time case. However, the noise level at the output is reduced because of the closed loop characteristics and has a noise-to-signal ratio of 0.22.

At the first stage in a typical single run of the three-stage algorithm, the BIC clearly identifies a third order, [3 3 0 0 0] model between $r(t_k)$ and $u(t_k)$; and, as in the discrete-time case, the coefficient of determination between the model output and the actual noise-free input is $R_T^2 = 0.999$. The MCS results are summarized in Table 9.2, where the good performance obtained in the discrete-time situation is repeated here. In particular, the single run and associated standard error estimates match those obtained from the MCS analysis; and the estimation results also match the optimal results obtained in the equivalent open-loop situation using the rivcbj algorithm, demonstrating the optimality of the CLRIVC estimation. Figure 9.5 shows the ensemble of impulse responses obtained in this example: the uncertainty bounds are smaller than those for the discrete-time results in Figure 9.4 but this is partly due to the higher output noise level in this former case and the longer sample size used here.

On the negative side, if the noise level is doubled, there are some very isolated realizations where the model is estimated rather poorly, suggesting that the CLRIVC estimation is not quite as robust as the CLRIV, probably because of the greater complexity of the hybrid continuous-time estimation procedure. Also, the computation time required for each realization is 35 times greater than that required in the discrete-time case, although this is partly explained by the sample length being ten

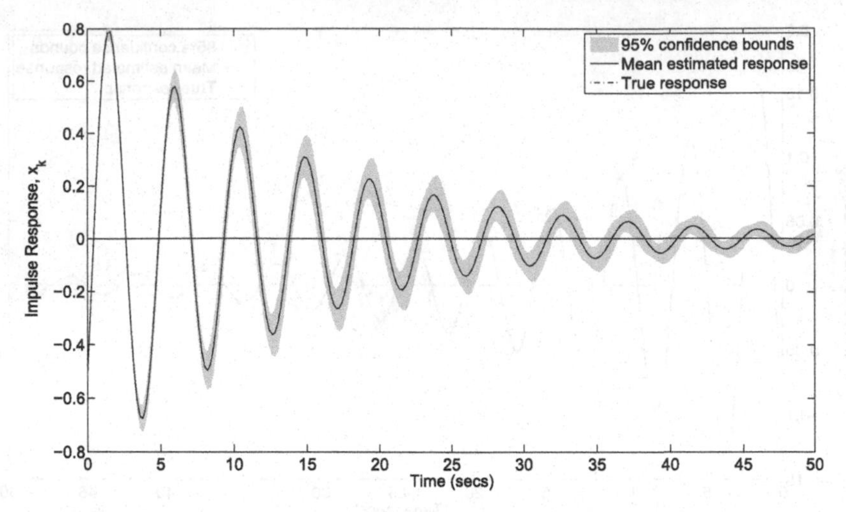

Fig. 9.5 MCS-based comparison of the true and CLRIVC estimated model impulse responses.

Table 9.2 3-Stage estimation and other results for the simulated continuous-time model.

Parameter		a_1	a_2	b_0	b_1	c_1	d_1
True	ϑ	0.141	2.0	-0.5	1.0	-0.85	0.5
3-stage RIVC	$\hat{\vartheta}$	0.141	2.002	-0.509	0.981	-0.851	0.499
(SR)	SE	0.0021	0.0028	0.0028	0.0055	0.0045	0.0075
3-stage RIVC	mean	0.140	2.000	-0.509	0.978	-0.856	0.487
(MCS)	SD	0.0025	0.0030	0.0030	0.0055	0.0047	0.0067
Open-loop	mean	0.140	1.999	-0.509	0.978	-0.852	0.501
(MCS)	SD	0.0022	0.0024	0.0030	0.0055	0.0047	0.0066
Simple†	mean	0.140	1.999	-0.509	0.983	ARMA	(2,2)
(MCS)	SD	0.0122	0.0154	0.0040	0.0093	-	-
Simple‡	mean	0.140	1.999	-0.509	0.984	-	-
(MCS)	SD	0.0122	0.0151	0.0047	0.0096	-	-

†Simple Method 1; ‡Simple Method 2

times larger to accommodate the smaller sampling interval used in the continuous-time estimation. To compensate for these minor disadvantages, however, CLRIVC estimation brings with it the many advantages that normally accrue from the estimation of continuous-time models, as discussed in chapter 8 and in chapters from the recent book on the subject (Garnier and Wang, 2008).

Finally, note that the variance of the estimates obtained with the simple two-stage algorithms is considerably higher than that obtained by adding the third stage, demonstrating the advantages of statistically optimal algorithm design in the case of stable systems. As we shall see in the next sub-section 9.4.3, however, these simple algorithms have advantages when we consider the case where the system contained in the closed-loop is unstable or marginally stable.

9.4.3 Example 9.3: Closed-loop, unstable and marginally stable system estimation

This section presents three simulation examples that demonstrate how the simple two-stage CLRIV and CLRIVC algorithms can work in the difficult but practically relevant case of unstable or marginally stable systems described by unstable, discrete or continuous-time TF models, although the performance will depend on the circumstances and convergence cannot be guaranteed. Although convergent estimates can be obtained at high levels of noise, the variance of the parameter estimates is naturally higher than in the case of a stable system because the estimates are more sensitive when the system is unstable. This means that associated properties of the model, such as the unstable impulse response, can have very high variances. Recognizing this, the basic results presented below are obtained with the additive noise variance σ^2 adjusted to give fairly low (*circa* 10%) noise levels, by standard deviation, so that the parametric error variance is similarly fairly low. However, the results obtained at higher noise levels are considered in less detail and we see that the algorithms remain potentially useful even at noise-to-signal ratios of unity.

A. Unstable discrete-time, oscillatory system

For this example, the damping parameter ζ in the continuous-time model (9.9) is set to -0.05, i.e. the negative of its value in the stable system analysis, and the discrete-time TF model used in the MCS analysis has the parameters shown in the second row of Table 9.3. Both of the simple methods perform well in this case, yielding very similar results. Also shown are the results obtained when the PEM algorithm from the SID Toolbox in Matlab is used, rather than the SRIV algorithm. The results are not so good but this is probably because the algorithm is not intended for use with unstable systems.

If the noise is increased to the level used in the stable system analysis, the results shown in Table 9.4 are obtained and the right panel of Figure 9.6 compares the associated ensemble of impulse responses in this case with those generated in the lower noise case, shown in the left panel. It is clear that, as the noise increases, so the variance of related properties increases even more markedly because of the sensitivity of the impulse response to uncertainty when the system is unstable.

If the noise is increased further to give a noise-to-signal ratio at the output of unity, then the ensemble averages obtained from the MCS analysis are as follows:

$$-1.946(0.058)\;1.027(0.067)\;-0.078067(0.014)\;0.119(0.010) \tag{9.13}$$

Table 9.3 Simple closed-loop method estimation results for the simulated discrete-time, unstable oscillatory model: low noise.

Parameter		a_1	a_2	b_0	b_1	c_1	d_1
True (hybrid)	ϑ	-0.1414	2.0	-0.5	1.0	-0.85	0.5
True (discrete)	ϑ	-1.948	1.029	-0.080	0.120	-0.85	0.5
Simple Method†	$\hat{\vartheta}$	-1.948	1.029	-0.080	0.121	-	-
(MCS)	SD	0.0027	0.0023	0.0039	0.0035	-	-
PEM	$\hat{\vartheta}$	-1.884	0.958	-0.051	0.0835	-	-
(MCS)	SD	0.0028	0.0021	0.0015	0.0021	-	-
Simple method‡	$\hat{\vartheta}$	-1.948	1.028	-0.0791	0.119	-	-
(MCS)	SD	0.0026	0.0024	0.0110	0.0105	-	-

†Simple Method 1; ‡Simple Method 2

Table 9.4 Simple closed-loop method estimation results for the simulated discrete-time, unstable oscillatory model: high noise.

Parameter		a_1	a_2	b_0	b_1	c_1	d_1
True (hybrid)	ϑ	-0.1414	2.0	-0.5	1.0	-0.85	0.5
True (discrete)	ϑ	-1.948	1.029	-0.080	0.120	-0.85	0.5
Simple Method†	$\hat{\vartheta}$	-1.947	1.027	-0.080	0.120	-	-
(MCS)	SD	0.0070	0.0074	0.0053	0.0049	-	-
PEM	$\hat{\vartheta}$	-1.885	0.959	-0.050	0.0828	-	-
(MCS)	SD	0.0075	0.0063	0.0042	0.0055	-	-
Simple method‡	$\hat{\vartheta}$	-1.952	1.032	-0.083	0.123	-	-
(MCS)	SD	0.0082	0.0073	0.0087	0.078	-	-

†Simple Method 1; ‡Simple Method 2

where the standard deviations are shown in parentheses. We see now that the estimates are still consistent and asymptotically unbiased but the variance is very high, implying that the sample size in such high noise situations would have to be substantially increased to obtain practically useful results. However, these are very high noise levels for a closed-loop system and it is likely, in practice, that efforts would be made to reduce the noise levels both by good experiment design and tuning control system gains to enhance disturbance rejection.

B. Discrete-time, double integrator system

The continuous-time model (9.9) is converted to a double integrator system if $\zeta = 0$ and the discrete-time TF model used in the MCS analysis then has the parameters shown in the second row of Table 9.5. All other settings in the simulations are the same as for the stable system example in section 9.4.1. In this double integrator case, the open-loop system is, of course, violently unstable, and so this provides an acute

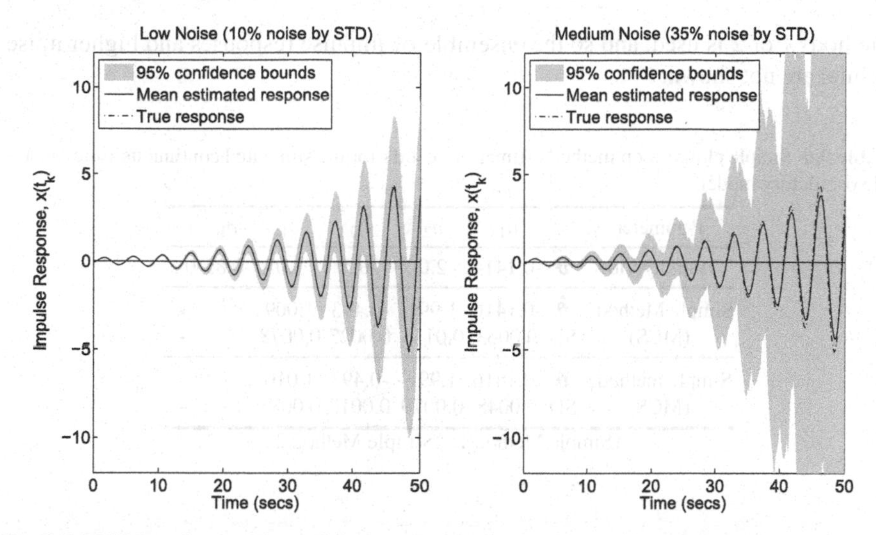

Fig. 9.6 MCS-based comparison of the impulse responses generated by the discrete-time, oscillatory model estimated by the simple closed-loop method 1., in low (left-hand panel) and medium (right-hand panel) noise level situations.

Table 9.5 Simple closed-loop method estimation results for the simulated discrete-time, double integrator model.

Parameter		a_1	a_2	b_0	b_1	c_1	d_1
True (hybrid)	ϑ	0.0	0.0	-0.5	1.0	-0.85	0.5
True (discrete)	ϑ	-2	1.0	-0.08	0.12	-0.85	0.5
Simple Method†	$\hat{\vartheta}$	-1.999	0.999	-0.080	0.120	-	-
(MCS)	SD	0.0031	0.0029	0.0024	0.0018	-	-
Simple method‡	$\hat{\vartheta}$	-1.960	0.960	-0.137	0.162	-	-
(MCS)	SD	0.147	0.147	0.186	0.183	-	-

†Simple Method 1; ‡Simple Method 2

test for any closed-loop identification algorithm. The MCS results are summarized in Table 9.5 and we see that, unlike the situation in the case of the unstable oscillatory system considered in example A., the results obtained for the Simple Method 1 are now considerable better than those obtained with Simple Method 2, suggesting that the former has advantages in this explosively unstable system context.

C. Unstable continuous-time, oscillatory system

This example is the continuous-time equivalent of the discrete-time example A., above, and the MCS results are presented in Table 9.6. Overall, the results are quite similar to those obtained in the discrete-time case, whatever simple two-stage methods 1 or 2 is used, and so the ensemble of impulse responses and higher noise results are not shown.

Table 9.6 Simple closed-loop method estimation results for the simulated continuous-time, unstable oscillatory model.

Parameter		a_1	a_2	b_0	b_1	c_1	d_1
True (hybrid)	$\boldsymbol{\vartheta}$	-0.1414	2.0	-0.5	1.0	-0.85	0.5
Simple Method†	$\hat{\boldsymbol{\vartheta}}$	-0.1416	1.991	-0.493	1.009	-	-
(MCS)	SD	0.0063	0.0115	0.0027	0.0078	-	-
Simple method‡	$\hat{\boldsymbol{\vartheta}}$	-0.1416	1.992	-0.493	1.010	-	-
(MCS)	SD	0.0048	0.0079	0.0016	0.0085	-	-

†Simple Method 1; ‡Simple Method 2

9.5 Exercises

1. Referring to the example considered in section 9.4.1, use the Matlab c2dm routine, with the sampling interval set to $\Delta t = 0.02$, in order to obtain the discrete-time equivalent of the continuous-time model in equation (9.9) with $\zeta = 0.05$ and $\omega_n = 2$ rad/sec. What do you note about this TF model (hint: apply the Matlab roots routine to the denominator polynomial)?
2. Using discrete-time (z^{-1} polynomial) block diagram algebraic manipulation, compute the deterministic closed-loop transfer function (i.e. with $\xi(t) = 0$) between $y_d(t_k)$ and $y(t_k)$ for the model in exercise 1 (see equations (B.45) and (B.47) of section B.3.7, Appendix B, with the s operator replaced by the z^{-1} operator).
3. Why do you think it might be difficult to estimate the closed-loop TF model in exercise 2., even if the data were noise-free?
4. Write a Matlab m-file script to implement the simple two-stage method 1 for discrete-time models. As pointed out, this only requires appropriate calls to the rivbj routine in CAPTAIN.
5. Write a Matlab m-file script to implement the three-stage CLRIV method for discrete-time models. Once again, only calls to the rivbj routine in CAPTAIN are required. Test this, and the simple algorithm developed in exercise 1., on the example in section 9.4.1, or preferably one of your own choice.
6. When the clrivbj and clrivcbj implementations of the two and three-stage closed-loop estimation algorithms become available in the CAPTAIN Toolbox (see Appendix G), run the CLdemo.

9.6 Summary and Conclusions

This chapter has demonstrated how the existing refined and simplified refined instrumental variable algorithms, as developed for the identification and estimation of discrete or continuous-time transfer function models in an open-loop situation, can be combined to allow for similar estimation in a closed-loop setting. Indeed, within this more complex environment, it is possible to apply the simplified versions to situations where the enclosed system is open-loop unstable, although convergence cannot be guaranteed. These generalized closed-loop algorithms can be used in either a two or a three-stage form and the three-stage implementation yields statistically optimal estimates in the case of stable open-loop systems. The statistical properties of the estimates in the case where the open-loop system is unstable are difficult to evaluate and further research is required in this situation. This will need to investigate whether it is possible to reduce further the variance of the estimates in the case on unstable open-loop systems, particularly in the case of continuous-time systems subject to high levels of noise.

Finally, it should be emphasized that the two and three-stage estimation algorithms described in this chapter have been developed only recently and, so far, have been evaluated only by simulation analysis, such as that described in section 9.4 of this chapter. It still remains to seen, therefore, how they perform with real data. They have been presented here, as we approach the the last three chapters of the book, in the hope that the reader is now in a mood to consider research orientated topics. This is continued in the last three chapters, which deal with more advanced and challenging aspects of research and development.

Chapter 10
Real-Time Recursive Parameter Estimation

One primary purpose of a recursive estimation algorithm is to process data as it is received from some source, with the aim of updating the estimates of parameters and states in a model that is being used for a practical purpose, such as model evaluation, forecasting or automatic control. 'Real-time' or 'on-line' recursive estimation of this type has received a lot of attention in different areas of science and engineering, initially in an automatic control context and latterly in relation to the topic of 'data-assimilation', in areas such as oceanography and hydrology.

The distinguished automatic control academic Karl Aström once said that recursive estimation was a 'fiddler's paradise' and this is certainly the case. To attempt to review all of the real-time recursive estimation algorithms that have been suggested since the publication of Kalman's seminal papers in 1960 would be a daunting task. Consequently all that this chapter will do is consider briefly several real-time recursive estimation algorithms that have been developed for the estimation of parameters in stochastic, dynamic systems; algorithms that are either of historical importance, such as the Extended Kalman Filter (EKF), or provide available tools that appear to have the most practical potential. In these days of the Internet with powerful search engines, the reader will then be able to follow this up and carry out more focussed and discerning research on such algorithms in the particular area of application that relates best to his or her area of interest.

In an attempt to aid the introductory reader in understanding a difficult subject, the previous chapters have followed sequentially from simple beginnings in deterministic, linear least squares methods; through the more complicated but still straightforward procedures of statistical regression analysis; to the quite complex techniques for solving more general problems of time series analysis. And in this latter case, we have been careful to look in detail only at those approaches which are related most closely to the simpler linear regression methods and can, to some extent, be regarded as logical 'pseudo-linear regression' developments of these methods in the context of stochastic, dynamic system estimation, particularly as they relate to TF model identification and estimation.

But there are other important recursive methods of time series model estimation which, while closely related to the procedures described in the previous chapters, represent somewhat different approaches to the problem. These procedures have been referred to briefly in previous chapters and, in the present chapter, we will discuss them in sufficient detail for the reader to become acquainted with their basic form and computational nature. In this introductory text, however, no attempt will be made to evaluate them fully and the potential user will need to follow up some of the references in order to obtain the requisite understanding demanded for practical application of the various techniques.

For much of the past thirty years, real-time recursive parameter estimation has been influenced most profoundly by publications on the theory and practice of *Recursive Prediction Error Minimization* (RPEM) emanating from the 'Swedish School' of control theory academics, who were strongly influenced by their teacher and mentor Karl Åström: most notably, Lennart Ljung and Torsten Söderström, in alliance with other academics, such as Petre Stoica (see e.g. Ljung and Söderström, 1983; Söderström and Stoica, 1989; and Ljung, 1999 and the prior references threin). This is a general approach to the design of real-time recursive estimation algorithms that can be applied to models of all types, including both transfer function and state space representations. And when applied to the estimation of parameters in transfer function models, it has a very close relationship to the recursive RIV algorithm discussed in chapter 7 of this book. The RPEM algorithm will be considered in the next section and the *Real-time Recursive Refined Instrumental Variable* (RRIV) extension of the RIV/RIVC algorithms will be considered in section 10.2. Following this, in section 10.3, we will step back in time and consider a real-time recursive parameter estimation algorithm that, since the early 1960s, has probably received more attention than any other, the *Extended Kalman Filter* (EKF). And finally, section 10.4 returns to the present and reviews briefly approaches to real-time recursive estimation that have been heavily influenced by the EKF but try to avoid the difficulties associated with nonlinear state space models by exploiting computationally intensive methods based on Monte Carlo simulation.

10.1 Prediction Error (PE) Methods and the RPEM Algorithm

Once the reader has come to terms with the statistical basis of time series models, the prediction error approach to recursive estimation becomes rather obvious: in simple terms, and based on the discussion in chapters 6 to 8, it implies the choice of a cost function such that the resultant recursive algorithm has a particular 'least squares' like form and the one sample-ahead-prediction errors (innovations process or recursive residuals) in this recursive algorithm, which represents only that part of $y(k)$ that cannot be predicted from the past data, are minimized. Usually, and certainly within the main context of the present book, stronger conditions apply and the normalized prediction errors (i.e. the normalized innovations sequence) should also have the properties of white noise with constant variance. In this way, the estimation problem is simplified in statistical terms and the resultant algorithm, if chosen correctly, is convergent and has certain desirable statistical properties, such as consistency and asymptotic statistical efficiency (See Appendix B).

Although the PE approach is very general and can be applied to any stochastically defined model form, including the ARX, ARMAX and DA, mentioned in chapter 6, as well as state space models (see later), let us consider it in the context of the model that has underlain most of the analysis in the previous chapters of this book, the discrete-time Box-Jenkins TF model (6.18), which is repeated here for convenience:

$$y(k) = \frac{B(z^{-1})}{A(z^{-1})}u(k) + \frac{D(z^{-1})}{C(z^{-1})}e(k) \tag{10.1}$$

where, the pure time delay δ on the input $u(k)$ has been dropped without any loss in generality.

In simple terms, the basic stimulus of the PEM approach to estimation is the definition of an error function that is usually defined in terms of the error between the noisy measured output $y(k)$ and the 'best prediction' $\hat{y}(k)$ of $y(k)$. In the case of the BJ model this error is clearly defined by the white noise process $e(k)$, since it is defined as the one-step-ahead prediction error and, if the theoretical statistical assumptions are valid, represents the irreducible uncertainty in the model. As a result, if the unknown parameter vector containing all the parameters in the BJ model is denoted by $\boldsymbol{\vartheta}$, then the PEM estimation problem can be posed in terms of minimizing the following nonlinear least squares cost function $\mathfrak{J}_2(\boldsymbol{\vartheta})$ with respect to $\boldsymbol{\vartheta}$, i.e.,

$$\hat{\boldsymbol{\vartheta}} = \arg\min_{\boldsymbol{\vartheta}} \mathfrak{J}_2(\boldsymbol{\vartheta},N) \qquad \mathfrak{J}_2(\boldsymbol{\vartheta},N) = \sum_{k=1}^{N} \varepsilon^2(\boldsymbol{\vartheta},k) \tag{10.2}$$

where,

$$\varepsilon(\boldsymbol{\vartheta},k) = \frac{C}{D}\left\{y(k) - \frac{B}{A}u(k)\right\} \tag{10.3}$$

and the z^{-1} arguments in the TF polynomials have been removed, here and subsequently, for convenience.

The optimization problem posed by the cost function (10.2) is clearly nonlinear in the parameters and one approach to its solution is the iterative RIV procedure described in chapter 7 based on the formulation of the pseudo-linear estimation equation (7.15). Another, more direct, approach which has dominated this form of time series analysis since the 1960s (see Box and Jenkins, 1970, in relation to the BJ model; and Aström and Bohlin, 1966, in relation to the ARMAX model), is to use some form of gradient-based optimization that adjusts an estimate $\hat{\boldsymbol{\vartheta}}$ of the parameter vector $\boldsymbol{\vartheta}$ until it minimizes the above cost function. This approach is termed the *Maximum Likelihood* (ML) method when $\varepsilon(k)$ can be assumed to have a Gaussian amplitude distribution; and the *Prediction Error Minimization* (PEM) method in more general contexts.

Of course, the ML and PEM approaches are based on *en bloc* processing and we are most concerned in this book with recursive estimation. Almost all research on recursive PEM methods has been in relation to purely discrete-time problems (e.g. early publications of this type are Ljung (1976); Ljung (1977b,a); and Ljung (1978), although Solo (1978, 1980) has considered both continuous and discrete-time implications). A later, comprehensive text on PEM methods, both *en bloc* and recursive, is the excellent book by Ljung (1999). However, the introductory reader to whom the present book is addressed may sometimes find these rather theoretical treatments of PEM methods a little difficult and analytically rather dense. Nevertheless, they represent very important contributions to the subject of recursive estimation since they deal with the important topics of algorithmic convergence and the asymptotic statistical properties of the estimates. Because of the difficult nature of the problem, however, we have purposefully omitted detailed discussion of these theoretical questions in this introductory text; rather we have let the reader assume that the algorithms described have generally good convergence properties unless otherwise stated.

This semi-theoretical approach does not imply that the convergence analysis is not important; merely it is intended to shield the reader new to the subject from rather esoteric theory before it is absolutely necessary. It also serves to emphasize that alternative experimental methods of evaluating convergence, such as the Monte Carlo procedures that have become so popular with the advent of fast desktop computers, are also important in assessing algorithmic performance. For instance, most purely theoretical analysis to date is based largely on asymptotic behaviour; small sample performance must, therefore, be considered by carefully conducted stochastic simulation experiments of the Monte Carlo type.

In relation to the theoretical analysis, it will suffice here to say that the theory can be based either on: (i) relating the behaviour of recursion to the solution of an ordinary, continuous-time differential equation (Ljung, 1977b,a and Ljung, 1999); (ii) utilizing probability theory to reduce the analysis to the consideration of a deterministic Lyapunov function (Hannan, 1976); or (iii) analysis of the behaviour as a stochastic difference equation (Solo, 1978, 1980). This latter approach allows for second order analysis and can indicate the asymptotic covariance behaviour of the recursive PE estimators.

In general, the *Recursive* PEM (RPEM) algorithm is constructed, by analogy with the recursive least squares and in accordance with the theoretical requirements for convergence, to take the following form (cf. the RLS algorithm in chapter 3 and the associated equation (3.20)),

$$\begin{aligned}
\hat{\boldsymbol{\vartheta}}(k) &= \hat{\boldsymbol{\vartheta}}(k-1) + \mathbf{P}(k)\boldsymbol{\pi}(k)\varepsilon(k) && \text{(RPEM-1)}\\
\mathbf{g}(k) &= \mathbf{P}(k-1)\boldsymbol{\pi}(k)[1+\boldsymbol{\pi}^T(k)\mathbf{P}(k-1)\boldsymbol{\pi}(k)]^{-1} && \text{(RPEM-2)}\\
\mathbf{P}(k) &= \mathbf{P}(k-1) - \mathbf{g}(k)\boldsymbol{\pi}^T(k)\mathbf{P}(k-1) && \text{(RPEM-3)}
\end{aligned}$$

where,

$$\varepsilon(k) = y(k) - \hat{y}(k|k-1)$$

is the recursive residual representing the latest one-step-ahead prediction (innovation) error; $\hat{\boldsymbol{\vartheta}}(k)$ is the estimate of $\boldsymbol{\vartheta}(k)$ at the kth sampling instant; and $\boldsymbol{\pi}(k)$ is the following vector based on the negative of the gradient of $\varepsilon(k)$ with respect to $\boldsymbol{\vartheta}(k)$ (compare equation (RPEM-1) with equation (3.20) of chapter 3)

$$\boldsymbol{\pi}(k) = -\frac{\partial\,\varepsilon(k)}{\partial\,\hat{\boldsymbol{\vartheta}}} \tag{10.4}$$

The motivation for this algorithm is clear when it is compared with both the recursive least squares algorithm RLS of Chapter 3 and the SA algorithm (C.5) in Appendix C: it is simply a matrix gain SA or stochastic gradient algorithm, whose matrix gain is generated by analogy with the recursive least squares theory and whose gradient term can be interpreted as being proportional to the negative gradient of the instantaneous least squares cost function (cf. equations (3.18) and (3.19)), i.e.,

$$J_I = \varepsilon^2(k)$$

$$\text{so,} \quad -\frac{1}{2}\nabla_{\hat{\boldsymbol{\vartheta}}}J_I = -\frac{\partial\,\varepsilon(k)}{\partial\,\hat{\boldsymbol{\vartheta}}}\varepsilon(k) = \boldsymbol{\pi}(k)\varepsilon(k)$$

In the case of the BJ model, the parameter vector is

$$\boldsymbol{\vartheta} = [a_1,\ldots,a_n,b_0,\ldots,b_m,c_1,\ldots,c_p,d_1,\ldots,d_q]^T \tag{10.5}$$

and the gradient vector $\boldsymbol{\pi}(k)$, for any value of $\boldsymbol{\vartheta}$, can be evaluated quite simply in this case by differentiating $\varepsilon(k)$ with respect to each parameter of $\boldsymbol{\vartheta}$ in turn. Thus, since

$$\varepsilon(k) = \frac{C}{D}y(k) - \frac{BC}{DA}u(k)$$

we obtain:

$$\begin{aligned}
\frac{\partial}{\partial a_i}[\varepsilon(k)] &= \frac{C}{DA}.\frac{B}{A}u(k-i) = \frac{C}{DA}\hat{x}(k-i) = \hat{x}_{f_1}(k-i)\\
\frac{\partial}{\partial b_i}[\varepsilon(k)] &= -\frac{C}{DA}u(k-i) = -u_{f_1}(k-i)\\
\frac{\partial}{\partial c_i}[\varepsilon(k)] &= \frac{1}{D}[y(k-i) - \frac{B}{A}u(k-i)] = \frac{1}{D}\xi(k-i) = \xi_{f_2}(k-i)\\
\frac{\partial}{\partial d_i}[\varepsilon(k)] &= -\frac{1}{D}[\frac{C}{D}y(k-i) - \frac{CB}{DA}u(k-i)] = -\varepsilon_{f_2}(k-i)
\end{aligned} \tag{10.6}$$

where $\hat{x}(k)$ is defined as

$$\hat{x}(k) = \frac{B}{A}u(k) \tag{10.7}$$

As in chapter 7, the f_1 subscript denotes that the variable is filtered by a prefilter with transfer function $f_1 = C/DA$; while the the f_2 subscript denotes prefiltering with the transfer function $f_2 = 1/D$.

When the expressions in (10.6) are implemented in the RPEM algorithm, the polynomials A, B, C and D need to be based on the latest recursive estimates of the parameters. As a result, $\boldsymbol{\pi}(k)$ is defined as follows:

$$\begin{aligned}\boldsymbol{\pi}(k) = [&-\hat{x}_{\hat{f}_1}(k-1), \ldots, -\hat{x}_{\hat{f}_1}(k-n), u_{\hat{f}_1}(k), \ldots, u_{\hat{f}_1}(k-m), \\ &-\hat{\xi}_{\hat{f}_2}(k-1), \ldots, -\hat{\xi}_{\hat{f}_2}(k-p), \hat{\varepsilon}_{\hat{f}_2}(k-1), \ldots, \hat{\varepsilon}_{\hat{f}_2}(k-q)]^T\end{aligned} \tag{10.8}$$

where the prefilters $\hat{f}_1$ and $\hat{f}_2$ are the latest estimated prefilters, defined as follows:

$$\hat{f}_1 = \frac{\hat{C}}{\hat{D}\hat{A}}; \quad \hat{f}_2 = \frac{1}{\hat{D}} \tag{10.9}$$

with the estimated polynomials $\hat{A}$, $\hat{C}$ and $\hat{D}$ defined by the latest recursive estimate $\hat{\boldsymbol{\vartheta}}$. Similarly, $\hat{x}(k)$ in (10.7) is generated as follows, based on the latest estimated polynomials $\hat{A}$ and $\hat{B}$, i.e.,

$$\hat{x}(k) = \frac{\hat{B}}{\hat{A}}u(k) \tag{10.10}$$

This will be recognized as the 'auxiliary model' that is used to generate the instrumental variable in the RIV algorithm of chapter 7, so that the RPEM algorithm is, in fact, using the same variable, although not exploiting it in IV terms. It is also exploiting exactly the same adaptive prefiltering as the RIV algorithm.

We see, therefore, that the similarities between the RPEM algorithm, with $\boldsymbol{\pi}(k)$ defined as in (10.8), and the RIV algorithm of chapter 7 are considerable (see also Appendix F, equations (F.4) *et seq.*). There are, however, three main differences.

1. First, whereas, in the RPEM algorithm, all elements of $\boldsymbol{\vartheta}$ (i.e. a_i, b_i, c_i and d_i) are estimated simultaneously in a single recursive algorithm with a $\mathbf{P}(k)$ matrix of dimension $(n+m+p+q+1) \times (n+m+p+q+1)$, in the RIV case the parameters are estimated in two separate but coordinated algorithms; i.e. RIV algorithm with $\mathbf{P}(k)$ of dimension $(n+m+1) \times (n+m+1)$ and IVARMA with $\mathbf{P}(k)$ of dimension $(p+q) \times (p+q)$. So, in effect, the total $\mathbf{P}(k)$ is block-diagonal, with the off-diagonal blocks empty.
2. Second, the RPEM algorithm does not incorporate the IV modifications that are inherent in the RIV algorithm in order to ensure robustness to the violation of the statistical assumptions of the BJ model (although we have seen that it incorporates the similarly defined instrumental variable $\hat{x}(k)$, demonstrating the importance of this variable in theoretical terms).
3. Finally, the RPEM algorithm is fully recursive and does not incorporate any iteration, whereas the iteration in the RIV algorithm is essential.

Indeed, as pointed out by Ljung and Söderström (1983) in Appendix 4.E of their excellent theoretical text on recursive estimation, the similarities are even greater if the RPEM algorithm is compared with the real-time recursive implementation of the RIV algorithm suggested by Young and Jakeman (1979b) and described in the first edition of the present book. This will not be pursued here, however, since an improved real-time recursive version of RIV is discussed later in section 10.2.

The RPEM algorithm above is derived heuristically by analogy with the RLS algorithm and, as such, it can be considered as a recursive approximation to the *en bloc* PEM algorithm. Indeed, exact equivalence only occurs when the RPEM algorithm is formulated for the ARX model (see exercise 1, section 10.6), where it becomes exactly equivalent to the RLS algorithm. This approximation is investigated by Söderström and Stoica (1989) (page 328 *et seq.*), who show that it can be derived by assuming that the estimate $\hat{\boldsymbol{\vartheta}}(k-1)$ minimizes the *en bloc* PEM cost function $\mathcal{J}(\boldsymbol{\vartheta},k-1)$ in equation (10.2) after $k-1$ samples have been processed and that the minimum point of $\mathcal{J}(\boldsymbol{\vartheta},k)$ is close to $\hat{\boldsymbol{\vartheta}}(k-1)$. Then $\mathcal{J}(\boldsymbol{\vartheta},k)$ is approximated by a second order Taylor series expansion about $\hat{\boldsymbol{\vartheta}}(k-1)$. Proceeding in this manner and introducing more approximations, they show that $\hat{\boldsymbol{\vartheta}}(k)$ can be obtained as an update to $\hat{\boldsymbol{\vartheta}}(k-1)$ by a Gauss-Newton algorithm. Then, exploiting the matrix inversion lemma (see equation (3.12), section 3.1 of chapter 3), they arrive at the RPEM algorithm.

The RPEM algorithm and the recursive implementation of the RIV algorithm (RRIV: see next section 10.2) can both be modified easily to allow for time variable parameter (TVP) estimation: indeed this is a prime reason for their development and, with this objective in mind, the Söderström and Stoica (1989) analysis is formulated in terms of an an exponentially-weighted-past (EWP: see section 4.1.2 of chapter 4) version of the *en bloc* PEM cost function. Noting the discussion about time variable parameter estimation in chapters 4 and 5, all that is necessary is the addition of modifications similar to those used in the RLS algorithms: either the addition of some procedure, such as the EWP 'forgetting factor', for curtailing the memory of the estimation; or modelling the evolution of the parameter vector as a GRW type process (although the `rpem` routine in the Matlab SID Toolbox only allows for a RW model). In both cases, FIS estimation could be used to obtain optimally smoothed estimates, using a similar approach to that described fully in chapters 4 and 5. As far as I am aware, however, RPEM has not been modified or evaluated yet in this form. The RRIV algorithm is still being developed but will eventually have an FIS option; and the sub-optimal DTF algorithm (the `dtfm` routine in CAPTAIN: see next section 10.2) already has such an option but is not designed for true real-time application.

10.1.1 Statistical Properties of the PEM Estimates for the BJ Model

Referring back to chapter 7, the IV vectors associated with the system and noise models model are $\hat{\boldsymbol{\phi}}(k)$ and $\hat{\boldsymbol{\psi}}(k)$, respectively. If the definition of the latter data vector in (7.28) is slightly modified by the removal of the initial $\hat{\xi}_{f_2}(k)$ element, so that it fits within the present RPEM context, i.e. the two vectors are defined as follows:

$$\begin{aligned}\hat{\boldsymbol{\phi}}(k) &= [-\hat{x}_{f_1}(k-1),\ldots,\ -\hat{x}_{f_1}(k-n),\ u_{f_1}(k-\delta),\ldots,\ u_{f_1}(k-\delta-m)]^T\\ \hat{\boldsymbol{\psi}}(k) &= [\hat{\xi}_{f_2}(k-1),\ldots,\ \hat{\xi}_{f_2}(k-p),\ -\hat{e}_{f_2}(k-1),\ldots,\ -\hat{e}_{f_2}(k-q)]^T\end{aligned} \tag{10.11}$$

then $\hat{\boldsymbol{\pi}}(k) = [\hat{\boldsymbol{\phi}}^T(k)\ \hat{\boldsymbol{\psi}}^T(k)]^T$ and the asymptotic Information Matrix (see Appendix B, section B.2.10) takes the form[1]:

$$\mathbf{I}(\boldsymbol{\vartheta}) = E\{\boldsymbol{\pi}(k)\boldsymbol{\pi}^T(k)\} = \left[\begin{array}{c|c} E\{\hat{\boldsymbol{\phi}}(k)\hat{\boldsymbol{\phi}}^T(k)\} & E\{\hat{\boldsymbol{\phi}}(k)\hat{\boldsymbol{\psi}}^T(k)\} \\ \hline E\{\hat{\boldsymbol{\psi}}(k)\hat{\boldsymbol{\phi}}^T(k)\} & E\{\hat{\boldsymbol{\psi}}(k)\hat{\boldsymbol{\psi}}^T(k)\} \end{array}\right] \tag{10.12}$$

The asymptotic error covariance matrix is given by the inverse of $\mathbf{I}(\boldsymbol{\vartheta})$ but, from the definition of $\hat{\boldsymbol{\phi}}(k)$ as an instrumental variable vector, we know that it is statistically independent of the noise terms $\xi(k)$ and $e(k)$ and is, therefore, independent of the vector $\hat{\boldsymbol{\psi}}(k)$. Consequently, the off-diagonal sub-matrices in the partitioned matrix $\mathbf{I}(\boldsymbol{\vartheta})$ are identically zero and the estimated noise parameter vector $\hat{\boldsymbol{\eta}}$ is asymptotically independent of the estimated system parameter vector $\hat{\boldsymbol{\rho}}$. This is, of course, precisely the result of Pierce's theorem given in section 7.2.3 of chapter 7; indeed the above analysis can be viewed as a simplistic proof of his theorem (although the interested reader should be directed to Pierce's paper for his more rigorous proof).

The asymptotic independence of the noise and system parameter estimates has important implications on both the PEM and RIV estimation procedures. We see that, for large sample size N at least, little is accomplished by estimating the two sets of parameter estimates, $\boldsymbol{\rho}$ and $\boldsymbol{\eta}$ simultaneously, since the off-diagonal blocks of the $\mathbf{P}(k)$ matrix in the RPEM algorithm will be close to zero. Indeed, from the numerical standpoint, it will not only be inefficient to compute these off diagonal blocks but it could be dangerous and lead to numerical problems, with the numbers composing these blocks being insignificantly different from zero and possibly going negative unless the computationally more expensive square-root formulation (see e.g. Gelb, 1974; or Ljung and Söderström, 1983) is utilized.

For lower sample size, it could be argued that these off-diagonal terms may be significant and that their retention will improve the estimation. It is difficult to prove theoretically that this will be the case since, as we have indicated previously, no small sample theory has been developed for this problem. Recourse to Monte Carlo simulation could provide an indication of the relative importance of these terms as regards estimation accuracy but, as far as I am aware, no thorough numerical evaluation of RPEM algorithm in the BJ model case has yet taken place. Comprehensive simulation results do suggest, however, that there is little to choose between the *en bloc* PEM and RIV algorithms as regards estimation performance and the PEM algorithm implemented in the Matlab SID Toolbox, `pem`, can exhibit convergence problems in some circumstances: for example, when the system is 'stiff' in dynamical terms, so that the TF model has widely spaced eigenvalues (see the numerous examples in previous chapters of the book). On the other hand, the PEM approach is superior in the case of MISO TF models with different denominators in the constituent TF models, as discussed in section 7.6 of chapter 7.

In contrast to the RPEM algorithm, the recursive implementation of the RIV takes place as an option in the iterative procedure so that, when the convergence

[1] Do not confuse $\mathbf{I}(\boldsymbol{\vartheta})$ here with the identity matrix.

of the iterations is complete, the final recursive pass through the data recovers the complete recursive equivalent of the *en bloc* solution and the recursive estimate $\boldsymbol{\vartheta}(N) = [\hat{\boldsymbol{\rho}}^T(N)\ \hat{\boldsymbol{\eta}}^T(N)]^T$ at the last sample N is exactly the same as the the *en bloc* estimate based on the N samples. But this equivalence comes at a cost: the recursive estimates are off-line not real time, on-line estimates.

10.2 Real-Time Recursive RRIV Estimation

Of course, it is possible to convert the RIV algorithm to an on-line form, either by the procedures discussed in the first edition of this book or by using an approach similar to that of RPEM (but constraining the off-diagonal blocks of $\mathbf{P}(k)$ to be null matrices, so that the estimates of the system and noise model parameter vectors are independent: see above discussion). However, there is an alternative method based on introducing iteration at each recursive update of the model parameters. This is motivated by the off-line DTF algorithm developed in section 6.10 of chapter 5 which, as shown there, is able to track TVPs very well. However, its requirement for iteration using the whole data set means that it cannot be implemented on-line in real-time: rather it is an algorithm for exploring the possibility of time variable parameters in a model either to gain more knowledge about a dynamic system, or to prepare for real-time TVP estimation. The *Recursive Refined Instrumental Variable* (RRIV) algorithm, as described below, is intended to provide a tool for such real-time applications. It must be stressed, however, that this RRIV algorithm is not yet fully developed and the details below are simply the latest research and development results.

10.2.1 The recursive algorithms

The extension of both the RIV and RIVC algorithms in chapter 7 and 8 to allow for TVP estimation follows straightforwardly from their recursive formulation and uses precisely the same procedures employed in chapters 4 and 5 for TVP estimation. First, however, consider the RIV recursive equations in section 7.1.2 of chapter 7, in a form where σ^2 in equation (RIV-2) is set to unity, i.e.,

$$
\begin{aligned}
\hat{\boldsymbol{\rho}}(k) &= \hat{\boldsymbol{\rho}}(k-1) + \mathbf{g}(k)[y_{f1}(k) - \boldsymbol{\phi}^T(k)\hat{\boldsymbol{\rho}}(k-1)] \\
\mathbf{g}(k) &= \hat{\mathbf{P}}_\rho(k-1)\hat{\boldsymbol{\phi}}(k)[1 + \boldsymbol{\phi}^T(k)\hat{\mathbf{P}}_\rho(k-1)\hat{\boldsymbol{\phi}}(k)]^{-1} \\
\hat{\mathbf{P}}_\rho(k) &= \hat{\mathbf{P}}_\rho(k-1) - \mathbf{g}(k)\boldsymbol{\phi}^T(k)\hat{\mathbf{P}}_\rho(k-1)
\end{aligned}
\qquad (10.13)
$$

This has the advantage that, in TVP estimation based on modelling the parameter variations, it introduces the NVR matrix $\mathbf{Q}_{nvr}$ that then can be optimized by maximum likelihood, as described in section 4.5.3 of chapter 4. For TVP estimation purposes, this form of the recursive RIV equations can now be considered in the context of the KALMSMO filtering equations in section 4.5.2 of chapter 4, with the above equations considered as the 'correction' equations.

The simplest case is where there is additive white noise and the parameter vector $\boldsymbol{\rho}(k)$ is assumed to evolve stochastically as a vector random walk process. This is probably the most generally useful setting in practical situations and the *Recursive Simplified Refined Instrumental Variable* (RSRIV) algorithm obtained in this manner, takes the following form:

Prediction:

$$\hat{\mathbf{P}}_{\rho}(k|k-1) = \hat{\mathbf{P}}_{\rho}(k-1) + \mathbf{Q}^{\rho}_{nvr} \qquad \text{(RSRIV-1)}$$

Correction:

$$\begin{aligned} \hat{\boldsymbol{\rho}}(k) &= \hat{\boldsymbol{\rho}}(k-1) + \mathbf{g}(k)[y_{f1}(k) - \boldsymbol{\phi}^T(k)\hat{\boldsymbol{\rho}}(k-1)] && \text{(RSRIV-2)} \\ \mathbf{g}(k) &= \hat{\mathbf{P}}_{\rho}(k|k-1)\hat{\boldsymbol{\phi}}(k)[1 + \boldsymbol{\phi}^T(k)\hat{\mathbf{P}}_{\rho}(k|k-1)\hat{\boldsymbol{\phi}}(k)]^{-1} && \text{(RSRIV-3)} \\ \hat{\mathbf{P}}_{\rho}(k) &= \hat{\mathbf{P}}_{\rho}(k|k-1) - \mathbf{g}(k)\boldsymbol{\phi}^T(k)\hat{\mathbf{P}}_{\rho}(k-1) && \text{(RSRIV-4)} \end{aligned}$$

where $\hat{\mathbf{P}}_{\rho}(k|k-1)$ is the one sample ahead prediction of the $\hat{\mathbf{P}}_{\rho}(k)$ matrix, so that the covariance matrix can be recovered from $\hat{\mathbf{P}}^*_{\rho}(k) = \hat{\sigma}^2\hat{\mathbf{P}}_{\rho}(k)$, where $\hat{\sigma}^2$ is an estimate of σ^2 obtained by a separate recursive algorithm, as discussed in section 4.5.4 of chapter 4 (which would also allow for heteroscedastic noise with changing variance over time). Of course, in this simple form, there is no need to explicitly predict the evolution of $\hat{\boldsymbol{\rho}}(k)$ between samples because the random walk assumption means that the prediction is simply $\hat{\boldsymbol{\rho}}(k|k-1) = \hat{\boldsymbol{\rho}}(k-1)$.

In the RSRIV algorithm, the parameters of the auxiliary model and prefilters have to be continually updated based on the TVP estimates: i.e.,

$$\begin{aligned} &\hat{x}(k) = \frac{\hat{B}(k,z^{-1})}{\hat{A}(k,z^{-1})}u(k) \\ &\hat{x}_{f_1}(k) = f_1(k,z^{-1})\hat{x}(k); \;\; y_{f_1}(k) = f_1(k,z^{-1})y(k) \\ &u_{f_1}(k) = f_1(k,z^{-1})u(k); \;\; \text{where, } f_1(k,z^{-1}) = \frac{1}{\hat{A}(k,z^{-1})} \end{aligned} \qquad (10.14)$$

and the k in the arguments indicates the TVP updating. In the full RRIV case with coloured additive noise, the TVP prefilter $f_1(k,z^{-1})$ is expanded to the form:

$$f_1(k,z^{-1}) = \frac{\hat{C}(k,z^{-1})}{\hat{D}(k,z^{-1})\hat{A}(k,z^{-1})} \qquad (10.15)$$

which requires the concurrent use of the recursive TVP version of the IVARMA noise model estimation algorithm. This takes the following form, again if the noise model parameter vector $\boldsymbol{\eta}(k)$ is assumed to evolve stochastically as a vector random walk process:

Prediction:

$$\hat{\mathbf{P}}_{\eta}(k|k-1) = \hat{\mathbf{P}}_{\eta}(k-1) + \mathbf{Q}_{nvr}^{\eta} \qquad \text{(RIVARMA-1)}$$

Correction:

$$\hat{\boldsymbol{\eta}}(k) = \hat{\boldsymbol{\eta}}(k-1) + \mathbf{g}_{\eta}(k)[\tilde{e}_{f_2}(k) - \boldsymbol{\psi}^T(k)\hat{\boldsymbol{\eta}}(k-1)] \qquad \text{(RIVARMA-2)}$$

$$\mathbf{g}_{\eta}(k) = \hat{\mathbf{P}}_{\eta}(k|k-1)\hat{\boldsymbol{\psi}}(k)[1 + \boldsymbol{\psi}^T(k)\hat{\mathbf{P}}_{\eta}(k|k-1)\hat{\boldsymbol{\psi}}(k)]^{-1} \qquad \text{(RIVARMA-3)}$$

$$\hat{\mathbf{P}}_{\eta}(k) = \hat{\mathbf{P}}_{\eta}(k|k-1) - \mathbf{g}_{\eta}(k)\boldsymbol{\psi}^T(k)\hat{\mathbf{P}}_{\eta}(k|k-1) \qquad \text{(RIVARMA-4)}$$

Clearly, the above algorithms could be modified to allow for modelling the parameter variations using other GRW stochastic model forms. Often, however, it can be assumed that the noise model parameters do not vary, so $\mathbf{Q}_{nvr}^{\eta} = \mathbf{0}$ and the constant parameter recursive IVARMA algorithm will be used intead. Finally, provided the data are being stored (see later), FIS smoothing (or even fixed lag smoothing) could be used at any sample to provide improved, off-line smoothed estimates of the parameters. However this will not be discussed here.

10.2.2 Implementation

The above RRIV algorithmic equations are quite complex to implement, except when they are applied to the simple linear ARX model, where they revert automatically to the SIVTVP estimation equations developed in chapter 6 that are used for DARX model estimation in chapter 5. In all other cases, the $\boldsymbol{\phi}(k)$ and $\boldsymbol{\psi}(k)$ data vectors are associated with a pseudo-linear estimation models that all involve adaptive prefilters.

The RPEM algorithm, as well as the RIVAML algorithm described in the first edition of this book (see also Young and Jakeman, 1979), update all the prefilters continually and this would be possible with the RRIV algorithm. Recognising the value of the iterative procedure in yielding robust performance and good practical results, however, an alternative approach has been developed recently that processes the data recursively in real-time *but also carries out iteration at each recursive step*. There are various ways in which this RRIV algorithm can be implemented and research is proceeding on this. However, the details of the implementation given below are those used in the subsequent simulation example and it seems to be a promising implementation.

The Real-Time Recursive-Iterative RRIV Algorithm

1. **Initialization** Optimize the diagonal elements of the $\mathbf{Q}_{nvr}$ matrix over an existing set of data containing *iwh* samples of the input and output variables. *Just prior to initiation of real-time operation*, apply the RIV constant parameter estimation algorithm over *iwp* data samples and use the estimation results to define initial estimates of the parameter vectors $\hat{\boldsymbol{\rho}}(0)$, and $\hat{\boldsymbol{\eta}}(0)$, together with the associated $\hat{\mathbf{P}}_{\eta}(0)$ and $\hat{\mathbf{P}}_{\rho}(0)$ matrices.
2. **First Recursion** Following the initialization, import the next *ruw* samples of input-output data, where *ruw* is the recursive update window which can be any size defined by the user. Then apply the RRIV algorithm for *it* iterations over these latest *ruw* samples in order to estimate the time variable system and noise model parameters over this data window up to sample k, starting the first iteration recursions with the prior estimates $\hat{\boldsymbol{\rho}}(0)$, $\hat{\boldsymbol{\eta}}(0)$ and matrices $\hat{\mathbf{P}}_{\eta}(0)$ and $\hat{\mathbf{P}}_{\rho}(0)$. During each iteration use the last pfw samples of data $\{u(k-pfw+1), \ldots, u(k); y(k-pfw+1), \ldots, y(k)\}$ (the auxiliary model/prefilter data window), where pfw may be larger than *ruw*, to produce the prefiltered input, output and instrumental variables required for RRIV estimation. A similar iterative processing strategy to that employed by the standard RIV algorithm of chapter 7 is employed here but with the (now time variable) parameters of the auxiliary model and prefilters at the jth iteration based on the estimated TVP parameters from the previous $(j-1)^{th}$ iteration.
3. **Subsequent Recursions** Read in the next *ruw* data samples and recursively update the parameters over the *ruw* interval in the same way as in step 2., but using initial parameter estimates $\hat{\boldsymbol{\rho}}$, $\hat{\boldsymbol{\eta}}$ and associated $\hat{\mathbf{P}}_{\rho}$, $\hat{\mathbf{P}}_{\eta}$ matrices based on the values obtained at the end of the RRIV iterations in step 2. Continue this procedure, each time reading in the latest *ruw* data samples and recursively updating the parameter estimates from initial parameter estimates $\hat{\boldsymbol{\rho}}$, $\hat{\boldsymbol{\eta}}$ and associated $\hat{\mathbf{P}}_{\rho}$, $\hat{\mathbf{P}}_{\eta}$ matrices defined by those that are obtained at the end of the immediately previous recursive update.

Comments

1. The NVR hyper-parameter optimization interval *iwh* should be large enough to provide information on possible parameter variation, so that the optimization is well-informed in this regard. On the other hand, the estimation initiation interval *iwp* should be fairly short, so that the RIV constant parameter estimates reflect the current conditions and are not too distorted by parameter variation over this sampling interval.
2. The recursive update interval *ruw* will depend upon the application requirements but will normally be quite small. The number of iterations, *it*, should be selected as small as possible consistent with good estimation performance. However, long experience with RIV estimation, as well as simulation results such as those reported below, suggest that *it* = 3 is best. The auxiliary model/prefilter data window of pfw samples should normally be set equal to *ruw*. However, in the case

where $pfw = k$, $\forall\ k$, the data are continually accumulated so this is the most appropriate setting in data assimilation situations, where it is required to access and store the data during real-time operation.

3. Since the refined instrumental variable approach to estimation is unified and can be applied to discrete-time of hybrid continuous-time systems, RRIVC and RSRIVC algorithms can be developed along the same lines as those discussed in this section.

10.2.3 Example 10.1: RSRIV estimation of TVPs in a simulated 2nd order TF model

This example considers the following second order, discrete-time TVP model, with two variable parameters and one constant parameter:

$$\begin{aligned}
x(k) &= \frac{b_0(k)}{1+a_1(k)z^{-1}+0.7z^{-2}}u(k) \\
\xi(k) &= \frac{1+0.2z^{-1}}{1-0.85z^{-1}}e(k) \qquad e(k) = \mathcal{N}(0,1) \\
y(k) &= x(k)+\xi(k) \\
u(k) &= \frac{0.5}{1-0.5z^{-1}}e_u(k) \qquad e_u(k) = \mathcal{N}(0,1)
\end{aligned} \tag{10.16}$$

where $a_1(k)$ is a pseudo random binary sequence of ± 0.5 centered on -1.4, while $b_0(k) = 1.5+0.5sin(0.02k)$. The RSRIV estimation results were obtained over a data set of 3000 samples and the noise $\xi(k)$ yields a noise/signal ratio by standard deviation of 0.22. Although the noise $\xi(k)$ is coloured, RSRIV estimation is used to demonstrate that the algorithm is robust in this regard. Indeed, the results obtained when $\xi(k)$ is replaced by white noise with the same noise/signal ratio are quite similar.

The ML optimized diagonal elements of the $\mathbf{Q}_{nvr}$ matrix, based on a window of $iwh = 1000$ prior data samples, are $\text{NVR}_{a_1} = 0.000035$, $\text{NVR}_{a_2} = 5\times 10^{-18}$ and $\text{NVR}_{b_0} = 0.00092$, showing that the ML optimization has correctly inferred that the a_2 parameter is constant. The constant parameter SRIV initialization estimates, based on the iwp window of 100 data samples, provide the following initial values for the elements of $\hat{\boldsymbol{\rho}}(0)$ and $\mathbf{P}_\rho(0)$:

$$\hat{a}_1(0) = -1.282;\ \hat{a}_2(0) = 0.699;\ \hat{b}_0(0) = 1.816$$

$$\mathbf{P}_\rho(0) = 10^{-4}\begin{bmatrix} 4.3 & -3.8 & 0.002 \\ -3.8 & 4.16 & -0.002 \\ 0.001 & -0.002 & 102.1 \end{bmatrix}$$

Figure 10.1 shows the RSRIV results obtained from these starting values, with a recursive update interval $ruw = pfw = 2$ and $it = 3$ iterations for each recursive update. Clearly, the two TVPs and the constant parameter are all estimated well; and the TVP model output closely matches the noise-free output $x(k)$, with $R_T^2 = 0.97$.

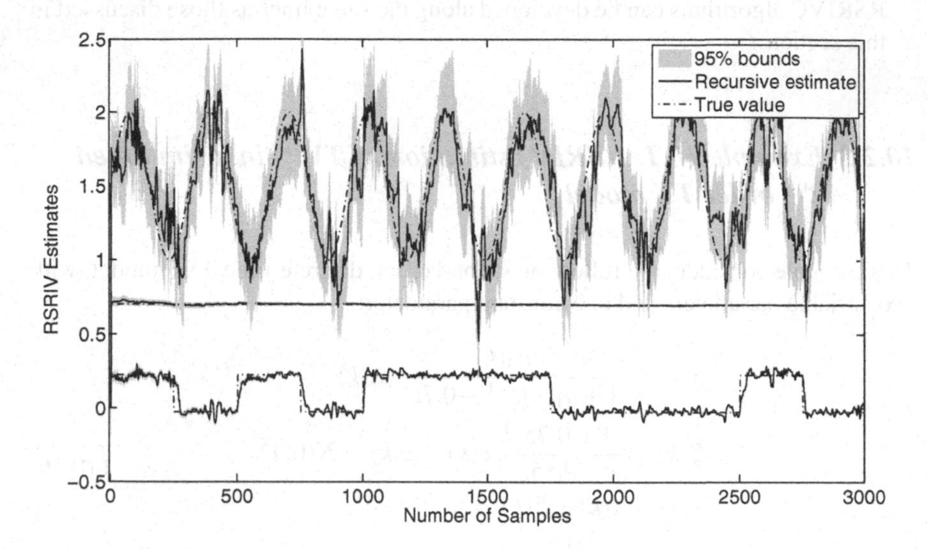

Fig. 10.1 RRIV estimation results for the simulated time variable parameter system (10.16).

If the rpem routine in the Matlab SID Toolbox is applied in a similar manner to the same data (i.e. without concurrent noise model parameter estimation and with the appropriate 'kf' option selected), then it produces rather poor results if the same optimized NVR values are utilized. However, if the optimized NVR values are multiplied by 0.1, then quite similar results are obtained, as shown in Figure 10.2, where we see that the RSRIV estimates of the sinusoidal b_0 parameter variations are a little more volatile than those obtained by RPEM. Table 10.1 compares the variances of the errors between the parameters estimated by the two methods and the true parameter variations. The RSRIV results are marginally better the RPEM, but the differences are not very large and, of course, some 'tuning' of the RPEM NVR values might improve the estimation results.

Table 10.1 Variances of the errors between the RSRIV and RPEM estimated parameters and true parameter variations.

Algorithm	$\hat{a}_1$	$\hat{a}_2$	$\hat{b}_0$
RSRIV	0.0031134	0.00051298	0.033316
RPEM	0.005744	0.0039718	0.038928

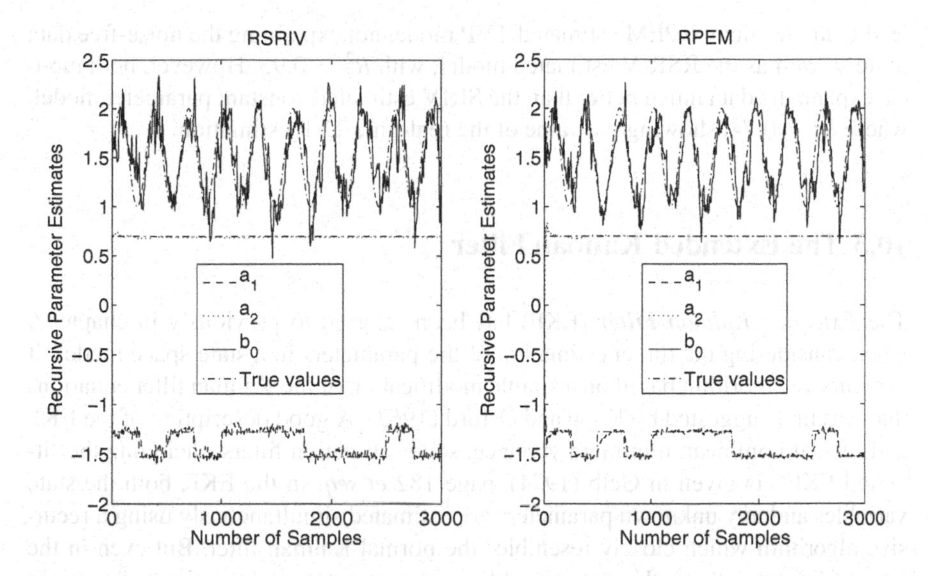

Fig. 10.2 Comparison of RRIV (left panel) and RPEM (right panel) estimates of the model parameter variations.

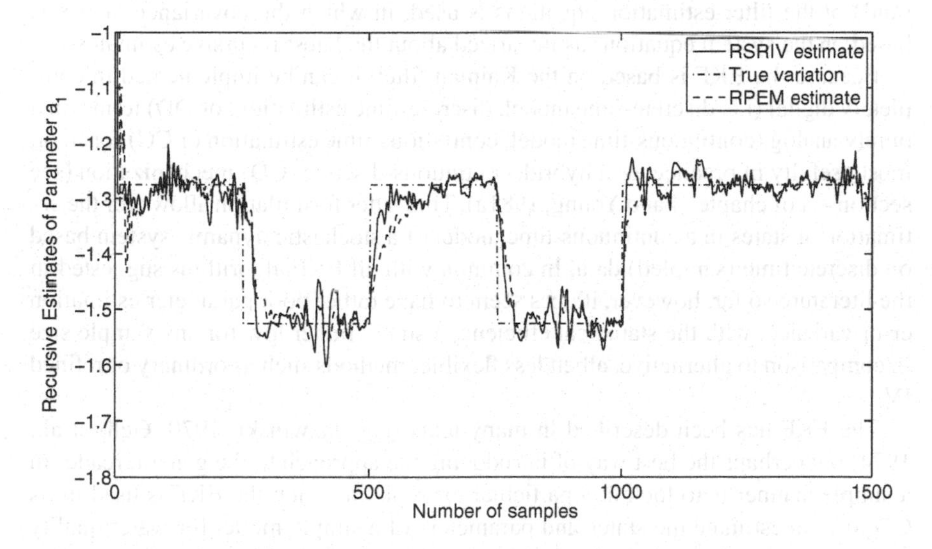

Fig. 10.3 Comparison of RRIV and RPEM estimates of the a_1 parameter variations in Figure 10.2 over the initial segment of the data.

A more detailed comparison of the $a_1(k)$ parameter estimates in Figure 10.3 reveals that the tracking of the step change by RPEM is rather slower than RSRIV and this is may be the reason for the results in Table 10.1. These differences also lead to the resultant RPEM estimated TVP model not explaining the noise-free data quite so well as the RSRIV estimated model, with $R_T^2 = 0.93$. However, both models explain the data much better than the SRIV estimated constant parameter model, where $R_T^2 = 0.74$, showing the value of the real-time TVP estimation.

10.3 The Extended Kalman Filter

The *Extended Kalman Filter* (EKF) has been referred to previously in chapter 6 when considering the direct estimation of the parameters in a state space model. It is recursive algorithm based on a simple modification of the Kalman filter equations that was first suggested by Kopp and Orford (1963). A good description of the EKF and related nonlinear, minimum variance, state estimation filters, including the 'iterated EKF', is given in Gelb (1974), page 182 *et seq*. In the EKF, both the state variables and any unknown parameters are estimated simultaneously using a recursive algorithm which closely resembles the normal Kalman filter. But even in the case of linear systems, the state variables and parameters to be estimated occur in product form, so that the estimation problem is always one of direct nonlinear estimation with its attendant practical difficulties. In order to allow for the nonlinearity, a form of linearization (sometimes termed 're-linearization' or 'statistical linearization') of the filter-estimation equations is used, in which the covariance matrix is based on the system equations as linearized about the latest recursive estimates.

Because the EKF is based on the Kalman filter, it can be implemented in completely digital (i.e. discrete-time model, discrete-time estimation, or DD) terms; in a purely analog (continuous-time model, continuous-time estimation or CC) form; or, most usefully in practice, as a hybrid, continuous-discrete (CD) mechanization (see section 4.4 of chapter 4 and Young, 1981a). This latter formulation allows for the estimation of states in a continuous-time model of a stochastic dynamic system based on discrete time (sampled) data. In common with all EKF algorithms suggested in the literature so far, however, it does seem to have rather poor parameter estimation error variance, with the statistical efficiency usually rather low for any sample size in comparison to alternative, albeit less flexible, methods such as ordinary or refined IV.

The EKF has been described in many texts (e.g. Jazwinski, 1970; Gelb et al., 1974) but perhaps the best way of introducing the approach to the general reader in a simple manner is to look at a particular example in which the EKF is used in its CD form to estimate the states and parameters of a simple model for water quality in a non-tidal river system. Preparatory to this, let us consider a continuous-time system described by the following state space differential equation

$$\frac{d\mathbf{x}(t)}{dt} = \mathbf{F}\mathbf{x}(t) + \mathbf{G}\mathbf{u}(t) + \boldsymbol{\eta}_1(t) \tag{10.17}$$

where $\mathbf{x}(t)$ is an n dimensional state vector $\mathbf{u}(t)$ is a q dimensional vector of deterministic forcing functions; $\boldsymbol{\eta}_1(t)$ is an n dimensional vector of zero mean white noise disturbances to the system; $\mathbf{F}$ and $\mathbf{G}$ are, respectively, $n \times n$ and $n \times q$ matrices that characterize the system.

Suppose that some of the elements in $\mathbf{F}$ and $\mathbf{G}$ are unknown parameters and that we wish to estimate both the state $\mathbf{x}(t)$ and the unknown parameters on the basis of our knowledge of the input forcing vector $\mathbf{u}(t)$ and *discrete-time* observations of the system behaviour made by means of an observation process,

$$\mathbf{y}(t_k) = \mathbf{H}\mathbf{x}(t_k) + \boldsymbol{\xi}(t_k) \tag{10.18}$$

where $\mathbf{y}(t_k)$ is the pth order observation vector at the kth time instant, and $\mathbf{H}$ is a $p \times n$ observation matrix[2]. One approach to this problem is to augment the state vector $\mathbf{x}(t)$ with a parameter vector $\mathbf{a}$ composed of the unknown parameters and then formulate additional state equations for this parameter vector: e.g., if the parameters are all assumed time-invariant, a suitable state representation would be simply

$$\frac{d\mathbf{a}(t)}{dt} = 0 \tag{10.19}$$

If the augmented state vector is defined as

$$\mathbf{X}(t) = \begin{bmatrix} \mathbf{x}(t) \\ \mathbf{a}(t) \end{bmatrix} \tag{10.20}$$

the augmented state equations will be of the following general form.

$$\frac{d\mathbf{X}(t)}{dt} = f\{\mathbf{X}(t), \mathbf{u}(t)\} + \boldsymbol{\eta}(t); \quad \boldsymbol{\eta}(t) = \left[\, \boldsymbol{\eta}_1(t) \,\vdots\, \mathbf{0} \,\right] \tag{10.21}$$

with a linear observation equation,

$$\mathbf{y}(t_k) = \mathbf{C}\mathbf{X}(t_k) + \boldsymbol{\xi}(t_k); \quad \mathbf{C} = \left[\, \mathbf{H} \,\vdots\, \mathbf{0} \,\right] \tag{10.22}$$

In equation (10.21), $f\{\mathbf{X}(t), \mathbf{u}(t)\}$ is a non-linear vector function, where the non-linearities arise from the product terms involving elements of $\mathbf{x}(t)$ and $\mathbf{a}$. If any elements of the parameter vector $\mathbf{a}$ are assumed to vary stochastically over the observation interval, then (10.19) will need to be modified to allow for this. For example, in the simplest case of a random walk, the rate of change will be equal to a random white noise vector, say $\boldsymbol{\eta}_2(t)$ that will then replace the zero vector of the extended $\boldsymbol{\eta}(t)$ vector defined in (10.21).

A simple example of this kind of system representation can be obtained by considering the following equation for biochemical oxygen demand (BOD) at the output of a reach in a non-tidal river system (Young and Beck, 1974; Beck and Young, 1975, 1976)

$$\frac{dx_1(t)}{dt} = -(k_1 + \frac{Q_f}{V_m})x_1(t) + \frac{Q_f}{V_m}u_1(t) + L_A(t) + \eta_1(t) \tag{10.23}$$

[2] As in chapter 8, the argument t_k is used to emphasize that this is a hybrid model.

Here $x_1(t)$ is the BOD, which can be considered as an aggregate or macro-measure of the oxygen absorbing potential of substances in the stream (such as decaying organic material from effluent discharges); it is defined as the oxygen absorbed (mg/l) over a 5 day period by a sample of the river water, in the absence of light, and at a constant temperature of 20^oC. Other terms in equation (10.23) are defined as follows: k_1 is the BOD decay rate constant (day^{-1}); Q_f is the volumetric flow rate in the stream (m^3 day^{-1}); V_m is the mean volume of the reach (m^3); $u_1(t)$ is the input BOD from the previous upstream reach of the river system (mg^{-1}); $L_A(t)$ is the rate of addition of BOD to the reach by local run-off from land adjoining river (day^{-1}); and $\eta_1(t)$ is the lumped effect of all stochastic disturbances to the BOD in the system (mg^{-1} day^{-1}).

Suppose now that the parameter k_1 and the variable $L_A(t)$ are to be estimated on the basis of noisy, daily observations of the BOD. In this situation, the observation equation (10.22) will be simply a scalar equation of the form

$$y_1(t_k) = x_1(t_k) + \xi_1(t_k) \tag{10.24}$$

where $\xi_1(t_k)$ is the measurement noise of the kth day. In line with the above formulation, k_1 and $L_A(t)$ can be considered as additional states $x_2(t)$ and $x_3(t)$. If it is now assumed that k_1 is time invariant while $L_A(t)$, which may vary, is a continuous random walk (i.e. the continuous-time equivalent of the discrete-time random walk model), then the augmented state equations can be written in the form (10.21) with

$$f\{\mathbf{X}(t),\mathbf{u(t)}\} = \begin{bmatrix} -\left[x_2(t)+\frac{Q_f}{V_m}\right]x_1(t)+\frac{Q_f}{V_m}u_1(t)+x_3(t) \\ 0 \\ 0 \end{bmatrix}; \quad \boldsymbol{\eta}(t) = \begin{bmatrix} \eta_1(t) \\ 0 \\ \eta_3(t) \end{bmatrix}$$

Here, $\mathbf{X}(t) = [x_1(t)\ x_2(t)\ x_3(t)]^T$ and $\eta_3(t)$ is white noise disturbance introduced so that the final equation, i e. $\frac{dx_3(t)}{dt} = \eta_3(t)$ represents the required continuous random walk process for L_A.

The basic idea of the continuous-discrete EKF can be explained most easily by referring back to the ordinary discrete-time Kalman filter equations KF in section 4.4 of chapter 4. In the present situation, the *a priori* prediction of the state between samples can be generated by replacing equation KF-1 by an equivalent expression obtained by reference to the state equation (10.21), with the above definitions. Since $\boldsymbol{\eta}(t)$ is assumed to be a vector of serially uncorrelated white noise disturbances which are, therefore, predicted to have zero value, it seems reasonable that this new prediction equation can be approximated by the deterministic nonlinear differential equation

$$\frac{d\hat{\mathbf{X}}(t)}{dt} = f\{\hat{\mathbf{X}}(t),\mathbf{u}(t)\} \tag{10.25}$$

which has to be solved over the sampling interval $\Delta t = t_k - t_{k-1}$ from initial conditions defined by the current estimate $\hat{\mathbf{X}}(t_{k-1})$ at the $(k-1)^{th}$ sampling instant. The justification for (10.25) involves a Taylor series expansion of $f\{\mathbf{X}(t),\mathbf{u}(t)\}$ about the current estimate of the state vector and dropping all but the first term in this expansion (see Gelb et al, 1974): i.e,

$$f\{\mathbf{X}(t),\mathbf{u}(t)\} = f\{\hat{\mathbf{X}}(t),\mathbf{u}(t)\} + \left[\frac{\partial f}{d\mathbf{X}}\right]_{\mathbf{X}(t)=\hat{\mathbf{X}}(t)} \{\mathbf{X}(t) - \hat{\mathbf{X}}(t)\} + \ldots \tag{10.26}$$

The first order approximation $\hat{f}\{\mathbf{X}(t),\mathbf{u}(t)\}$ to $f\{\mathbf{X}(t),\mathbf{u}(t)\}$ is then obtained by dropping all but the first term in the expansion, i.e.,

$$\hat{f}\{\mathbf{X}(t),\mathbf{u}(t)\} = f\{\hat{\mathbf{X}}(t),\mathbf{u}(t)\} \tag{10.27}$$

which is the approximation used in (10.25).

An approximate differential equation for the evolution of the error covariance matrix $\mathbf{P}(t)$ associated with the estimated state vector[3] can be obtained using similar reasoning, this time retaining the first two terms in the Taylor series expansion, and the result takes the form

$$\frac{d\mathbf{P}(t)}{dt} = \boldsymbol{\Phi}(\hat{\mathbf{X}}(t),\mathbf{u}(t))\mathbf{P}(t) + \mathbf{P}(t)\boldsymbol{\Phi}^T(\hat{\mathbf{X}}(t),\mathbf{u}(t)) + \mathbf{Q} \tag{10.28}$$

where, $\boldsymbol{\Phi}(\hat{\mathbf{X}}(t),\mathbf{u}(t))$ is a matrix whose elements ϕ_{ij} are defined in the following manner.

$$\phi_{ij}(\hat{\mathbf{X}}(t)) \equiv \left[\frac{\partial f_i(\mathbf{X}(t))}{\partial \mathbf{X}_j(t)}\right]_{\mathbf{X}(t)=\hat{\mathbf{X}}(t_{k-1})} \quad ; \quad \text{for } i,j = 1,2,3 \tag{10.29}$$

and $\mathbf{Q}$ is the covariance matrix of the stochastic disturbances (see Gelb et al., 1974, page 182 *et seq.*). Note that, in the present simple example, there are no parameters in the coefficient of the input variable u_1 that need to be estimated, so $\boldsymbol{\Phi}(\hat{\mathbf{X}}(t),\mathbf{u}(t))$ is independent of $u_1(t)$. Gelb et al (1974) pursue this approach and derive a continuous-discrete EKF based on these continuous-time prediction equations and associated, discrete-time correction equations. Actually, they assume that the observation equation is also nonlinear, so a similar approach has to be applied to this. In the present simple example, the observation equation (10.22) is linear, so this is not required.

The advantage of the above formulation is that both of the nonlinear prediction equations (10.25) and (10.28) are continuous-time differential equations and so can handle non-uniform sampling (i.e. a changing sampling interval Δt) because they can be solved over any time interval while waiting for the receipt of the next set of sampled data. However, a simpler alternative to this approach is possible if uniform data sampling is being used: then the nonlinear equation (10.28) can be converted to a discrete-time form.

In order to examine this second approach, consider again the river water quality example, where the prediction of the augmented state vector estimates between observations is obtained by the solution of the following equations between sampling

[3] Note that in the discrete-time KF algorithm of section 4.4 in chapter 4, the covariance matrix is denoted by $\mathbf{P}^*(k)$ but the superscript * has been dropped here for convenience.

instants:

$$\begin{aligned}
\frac{dx_1(t)}{dt} &= -(x_2(t) + \frac{Q_f}{V_m})x_1(t) + \frac{Q_f}{V_m}u_1(t) + x_3(t) \\
\frac{dx_2(t)}{dt} &= 0 \\
\frac{dx_3(t)}{dt} &= 0
\end{aligned} \tag{10.30}$$

The transition matrix $\mathbf{A}(t_k) = \mathbf{A}(t_k, t_{k-1}, \hat{\mathbf{X}}(t_{k-1}), \mathbf{u}(t_{k-1}))$ for the discrete time, linear, approximation to these equations is obtained by reference to the matrix $\boldsymbol{\Phi}(\hat{\mathbf{X}}(t), \mathbf{u}(t))$, whose elements f_1, f_2 and f_3, in this case, are obtained by reference to equation (10.30) as

$$f_1 = -(\hat{x}_2(t) + \frac{Q_f}{V_m})\hat{x}_1(t) + \frac{Q_f}{V_m}u_1(t) + \hat{x}_3(t); \quad f_2 = 0; \quad f_3 = 0$$

In the present case, therefore, $\boldsymbol{\Phi}(\hat{\mathbf{X}}(t), \mathbf{u}(t))$ is given by[4]

$$\boldsymbol{\Phi}(\hat{\mathbf{X}}(t), \mathbf{u}(t)) = \begin{bmatrix} -[\hat{x}_2(t_{k-1}) + \frac{Q_f(t_{k-1})}{V_m}] & -\hat{x}_1(t_{k-1}) & 1 \\ 0 & 0 & 0 \\ 0 & 0 & 0 \end{bmatrix}$$

Under certain minor and, in practice, non-limiting assumptions $\mathbf{A}(t_k)$ can now be derived as follows, using standard state space techniques (see, for example, Dorf, 1965; Ogata, 1967) i.e.,

$$\mathbf{A}(t_k) = \mathbf{A}[\hat{\mathbf{X}}(t_{k-1})] = e^{\boldsymbol{\Phi}\Delta t} \tag{10.31}$$

where $e^{\boldsymbol{\Phi}\Delta t}$ is the 'matrix exponential' function: see Appendix B, equation (B.53).

Having obtained a discrete-time, linearized representation of the system at the $(k-1)^{th}$ sampling instant, it is now simple to substitute in the remaining equations of the linear KF algorithm, recognising that $\mathbf{A}(t_k) = A[\hat{\mathbf{X}}(t_k - 1), \mathbf{u}(t_k - 1)]$ will need to be re-evaluated at each recursive update as the $\boldsymbol{\Phi}(\hat{\mathbf{X}}(t), \mathbf{u}(t))$ matrix changes in sympathy with the changes in the recursive estimates. Of course, having obtained this linearized model of the system, it would be possible to use this to predict the state vector, rather than solving the nonlinear differential equation by some form of numerical integration. However, experience has shown that, in general, it is better to retain the continuous-time nonlinear equation for prediction. By analogy with this simple example, the general continuous-discrete time EKF algorithm can be written in the following general form.

First, the predictive step involving integration of the nonlinear differential equations (10.25) over the sampling interval, using by some robust form of numerical integration:

[4] The flow Q_f is a variable quantity and so its sampled value at the $(k-1)^{th}$ instant is used in the evaluation of $\boldsymbol{\Phi}(\hat{\mathbf{X}}(t), \mathbf{u}(t))$. Nominally V_m is also time variable but is assumed constant here for simplicity.

Prediction:

$$\hat{\mathbf{X}}(t_k|t_{k-1}) = \hat{\mathbf{X}}(t_{k-1}) + \int_{t=t_{k-1}}^{t=t_k} f[\hat{\mathbf{X}}(t), \mathbf{u}(t)]\, dt \qquad \text{(EKF-1)}$$

$$\mathbf{P}(t_k|t_{k-1}) = \mathbf{A}(t_k)\mathbf{P}(t_{k-1})\mathbf{A}^T(t_k) + \mathbf{Q} \qquad \text{(EKF-2)}$$

Followed by discrete-time correction of these predictions:

Correction:

$$\hat{\mathbf{x}}(t_k) = \hat{\mathbf{x}}(t_k|t_{k-1}) + \mathbf{G}(k)\{\mathbf{y}(k) - \mathbf{C}(t_k)\hat{\mathbf{x}}(t_k|t_{k-1})\} \qquad \text{(EKF-3)}$$

$$\mathbf{G}(t_k) = \mathbf{P}(t_k|t_{k-1})\mathbf{C}^T(t_k)[\mathbf{R} + \mathbf{C}(t_k)\mathbf{P}(t_k|t_{k-1})\mathbf{C}^T(t_k)]^{-1} \qquad \text{(EKF-4)}$$

$$\mathbf{P}(t_k) = \mathbf{P}(t_k|t_{k-1}) - \mathbf{G}(t_k)\,\mathbf{C}(t_k)\mathbf{P}(t_k|t_{k-1}) \qquad \text{(EKF-5)}$$

In this form of the continuous-discrete EKF algorithm, $\mathbf{Q}$ is the covariance matrix of the discrete equivalent $\boldsymbol{\eta}(t_k)$ of the white noise disturbance vector $\boldsymbol{\eta}(t)$; while $\mathbf{R}$ is the covariance matrix of the measurement noise vector $\boldsymbol{\xi}(t_k)$; i.e. both of the stochastic disturbance vectors are assumed to have the same kind of statistical properties as in the ordinary KF and, in this case, are assumed constant (i.e. the statistical properties of the noise are assumed stationary). Clearly the system equations must be defined such that these assumptions are reasonably valid, although in practice strict adherence to these requirements is rarely necessary.

In this latter connection, it is important to realize that, while algorithm EKF resembles the ordinary KF algorithm it is, of course, a non-linear estimation procedure: thus unlike the KF, even if the theoretical assumptions are fully satisfied, there is no guarantee of convergence, nor is the $\mathbf{P}(t_k)$ matrix necessarily any accurate indication of the covariance matrix of the estimation errors. Nevertheless, if care is taken in its utilization and provided that the statistical re-linearization procedure is valid within the range of confidence associated with the estimates at each recursive step, then the algorithm can work quite well in practice and has proven popular over many years.

It should also be stressed that the above algorithm is not the only form of the EKF. In fact the reader who carries out a web search will soon discover how many publications there are available on this and related topics. For the present purposes, however, it is sufficient to note the following comments.

1. The EKF is not limited to the kind of parameter estimation problem considered here: potentially, it can be applied to any nonlinear, stochastic, dynamic system (continuous, discrete or hybrid continuous-discrete) that can be considered in this manner.

2. Being an approximate realization of nonlinear estimation equations, it is clearly possible to formulate other higher order approximations (e.g. based on higher order series expansions), as shown, for example, by Jazwinski (1970) and Gelb et al (1974).
3. As mentioned earlier, Gelb et al describe an iterated version (IEKF) whose motivation for iteration is not unlike that used as the basis for RIV and RRIV estimation. Here, the paper by Haupt et al. (1996) is an interesting one that compares the performance of the EKF and IEKF with an alternative 2-step algorithm.
4. Yet another alternative is to decompose the estimation problem into a sub-problem that estimates the states, given the parameter estimates obtained from a parameter estimation iteration; and a second, coordinated, sub-problem, that estimates the parameters given the estimates obtained from a state estimation iteration. A fully recursive approach of this type which does not use iteration to enhance the estimation accuracy, has been suggested by Todini (1978) and this seems to yield quite good results.

Consequently, the EKF in these and other forms, has wide application potential, as demonstrated by the enormous number of papers that have utilized the EKF, in one form or another, since the publication of Kopp and Orford's seminal paper in 1963.

10.3.1 The EKF in a recursive prediction error form

Since the EKF is nominally a statistical approach (although the statistical implications are not too clear because of the largely heuristic derivation of the algorithm), the reader might wonder whether the EKF can be considered in PEM terms. This has been discussed by Ljung (1979a) in the context of the completely discrete-time EKF, and he shows that it is only an approximate PEM method because linearization is applied to the normal stochastic state space representation rather than the innovations or Kalman filter representation (see section 6.2 of chapter 6), as required in the PEM case. He has suggested modifications to the discrete-time EKF to make it a PEM-like algorithm and Beck, for instance, has considered this and developed the idea further (see below).

This latter point probably accounts, in part, for the rather poor performance of the EKF in certain circumstances: problems of convergence and low statistical efficiency are often reported in the literature. Beck and Young (1976) suggested that the limitations of the EKF in statistical efficiency terms, coupled with its inherent flexibility, particularly when used in a CD mode with non-uniform sampling intervals, make it more appropriate for use in model structure identification studies than in final parameter estimation. Here, the EKF is used merely to test the dynamic behaviour of the continuous-time model (which is embedded within the algorithm) and to evaluate structures which lead either to constant parameter representations or representations with variable parameters that have direct physical significance. In these applications the EKF is, in effect, being used as a model structure identification procedure and human interaction is essential. However, the worst aspect of the

EKF is the rather poor parameter estimates, which led Young and Whitehead (1977) to move away from the EKF and utilize an alternative IV-based approach when considering water quality models such as that considered in the previous sub-section.

This use of the EKF in a model structure identification context has been considered by Beck and his co-workers over subsequent years. In the latest research on this topic (see Lin and Beck, 2007, and the prior references therein), they have now developed a *Recursive Prediction Error* (RPE) algorithm, inspired by the ideas of Ljung but formulated in a rather different manner that, to some extent, resembles the RRIV algorithm considered in the previous section 10.2 of this chapter. Both are optimal maximum likelihood estimation algorithms but are not limited in this sense; both can be applied to both discrete-time and continuous-time models; and finally, both model the potentially time variable parameters by a stochastic GRW process. Indeed, the RPE and RRIV algorithms are quite complementary so that, taken together, they constitute a powerful general approach to TVP estimation.

However, there are differences: the RPE algorithm is formulated within a 'hypothetico-deductive' framework, where the model structure is postulated on the basis of assumptions regarding the physical nature of the system and then the constant or time variable parameters that characterize this structure are estimated from the available time-series data. On the other hand, as we have seen in this and previous chapters, the RIV and RRIV algorithms are formulated within an 'inductive' framework, where no prior assumptions about the model structure are made, other than that the system can be modelled by a fairly general set of differential equations or a discrete-time equivalent of this. In this inductive approach, the interpretation of the model in physically meaningful terms *follows* the parameter estimation phase, which may include both TVP and the related *State-Dependent Parameter* (SDP) estimation methods described in the chapter 11. This inductive philosopy is also the cornerstone of an approach to model identification and estimation that I have called *Data-Based Mechanistic* (DBM) modelling and which is outlined in chapter 12.

10.4 Computationally Intensive Methods of Recursive Estimation

The EKF is a simple, approximate solution to the optimal, nonlinear, estimation and filtering problem, the general solution to which is infinite dimensional (Kushner, 1967). Unfortunately, as an approximation, the EKF has various limitations, such as problems with covariance estimation and convergence for multi-dimensional, nonlinear models. As Julier et al. (2000) conclude

> *Although the EKF (in its many forms) is a widely used filtering strategy, over 30 years of experience with it .... has led to a general consensus that it is difficult to implement, difficult to tune, and only reliable for systems that are almost linear on the time scale of the update intervals structure.*

Not surprisingly, other approximate solutions are possible and some of these, as pointed out previously, are discussed in the early but very influential book by

Jazwinski (1970). However, such algorithms are normally based on higher order expansions and, while often theoretically superior, they do not possess the attractive, practical simplicity of the KF and EKF.

Recently, however, the easy availability of fast desktop computers, particularly of the multi-processor and parallel processing variety, has led to the development of new, computationally intensive approaches to recursive estimation for nonlinear stochastic systems based on stochastic simulation methods that exploit random sampling of various kinds, including Monte Carlo-based approaches (see section B.3.9 of Appendix B). In part, these have been prompted by the limitations of the EKF, but mostly they have attracted attention because they provide a way of solving recursive nonlinear and non-Gaussian filtering problems for which there is no analytical solution. These computationally intensive methods are often formulated in Bayesian estimation terms and so referred to as 'numerical Bayesian' methods. An early example is the Monte Carlo method for nonlinear, non-Gaussian state space filtering and smoothing (Gordon et al., 1993, 2000; Kitagawa and Gersch, 1996; Doucet et al., 2001). The book by Kitagawa and Gersch, in particular, provides a very readable introduction to this subject and the reader interested in finding more about this approach should start by consulting this text.

Such random sampling approaches to recursive estimation and filtering are now a popular topic for research and development in statistics, engineering and environmental science and there are a large and growing number of papers and texts on the subject. Amongst the other books the reader is recommended to consult are those of Ruanaidh and Fitzgerald (1996); Gammerman (1997); Durbin and Koopmans (2001) and some of the chapters in Fitzgerald et al. (2000). However, it is too broad and involved a topic to address in the present book and all that I will do here is to briefly outline three of the methods that I have found interesting and potentially useful in the environmental context.

(A) **The Ensemble Kalman Filter (EnKF)**

The EnKF is probably the simplest and, for this reason, one of the most popular of the recursive methods that exploit Monte Carlo-based methods. It is an adaptation of the standard, analytic KF algorithm to non-linear systems using Monte Carlo sampling and ensemble averaging in the prediction step and linear updating in the correction step; and it can be seen as a simple but computationally intensive alternative to the EKF. A definitive account of the EnKF appeared with the publication of Evensen's book on the subject (Evensen, 2007) and the recent paper by Clark et al. (2008) provides a comprehensive and critical evaluation of both the EnKF and the related *Ensemble Square Root Filter* (EnSRF) when applied to the distributed hydrological model *TopNet* of the Wairau River basin in New Zealand. There are various ways in which joint state/TVP estimation can be carried out within an EnKF framework but a relevant one in the present context is that suggested by Moradkhani et al. (2005) and tested on data from the Leaf River in the USA (the same data as those used for the example discussed later).

The basic implementation of the EnKF for state estimation and forecasting is quite simple because the correction step in the recursions is the same as the standard KF. The Monte Carlo sampling and ensemble averaging is only required in the prediction step, which simply involves the computation of the ensemble mean and its associated covariance matrix, computed from the deviations of the ensemble members from the mean (acting as a surrogate for the true state, which is unknown, of course). However, Moradkhani et al develop a 'dual EnKF' which requires separate state space representation for the state variables and parameters through two linked algorithms (filters) running in parallel. Here, the parameters are treated in a similar manner to the state variables, with the parameters assumed to follow a stochastic RW process, exactly the same as that used in earlier chapters of the present book. However, the implementation of the dual recursions could be accomplished in various ways and this is not all that clear from the description in the paper.

The EnKF results obtained with the Leaf River data are promising but they suggest the need for further research on the practical implications of the filter in relation to real-time state/parameter updating. For example, the flow forecasts appear good but it may be that these are estimates based on all the data up to the latest day, rather than one-day-ahead forecasts. Also the parameter estimation convergence is fairly slow when compared to that of the recursive RIV estimation algorithm applied to the same data (see example below and Young, 2010b,c) and so some questions remain about how useful an EnKF implementation, such as this, would be in tracking time variable parameters. Finally, the HYMOD model used in the study is quite small and simple, so it is not clear how well this EnKF approach would work in the case of a large, highly nonlinear model for which the EnKF is really intended (see Weerts and Serafy, 2006; Clark et al., 2008).

(B) **The Particle Filter (PF)**

When interpreted in Bayesian terms, the KF can be considered as a very special, analytically tractable version of the general recursive Bayesian filter, as obtained when the state space model and observation equations are linear and the additive stochastic disturbance inputs have Gaussian amplitude distributions. The PF, on the other hand, is a sequential, Monte Carlo-based approximate mechanization of the prediction and correction stages of a fairly general recursive Bayesian filter (Gordon et al., 2000; Doucet et al., 2001; Moradkhani et al., 2005; Smith et al., 2006) and so it applies to general nonlinear models with nonlinear observations and non-Gaussian stochastic disturbances. In the PF, the underlying posterior probability distribution function is represented by a cloud of particles in the state space and the samples automatically migrate to regions of high posterior probability. Moreover, in theoretical terms, convergence is not particularly sensitive to the size of the state space.

On the basis of this description, the PF seems extremely flexible and potentially very attractive. As so often with general methods such as this, however, there are practical drawbacks. It is naturally very expensive in computational terms and practical restrictions on the number of particles that can be used in sampling the prior distributions often lead to posterior distributions that are dominated by only a few particles. This can introduce a need for modifications, such as the use of techniques that include residual re-sampling, *Markov Chain Monte Carlo*

(MCMC) analysis, sequential importance sampling and sampling importance resampling: see e.g. Moradkhani et al. (2005). The latter reference and the one by Smith et al. (2006) are interesting because they also analyze the Leaf River data, although only in an estimation, rather than a forecasting sense.

The results obtained by Moradkhani et al are rather mixed: for example, some of the high flows fall outside of the estimated uncertainty intervals and there is high interaction between the estimated states and parameters, to the detriment of the parameter estimates (also a common characteristic of the EKF and perhaps an argument, once again, for the separation of state and parameter estimation). Similar comments to those made about the EnKF in the last sub-section apply to the PF: namely, (i) it is not clear how well the PF approach would work in the case of a large, highly nonlinear model rather than the simple HYMOD model used by Moradkhani et al; and (ii) the parameter estimation results show fairly slow convergence and its time variable parameter tracking ability is questionable.

On the other hand, the results obtained by Smith et al make good sense and they allow the authors to investigate the shortcomings in the HYMOD model structure. Also, the time variable parameter tracking results seem quite reasonable, although some of the estimated variations are rather volatile when compared with the recursive estimates that I obtained using an adaptive Kalman Filter applied to the same data with a similar complexity model (see example below and Young, 2010b,c).

Finally as regards the comparison of the EnKF and PF, Weerts and Serafy (2006) (see earlier) conclude that "For low flows, {the} EnKF outperforms both particle filters (the *Sequential Importance Resampling* (SIR) filter; and *Residual Resampling* filter (RR) variations), because it is less sensitive to mis-specification of the model and uncertainties."

(C) **The Unscented Kalman Filter (UKF)**

The UKF operates on the premise that it is easier to approximate a Gaussian distribution than it is to approximate an arbitrary nonlinear function (see e.g. Julier et al. (2000) and the prior references therein). Instead of linearizing using Jacobian matrices, as in the EKF, the UKF uses a *deterministic* 'sigma point filter' sampling approach to capture the mean and covariance estimates with a minimal set of sample points. It has some similarities with the EnKF but the random sampling strategy of the EnKF is replaced by this deterministic approach, which will be more efficient both computationally and statistically when its assumptions are satisfied. The UKF appears to be a powerful nonlinear estimation technique and has been shown to be a superior alternative to the EKF in a variety of applications including state estimation and parameter estimation for time series modeling. A dual-UKF method involving state and parameter estimation is described by Tian et al. (2008) in relation to the design of a system for assimilating satellite observations of soil moisture using the NCAR Community Land Model.

These computationally intensive, numerical approaches to recursive estimation have great potential for practical application to nonlinear stochastic systems of all kinds because they are inherently so flexible and free from the kind of assumptions that are necessary to formulate the 'analytic' algorithms that have been described in the earlier chapters of present book. Nevertheless, in any particular application, it is necessary to evaluate whether the more computationally efficient and easily accessible algorithms might not suffice (or, indeed, may have practical advantages). After all, these new methods are still only approximate solutions that depend very much on how well they are implemented, so they are not guaranteed to produce any better results in any specific application, as illustrated by the example considered in the next section. Moreover, they are not easy to program into a general form that would allow them to be as easily accessible for day-to-day application as the recursive routines currently available in toolboxes such as CAPTAIN and SID. On the contrary, each application normally needs to be evaluated carefully and a customized algorithm has to be written, normally by a specialist with some considerable experience of these methods.

10.5 Example 12.1: Data Assimilation and Adaptive Forecasting

This example is concerned with daily rainfall-flow data from the Leaf River catchment, a humid watershed with an area of 1944 km^2 located north of Collins, Mississippi, USA. These data have been selected because, as mentioned in the previous section, they have been used as the basis for the recent research on the application of some of the newest methods of recursive data assimilation: the EnKF and the PF, as described in Moradkhani, Hsu et al and Moradkhani, Sarooshian et al., both 2005; and Smith et al. (2006). Consequently, the results presented below can be viewed in the context of these previous studies and the methods that they describe[5]. The full details of this example are available in Young (2010c).

For simplicity of presentation and in order that more detailed aspects of the model estimation and forecasting system design process can be emphasized, the example concerns only a single model relationship between rainfall and flow at a single site on the Leaf River (as in the other studies of the Leaf River mentioned previously). When this methodology is applied to a whole river basin, however, a simple model such as this would form just one element in a quasi-distributed model of the total catchment network. A typical example is the Lancaster forecasting system for part of the River Severn (Romanowicz et al., 2006; Young et al., 2006) which contains additional rainfall-flow models of this type, as well as linear and nonlinear flow routing models. The same methodological approach to that described below is used for each of these sub-models, prior to their assembly in the complete catchment model and forecasting system. This approach is currently being used in the design of a forecasting system for the River Eden catchment in the UK that is being incorporated into the Delft-FEWS scheme (Leedal et al., 2010) as part of a project funded by the UK's Environment Agency in connection with the development of its

[5] Note that, in these papers, the term 'residence time' is used incorrectly when referring to the *inverse* of the residence time, with inverse time units ($days^{-1}$); the residence time referred to in the present section is defined in the standard manner, with daily time units.

National Flood Forecasting System (NFFS: see also Beven et al., 2008; Young et al., 2009)

Despite its simplicity, the example presents a difficult flow forecasting exercise. Indeed, it has been selected because these difficulties help to illustrate various important aspects of forecasting system design and the use of recursive estimation methods. For instance, the example is concerned with daily data and yet there is no advective delay δ between the occurrence of rainfall and its effect on flow. Indeed, there is a significant *instantaneous* effect (i.e. a flow effect resulting from rainfall falling within the same day). Clearly, therefore, one-day-ahead flow forecasting, without an accompanying one-day-ahead forecast of the rainfall (rather difficult, particularly in the UK!), presents quite a challenge.

10.5.1 The DBM model

The structure of the *Data-Based Mechanistic* (DBM) model (see chapter 12) between rainfall $r(k)$ and flow $y(k)$ is identified, initially, from an estimation data set of 366 days over 1952-1953. This structure identification phase is important and, in this example, it requires the use of *State-Dependent Parameter* (SDP) estimation, as described in the next chapter 11. Suffice it to say that, in this case, such SDP analysis suggests a nonlinear model of the 'Hammerstein' type, with the rainfall input $r(k)$ being converted by an input SDP nonlinearity $\mathcal{F}\{s(k)\}$ into an unobserved 'effective rainfall' $u(k)$ (i.e. the rainfall that is effective in causing the changes in flow). This then enters a linear, 3rd order, discrete-time TF model, with three real eigenvalues, two of which are repeated: i.e.,

$$\begin{aligned} y(k) &= \frac{b_0 + b_1 z^{-1} + b_2 z^{-2} + b_3 z^{-3}}{1 + a_1 z^{-1} + a_2 z^{-2} + a_3 z^{-3}} u(k) + e(k) \\ &= \frac{b_0 + b_1 z^{-1} + b_2 z^{-2} + b_3 z^{-3}}{(1 + f_1 z^{-1})(1 + f_2 z^{-1})^2} u(k) + e(k) \\ u(k) &= \mathcal{F}\{s(k)\} r(k) \end{aligned} \tag{10.32}$$

where $y(k)$ here is the measured flow and the additive observation noise $e(k)$ is normally heteroscedastic with changes in variance that are a function of the flow magnitude. Here, $\mathcal{F}\{s(k)\}$ reflects the effects of the 'soil moisture' changes $s(k)$ going on in the catchment: when it has not rained for some time, the soil dries out so that, when it rains again, some of the rainfall is intercepted by the soil and does not cause changes in flow. However, if there is then further rain so that $s(k)$ (sometimes called 'catchment storage') increases to saturation, so less rainfall is intercepted in this manner and the effects of the rainfall on flow become larger. In fact, $s(k)$ is difficult to measure directly, so the scaled flow $y(k)$ is used as a '*surrogate*' measure of soil moisture, i.e. $u(k) = \mathcal{F}\{y(k)\} r(k)$. Note that this does not imply an actual feedback of flow to the nonlinearity because $y(k)$ is only functioning in this surrogate role. As a result, the model (10.32) can only be used for the present forecasting purposes and not for simulation.

The SDP nonlinearity $\mathcal{F}\{y(k)\}$ is initially estimated nonparametrically by the sdp routine in CAPTAIN (see next chapter 11). The resulting nonparametric estimate is shown in Figure 10.4 and it is found that an exponential function,

$$\mathcal{F}\{y(k)\} = 1 - e^{-\gamma y(k)} \tag{10.33}$$

provides a reasonable parameterization of this nonparametric SDP.

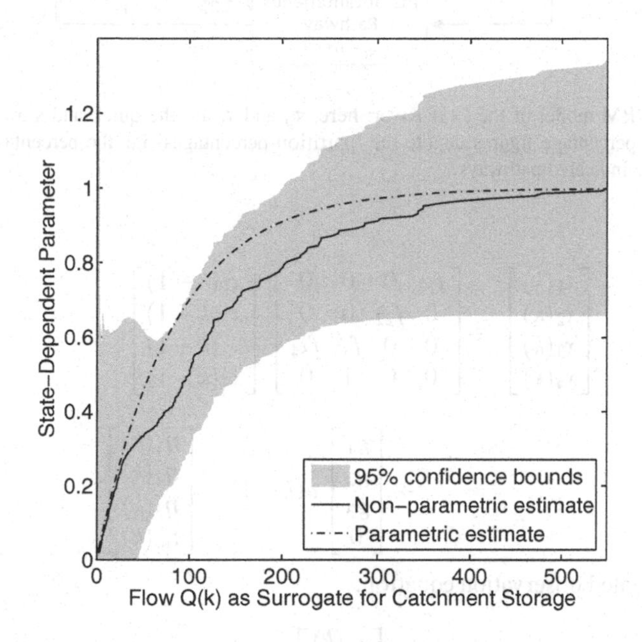

Fig. 10.4 Nonparametric (black line) and parametric (dash-dot) estimates of the estimated SDP effective rainfall nonlinearity, with $\gamma = 0.0121$.

At this stage, the TF model (10.32) is converted into a suitable state space form for use within a KF-based forecasting engine. The model is a stiff dynamic system with three real eigenvalues, two of which are identical (0.445) and very different from the third (0.961). These are associated with time constants (residence times) of 1.23 days ($\times 2$) and 25.4 days, respectively. Consequently, the TF model can be decomposed by partial fraction expansion (made more difficult in this case because of the repeated eigenvalues) into a parallel connection of first order processes, as shown diagrammatically in Figure 10.5. The associated state space model takes the form:

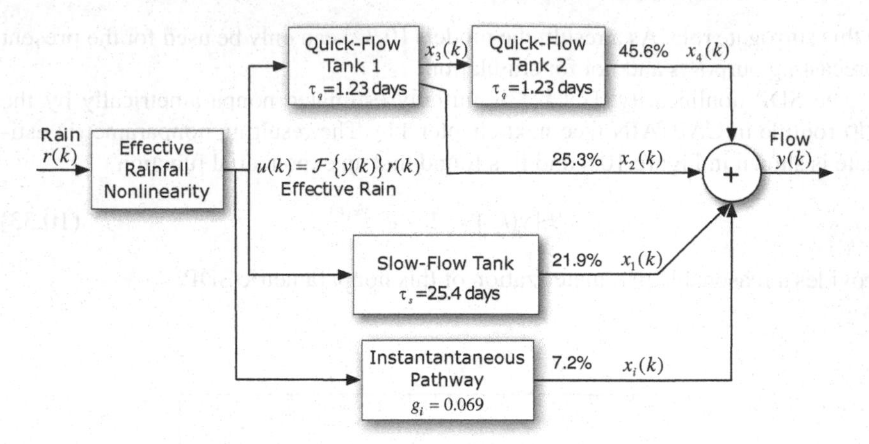

Fig. 10.5 The DBM model of the Leaf River: here, τ_q and τ_s are the quick and slow residence times, while the percentage figures denote the 'partition percentages' i.e. the percentage of flow passing down the indicated pathways.

$$
\begin{bmatrix} x_1(k) \\ x_2(k) \\ x_3(k) \\ x_4(k) \end{bmatrix} = \begin{bmatrix} f_{11} & 0 & 0 & 0 \\ 0 & f_{22} & 0 & 0 \\ 0 & 0 & f_{33} & f_{34} \\ 0 & 0 & 1 & 0 \end{bmatrix} \begin{bmatrix} x_1(k-1) \\ x_2(k-1) \\ x_3(k-1) \\ x_4(k-1) \end{bmatrix} + \begin{bmatrix} g_1 \\ g_2 \\ g_3 \\ 0 \end{bmatrix} u(k-1) + \begin{bmatrix} \eta_1(k) \\ \eta_2(k) \\ \eta_3(k) \\ \eta_4(k) \end{bmatrix} \tag{10.34}
$$

with the associated observation equation:

$$
y(k) = \begin{bmatrix} 1 \; 1 \; 0 \; 1 \end{bmatrix} \begin{bmatrix} x_1(k) \\ x_2(k) \\ x_3(k) \\ x_4(k) \end{bmatrix} + g_i u(k) + e(k) \tag{10.35}
$$

Here, the state variables $x_i(k)$, $i = 1,2,4$ are, respectively (see Figure 10.5), the 'slow flow' $x_1(k)$, normally associated with the groundwater processes; and the two 'quick' flows, $x_2(k)$ and $x_4(k)$, normally associated with the surface and near surface processes. The fourth state, $x_3(k)$, is an intermediate state arising from the transfer function decomposition. These are all 'unobserved state variables' that have to be estimated by the KF algorithm when the model is used in forecasting. As we see in the observation equation (10.35), the flow measurement is the sum of the first, second and fourth of these state variables, plus an instantaneous term dependent on rainfall occurring during the same day. The steady state gains of the parallel pathways can be used to define the 'partition percentages' shown in Figure 10.5. Finally, the heteroscedastic observation noise $e(k)$ is assumed to have variance $\sigma_k^2 = \mathcal{F}_n\{y(k)\}\sigma^2$ that is a SDP function of the flow $y(k)$ and the stochastic input covariance matrix $\mathbf{Q}$ is assumed to be purely diagonal, so that the hyper-parameters that need to be specified are the NVR parameters $\mathbf{Q}/\sigma^2$.

The parameters in the state space model (10.34) are obtained from the estimated nonlinear TF model parameters

$$\boldsymbol{\rho} = [a_1\ a_2\ a_3\ b_0\ b_1\ b_2\ b_3\ \gamma] \tag{10.36}$$

The estimates of these parameters are based on an estimation data set over 1952-1953 and are obtained by constrained optimization, in order to ensure that the model has real eigenvalues, similar to that described in example 7.2, section 8.6.2 of chapter 7. These are used to define the following state space model parameter estimates:

$$\begin{aligned} \hat{f}_{11} &= 0.9613(0.008);\ \hat{f}_{22} = 0.4449(0.010);\ \hat{f}_{33} = 0.8898(0.020)\\ \hat{f}_{34} &= -0.1979(0.009);\ \hat{g}_1 = 0.0081(0.006);\ \hat{g}_2 = 0.13467(0.007)\\ \hat{g}_3 &= 0.13472(0.003);\ \hat{g}_i = 0.0691(0.004);\ \ \hat{\gamma} = 0.0121(0.0005) \end{aligned} \tag{10.37}$$

Of course, the directly estimated parameters in this case are the TF model parameters. As a result, the figures shown in parentheses here are the standard errors for these derived state space model parameters, as obtained by Monte Carlo analysis (see section B.3.9 of Appendix B) using the parametric error covariance matrix from the TF model estimation. The model (10.34)–(10.35) with the above parameter estimates constitutes the prior, 'nominal' DBM model of the Leaf River data.

10.5.2 Parameter updating by RRIV estimation

In order to allow for parameter updating, we need to introduce a capacity for updating the parameters of this model in real-time as additional daily rainfall-flow data arrive. This exploits the real-time recursive RRIV algorithm described in section 10.2, with suitably optimized hyper-parameters, to update the parameters of the DBM transfer function model; the parameters in the state space model are then obtained from these by transformation, on a continuing basis. The recursive TVP estimation could include the continual updating of the parameter γ in the SDP effective rainfall nonlinearity (10.33) but this did not affect the forecasting ability very much and was maintained at its nominal value. However, the SDP model for the heteroscedasticity is identified as $\sigma_k^2 = 0.01y(k)^2$ and this is incorporated in the estimation algorithm to account for the changes in the observation noise variance.

10.5.3 State updating by the Kalman Filter

Given the model (10.34)–(10.35), the discrete-time KF algorithm provides an obvious starting point for the design of a real-time forecasting engine. In order to utilize this, it is necessary to quantify the various hyper-parameters that control the KF forecasting performance.

Although these hyper-parameters could be optimized, this was not attempted in this case so that the ease of manual selection could be demonstrated. The NVR hyper-parameters were selected on the basis of the empirically estimated diagonal elements of the state variable covariance matrix and the estimate of the residual variance σ^2, computed over the same data as that used for the nominal model estimation. The diagonal elements of the resulting NVR matrix $\mathbf{Q}_{nvr}$ were then normalized around the second state element and defined as follows:

$$\mathbf{Q}_{nvr} = \text{diag}(\delta_q\,[0.634\;\;1.0\;\;3.245\;\;0\,]) \tag{10.38}$$

where δ_q is now the only hyper-parameter to be determined. This was selected by a systematic procedure to yield the best overall one-day-ahead forecasting performance.

10.5.4 Typical adaptive forecasting results

Typical adaptive forecasting results are presented in Figures 10.6 and 10.7. which shows three years of real-time updating, following initiation after 50 days. Here, the initial covariance matrix for the RRIV parameter estimation is set to reflect some considerable uncertainty in the parameters and so the estimates are rather volatile when the first large rainfall and flow events occur.

The recursively updated parameter estimates, in this case, are the coefficients of the numerator polynomial in the transfer function since these reflect well the changes in the model characteristics that, when they occur, are linked mainly with the gain of the model. In previous projects of this type, only a single gain parameter has been updated (Lees et al., 1993, 1994; Romanowicz et al., 2006) but this has been modified in this example to illustrate how multiple parameter estimates can be updated if this is necessary. The resulting estimates, as plotted in the upper middle panel, vary quite a lot while the RRIV estimation algorithm is 'learning' the model parameters from the rainfall-flow data. However, after this is completed early in 1953, they then settle down to become fairly stable when sufficient information has been processed to engender confidence in the estimates. Note that the associated changes in the parameters of the state space model (10.34) can be inferred straightforwardly from the changes in these estimated transfer function parameters and the same estimation behaviour is reflected in the associated changes in the partition percentages, shown in the lower middle panel. Here, for clarity, the quick-flow percentage is obtained by aggregating the percentages of the two estimated quick-flow

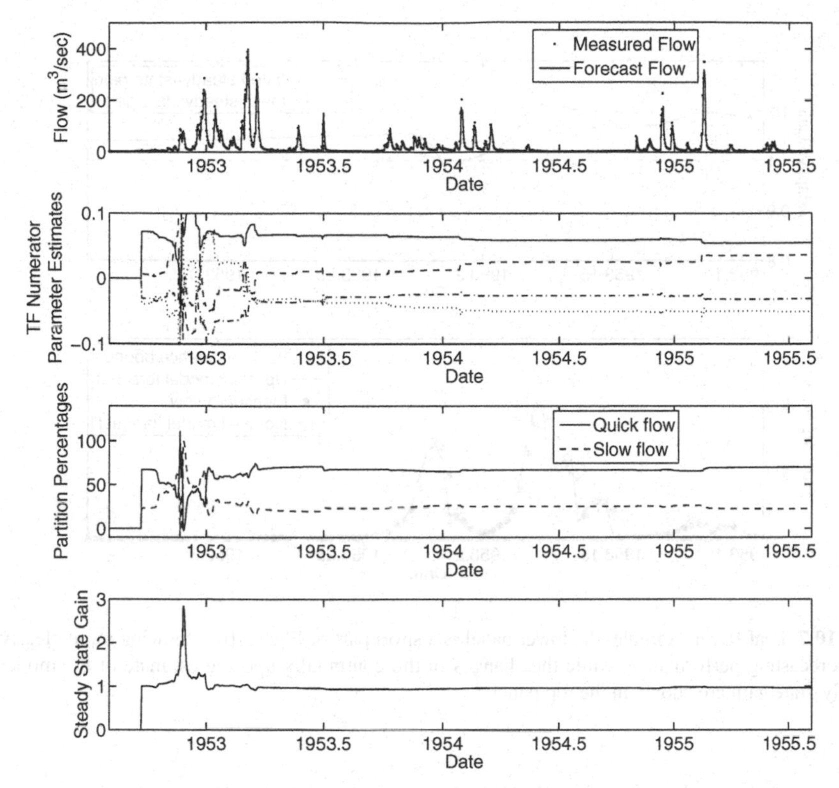

Fig. 10.6 Leaf River example: three years of real-time updating following initiation after 50 days: measured and forecast flow (upper panel); recursive estimates of TF numerator parameters (upper middle panel); partition percentages (lower middle panel); and overall steady state gain (lower panel).

pathways. The overall steady state gain, plotted in the lower panel, also shows little change after the learning period is complete.

Figure 10.7 presents a more detailed view of the adaptive forecasting performance. The lower panel is a short segment of Figure 10.6 showing more clearly the effect that real-time model parameter updating has on the forecasting performance. The estimated 95% confidence interval is consistent with the forecasts and captures the heteroscedastic behaviour of the forecasting errors, except during the upward part of the hydrograph, where the flow measurement is sometimes marginally outside this interval. This is a direct consequence of the forecasting problems, mentioned previously, caused by the absence of any advective time delay between the rainfall and flow in this example: since the rainfall is not being predicted one-day-ahead, the forecasting system has no prior warning of the rainfall and it is impossible for it to forecast the associated change in flow unless such a rainfall forecast is also available.

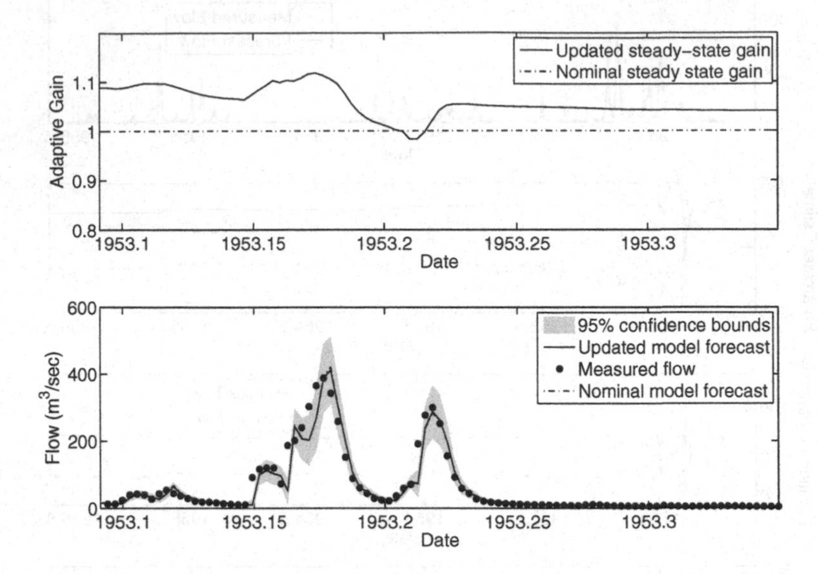

Fig. 10.7 Leaf River example: the lower panel is a short part of Figure 10.6 showing more clearly the forecasting performance; while the changes in the continually updated estimate of the model steady-state gain are shown in the top panel.

Finally, note that the computational cost of the adaptive forecasting implementation described here is very small: each daily state/parameter update takes only a few microseconds. In contrast, the numerically intensive alternatives, which do not perform any better than the system described here, are much more computationally expensive: e.g. in order to obtain similar performance, Moradkhani et al. (2005) require a minimum Monte Carlo ensemble of 40 realizations at each recursive update.

10.6 Exercises

1. Show that the RPEM algorithm reverts to the RLS algorithm when applied to the ARX model.
2. Show how the state space model (10.34) can be derived from the factored TF model (10.32).

10.7 Summary

This chapter has discussed briefly some of the major alternatives to the recursive methods of time series analysis described in previous chapters of the book and introduced the real-time recursive version RRIV of the RIV algorithm. The first of these alternatives, the real-time RPEM algorithm, as described in section 10.1 of the chapter, is closely related to the RRIV approach introduced in section 10.2. RPEM estimation applied to discrete-time TF models is similarly posed in the observation space and it possesses the same asymptotic statistical properties as RRIV. Also it is very flexible in the sense that, like RRIV, it can be modified to handle other candidate time series model forms, such as the ARMAX and DA representations (see e.g. Jakeman and Young, 1981, 1983). As we have seen, however, while the two algorithms exhibit similar performance, they are implemented in quite different ways and so the results will not be identical.

The major advantage of the RPEM approach at this time is its elegant theoretical background, which is helpful not only in assessing the convergence and asymptotic statistical properties of PEM inspired algorithms in either the observation or state space, but also in similarly assessing algorithms like RRIV, that are based on pseudo-linear regression models (Solo, 1978, 1980), or state space based methods such as the EKF. In this regard, the reader is urged to consult Ljung and Söderström, (1983), Söderström and Stoica (1989) and Ljung (1999) which deal with the more theoretical aspects of PEM estimation in considerable detail.

The chapter also introduces the reader to the EKF algorithm. This is attractive in some applications because of its relative simplicity and because it tackles the problem of time series analysis from a state space standpoint. However, its performance is sub-optimal and can be problem dependent; in particular it is known to generate parameter estimates which can have low statistical efficiency. Of more contemporary importance are the computationally intensive algorithms based on Monte Carlo simulation and ensemble averaging that have emerged as alternatives to the KF and EKF in recent years. These can fulfil a similar role to the real-time recursive RRIV and RPEM algorithms but, while more flexible and widely applicable, they are computationally intensive and normally need to be customized to any specific application. The example of flow forecasting in the Leaf River illustrates well how it is necessary to carefully evaluate the need for such algorithms before deciding whether they are really required

Finally, it should be emphasized that other, alternative, recursive methods of time series analysis, for both TF and state space models, have been suggested over many years and these are not considered in this chapter. But it is felt that the RPEM, RRIV and the other methods that are discussed here provide what are currently, at least, the most attractive approaches, both in theory and practice. This does not mean, however, that the other approaches, such as *extended least squares*, *generalized least squares* and *correlation methods* (as reviewed and compared by Isermann et al., 1973), should not be considered as alternatives, so the reader is urged to supplement the reading of this book by acquainting himself or herself with these other procedures, which may have advantages in particular practical circumstances.

Part III
Other Topics

Chapter 11
State-Dependent Parameter (SDP) Estimation

11.1 Introduction

Chapters 4, 5 and 10 have shown that the estimation of time variable parameters can help to model 'non-stationary' systems whose static and dynamic behaviour changes over time. On the other hand, if the changes in the parameters are functions of the state or input variables (i.e. the parameters and/or their changes actually constitute stochastic state variables), then the system is truly nonlinear and likely to exhibit severe nonlinear behaviour. Normally, this cannot be approximated in a simple TVP manner; in which case, recourse must be made to the alternative, and more powerful *State-Dependent Parameter* (SDP) modelling methods that are the main topic of this chapter.

The extension of the TVP estimation methods to allow for state dependency can be traced back to a little known Conference paper by Hoberock and Kohr (1966), who used it in connection with simulated deterministic differential equation models. They employed a continuous-time steepest descent algorithm, implemented within an analog computer, to estimate the parameters and, as a result, their method is rather sensitive to noise on the data. As far as I am aware, the idea of SDP modelling, within a more robust stochastic setting and exploiting recursive estimation, was originated by myself (Young, 1969b, 1981b) and Jerry Mendel (Mendel, 1969, 1970). Both approaches enhanced recursive estimation performance by assuming that the model parameters could vary because of their dependence on the variations in other measured variables. I then explored these ideas within a broader SDP setting (Young, 1978a) and Priestley (1988) took them up in a series of papers and a book on the subject. These earlier publications do not, however, exploit the power of recursive fixed interval smoothing (FIS) that provides the main engine for the developments described in the present chapter and which was first introduced in Young (2000).

SDP estimation of this type involves the non-parametric *identification* of the state dependency (i.e. identifying the nature and location of SDP nonlinearities, rather than their final estimation), using recursive methods of time variable parameter estimation which allow for rapid (state dependent) parametric change. As we shall see, the standard methods of TVP estimation developed previously for nonstationary time series analysis need to be modified considerably in this SDP setting to allow for the much more rapid temporal changes that arise from the state dependency. I must admit that discovering how this difficult task might be accomplished took me some time and I suggested one rather limited approach (Young, 1993b; Young and Beven, 1994), before realising that a much better alternative was possible. This conveniently exploits recursive smoothing in the form of the recursive FIS algorithms discussed in chapters 4 and 5, combined with special data re-ordering and 'back-fitting' procedures, to obtain estimates of any state dependent parameter variations. The re-ordering procedure, which is defined by the variables (states) on which the parameters are found to be dependent, is crucial because it converts the estimation of the rapidly variable SDPs in the normal temporal observation space into the estimation of slowly time variable parameters in the re-ordered observation space, thereby allowing for the use of the FIS/TVP algorithms considered in chapters 4 and 5. The reason for the back-fitting is discussed later.

The above approach yields recursive FIS estimates that are defined at every sample in the re-ordered observed series. This means that the state dependencies are, in effect, estimated in the form of non-parametric relationships (graphs) between this estimated smooth parameter variation and the associated state variable(s). Indeed, the re-ordering for each parameter is normally based on the ascending values of the associated state on which it is found to be dependent and this means that resulting estimated parameters and states are in an ideal form for immediate plotting against each other (since this is precisely what is do when plotting line graphs).

The data on which the graph is based can be considered as a 'look-up table' for the SDP variations and the results could be used directly in this form (e.g. Simulink, the very useful 'icono-graphic' simulation facility in Matlab, allows the modeller to use such objects). However, the graph often suggests some convenient parameterization of the non-parametric relationships (such as the exponential function used in the flow forecasting example described in section 10.5 of the previous chapter). This can be accomplished in various ways, from simple curve fitting, based on ordinary or weighted least squares methods, to the use of neural/neuro-fuzzy networks (e.g. Jang et al, 1997), or *Radial Basis Functions* (RBF) and wavelets (Buhmann, 2003; Strang and Nguyen, 1996; Truong et al., 2006). The advantage of such parameterization is that it makes the model fully self-contained and parametric in form, so revealing more clearly the nature of the identified nonlinear dynamic system.

It should be noted that, having reached this parameterization stage of the SDP modelling, the SDP models can be linked with what are referred to as *Non-Linear Parameter Varying* (NLPV) models in the control and systems literature (see e.g. Previdi and Lovera, 2004); a link that is demonstrated in Young (2005). This name derives from the earlier *Linear Parameter Varying* (LPV) models that are discussed briefly in section 4.8 of chapter 4 and the state on which the parameters are assumed to depend is called the 'scheduling' variable because of the link with scheduled gain control systems. The main difference between the SDP and NLPV modelling approaches, in addition to the fact that the latter are considered mainly within an automatic control context, is that the initial nonparametric identification stage is omitted in NLPV (and LPV) modelling. The parametric forms and the location of the SDPs appear to be selected either by assuming that all parameters are state-dependent and modelling them using some of the functions mentioned above (quite often polynomials), or by selecting them based on prior knowledge of the physical system being modelled.

Returning to SDP estimation as considered here, having identified a structural form for the nonlinear model of the system based on the parameterized nonlinear relationships, the model can be used in the nonlinear SDP transfer function form, or converted into a stochastic state space form. The final constant parameter estimation phase of the nonlinear modelling then exploits some method of nonlinear optimization, such as straightforward *en bloc* nonlinear least squares optimization; or more sophisticated *Maximum Likelihood* (ML) methods of estimation based on Gaussian assumptions for the stochastic disturbances and the application of *Prediction Error Decomposition*: see section 4.5.3 of chapter 4. The reader might be surprised to see that I have limited this final stage to Gaussian normal stochastic disturbances. This is not so much a limitation as it might seem at first sight, however, because such normally distributed inputs are propagated *within* the *nonlinear* system as non-normal disturbances to the system states. And, of course, the assumption of normality for the inputs considerably simplifies the process of parameter estimation.

The final model identified and estimated in the above manner should provide a parametrically efficient representation of the stochastic, nonlinear system that has considerable potential for use in subsequent signal processing, time series analysis, forecasting and automatic control system design. For example, although the SDP methodology, as described in the next sections of this chapter, exploits recursive estimation in an *off-line* manner, this sequential processing of the data facilitates the development of related *on-line adaptive* methods of signal processing, forecasting and control using, for example, real-time recursive procedures, such as those described in the previous chapter 10. This kind of adaptive implementation is potentially advantageous because it introduces a mechanism for on-line modifications to the parameters that allow for the inherent approximations that are inherent in the SDP model synthesis.

In order to illustrate the practical application and utility of the SDP approach, the chapter contains simulation examples, as well as two practical studies involving macro-economic data from the USA and the analysis of electrical signals obtained from experiments carried out on the axon of a squid. Other simulation and practical examples cited in the references cover a variety of application areas from the environment through engineering to economics.

11.2 SDP Identification of Nonlinear Input-Output Systems

While the DARX model discussed in section 5.4 of chapter 5 can produce fairly complex response characteristics, it is only when the parameters are functions of the system variables, and so vary at a rate commensurate with these variables, that the resultant model can behave in a heavily nonlinear or even chaotic manner. A SDP model of this type is the *State Dependent Parameter ARX* (SDARX) model. In its simplest single input, single output form, the this model equation can be written most conveniently in the following form[1],

$$y(k) = \mathbf{z}^T(k)\boldsymbol{\rho}(k) + e(k); \qquad e(k) = \mathcal{N}(0, \sigma^2) \tag{11.1}$$

where,

$$\begin{aligned} \mathbf{z}^T(k) &= [-y(k-1)\ -y(k-2)\ \cdots\ -y(k-n)\ u(k-\delta)\ \cdots\ u(k-\delta-m)] \\ \boldsymbol{\rho}(k) &= [a_1\{\chi(k)\}\, a_2\{\chi(k)\} \cdots a_n\{\chi(k)\}\, b_0\{\chi(k)\} \cdots b_m\{\chi(k)\}]^T \end{aligned} \tag{11.2}$$

and $a_i\{\chi(k)\}$, $i = 1, 2, \ldots, n$, and $b_j\{\chi(k)\}$, $j = 0, 1, \ldots, m$, are the state dependent parameters, which are assumed to be functions of one of the variables in a non-minimal state vector $\chi^T(k) = [\mathbf{z}^T(k)\ \mathbf{U}^T(k)]$. Here $\mathbf{U}(k) = [U_1(k)\ U_2(k)\ \cdots\ U_r(k)]^T$ is a vector of other variables that may affect the relationship between these two primary variables but are not variables that appear in $\mathbf{z}(k)$: for example, the 'air data' variables used in the self-adaptive autostabilization example mentioned in section 4.8 of chapter 4. As usual, δ is a pure time delay on the input variable and $e(k)$ is a zero mean, white noise input with Gaussian normal amplitude distribution and variance σ^2 (although this assumption is not essential to the practical application of the resulting estimation algorithms). Finally, for convenience of notation, let $\boldsymbol{\rho}(k)$ be defined as follows,

$$\boldsymbol{\rho}(k) = [\rho_1\{\chi(k)\}\ \rho_2\{\chi(k)\}\ \ldots\ \rho_{n+m+1}\{\chi(k)\}]^T \tag{11.3}$$

with $\rho_i\{\chi(k)\}$, $i = 1, 2, \ldots, n+m+1$, relating to the $a_i\{\chi(k)\}$ and $b_j\{\chi(k)\}$ through (11.2).

Until recently, SDP estimation has been limited to SDP models in which each state dependent parameter is a function of only one 'state' variable. A typical, simple example of this is a SDARX model in the form of the following nonlinear 'forced logistic growth' equation:

$$y(k) \;=\; \alpha\, y(k-1) - \alpha\, y(k-1)^2 + u(k) + e(k); \quad e(k) = \mathcal{N}(0, \sigma^2) \tag{11.4}$$

or,

$$\begin{aligned} y(k) \;&=\; a_1\{y(k-1)\}\,.\,y(k-1) + b_0\{u(k)\}\,.\,u(k) + e(k) \\ \text{where, } a_1\{y(k-1)\} \;&=\; \alpha - \alpha\, y(k-1) \ \text{ and } \ b_0(u(k)) = 1.0 \quad \forall\, k \end{aligned} \tag{11.5}$$

Although it is simple, this model can exhibit rich behavioural patterns: from simple to chaotic response, depending on α (see Young, 2000).

---

[1] Note that it is possible to consider continuous-time models by exploiting FIS estimation applied to integrated random walk models (see chapter 5) in order to generate multiple derivatives of noisy data (Young et al., 1993), but this is not discussed here.

11.2.1 Full SDP estimation

In the special case of a SDP model such as (11.1) when each state dependent parameter is a function of only one variable, say $z_j(k)$, $i = 1,2, \ldots, n+m+1$, the rate of variation with k will be commensurate with the temporal variations in this variable, which could be the input variable $u(k)$, the output variable $y(k)$, one of their past values, $y(k-i)$, $i = 1,2, \ldots, n$, or $u(k-i), i = 1,2, \ldots, m$; or it could be one of the other variables $U_i(k)$, $i = 1,2, \ldots, r$. It cannot be assumed, therefore, that the simple GRW model for parameter variation used in chapters 4 and 5 is appropriate to describe such variations, which could be very rapid. Consequently, it would appear that the stochastic state space model should include prior information on the nature of the parameter variation if the TVP estimation methodology discussed in these previous chapters is to work satisfactorily. Fortunately, it is possible to remove this requirement if we resort to the rather unusual procedure of sorting the data in a *non-temporal order*. Then, if this ordering is chosen so that the SDP variations associated with the *sorted* series are smoother and less rapid than in the normal temporal observation space, it is more likely that a simple GRW process can be utilized to describe their evolution *in this transformed observation space*.

For example, if the time series are sorted in some common 'ascending order of magnitude' manner (i.e. the *sort* operation in Matlab), then the rapid natural variations in the 'state variables', on which the parameters are dependent, are effectively eliminated from the data and replaced, in the transformed data space, by much smoother and less rapid variations. An example of this is shown in Figure 11.1 for the model (11.5) with $\alpha = 2.0, \sigma^2 = 0.01$ and the input $u(k)$ generated randomly from a normal distribution between zero and 0.08. Table 11.1 lists a sample of ten normal ($y(k-1)$, second column) and re-ordered ($y^o(k-1)$, fourth column) delayed output samples, with the normal time index in the first column and the re-ordered index in the third column. It is clear that, if the SDP $a_1\{y(k-1)\}$ is, indeed, related

Table 11.1 Forced logistic growth model: example of normal and sorted data.

Time Index	$y(k-1)$	Time Index	$y^o(k-1)$
35	0.658	669	0.3249
36	0.438	848	0.3253
37	0.476	51	0.3256
38	0.513	734	0.3280
39	0.675	703	0.3281
40	0.466	858	0.3282
41	0.877	814	0.3285
42	0.184	531	0.3285
43	0.400	234	0.3301
44	0.647	322	0.3307
45	0.366	764	0.3325

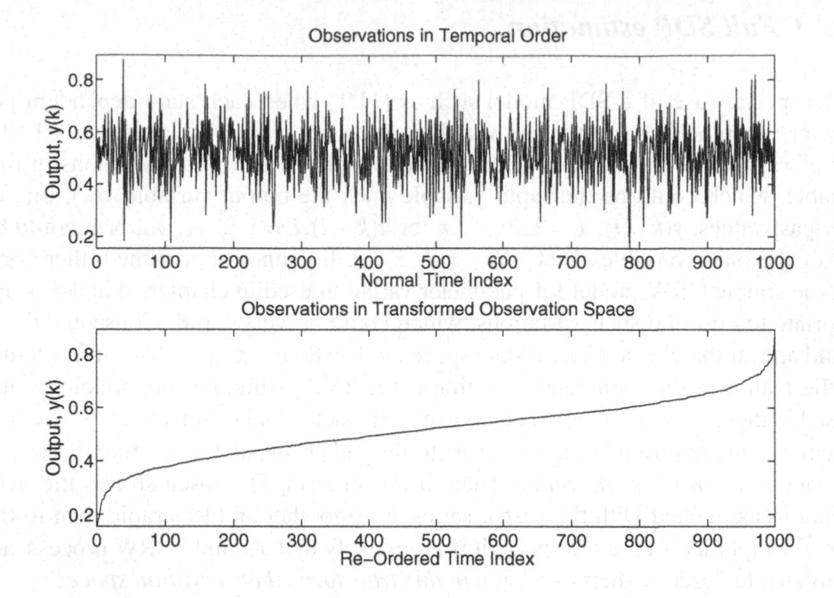

Fig. 11.1 Forced logistic growth model: output data plotted in normal time (upper panel) and re-ordered in ascending order of magnitude (lower panel).

to the user-specified variables $y(k-1)$, then it will be similarly affected by this same sorting.

So let us summarize the estimation procedure in this example. At each recursive step in the transformed data space, the data vector $\mathbf{z}(k)$ in (11.2) is selected with k defined by the latest ordered index but with natural temporal order retained within the vector, so that the recursive processing is the same as normal, except that the order in which this is carried out is defined by the sorting of $y(k-1)$. Following FIS estimation in this transformed data space, however, these SDP estimates can be 'unsorted' (the *unsort* routine in CAPTAIN) and their true, rapid variation will become apparent. Of course, the nature of the sorting will affect the estimation and it seems likely that there will be an optimum sorting which results in minimum variance estimates. However, such optimum sorting will naturally depend upon the nature of the state dependency and its definition would require some sort of iterative estimation procedure. In practical terms, therefore, the common ascending order sorting and un-sorting operations seem the most straightforward and are utilized here.

One obvious requirement of this approach to SDP estimation is that the sorting of data, prior to FIS estimation, must be *common to all of the variables in the relationship* (11.1). If a single, ascending order strategy is selected, therefore, it is necessary to decide upon which variable in the model the sorting should be based. The simplest strategy is to sort according to the ascending order of the 'dependent' variable $y(k)$. Depending upon the nature of each SDP in the vector $\boldsymbol{\rho}(k)$, however, a single variable sorting strategy, such of this, may not produce satisfactory results since each SDP in a multi-order model may be dependent on a different variable. If this is the case, then a more complicated, but still straightforward, iterative 'back-fitting' procedure can be exploited. Here, each parameter is estimated *in turn*, based on the *Modified Dependent Variable* (MDV) series obtained by subtracting all the other terms on the right hand side of (11.1) from $y(k)$, using the values of the other parameter estimates from the previous iteration. At each such back-fitting iteration, the sorting can then be based on the single variable associated with the current SDP being estimated.

Since the SDP estimates resulting from this back-fitting algorithm are themselves time series, it will be noted that the algorithm constitutes a special form of non-parametric estimation and, as such, can be compared with other non-parametric methods, such as the *Generalized Additive Modelling* (GAM) approach of Hastie and Tibshirani (1996). However, in both conceptual and algorithmic terms, the SDP approach described here is significantly different from this earlier approach and seems more appropriate to the estimation of nonlinear, stochastic, dynamic models. Moreover, the recursive methodology, on which SDP estimation is based, is couched in optimal maximum likelihood terms that seem more elegant and flexible than the 'scatter-plot smoothing' procedures used by Hastie, Tibshirani and others.

The back-fitting algorithm for the SDARX model (11.1) takes the following form where, as defined above, the $z_j(k),\ j = 1,2,\ \ldots, n+m+1$, are the state variables associated with each of the SDPs in the model:

Back-fitting Algorithm for SDP Models

1. Assume that FIS estimation has yielded prior TVP estimates $\hat{\rho}_i^{k=0}, i = 1,2,\cdots, m+n+1$ of the SDPs[2].
2. Iterate: $i = 1,2,\cdots,m+n+1; k = 1,2,\cdots,k_c$
   (i) form the MDV $y^i(k) = y(k) - \sum_{j\neq i} z_j(k).\hat{\rho}_j^k(k|N)$;
   (ii) sort both $y^i(k)$ and $z_i(k)$ according to the ascending order of $z_i(k)$;
   (iii) obtain an FIS estimate $\hat{\rho}_i^k(k|N)$ of $\rho_i(k)$ in the MDV relationship $y^i(k) = \rho_i(k).z_i(k)$.
3. Continue 2. (each time forming the MDV and then sorting according to the current right-hand side variable $z_j(k)$, prior to FIS estimation), until iteration k_c, when the individual SDPs (each of which are time-series of length N) have not changed significantly according to some chosen criterion. The smoothing hyper-parameters required for FIS estimation at each stage are optimized by ML, as explained in chapters 4 and 5 and discussed further below.

Note that the ML optimization can be carried out in two ways: after every complete iteration (each involving $m+n+1$ FIS operations) either until convergence is achieved or for a user-specified number of iterations; or only at the initial complete iteration, with the hyper-parameters maintained at these values for the rest of the back-fitting. The latter choice seems most satisfactory in practice, since very little

[2] As a default, these can be simply the constant least squares parameter estimates, since the convergence of the back-fitting procedure is not too sensitive to the prior estimates, provided they are reasonable.

improvement in convergence occurs if optimization is continued after this stage. Normally, convergence is completed after only a few iterations, although it can be more lengthy in some circumstances (see Conclusions section 11.7). This SDP algorithm is available as the sdp routine in the CAPTAIN Toolbox.

When the sdp routine is applied to the data generated by the forced logistic growth model (11.5), the ML optimized NVR matrix (see section 4.2 of chapter 4) has the diagonal values $\mathbf{Q}_{nvr} = \text{diag}[8.9 \times 10^{-8} \;\; 2.5 \times 10^{-17}]$ and it is clear that the optimization has successfully identified that the potential state dependency resides in the parameter $a_1\{y(k-1)\}$, while the input parameter $b_0\{y(k-1)\}$ is effectively time-invariant (i.e. not a function of $y(k-1)$), with the optimized NVR value virtually zero. The associated graphical estimation results are shown in Figure 11.2, where the two estimated parameters are plotted (full lines) against $y(k-1)$ and compared with the true parameters (dash-dot lines), with the estimated 95% confidence bounds shown in grey. Despite the high level of noise, the true SDPs lie within the 95% confidence bounds of the estimated SDPs over most of the $y(k-1)$ range, as required: the poorer estimates of $a_1\{y(k-1)\}$ at the extremes of this range are because there are relatively few samples occurring in these regions (i.e. very low and high values of $y(k-1)$: see figure 11.1).

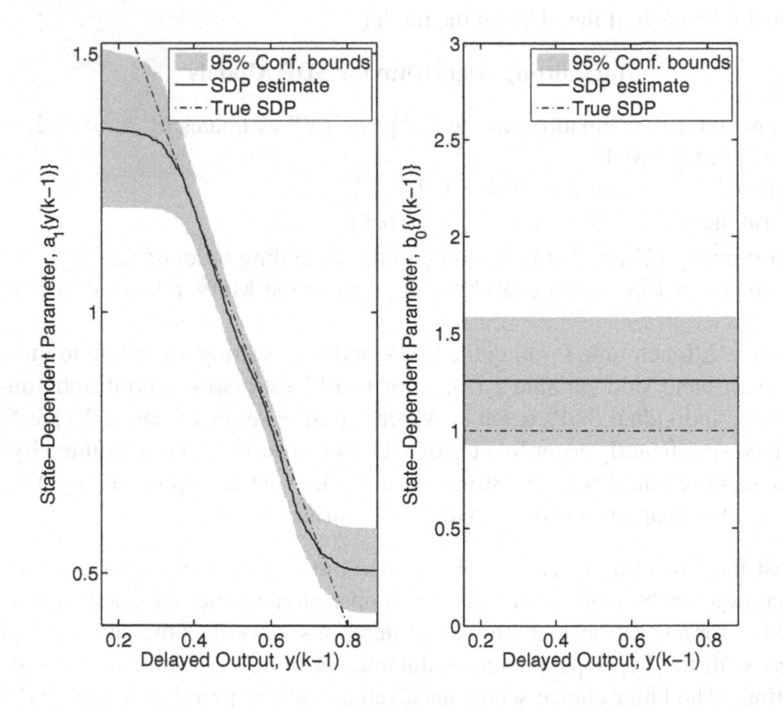

Fig. 11.2 Forced logistic growth model: estimated SDP $a_1\{y(k-1)\}$ (left panel) and $b_0\{y(k-1)\}$ (right panel) which is identified as constant (i.e. not dependent on $y(k-1)$).

11.2.2 Parameterization and final nonlinear model estimation

Parameterization of the identified SDP functions depends, of course, on the nature of the nonparametric SDP estimate, which is normally plotted as a graph of the SDP estimate $\rho_j\{z_j(k)\}$ versus the state variable on which it has been assumed dependent, $z_j(k)$. The parameterization problem is, therefore, to find a low dimensional parametric relationship for $\rho_j\{z_j(k)\}$ that is able to characterize the graphical relationship as well as possible, without over-fitting. Sometimes a simple, well known function, such as $\tanh(\gamma z_j(k))$, an inverse function $z_j^{-1}(k)$, $\cos(\gamma z_j(k))$, a power law $z_j^{\gamma}(k)$, or an exponential-type function, $1-e^{-\gamma z_j(k)}$, will suffice. The first three of these are used in later examples and the latter two have been used in hydrological modelling (see later, section 11.4, and the example in section 10.5.1 of the previous Chapter 10). However, a more general solution is to let each nonlinear SDP $\rho_j\{z_j(k)\}$ be approximated as a linear sum of basis functions $\mathscr{B}_{j,i}\{z_j(k)\}$ in $z_j(k)$, i.e.,

$$\rho_j\{z_j(k)\} = \sum_{i=1}^{n_j} \gamma_{j,i}\mathscr{B}_{j,i}\{z_j(k)\} \quad j = 1,2,\ldots,n+m+1 \tag{11.6}$$

where $\mathscr{B}_{j,i}\{z_j(k)\}$ can take on a number of different forms: for example, terms in a polynomial; radial basis functions; splines; wavelets; etc. The number of these basis functions n_j may be different for each SDP, so the parameterization problem is to find the n_j and the associated weights $\gamma_{j,i}$ that best reproduce the shape of each nonparametric estimate $\rho_j\{z_j(k)\}$ provided by the SDP algorithm. There are various toolboxes in Matlab, such as the 'Curve Fitting Toolbox' (e.g. Ochieng and Otieno (2009) use this in a hydrological modelling example) and the 'Neural Network Toolbox' (see later, example 11.3), that can be helpful in this task. And other approaches, such as 'neuro-fuzzy' modelling (Jang et al., 1997), also could be employed.

Having parameterized the SDP in any of these ways, the final parametric optimization of the whole model can then be carried out in various ways. For instance, as mentioned earlier in the Introduction, it could utilize some method of deterministic or stochastic nonlinear optimization. However, simpler approaches are possible in special cases. For example, if the SDP's only occur in the numerator polynomial of the TF and all have the same functional representation $\rho\{z(k)\}$, so that they can be factored out, then the numerator can be written in the form:

$$B(z^{-1}) = \{b_0 + b_1 z^{-1} + b_2 z^{-2} + \ldots + b_m z^{-m}\}\rho\{z(k)\} \tag{11.7}$$

This means that the *single* identified SDP nonlinearity is then applied to the input variable $u(k)$ and the model takes the special 'Hammerstein' form:

$$y(k) = \frac{B(z^{-1})}{A(z^{-1})}v(k); \quad v(k) = \rho\{z(k)\}u(k) \tag{11.8}$$

In other words, the model becomes a serial connection of the nonlinear SDP function and the linear TF model. A practical example of this is the rainfall-flow model

(10.32) in the previous Chapter 10, where the input rainfall is transformed to the rainfall that is 'effective' in causing flow variations (see also the later section 11.4).

The parameters of the Hammerstein nonlinear model (11.8) can be estimated using a quite simple constrained optimization procedure similar to that described earlier in section 8.6.2 of Chapter 8. Here, if a parameterization such as (11.6) is used, the $\gamma_{j,i}$ defining the single SDP that controls the input transformation are optimized iteratively, using an optimization routine such as lsqnonlin in Matlab. Here, the linear TF part of the model is estimated, at each iteration of the optimization routine, by the rivbj algorithm in the CAPTAIN Toolbox, with the nonlinearly transformed input $v(k)$ defined by the currently estimated SDP nonlinearity, defined by the latest $\hat{\gamma}_{j,i}$ estimates computed at the previous iteration. Such optimization is normally quite straightforward because reasonable estimates $\hat{\gamma}_{j,i}$ of the $\gamma_{j,i}$ have been defined in the previous parameterization exercise and can be used as starting values for the optimization, thus ensuring convergence.

Of course this approach is not limited to SDPs parameterized by (11.6). However, a simpler *but less statistically efficient*, RIV-based approach to Hammerstein model estimation, which is limited in this way, has been suggested recently for open (Laurain et al., 2008) and closed (Laurain et al., 2010) loop systems. In this alternative solution, the RIV algorithm is applied directly to the model after it has been reformulated in multi-input terms. For instance, consider the following Hammerstein model with a polynomial-type input nonlinearity:

$$
\begin{aligned}
y(k) &= \frac{b_0 + b_1 z^{-1}}{1 + a_1 z^{-1} + a_2 z^{-2}} v(k) + \xi(k) \\
v(k) &= \alpha_1 u(k-1) + \alpha_2 u^2(k-1) + \alpha_3 u^3(k-1)
\end{aligned} \tag{11.9}
$$

This can be written in the alternative multi-input, common denominator form

$$
y(k) = \frac{B_1(z^{-1})}{A(z^{-1})} u(k-1) + \frac{B_2(z^{-1})}{A(z^{-1})} u^2(k-1) + \frac{B_3(z^{-1})}{A(z^{-1})} u^3(k-1) + \xi(k) \tag{11.10}
$$

where $A(z^{-1}) = 1 + a_1 z^{-1} + a_2 z^{-2}$, $B_1(z^{-1}) = b_{1,0} + b_{1,1} z^{-1}$, $B_2(z^{-1}) = b_{2,0} + b_{2,1} z^{-1}$ and $B_3(z^{-1}) = b_{3,0} + b_{3,1} z^{-1}$. This only requires measurements of the output $y(k)$ and the basis functions associated with the input nonlinearity, in this case the polynomial terms $u(k)$, $u^2(k)$ and $u^3(k)$. The standard RIV approach can be applied to this multi-input model to yield estimates of the TF model parameters $\hat{a}_1$, $\hat{a}_2$, $\hat{b}_{1,0}$, $\hat{b}_{1,1}$, $\hat{b}_{2,0}$, $\hat{b}_{2,1}$, $\hat{b}_{3,0}$ and $\hat{b}_{3,1}$; and then, by simple substitution, we find that:

$$
\hat{b}_{1,0} = b_0 \alpha_1;\ \hat{b}_{1,1} = b_1 \alpha_1;\ \hat{b}_{2,0} = b_0 \alpha_2;\ \hat{b}_{2,1} = b_1 \alpha_2;\ \hat{b}_{3,0} = b_0 \alpha_3;\ \hat{b}_{3,1} = b_1 \alpha_3
$$

Although this provides a model of the system, the estimates of the input nonlinearity parameters $\hat{\alpha}_1$, $\hat{\alpha}_2$ and $\hat{\alpha}_3$ cannot be recovered unless some prior knowledge is assumed. For example, if the first parameter α_0 is known or, as may be the case in practice, it is equal to unity, then 'average' estimates of the other nonlinearity parameters can be recovered from these RIV estimates using the general expression:

$$\bar{\hat{\alpha}}_i = \frac{1}{m}\sum_{j=0}^{m-1}\frac{\hat{b}_{i,j}}{\hat{b}_{1,j}};\ i=2,\ldots,\ n_1 \tag{11.11}$$

where m is the order of the TF numerator polynomial and n_1 the number of terms in (11.6) with $j=1$, since there is only one SDP in this Hammerstein model (in the case of (11.9), $m=2$ and $n_1=3$). In statistical terms, this is not an altogether satisfactory solution, but it is simple to obtain and it could provide a good initiation for the gradient optimization procedure outlined above.

As an example, Monte Carlo simulation results for the model in (11.9) are reported in Table 11.2. These are based on 100 stochastic realizations where, for each realization, the input $u(k)$ is 1000 samples of a normally distributed white noise variable with unity variance passed through a first order, discrete-time filter with numerator 0.4 and a denominator $1-0.8z^{-1}$. The additive noise $\xi(k)$ for each realization is a random sample from a zero mean, white noise sequence with variance about 18, which results in a noise/signal ratio, based on standard deviations, of 0.5. Table 11.2 compares the results obtained using the constrained nonlinear optimization and the multi-input linear estimation procedures. The initial values of the nonlinearity parameter estimates used in the nonlinear optimization are based on fitting a 3rd order polynomial to the the nonparametric SDP estimation results generated by the sdp routine in the CAPTAIN Toolbox. The standard error (SE) estimates in Table 11.2 are obtained from the rivbj routine in the CAPTAIN Toolbox and, in the case of the nonlinear optimization, supplemented by the lsqnonlin uncertainty estimation results for $\hat{\alpha}_2$ and $\hat{\alpha}_3$ obtained from the confint routine in Matlab. The $\hat{\alpha}_2$ and $\hat{\alpha}_3$ estimates in the case of the multi-input linear estimation method are computed from the rivbj estimated numerator polynomial coefficients using (11.11). As a result, their estimated SEs are not available directly and so are not reported in Table 11.2. While the mean values of the MCS parameter esti-

Table 11.2 Comparative Monte Carlo Simulation results for the Hammerstein model (11.9): SD is the standard deviation from the MCS; SE is the standard error from a single run of the RIV algorithm.

Parameter		a_1	a_2	b_0	b_1	α_2	α_3
TrueValues	ρ	-1.6	0.8	0.85	0.65	0.5	0.25
Constrained Nonlinear	$\hat{\rho}$	-1.599	0.799	0.826	0.667	0.504	0.254
(MCS)	SD	0.009	0.007	0.097	0.118	0.030	0.031
(SR)	SE	0.008	0.007	0.099	0.121	0.028	0.029
Multi-input Linear	$\hat{\rho}$	-1.601	0.801	0.819	0.674	0.538	0.295
(MCS)	SD	0.009	0.007	0.172	0.188	0.133	0.102
(SR)	SE	0.008	0.007	0.170	0.185	–	–

mates produced by the multi-input linear approach are quite reasonable when compared with the equivalent constrained nonlinear optimization estimates, the standard deviations are between 1.6 and 4.4 times as large for the TF numerator and nonlinearity parameters.

So, to summarize, the multi-input linear method is simple and straightforward but it does have some limitations: it can only be used with Hammerstein models where the nonlinearity is modelled by a linear sum of basis functions; prior knowledge is required to fully infer estimates of the input nonlinearity parameters; and the statistical efficiency of the parameters associated with the input is impaired by the need to estimate more than the minimum number of parameters, so that the relative estimation uncertainties can be much higher.

11.2.3 The problem of Errors-in-Variables (EIV) estimation

When there are errors-in-variables (see section 7.4 of chapter 7), the SDP estimates are 'asymptotically' biased away from their true values if the assumed, and rather special, signal topology of the SDARX model (11.1) is violated. This is a similar effect to that encountered in the constant parameter situation but, in this SDP situation, 'asymptotic' is being used in a loose sense, since concepts such as asymptotic bias are not strictly applicable in the TVP and SDP situations. For example, if $y(k)$ in (11.1) is measured in the presence of noise $\xi(k)$, in addition to the 'system' noise $e(k)$, then this kind of troublesome bias can affect *regions* of the parameter estimates (see examples in Young, 2000).

When considering the practical implications of any errors-in-variables bias on the estimates of the parameters in the SDARX model, it is important to stress that SDP estimation is being suggested here primarily as a method of identifying the nonlinear model structure and not necessarily as an end in itself. Moreover, the bias can be quite small even in quite high noise situations. For instance, consider again the forced logistic growth model (11.5) with $\alpha = 2$ but with an errors-in-variables (transfer function) structure and the input parameter now defined as being nonlinearly dependent on the input variable $u(k)$:

$$\begin{aligned} x(k) &= a_1\{x(k-1)\}x(k-1) + b_0\{u(k)\}u(k) \\ y(k) &= x(k) + \xi(k); \quad u(k) = \mathcal{U}(0,3.2); \quad \xi(k) = \mathcal{N}(0, 3\times 10^{-5}) \\ a_1\{x(k-1)\} &= 2.0 - 2.0\,x(k-1); \;\; b_0\{u(k)\}u(k) = 0.01(tanh(u(k)) + 0.01u(k)) \end{aligned}$$

or, in TF terms,

$$y(k) = \frac{b_0\{u(k)\}}{1 - a_1\{x(k-1)\}z^{-1}}u(k) + \xi(k)$$

Both SDP estimates are estimated well, as shown in Figure 11.3, so providing useful information on the location and nature of the nonlinearities in this case (the full details of this example are given in Young et al, 2001).

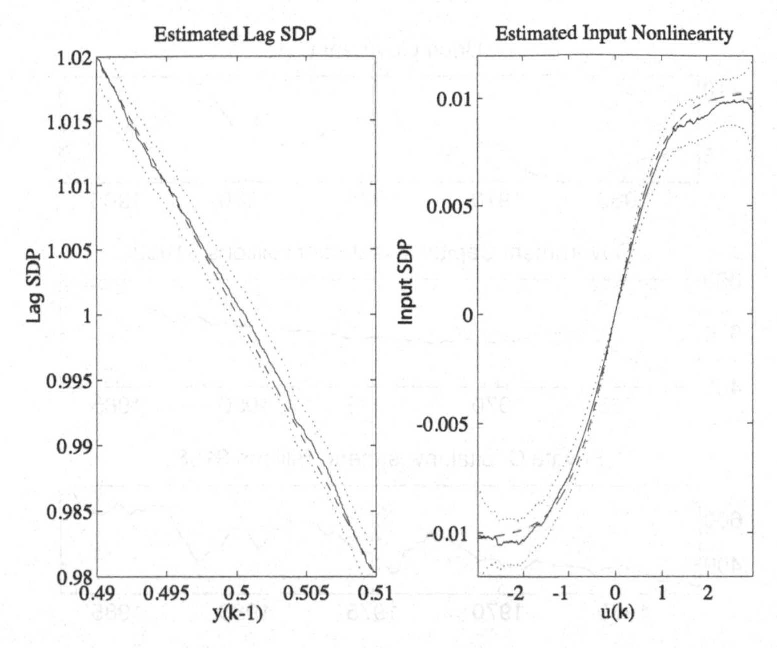

Fig. 11.3 SDP estimation results for a system with severe input and lag nonlinearities, showing the relative insensitivity, in this case, to a high level of errors-in-variables noise (noise/signal ratio=0.7 by standard deviation). This figure appeared originally in Young et al. (2001).

Example 11.2: Modelling Percentage unemployment in the USA 1962: 1987

This example is concerned with the analysis of the data concerning the USA economy for the 25 years from 1962-1987, as shown in Figure 11.4. This is part of a larger data set analysed by Young and Pedregal (1999a) but the present analysis concentrates more fully on how SDP estimation helps to identify the data transformations used in this previous study.

At first sight, there appear to be some tentative relationships between the variables plotted in Figure 11.4: not unexpectedly, for instance, the upward movements of unemployment $y(k)$ appear to be linked with similarly timed downward fluctuations in private capital investment $u_2(k)$. But it is not at all that clear how unemployment is being affected by either government capital investment $u_1(k)$ or the *Gross Domestic Product* (GDP) $g(k)$. However, because the variables have pronounced long term trends, it seems worthwhile to remove these trends using one of the unobserved component (UC) models discussed in Chapter 5.

For a quick UC analysis of this type, intended simply to evaluate whether it provides a reasonable approach to modelling, the IRWSMOOTH algorithm discussed in section 4.5.4 provides a useful starting point. Here, a 'reasonable' trend, judged by eye, can be extracted very quickly using either a manually selected NVR value

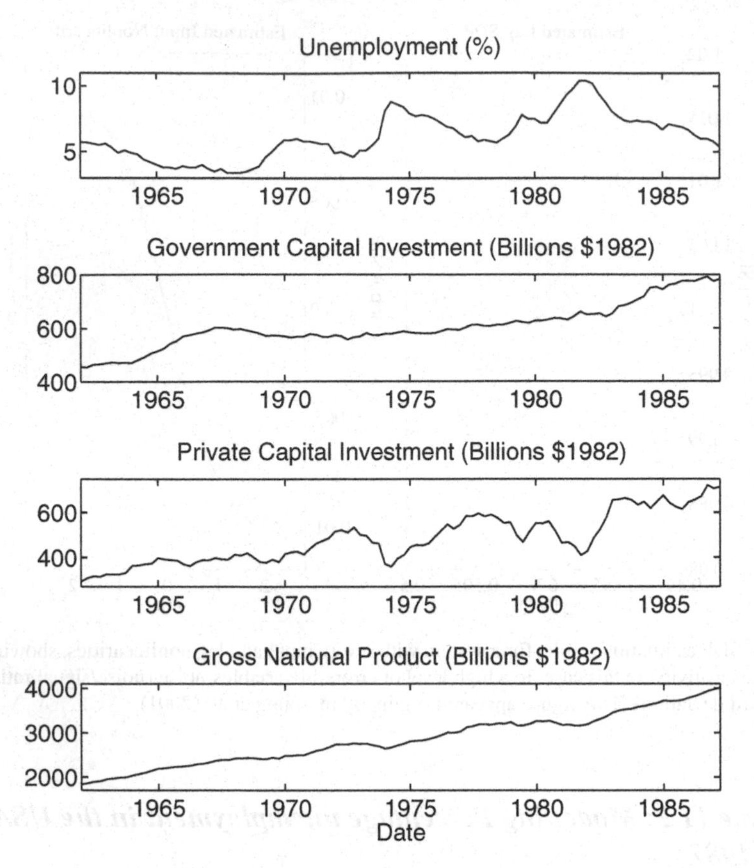

Fig. 11.4 USA unemployment example: unemployment and other major USA economic indicators 1962-1987.

or one selected by optimization using the irwsmopt routine in CAPTAIN. Here, it is best to employ optimization based on the f-step ahead prediction errors, rather than maximum likelihood, since this yields the smoother trend required in this initial analysis. In this case, $f = 6$, which makes sense because it is optimizing predictions 1.5 years ahead, produces NVR=0.00001, which is what one might well select by eye. This removes these trends very effectively to yield the detrended variables $\Delta y(k)$, $\Delta u_1(k)$ and $\Delta u_2(k)$. Then, SRIV identification and estimation applied to linear TF models, using the rivbjid and rivbj routines in CAPTAIN, with $\Delta y(k)$ as the output and the other two variables as inputs, shows that a [1 1 1 0 0] MISO model is well identified with $R_T^2 = 0.88$, while an alternative [1 1 2 0 0] model has an $R_T^2 = 0.91$.

This initial analysis produces promising results and it could be followed by a more rigorous analysis using the DHR model to estimate and remove the trend for each variable. However, the problem with such linear models is that they only explain the deviations about the long term trends. So it is reasonable to ask if there may be some underlying *nonlinear* relationship that will explain both the short and long term behaviour in a single model of similar form to the linear model, i.e. a first order MISO TF model with state-dependent parameters? One possibility is that the model parameters may be dependent on the GNP, which one would assume should have some part to play in the explanation of the unemployment changes.

In other words, let us consider a model of the form:

$$y(k) = \frac{b_{10}\{g(k)\}}{1+a\{g(k)\}(z^{-1})}u_1(k) + \frac{b_{20}(g(k))}{1+a\{g(k)\}(z^{-1})}u_2(k) + \xi(k) \quad (11.12)$$

where $y(k)$, $u_1(k)$ and $u_2(k)$ are the original measured variables (i.e not detrended) and in which all the parameters are potentially functions of the GNP, $g(k)$. In order to consider this relationship, where the variables have very different magnitudes and trends, it is convenient and computationally advantageous to standardize (or normalize) the variables by subtracting their mean value and dividing by the standard deviation about the mean value, using the stand routine in CAPTAIN, which conveniently returns the means and standard deviations for later use in de-standardization. Application of the sdp routine in CAPTAIN to these standardized data suggests that the $a\{g(k)\}$ parameter is constant and not dependent on $g(k)$ but that the two input parameters are strongly dependent on the *inverse* of $g(k)$, as shown in Figure 11.5. Here, the non-parametric SDP estimates are shown as a full lines and the dash-dot lines are the following inverse relationships $g_i(k)$, as estimated by simple linear least squares.

$$b_{10}(k) = 146.9\{g_i(k) - \bar{g}_i(k)\}; \quad b_{20}(k) = 180.2\{g_i(k) - \bar{g}_i(k)\} \quad (11.13)$$

where $g_i(k) = 1/g(k)$ and the over-bar indicates the mean value, which is removed because the other variables are standardized.

Having identified a SDP model form, the final stage in the SDP modelling is to estimate the following model in its parameterized form, recognising that the parameterized relationships in (11.13) are only being used to identify the *form* and *location* of the SDP nonlinearities, which will be re-estimated in the final model estimation, which is considered later:

$$y(k) = \frac{b_{10}}{1+a_1z^{-1}}\left\{\frac{u_1(k)}{g(k)}\right\} + \frac{b_{20}}{1+a_1z^{-1}}\left\{\frac{u_2(k)}{g(k)}\right\} + \xi(k) \quad (11.14)$$

Here, the $g_i(k)$ has been associated with the input variables recognising that the SDP relationships have identified the need for transformed inputs based on the ratios $u_1(k)/g(k)$ and $u_2(k)/g(k)$, respectively. But these nonlinearities are not abstract transforms; they have an immediate macro-economic meaning: the resulting inputs measure the fraction of economic output that is being devoted to government and private capital investment, respectively. In other words, SDP estimation has shown that it is possible to obtain a simple linear dynamic relationship between the three variables by considering the *Relative Government Investment*

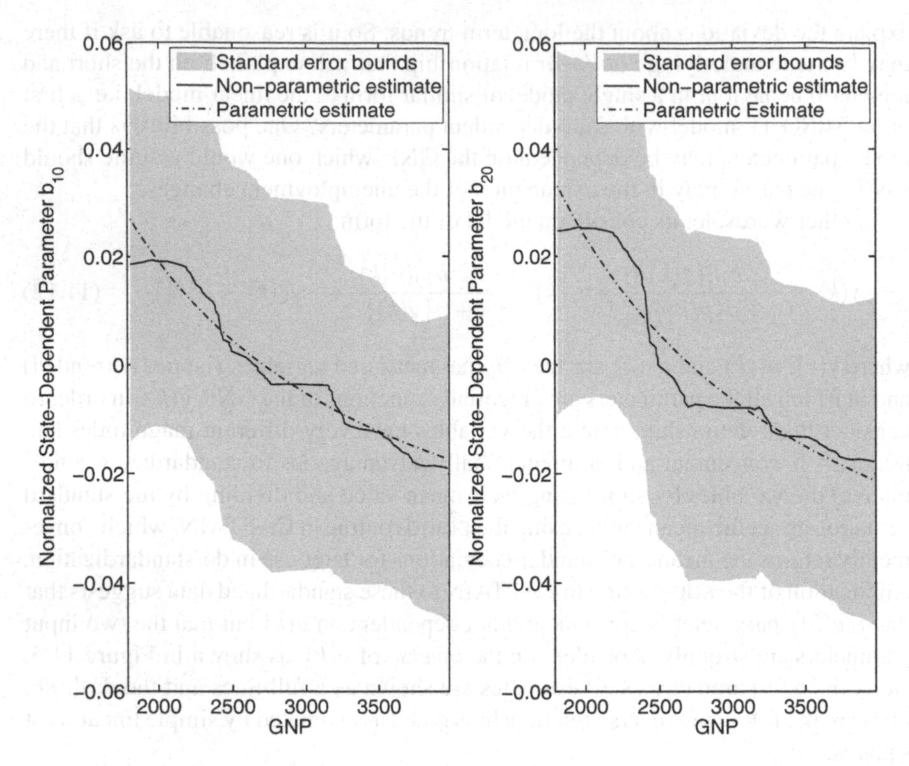

Fig. 11.5 USA unemployment example: standardized non-parametric and parametric state-dependent parameter estimates.

($\mathrm{RGI}(k) = u_1(k)/g(k)$) and *Relative Private Investment* ($\mathrm{RPI}(k) = u_2(k)/g(k)$) as inputs, rather than the measured investment levels $u_1(k)$ and $u_2(k)$. And the use of these relative investment measures not only makes great sense in macro-economic terms, it removes the trends that are present in $u_1(k)$ and $u_2(k)$, so removing the need for the prior trend removal procedure used in the initial linear modelling analysis mentioned previously.

Conveniently, with the introduction of $\mathrm{RGI}(k)$ and $\mathrm{RPI}(k)$ as input variables, the resulting linear dynamic model can be estimated using the RIV algorithm for MISO estimation in either its common denominator or different denominator form. This is carried out using the standardized variables, as obtained by application of the stand routine in CAPTAIN that subtracts the mean value and divides by the standard deviation. The resulting model parameter estimates and estimated standard errors are then modified to allow for this, in order to yield a model that relates the original measured variables. In the common denominator case, this latter model takes the form of equation (11.14), with the model for the additive noise $\xi(k)$ identified as an ARMA(2,1) process. The rivbj estimated parameters and standard errors (in parentheses) are as follows:

$$\hat{a}_1 = -0.739(0.04); \; \hat{b}_{10} = -17.0(3.16); \; \hat{b}_{20} = -32.0(3.14);$$
$$\hat{c}_1 = -1.863(0.05); \;\; \hat{c}_2 = 0.906(0.05); \;\; \hat{d}_1 = -0.975(0.01);$$
$$\hat{\sigma}_\xi^2 = 0.084; \; \sigma^2 = 0.021$$

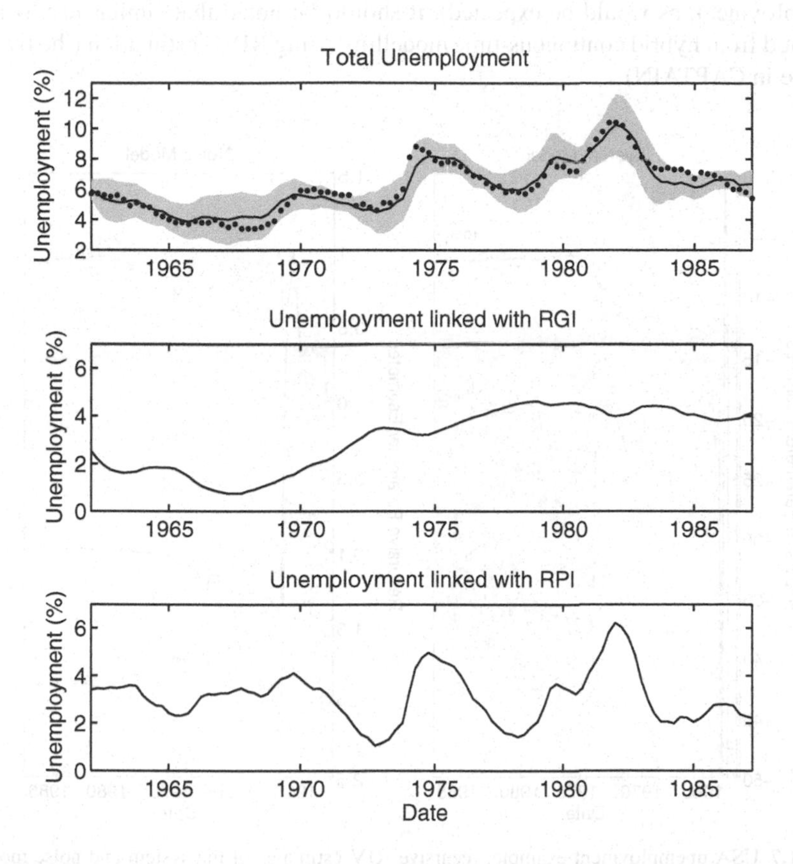

Fig. 11.6 USA unemployment example: RIV estimated MISO model output and 95% confidence bounds (grey) compared with the measured percentage unemployment (dots: top panel); percentage unemployment arising from the RGI input (middle panel); percentage unemployment arising from the RPI input (bottom panel).

This model explains the data well, as shown in Figure 11.6, with $R_T^2 = 0.92$ and $R^2 = 0.98$. Here the 95% uncertainty bounds, shown grey in the top panel, are obtained by MCS analysis using 1000 realizations (see section B.3.9, Appendix A) based on the full, stochastic, system plus noise model and the estimated covariance matrices. These uncertainty bounds are relatively large but nicely encompass all the measured data, as expected. This is reflected in the recursive estimates of the parameters, as shown in Figure 11.7, where the convergence is quite slow, with high volatility before 1959. Also, while the denominator parameter a_1 is well defined after this, the two numerator parameter estimates are still changing to some degree at the end of the data, with the 95% confidence bounds remaining very wide. Thus there remains some uncertainty about the effects of the relative investment inputs, although both are clearly required to explain the unemployment changes and the both contribute negatively to unemployment (i.e. any rise causes reductions in unemployment, as would be expected). It should be noted that similar results are obtained from hybrid continuous-time modelling using RIVC estimation (the rivcbj routine in CAPTAIN).

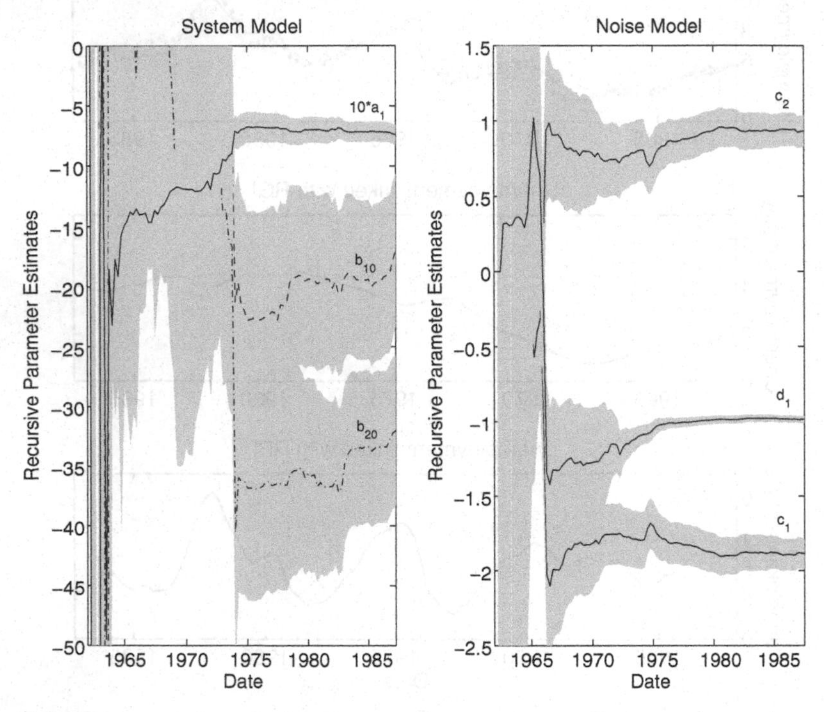

Fig. 11.7 USA unemployment example: recursive RIV estimates of the system and noise model parameters.

Despite these uncertainties which are, of course, an inherent part of stochastic modelling, the above MISO model can be interpreted quite well in economic terms, which increases our confidence in the model. For example, the steady state gains if the two TF components, which would be termed 'long-term multipliers' by economists, are -65.2 and -122.8, respectively, so that a rise in private capital investment has almost double the effect on reducing unemployment as government investment. And this remains the case even if the uncertainty is taken into account. Also, the time constant associated with each TF is 0.83 years, which makes reasonable economic sense. Other economically meaningful aspects of the model are discussed in Young and Pedregal (1999a), which considers models estimated on data from 1948 to 1988 and validated over the period 1988-1998. This concentration of the physical aspects of the model is an example of *Data-Based Mechanistic* (DBM) modelling, as discussed in the next chapter 12, where the estimated model is only deemed credible if it makes reasonable physical (here economic) sense.

Finally, when different denominator polynomial MISO models are estimated using either the pem algorithm in the Matlab SID Toolbox or the rivbjdd algorithm in CAPTAIN (see earlier comments in section 7.6 of chapter 7 and Appendix G), they do not explain the data as well as the MISO model considered above, with $R_T^2 = 0.895$. This is quite unusual given that this model form has one additional parameter, but it demonstrates the virtue of evaluating both the common and different denominator forms of the MISO TF model.

11.3 SDP Identification of Purely Stochastic Nonlinear Systems

A special example of the SDP model (11.1) is the following *State Dependent parameter Auto-Regressive* (SDAR) model:

$$y(k) = \mathbf{z}^T(k)\boldsymbol{\rho}(k) + e(k) \tag{11.15}$$

where

$$\begin{aligned} \mathbf{z}^T(k) &= [-y(k-1)\ \ -y(k-2)\ \ \cdots\ \ -y(k-n)] \\ \boldsymbol{\rho}(k) &= [a_1(\chi(k))\ a_2(\chi(k))\ \cdots\ a_n(\chi(k))]^T \end{aligned} \tag{11.16}$$

Clearly, the same SDARX estimation methods discussed above can be applied to this model. Typical simulation results for a well known nonlinear model are discussed in the following example and a real example is described after this.

Example 11.1: The Cosine Map model

The cosine map model (e.g. Zhan-Qian and Smith, 1998) takes the form

$$y(k) = \cos(2.8y(k-1)) + 0.3y(k-2) + e(k); \qquad e(k) = \mathrm{N}(0, 0.01) \tag{11.17}$$

A typical 2000 sample simulation of the model is presented in Figure 11.8, which shows the time response in the upper panel and the phase plane $\{y(k) \sim y(k-1)\}$ plot in the lower panel. In the latter graph, the noise free response is shown as a full thick line with the noisy response plotted as dots.

This a typical stochastic model that exhibits underlying chaotic response characteristics. It provides a testing example for the SDP approach, however, because it is not in the assumed affine form (Co and Ungarala, 1997): in particular, the SDP term $a_1\{y(k-1)\}.y(k-1)$ in the most appropriate SDAR model,

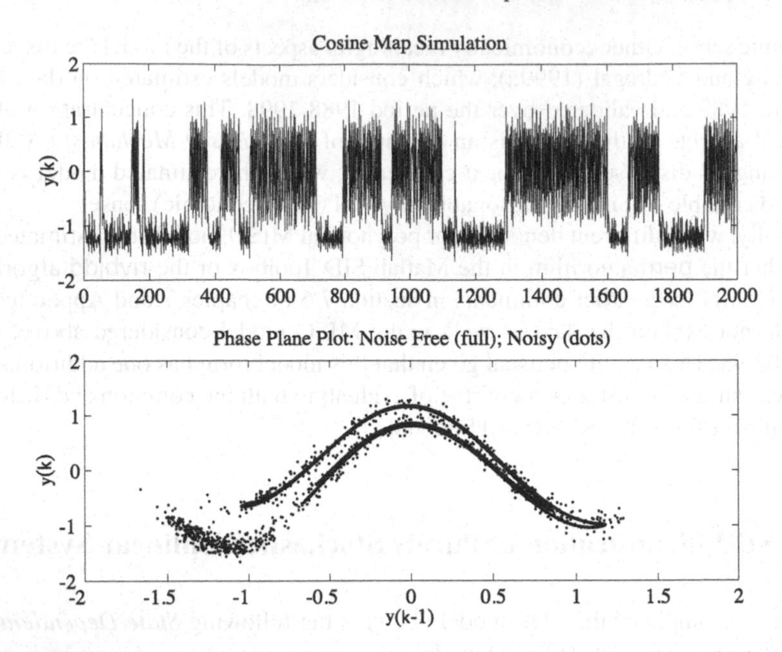

Fig. 11.8 Simulated cosine map model. Upper panel: measured output. Lower panel: phase plane (embedding) plot of $y(k)$ vs $y(k-1)$ (noise-free response shown as a full line). This figure appeared originally in Young (2000).

$$
\begin{aligned}
y(k) &= a_1\{y(k-1)\}.y(k-1) + a_2\{y(k-2)\}.y(k-2) + e(k) \\
e(k) &= \mathcal{N}(0, \sigma^2)
\end{aligned}
\tag{11.18}
$$

is not able to represent the equivalent term $\cos(2.8y(k-1))$ in (11.17) exactly, since $\cos(2.8y(k-1))/y(k-1)$ has a singularity at $y(k-1) = 0$. Despite this difficulty, SDAR estimation yields excellent results, as illustrated in Figures 11.9 and 11.10. These results were obtained using the sdp routine in CAPTAIN which yields an ML optimized NVR matrix (see section 4.2) with the diagonal values $\mathbf{Q}_{nvr} = \text{diag}[0.0057\ 1.28 \times 10^{-7}]$ and it is clear that the optimization has successfully identified that the potential state dependency resides in the first lag parameter $a_1(k) = a_1\{y(k-1)\}$, while the second lag parameter $a_2(k) = a_2\{y(k-2)\}$ is effectively time-invariant, with a very small optimized NVR value.

Figure 11.9 shows the FIS estimate of the cosine nonlinearity subsequent to the convergence of the back-fitting procedure, which took 6 iterations in this case. Except for the region around the singularity at $y(k-1) = 0$ (see the dip in the centre of the plot), the estimation is very good. The associated $\hat{a}_2(k|N) = 0.291(0.0038)$, for all k, is estimated as being time-invariant despite the fact that the NVR, $\mathbf{Q}_{nvr}(2,2) = 1.28 \times 10^{-7}$, is not too small in this case. Figure 11.10 compares the actual phase plane plot for the data used in the estimation (left hand panel) with a similar plot based on data from a typical random realization of the SDAR model (right hand

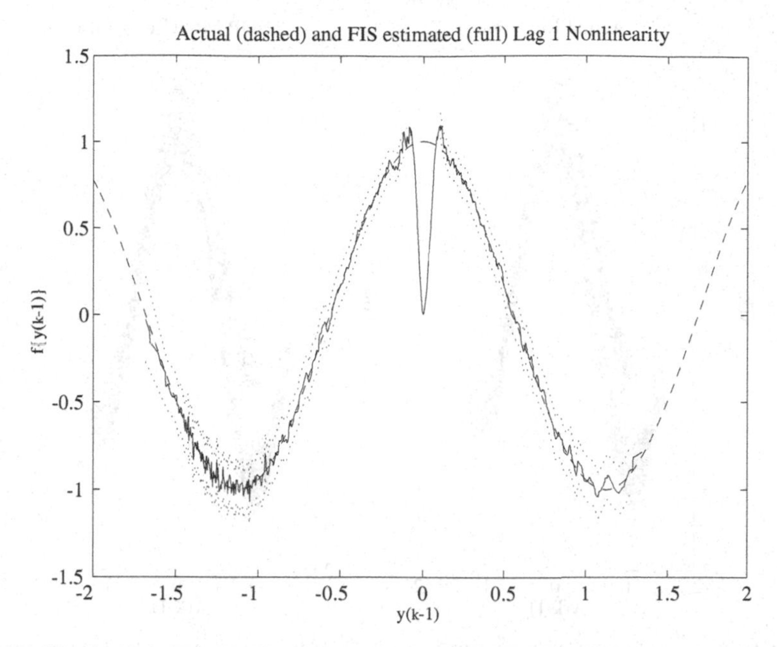

Fig. 11.9 Simulated cosine map model. Comparison of FIS estimated and actual cosine nonlinearity. This figure appeared originally in Young (2000).

panel). Similar agreement is found in both the time plots of the two series and the histograms.

Finally, on the basis of the above SDP results, the form of the nonlinear equation is identified correctly as:

$$y(k) = \alpha \cos(\beta y(k-1) + \gamma y(k-2) + e(k), \qquad (11.19)$$

based on the obvious cosine form of the SDP estimate in Figure 11.9; and the optimized ML estimates of the, now constant, parameters α, β and γ in this model are $\hat{\alpha} = 0.998(0.004)$, $\hat{\beta} = 2.797(0.004)$ and $\hat{\gamma} = 0.303(0.003)$.

Example 11.3: Analysis of squid data

This example is based on the analysis of the signal shown in Figure 11.11, which was obtained by Kazu Aihara and Gen Matsumoto[3] from experiments on the giant axon of a squid (Mees et al., 1992). The signal comprises voltage measurements made from a micro-pipette inserted into a giant axon. Squid are used for such experiments because they have large diameter axons: this is because the nerves are not

[3] I am most grateful to them for permission to use their data

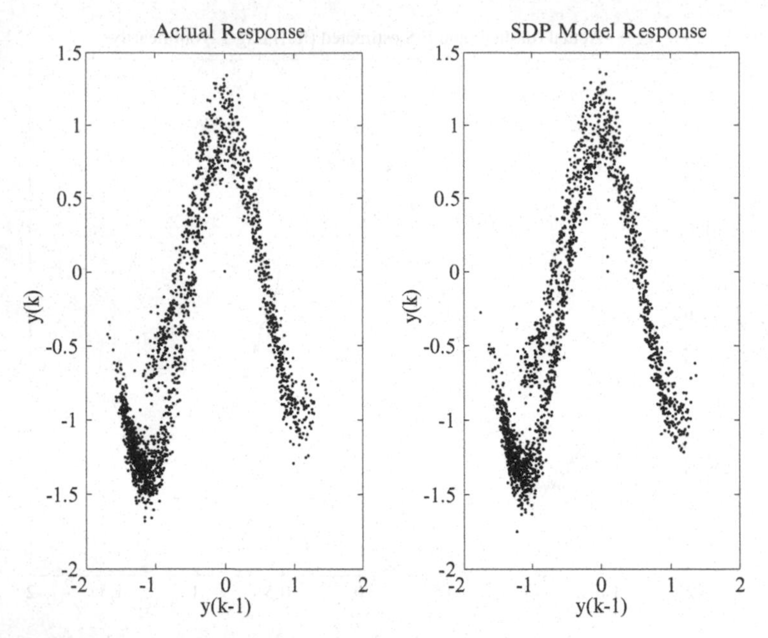

Fig. 11.10 Simulated cosine map model. Comparison of the phase-plane plot for the data used in the estimation (left panel) with the phase-plane plot obtained from a random realization of the estimated SDAR model. This figure appeared originally in Young (2000).

myelinated (insulated), so they need to have large diameter to reduce ion leakage and so maintain the transmission speed. The experiment is done in vitro (i.e. the nerves are surgically removed from the squid), with the membrane voltage clamped. This is normally referred to as the 'forced' response. In effect, therefore, Figure 11.11 shows a putative chaotic response to a periodic signal. It is analysed here, however, as a purely stochastic, unforced dynamic process. Only the first order SDAR model results will be discussed here: additional results based on the estimation of a second order SDAR model are presented in Young (2001d).

The simplest SDAR model that could produce the dynamics shown in Figure 11.11 is the following 1st order SDAR,

$$y(k) = a_1\{y(k-1)\}.y(k-1) + e(k) \tag{11.20}$$

The FIS estimate of the SDP obtained by application of the sdp routine is shown in Figure 11.12 and the associated nonlinear function in Figure 11.13. These results were obtained with the single NVR hyper-parameter optimized at $\text{NVR}(a_1) = 5.01 \times 10^{-6}$, under the assumption of an RW model for the SDP parameter variation (an IRW assumption provides a slightly more smoothed estimate but does not make any significant difference). The standard error band is shown as the dashed lines on both plots. This model explains 92.2% of the squid data (i.e. the Coefficient of Determination based on the model residuals $R^2 = 0.922$).

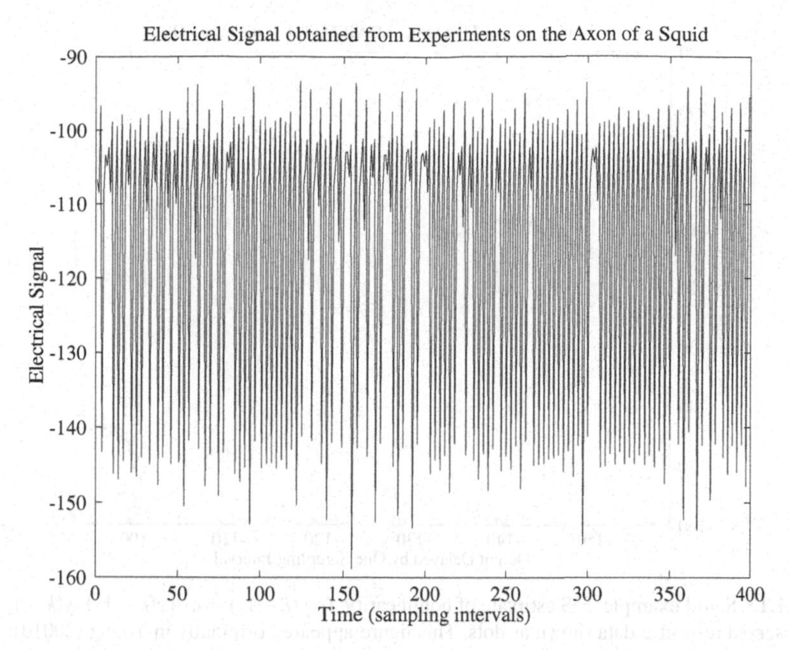

Fig. 11.11 Squid example: signal obtained from experiments on the giant axon. This figure appeared originally in Young (2001d).

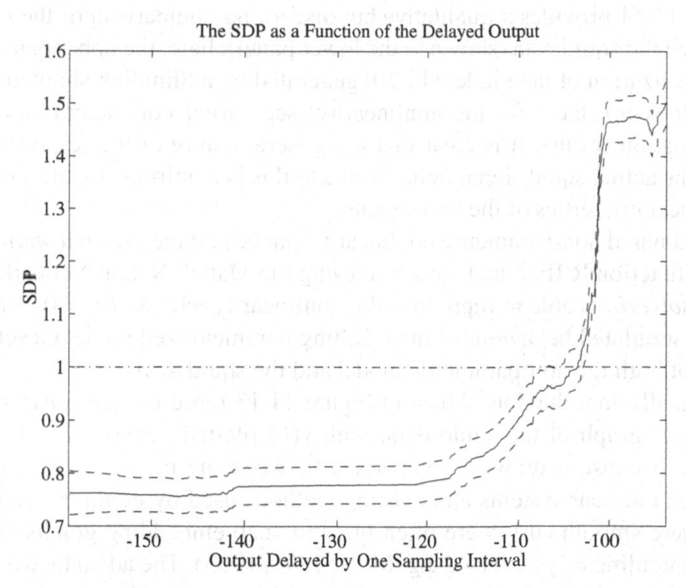

Fig. 11.12 Squid example: FIS estimate of the state dependent parameter $a_1\{y(k-1)\}$. This figure appeared originally in Young (2001d).

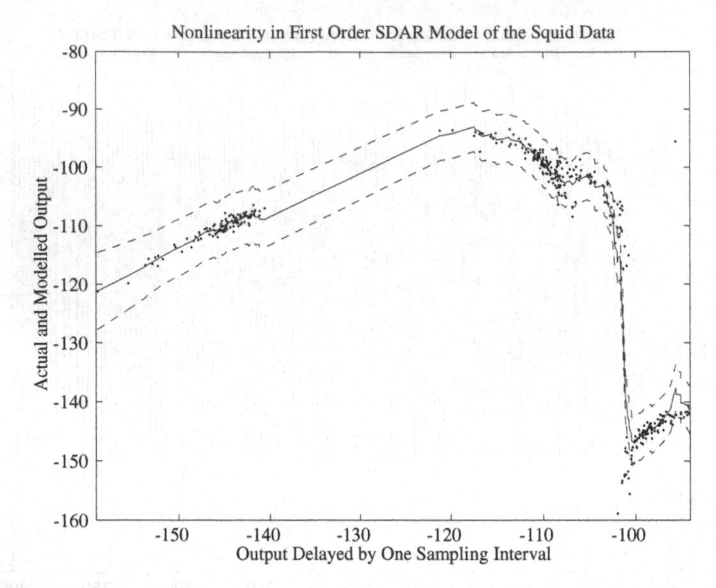

Fig. 11.13 Squid example: FIS estimate of nonlinearity $f_1(y(k-1)) = a_1\{y(k-1)\}.y(k-1)$ with the observed response data shown as dots. This figure appeared originally in Young (2001d).

Figure 11.14 provides a qualitative but discerning comparison of the model behaviour and the squid data (shown in the lower panel): here, the upper panel shows a random realization of the model (11.20) generated by a Simulink simulation model using a 'look-up' table for the nonlinearity (see earlier comments) based on the SDP estimation results. It is clear that the general nature of the response visually matches the actual squid signal rather well and this is confirmed by the similarity in the statistical properties of the two signals.

The estimated non-parametric nonlinearity can be parameterized in various ways. Here, a 9 function RBF model, designed using the Matlab Neural Network Toolbox function *solverb*, is able to reproduce the nonlinearity with an $R^2 = 0.997$, and the Simulink simulated behaviour of the resulting parameterized model closely resembles that of both the non-parametric model and the squid data.

Note finally that the dots shown in Figure 11.13 represent the 'phase-plane' or 'embedding' graph of the squid data, with $y(k)$ plotted versus $y(k-1)$. This is interesting because it draws a comparison between the present SDP approach to modelling nonlinear systems and existing methods used by nonlinear systems theorists, where smooth curves are often fitted to such embedding graphs in order to model the nonlinear system (see e.g. Mees, 1991, 1993). The advantage of the SDP approach is that the smoothing is optimal in a ML sense and is carried out within a stochastic, dynamic systems setting. Also, the smoothing is applied to the estimation of the SDP parameter, rather than the nonlinear function as a whole, and so it provides a more flexible and informative result.

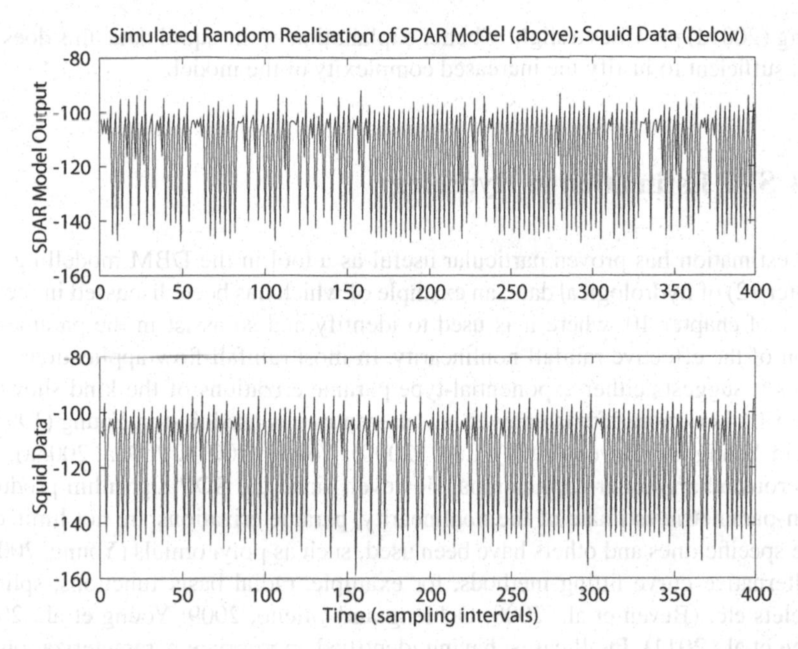

Fig. 11.14 Squid example: qualitative comparison of a random realization of the SDAR model behaviour (upper plot) and the squid data (lower panel). This figure appeared originally in Young (2001d).

For instance, the SDP in the model (11.20) can be considered, in an approximate sense, as the changing eigenvalue of a first order, discrete-time dynamic system (we might refer to this as a *virtual-eigenvalue*) and this provides some additional insight into the nature of the system. By reference to Figure 11.12, for example, we see that, for $y(k)$ less than about -120, this eigenvalue is approximately constant at a value of about 0.75, suggesting that the underlying behaviour over this part of the state space is quite stable, with a (virtual) time constant of approximately 3.5 sampling intervals. Between values of about -120 and -105, however, the eigenvalue increases steadily to unity, where the system is at a point of neutral stability (i.e. the system is acting transiently as an integrator). Thereafter, for values greater than -105, the eigenvalue rises sharply to around 1.5 and the underlying system is clearly exponentially unstable. However, if we now consider the temporal variation of the SDP, it is clear that, every time that the SDP exceeds unity, the instability of the system drives it immediately back to a location where the SDP is in the region where the eigenvalue is about 0.75 again and the system is stabilized. This is also well illustrated by a stacked plot of the changing impulse response associated with the AR model defined by the SDAR model at each instant of time instant.

So, to conclude, it would appear that the first order SDAR model provides a reasonable mechanism for characterising the behaviour of the electrical activity in the axon of the squid and that, although the second order SDAR model obtained in

Young (2001d) provides a slightly better explanation of the squid data, this does not seem sufficient to justify the increased complexity of the model.

11.4 SDP Estimation in Hydrology

SDP estimation has proven particular useful as a tool in the DBM modelling (see chapter 12) of hydrological data, an example of which has been discussed in section 10.5.1 of chapter 10, where it is used to identify and so assist in the parameterization of the effective rainfall nonlinearity. In most rainfall-flow applications such as this it suggests either exponential-type parameterizations of the kind shown in Figure 10.4 or, more often, a power-law alternative, as used first in Young (1993b), then in Young and Beven (1994); Lees (2000); Young (2001b, 2002a, 2003a), and numerous other similar applications. However, since the SDP algorithm produces a non-parametric estimate of the nonlinearity, parameterizations are not limited to these specific ones and others have been used, such as polynomials (Young, 2003a) or alternative curve fitting methods, for example, radial basis functions, splines, wavelets etc. (Beven et al., 2008; Ochieng and Otieno, 2009; Young et al., 2009; Beven et al., 2011). In all cases, having identified appropriate parameterization for the effective rainfall nonlinearity, the whole model including this input nonlinearity and the linear TF model, are optimized in the manner described earlier in section 11.2.2.

More recently, more complicated hydrological examples of SDP estimation have been reported: for example, the possibility of hysteresis effects in the case where the rainfall-flow dynamics are affected by snow-melt (Young et al., 2007); and more complex rainfall-flow dynamics that have been identified in rapidly sampled (10 min.) rainfall-flow data collected in the small Pontbren experimental catchment in Powys, mid-Wales, UK (McIntyre et al., 2011). This latter example is particularly interesting since it is the first time that the SDP algorithm has detected significant internal nonlinearity in a TF model of rainfall-flow dynamics: i.e. where the estimates of the TF denominator parameters are found to be state dependent.

11.5 SDP Estimation in Forecasting and Automatic Control

It is clearly possible to develop state estimation and control system design methods based on the new class of SDP nonlinear models discussed in this chapter. These can be considered in parametric or non-parametric form (the latter providing an alternative way of considering estimation, forecasting and control system design).

The combined KF and FIS algorithms, when used in a more conventional state estimation setting, can function with system matrices characterized by time, or even state dependent parameters. This extends their range of applicability to a considerable extent. Thus, the SDP models described here can provide the basis for rather novel, non-parametric or state dependent parameter, KF–FIS design, with implications for both modelling and nonlinear optimization based on prediction error decomposition. For example, the blowfly population modelling example in Young (2000) has shown how a SDP implementation of the KF can be used in the final stage of nonlinear model parameter optimization based on prediction error decomposition, as well as for mult-step-ahead forecasting. This blowfly example is a very interesting one and the reader might like to consult this reference. The data, collected laboriously by the Australian scientist A. J. Nicholson (1954), have received attention previously in the ecological literature (Gurney et al., 1980). Without any prior assumptions, the SDP analysis identifies the nature and location of the nonlinearity; and the resulting estimated model agrees well with the model, based on ecological concepts, that Gurney et al. fitted to the Nicholson data.

In the case of automatic control system design, Young (1996) adumbrated a SDP approach to nonlinear control system design by showing how a SDP version of the *Proportional-Integral-Plus* (PIP) controller (see earlier, chapter 9, and Appendix G) could be used to successfully control the chaotic forced logistic growth equation. Further research in this area has followed in recent years: see e.g. Taylor et al. (2000, 2008). SDP estimation has been applied to a variety of other nonlinear systems within an automatic control context, including a well known example (Sjöberg et al., 1995) of a nonlinear hydraulic servo with assymetric limiting and deadzone nonlinearities. For instance, Young (2001a) has shown how SDP identification and parameter estimation can produce a considerably reduced complexity model when compared with the Quasi-ARMAX nonlinear model obtained by Hu et al. (2001).

11.6 Exercises

1. Investigate other approaches to nonlinear modelling, such as the following models described in Granger and Teräsvirta (1993): the family of *Smooth Transition Regressive* (STR) and *Smooth Transition AutoRegressive* (STAR); *Nonlinear AutoRegressive* (NLAR); *Threshold AutoRegressive* (TAR); bilinear; *SiGn Nonlinear* (SGN) autoregressive; and *Rational Nonlinear AutoRegressive* (NAR). Show how these can be considered in the wider context of SDP modelling.
2. As in 1., read the paper by Chen and Billings (1989) and show how NARMAX and other models mentioned in the paper can be considered in SDP terms.
3. Read the interesting book by Priestley (1988) and compare the approach to SDP estimation described there with the SDP estimation methods discussed in the present chapter (the related chapter by Young (1993b) may help understanding).
4. Run the SDP simulation example in CAPTAIN (see instructions for running CAPTAIN demonstration examples in Appendix G).

11.7 Summary and Conclusions

In the last twenty years the advent of fast computers and the desire of theorists to extend the boundaries of time series analysis has led to an explosion of research on Monte Carlo-based numerical methods, from either classical (e.g. Durbin and Koopman, 2001) or Bayesian (e.g. Ruanaidh and Fitzgerald, 1996; Gamerman, 1997) perspectives. One motivation of this more recent research is clearly to extend the 'Gaussian' methods of standard recursive estimation to non-Gaussian and nonlinear time series, using models in which the *stochastic inputs* are non-Gaussian. But models of non-Gaussian and nonlinear processes do not necessarily require the assumption of non-Gaussian inputs. As this chapter has shown, a fairly wide class of non-Gaussian and nonlinear time series can be represented by SDP models with Gaussian inputs. When it is possible (and the methods do seem quite widely applicable), this is clearly advantageous, since it allows for the use of well tried and robust algorithms that are computationally much less demanding than even the 'classical' non-Gaussian methods (see Durbin and Koopman, 2001).

It is clear that the simulated and real examples presented in the present chapter, combined with those discussed in the other cited references, demonstrate the efficacy of the proposed SDP approach to modelling for a fairly wide and practically useful class of nonlinear stochastic systems. However, the proposed technique is relatively new and it raises a variety of interesting theoretical questions and possibilities for extending the approach to an even richer class of nonlinear stochastic systems. For example:

1. How can the approach be extended to handle multi-state dependencies, where the SDPs may be functions of several state variables? One approach currently being investigated is to model the SDPs with neuro-fuzzy functions of several variables (e.g. Hu *et al*, 2001) but with the structure of the nonlinear system identified first from the data using the methodology described in the present chapter. Another is to develop a multi-state SDP version of the current SDP algorithm. This extension is not straightforward because the techniques used in the current SDP algorithm do not translate easily to the situation where the SDP is a function of several states. However, very recent research has developed a particular approach to this problem which has produced promising results (Sadeghi et al., 2010).
2. What is the best method of handling the estimation bias that occurs when the proposed SDP modelling approach is applied in situations where *errors-in-variables* problems occur, particularly those that arise when the transfer function form of the SDP model is used (rather than those arising from input noise effects)? An IV approach could obviate such problems in the case of well-behaved nonlinear input-output models. But alternative approaches are required in the case of purely stochastic nonlinear systems without inputs (which, as we have seen in this book, are normally required in IV estimation to generate the IVs) and/or nonlinear systems that exhibit sensitive, chaotic behaviour.
3. Although no convergence problems have been encountered so far in the evaluation of the proposed SDP estimation procedure, what conditions are required

for convergence of the back-fitting procedure? Hastie and Tibshirani (1996) use a similar back-fitting procedure for the estimation of their GAM (see previously, section 11.2.1). It needs to be established whether their conclusions as regards convergence (which are not entirely persuasive, in any case) are applicable to the models and back-fitting procedure described in this chapter. Unlike the GAM, for instance, the nonlinear functions in the SDP models are factorized into the product of the SDP and the model variable (although this is not essential); and the SDP is estimated by optimal FIS smoothing (rather than the more conventional scatter-plot smoothing used by Hastie and Tibshirani).

4. The back-fitting procedure does not provide complete covariance information on the SDP estimates. Could this be distorting the standard errors on the estimates? In more general terms, what are the full *theoretical* statistical properties of the SDP estimates obtained by back-fitting?
5. Finally, what are the identifiability conditions on the SDP models? It is clear that problems analogous to collinearity in constant parameter model estimation can occur and that back-fitting convergence may be affected by such problems. Also, in the case of input-output models, the nature of the input signals will affect the identifiability of the model parameters. It is necessary to explore these factors further and establish what other factors may affect the identifiability of the model.

Regardless of the answers to these questions, however, the SDP approach to the identification of nonlinearities in stochastic systems appears to hold great promise and is an excellent example of how recursive estimation can be exploited in non-standard contexts. In contrast to other approaches, such as neural networks and NARMAX models (see e.g. Chen and Billings, 1989), SDP estimation attempts to identify the type of nonlinearity and, therefore, the form of the nonlinear model, *prior* to the estimation of the parameters in the finally identified model. This helps to ensure that the final nonlinear model is efficiently parameterized (parsimonious) and it should avoid the over-parameterization that normally accompanies neural network and, to a lesser extent, the 'black-box' NARMAX models. Indeed, the SDP approach has been developed as a primary tool in *Data-Based Mechanistic* modelling, as discussed in the next chapter 12.

SDP estimation also provides a non-parametric model that can be useful in its own right. As we have seen, the SDP model can be simulated easily in programs such as Simulink, thus removing the need for the final parametric estimation in some applications, such as simulation, forecasting and automatic control.

Finally, it must be stressed that, although SDP approach to modelling nonlinear, stochastic, dynamic systems from time series data is a reasonably general approach that encapsulates other nonlinear model forms (see Exercises 1 and 2), it is not a completely general solution to the problem of modelling nonlinear stochastic systems: indeed, because of their infinite variety, there is no completely unified theory of nonlinear systems. So the SDP modelling should be added to other approaches that have been proposed previously and will be proposed in the future. This chapter has not attempted to review the existing alternative methods since this would require another book of at least the present size.

Chapter 12
Data-Based Mechanistic (DBM) Modelling

So far in this book, I have concentrated on the development, implementation and evaluation of various recursive estimation algorithms. But such estimation algorithms, no matter how elegant and potentially powerful, are not an end in themselves: they are only worth developing if they perform a practically useful function in the wider context of mathematical modelling. In this final chapter, therefore, these estimation algorithms are considered in a more philosophical manner and linked together in terms of their ability to service the requirements of a general approach to stochastic, dynamic modelling from time series data that I have called *Data-Based Mechanistic* (DBM) modelling.

12.1 Introduction

DBM modelling is a 'method theory' that has been developed over many years. Its name emphasizes my contention that, while the model should be inferred from the analysis of data in as objective a manner as possible, it should also have a clear, scientifically acceptable, mechanistic interpretation. In a very real sense, therefore, although DBM modelling exploits some of the methodology used in 'black-box' modelling, it is actually a reaction against the notion of pure black-box models. While such models provide an excellent vehicle for activities such as signal and image processing, or some aspects of statistical forecasting and automatic control, I believe they are not sufficient for scientific and engineering research that is attempting to investigate the nature of dynamic systems in the natural and man-made world. In other words, a model should not just explain the time series data well, it should also provide a mechanistic description of the system under investigation; a description that further enhances our confidence in its ability to approximate reality in a meaningful manner.

In his excellent book *The Road to Reality*, Roger Penrose expresses my views on modelling very well. When discussing the mathematical nature of the quantum particle, he says:

> Are the states 'real'? .... quantum physicists tend not to be clear about this issue. Most of them are distinctly uncomfortable about addressing this issue of 'reality' at all ... All that we should ask of our formalism, they might claim, would be to give answers to appropriate questions that we may pose of a system, and that those answers agree with observational fact ...My own viewpoint is that the question of 'reality' *must* be addressed .... We need a notion of reality, *even if only a provisional or approximate one*[1], for without it our objective universe, and thence the whole of science, simply evaporates before our contemplative gaze!

Of course, Penrose is addressing fundamental questions of science and I am not suggesting that his comments are set within the more modest context of the present book, where I am addressing more down-to-earth concepts, such as how we can model real systems from observational data, such as those considered in various chapters in this book: e.g. the transport and dispersion of solutes in a water body (chapters 6, 7, 8 and later in the present chapter); the evaluation of a Global Circulation Model in chapter 8; the modelling and forecasting of river flow in chapter 10 and the characterization of both biological and macro-economic behaviour in chapter 11. But I believe the philosophical basis for his comments still applies: 'agreement with observational fact' is not enough in science, 'we need a notion of reality' that gives us confidence that our model is at least describing an approximation of what is happening in the real world.

The above questions about the nature of mathematical modelling are also influenced by the way in which scientific modelling is carried out. During the 20th Century, scientific philosophers such as Karl Popper (1959) and Thomas Kuhn (1962), looked at the philosophy of science in the context of the scientific research that they observed at the time; research that was dominated by enormous advances that occurred in physics during the latter part of the 19th Century and the first half of the 20th. Popper, in particular, was a proponent of what he termed the *hypothetico–deductive* method. Here, the model, or theory in more general terms, forms a hypothesis that is tested against data, usually obtained from experiments, often within a laboratory where experiments could be carefully planned. And, in Popper's view, the aim is not to 'prove' the hypothesis, but rather to attempt its 'falsification' and consider it to be 'conditionally valid' until falsified.

Whether Kuhn subscribed to the hypothetico-deductive concept is not clear. Rather, he viewed science from a 'paradigmatic' standpoint in which most 'ordinary science' worked within and embroidered defined paradigms; while the more fundamental achievements of science were those that questioned or even overturned these current paradigms (as Einstein's theories of relativity radically changed the Newtonian view of the World). In this regard, the hypothetico-deductive approach to scientific research used by ordinary scientists often tends to be too constrained by current paradigms: hypotheses are made within the current paradigm and do not often seek to question it.

One advantage of the hypothetico-deductive model is that it naturally relates to the 'physical reality', at least as it is perceived by the scientist. On the other hand, it often leads to the synthesis of large computer models that are over-parameterized and so not identifiable from the available data. Typical examples range from large

[1] My added emphasis.

hydrological and climatic models, such as the vast global circulation models and the smaller, but still very large, global carbon cycle models (see the discussion in Young et al., 1996; Parkinson and Young, 1998; Young and Parkinson, 2002), as used in climate change research; through some econometric models (e.g. Young and Ratto, 2011), to computer models of industrial processes (such as the large power system models of Åström and Bell, 2000, and Lu et al., 2011). In such examples, particularly those concerned with 'natural' systems, such as the environment, climate and the economy, the model contains a surplus content that often owes more to the perception of the scientist than to its evaluation against observational data.

By contrast, the methodology that we have considered in this book is *inductive*: the model is identified and estimated from the data in a minimally parameterized or parsimonious form. Such an inductive approach has a rich history in science and was, indeed, the normal approach to scientific research used during the era of the 'natural philosopher', the term used to describe the scientist prior to the twentieth Century (see Young, 2011). In the present world of science, however, inductive modelling is now more the domain of the statistician and time series analyst, as well as those engineers like myself who prefer to use efficiently parameterized, data-based models whose physical interpretation is inferred from the data-based model, taking a balanced but questioning account of any existing physical interpretations and models.

I believe, that this dichotomy between modelling approaches is a rather unsatisfactory state of affairs and so, as we shall see in this chapter, I have endeavoured to evolve the DBM approach so that it attempts to build a bridge between the hypothetic-deductive and the inductive modeller. In this way, the best aspects of both approaches are melded together into a unified procedure that tries to reach the right balance between parsimony, identifiability and realistic descriptive ability.

12.2 A brief Review of Data-Based Mechanistic Modelling

Although the term 'data-based mechanistic modelling' that was first used in Young and Lees (1993), the basic concepts of this DBM approach to modelling dynamic systems have been developed over many years. For example, they were first applied seriously within a hydrological context in the early 1970s, with application to the modelling of water quality and flow in rivers (Young and Beck, 1974; Young, 1974) and set within a more general framework shortly thereafter. Since then, they have been applied to many different systems in diverse areas of application from ecology, through engineering to economics (see e.g. Young, 1998, 2006 and the prior references therein)..

In its latest form, the DBM modelling strategy consists of the following seven major stages (Young and Ratto, 2009):

1. The important first stage in any modelling exercise is to define the objectives and to consider the types of model that are most appropriate to meeting these objectives. Since the concept of DBM modelling requires adequate data if it is

to be completely successful, this stage also includes considerations of scale and the likely data availability at this scale, particularly as they relate to the defined modelling objectives.

2. In the initial phases of modelling, it may well be that real observational data will be scarce, so that any major modelling effort will have to be centred on simulation modelling, normally based on largely deterministic concepts, such as the conservation laws (mass, energy momentum etc.). In the DBM simulation modelling approach, which is basically Bayesian in concept, these deterministic simulation equations are converted to a stochastic form by assuming that the associated parameters and inputs are inherently uncertain and can only be characterized in some suitable stochastic form, such as a probability distribution function (pdf) for the parameters and a continuous or discrete time-series model for the inputs. The subsequent stochastic analysis uses Monte Carlo Simulation (MCS), to explore the propagation of uncertainty in the resulting stochastic model, and sensitivity analysis of the MCS results to identify the most important parameters which lead to a specified model behaviour: e.g. Parkinson and Young (1998).
3. The initial exploration of the simulation model in stochastic terms is aimed at revealing the relative importance of different parts of the model in explaining the dominant behavioural mechanisms. This understanding of the model is further enhanced by employing *Dominant Mode Analysis* (DMA). This approach to dynamic model order reduction is applied to time-series data obtained from planned experimentation, not on the system itself, but on an often large, high order, computer simulation model of the system that, in effect, becomes a surrogate for the real system. In particular, the optimal methods of refined instrumental variable estimation (RIV/RIVC) described in chapters 7, 8, where necessary aided by the state-dependent parameter (SDP) method of nonlinear model identification described in chapter 11, are applied to these experimental data and yield low order approximations to the high order simulation model that are almost always able to explain its dynamic response characteristics to a remarkably accurate degree (e.g. greater than 99.99% of the large model output variance explained by the reduced order model output).
4. A more complete understanding of the links between the high order simulation model and its reduced order representation obtained in stage 3 is obtained by performing multiple DMA analysis over a user-specified range of simulation model parameter values. The mapping between the large and reduced order model parameters or responses then yields a full *Dynamic Emulation (or 'meta') Model* (DEM) that can replace the simulation model over a wide range of parameter values. This approach to high order model emulation is introduced in Young and Ratto (2009), while Young and Ratto (2011) describe in detail two methods of emulation: namely, 'stand-alone parameter mapping', which is used in the present chapter, and 'response mapping', with application to the emulation of the Nash-Cascade hydrological model and a large economic model.
5. Once experimental time series data are available, an appropriate model structure and order is identified by a process of statistical inference applied directly to these real time-series data and based on a generic class of dynamic models: although

discrete-time models could be utilized, linear, stochastic models described by continuous-time transfer functions (i.e. lumped parameter differential equations) are used in the present chapter since these have advantages when used in modelling physical systems (see section 8.1 of chapter 8). If such time series data are available at the start of the study, then this analysis will constitute the first stage in DBM modelling. The identification and estimation procedures used in the DBM modelling are the same optimal RIV/RIVC methods used in dominant mode analysis (see above 3 and 4).

6. If emulation modelling has been carried out prior to the acquisition of data then, if at all possible, the DBM model obtained at the previous stage 5 should be reconciled with the dynamic emulation version of the simulation model considered in stage 4. Otherwise, if time series data are available at the start of the study and a DBM model has been obtained at the previous stage 5, then any available and relevant emulation model that merits such investigation should be considered at this stage and reconciled with the DBM model. Although such reconciliation will depend upon the nature of the application being considered, the DBM model obtained from the real data should have strong similarities with the reduced order dynamic emulation model. If this is not the case, then the differences need to be investigated, with the aim of linking the reduced-order model with the high order simulation model via the parametric mapping of the dynamic emulation model (see later illustrative example).
7. The final stage of model synthesis should always be an attempt at model validation: see e.g Young (2001b). The word 'attempt' is important since validation is a complex process and even its definition is controversial. Some academics (e.g. Konikow and Bredehoeft, 1992), within a ground-water context; and Oreskes *et al.* (1994) in relation to the whole of the earth sciences) question even the possibility of validating models. However, statistical evaluation of the model by confirming that statistical diagnostics are satisfactory (e.g. no significant autocorrelation in the residuals or cross correlation between the residuals and input variables; no evidence of un-modelled nonlinearity etc.) is always possible and can engender greater confidence in the efficacy of the model. Also, one specific, quantitative aspect of validation is widely accepted; namely 'predictive validation' or 'cross-validation', in which the predictive potential of the model is evaluated on data other than that used in the identification and estimation stages of the analysis. When validated in this narrow sense, it can be assumed that the 'conditionally valid' model represents the best theory of behaviour currently available that has not yet been 'falsified' in a Popperian sense.

Although these are the seven major stages in the process of DBM model synthesis, they may not all be required in any specific application: rather, they are 'tools' to be used at the discretion of the modeller. Also, they are not the end of the modelling process. If the model is to be applied in practice (and for what other reason should it be constructed?) then, as additional data are received, they should be used to evaluate further the model's ability to meet its objectives. Then, if possible, both the model parameters and structure can be modified if they are inadequate in any way. This process, sometimes referred to as 'data assimilation', can be achieved in a variety of ways. Since most data assimilation methods attempt to mimic the Kalman Filter, however, it is likely to involve recursive updating of the model parameter and state estimates in some manner, as well as the use of the model in a predictive (forecasting) sense. This process of data assimilation is made simpler in the DBM case because the estimation methods used in DBM modelling, as described in the previous chapters of this book, are all inherently recursive in form and so can be used directly for on-line, Bayesian data assimilation (Young, 1984, 2002a; Romanowicz et al., 2006).

The only methods mentioned in the above DBM strategy that have not been described in the present book are 'model order reduction' and 'high order model emulation'. Before considering a specific example of DBM modelling, therefore, it is necessary to present briefly the main concepts involved in such emulation modelling. An important aspect of emulation is the definition of a reduced order model that reproduces the dynamic behaviour of its large model progenitor to a very accurate degree. Model order reduction has received a lot of attention in the control and systems literature and so, to begin with, let us consider briefly some of the mathematical techniques that have been used for such model order reduction.

12.3 Model Order Reduction

The linear model order reduction problem is to find a low order, linear model that mimics the dynamic behaviour of a much higher order model. Such order reduction can be used in various ways: for example, it can help to understand better the dominant modal behaviour of the high order model; or it can be used to simplify the problem of control and/or forecasting system design by allowing the designer to utilize the reduced order model in any design calculations in place of the high order model. Model order reduction has received a lot of attention in the last few decades and many different approaches have been suggested, from simple methods, such as the 'power reduction' approach of Liaw (1986), to more complex procedures, including 'proper orthogonal decomposition', 'balanced model order reduction' and 'structured total least squares' (Moore, 1981; Willcox and Peraire, 2002; Markovsky et al., 2005).

In the present chapter, *Dominant Mode Analysis* (DMA) (Young et al., 1996; Young, 1999a), which in this setting can be considered as a specific form of model reduction, is used for model order reduction and high order model emulation. Unlike some approaches, such as balanced model reduction, that require an analytical formulation of the problem and access to the model equations, this approach requires only that the high order model can be simulated with user-defined input perturbations. Planned experiments are performed on the high order model and reduced order TF models structures are then identified and estimated from these experimental data using either the rivbjid or rivcbjid algorithms in CAPTAIN, depending on whether the high order model is discrete or continuous-time in form. In the present context, where the data are normally noise-free, the SRIV and SRIVC options of these algorithms are used to search for a reduced model that is well defined, minimally parameterized and explains the high order model output to a high degree (e.g. with coefficients of determination normally greater than 0.99).

12.3.1 The ALSTOM gasifier example

This example is concerned with a 25th order, continuous-time model of a gasifier, a device that converts solid materials into gases. The model was issued as part of the ALSTOM benchmark challenge on gasifier control by ALSTOM Power UK and is used in Matlab as a demonstration example for the balred balanced model reduction routine. The reduced order estimation results are obtained using a 7th order continuous-time model identified using the SRIVC option of the rivcbj routine. These results are compared with those obtained using the balred routine and a *Proper Orthogonal Decomposition* (POD) algorithm.

The DMA was conducted using the unit impulse response data shown in the left hand panel of Figure 12.1 and the order identification results are shown in Table 12.1. Here, when rounded to four decimal places, the coefficient of determination based on the simulated reduced order model is $R_T^2 = 1.0$ for all models over 4th order, indicating a almost perfect explanation of the data in these cases. Consequently, it is necessary to consider the variance of the model error series and the other order identification measures reported in Table 12.1: namely, the YIC and BIC (see section 6.9 of chapter 6).

Table 12.1 Model Order Identification results for the Alstom gasifier example.

Model	R_T^2	YIC	BIC
[8 8 0 0 0]	1.0	-27.5	-1.88×10^5
[7 7 0 0 0]	1.0	-25.9	-1.22×10^5
[6 6 0 0 0]	1.0	-26.6	-0.61×10^5
[5 5 0 0 0]	1.0	-19.9	1.26×10^5
[4 4 0 0 0]	0.9999	-20.5	0.72×10^5
[3 3 0 0 0]	0.999	-15.6	2.47×10^5

Table 12.1 suggests that all order models from 3rd to 8th explain the data well, with the 6th and 8th order models having little to choose between them. From the point of view of a parsimony, the 6th order model looks best at first sight. However the later validation of the model suggests that the 7th order model is marginally superior. The SRIVC estimates of the parameters in this model are given below.

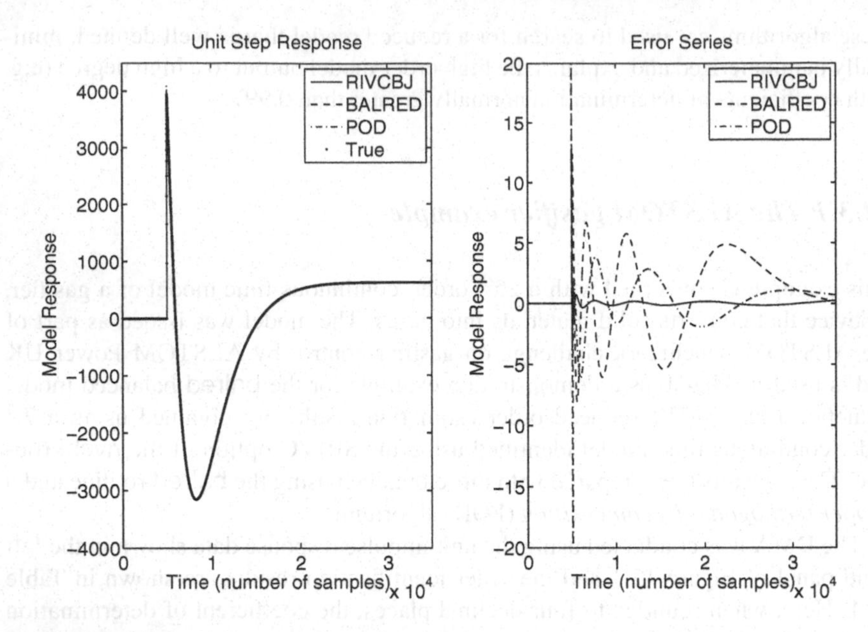

Fig. 12.1 Comparative estimation results for the model reduction exercise on the high order ALSTOM gasifier model.

$$\hat{\alpha}_1 = 0.5003;\ \hat{\alpha}_2 = 0.04584;\ \hat{\alpha}_3 = 8.709 \times 10^{-4}$$
$$\hat{\alpha}_4 = 1.841 \times 10^{-6};\ \hat{\alpha}_5 = 1.638 \times 10^{-9}$$
$$\hat{\alpha}_6 = 6.386 \times 10^{-13};\ \hat{\alpha}_7 = 9.31 \times 10^{-17}$$
$$\hat{\beta}_1 = 1142;\ \hat{\beta}_2 = 167.7;\ \hat{\beta}_3 = 3.395$$
$$\hat{\beta}_4 = 3.133 \times 10^{-3};\ \hat{\beta}_5 = -1.498 \times 10^{-6}$$
$$\hat{\beta}_6 = -1.647 \times 10^{-9};\ \hat{\beta}_7 = 5.761 \times 10^{-14}.$$

The left hand panel of Figure 12.1 confirms that this 7th order model, as well as the 7th order BALRED and POD models, all explain the impulse response data very well at first sight. However, the right hand panel reveals that the SRIVC estimated model has a much smaller error series than the other two, with the POD model being a little better, in this regard, than the BALRED estimated model. When the models are applied to the validation data set in Figure 12.2, the SRIVC estimated model remains the best but by a smaller margin. In terms of standard deviation, the model error series is now only 3 times better than the BALRED estimated model. However, the POD model now validates quite poorly and should be rejected.

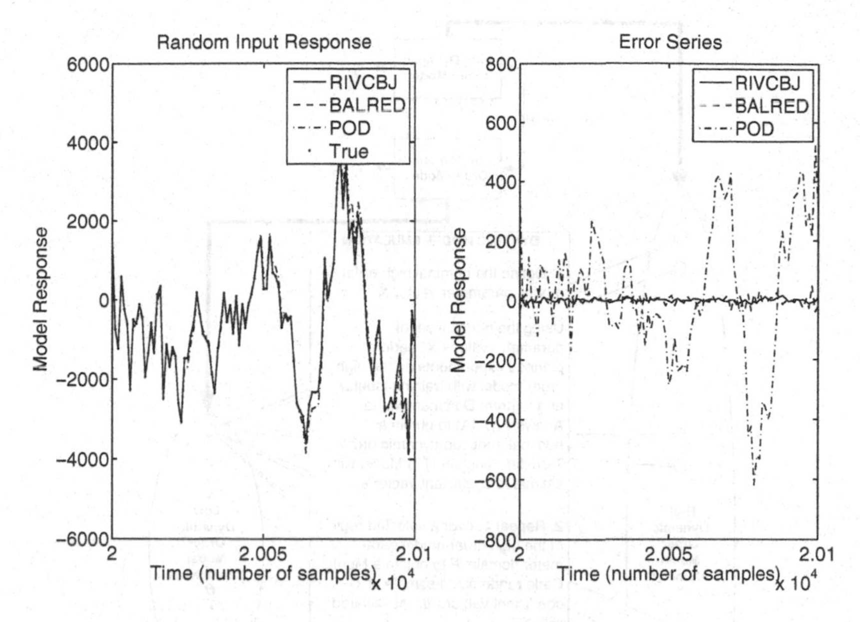

Fig. 12.2 ALSTOM gasifier model reduction: validation results.

12.4 Large Computer Model Emulation

The DMA approach to linear model reduction yields a minimally parameterized, low order model that often 'emulates' the dynamic behavior of the high order model to a remarkable degree. However, this is 'nominal' emulation: namely the dynamic behaviour is reproduced for a given nominal set of high order model parameters. In 'full' emulation, the objective is to emulate the high order model over a whole, user-defined range of parameter values, so that the user can employ the reduced order model as a complete surrogate for the high order model in applications that demand fast, repeated solution, as in Monte Carlo Simulation and Sensitivity analysis.

A proper dynamic emulator needs to reproduce the behaviour of the high order dynamic model in two simulation conditions: first, it should replicate the high order model behaviour at any time horizon and any realization of the forcing inputs; second, this replication should be for any value of its parameters $\mathbf{X} = [X_1, \ldots, X_p]$ over a user-specified range. The procedure for achieving these objectives suggested by Young and Ratto (2009), within the overall DBM modelling framework, involves the following four main stages, as shown diagrammatically in Figure 12.3:

1. **Identification and Estimation of a Nominal TF Model**: On a nominal set of high order model parameters $\bar{\mathbf{X}}$ (input factors), DMA is used to provide an identifiable, reduced order SISO or MISO transfer function model that reproduces the response of the high order model as well as possible.

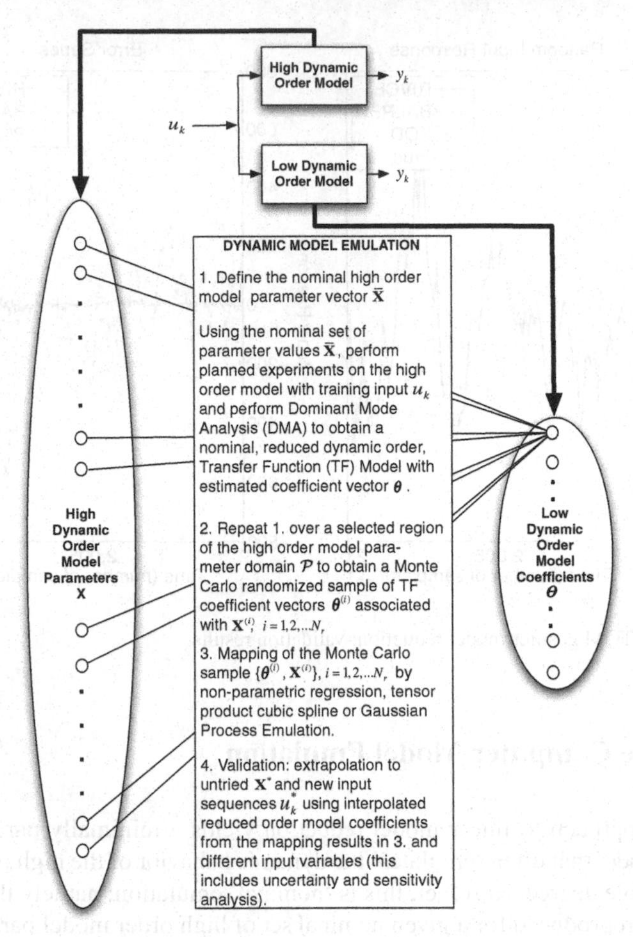

Fig. 12.3 A diagrammatic interpretation of full model emulation. This diagram appeared originally in Young and Ratto (2011).

2. **MCS Randomization**: Normally, *Monte Carlo Simulation* (MCS) analysis is carried out in which, for each set of independent, randomly chosen parameters $\mathbf{X}_i, i = 1, 2, \ldots, N$, from the high order simulation model, a dynamic SISO or MISO TF model is estimated, with the same reduced order obtained in step 1, to yield a set of TF model parameters $\boldsymbol{\theta}_i, i = 1, 2, \ldots, N$. Sometimes, this MCS randomization is not necessary because the DMA can be applied over a grid of the high order model covering the whole of the specified parametric range (Tych and Young, 2011).
3. **Mapping and Parameterization**: Based on the MCS-based results obtained in step 2, a technique such as non-parametric regression or tensor product cubic spline estimation (Wahba, 1990; Gu, 2002) is used to map the relationships between the high order model parameters and each parameter of the reduced order

TF model. If the whole grid of parameter values has been mapped, then the reduced order model parameters can be inferred by simple interpolation (Tych and Young, 2011).

4. **Validation, uncertainty and sensitivity analysis**: given the interpolated and extrapolated TF coefficients, the emulation model can be validated by generating dynamic simulations at any time-horizon and for any sequence of forcing inputs (this includes uncertainty and sensitivity analysis, the former using techniques such as smoothing spline analysis of variance (ANOVA): see Wahba (1990); Gu (2002).

The mapping in step 3, together with the associated TF models estimated in 2, constitute the full dynamic emulation model.

12.5 An Illustrative Example: DBM Modelling of Pollution Transport in a Wetland Area

This example relates to the same tracer data discussed and analysed in the earlier chapters 6, 7, 8 but sets this within the wider context of DBM modelling. One of the first models to be considered seriously in DBM terms was the *Aggregated Dead Zone* (ADZ) model for the transport and dispersion of solutes in river systems (see e.g. Beer and Young, 1983; Wallis et al, 1989; Green et al, 1994). The name 'dead zone' is not strictly true because such zones involve imperfect mixing processes: better terms are *Aggregated Mixing Zone* (AMZ) (Beven and Young, 1988) or *Active Mixing Zone* (AMV) (Young and Lees, 1993). These arose in subsequent uses of the ADZ-type model in relation to the modelling of imperfect mixing processes that characterize mass and energy flow processes in the wider environment: for instance, Beven and Young use the AMZ for modelling flow through porous media; Young and Lees generalize the concept to the AMV form and apply it to the DBM modelling of heat flow in soils; while Price et al. (1999) and Young et al. (2000) show how the AMV can be used very successfully for the DBM modelling of heat flow and the resultant temperature changes in buildings. This latter study is a very good example of DBM modelling because it shows how the DBM model can be related to the more classical equations of heat transfer and, in doing so, raises some interesting questions about the nature of these equations. It is clear, therefore, that the same ADZ/AMV modelling ideas have fairly wide applicability to flow processes involving mass and energy transfer in both the natural and built environment.

Research in this area of study is aided by the ability to conduct simple planned experiments using conservative tracer materials, such as the fluorescent red dye, Rhodamine WT, or potassium bromide, as discussed earlier in this book in connection with the tracer experiments of Chris Martinez and William R. Wise (see chapters 6, 7, 8). The Martinez-Wise data plotted in Figure 12.4 are used later as the basis for DBM modelling. However, the main objective of this illustrative example is to show how the complete DBM modelling procedure, as outlined section 12.2, can develop from an initial, fairly large simulation model, through emulation modelling, to modelling on the basis of real data, such as those in Figure 12.4. Although these data are already available and could be used directly for modelling analysis, let us assume, for illustrative purposes, that the data collection experiments have not yet taken place and, without access to data, we resort to more speculative simulation modelling.

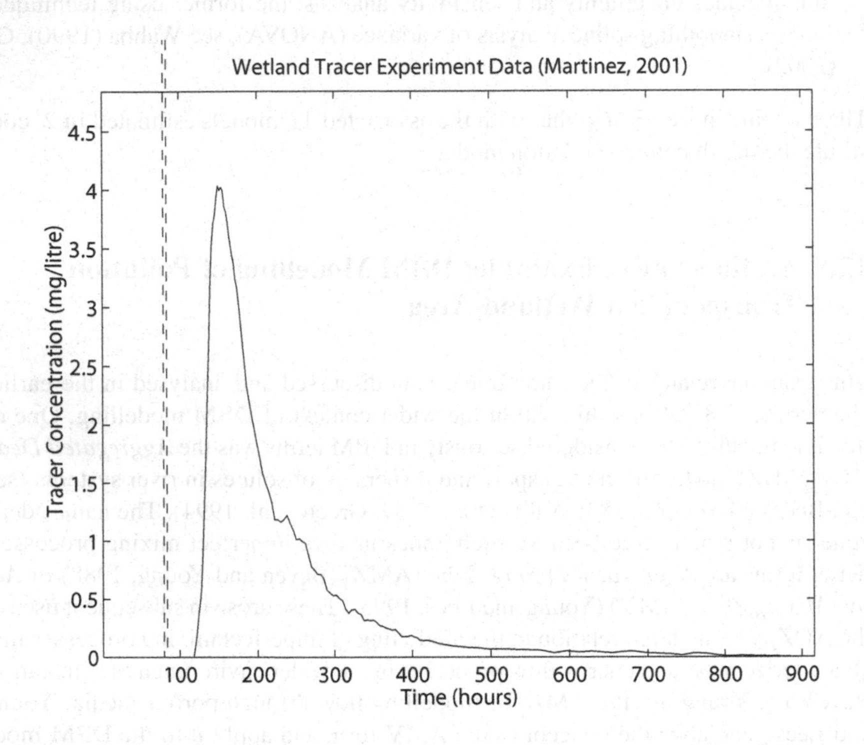

Fig. 12.4 Wetland tracer experiment data: the input $u(k)$ is an impulsive (or 'gulp') application of bromide tracer (dashed line) and the output $y(k)$ is the concentration of bromide measured every two hours at a downstream weir (full line).

12.5.1 The large simulation model

Although its efficacy as a description of solute transport and dispersion in rivers has been questioned, the *Advection Dispersion Equation* (ADE) has been the basis of numerous models that have been used in practice over many years. In the case of a conservative (non-decaying or non-reactive) solute, this model takes the form of the following partial differential equation for the changes in the concentration of the solute $c(t,s)$ in time (t) and space (s):

$$\frac{\partial c(t,s)}{\partial t} + V\frac{\partial c(t,s)}{\partial s} = D\frac{\partial^2 c(t,s)}{\partial s^2} \tag{12.1}$$

where D is the 'dispersion cefficient' and V is the velocity. Its derivation is discussed in many fluid dynamic texts and is outlined in Young and Wallis (1993).

The ADE derives originally from the work of the great fluid dynamicist G. I. Taylor (Taylor, 1954) who used a one dimensional Fickian diffusion equation to describe the random diffusion of a solute in a pipe, once sufficient time had elapsed to allow for full cross sectional mixing. Subsequently, although the same partial differential equation (PDE) model has been applied to solute dispersion in river channels, tracer studies such as that shown in Figure 12.4 show a longer 'tail' in the response that is redolent of non-Fickian behaviour. Two main solutions to this limitation have been suggested: the ADZ approach mentioned above, which replaces the PDE model by a lumped parameter, ordinary differential equation (ODE) relating solute concentration between spatial locations along the river; and a modified version of the ADE that is now termed the 'transient storage ADE' model (see e.g. Valentine and Wood, 1977; Martinez and Wise, 2003) and takes the following form:

$$\begin{aligned} \frac{\partial c(t,s)}{\partial t} + V\frac{\partial c(t,s)}{\partial s} &= D\frac{\partial^2 c(t,s)}{\partial s^2} + K\Gamma_c\{u(t,s) - c(t,s)\} \\ \frac{du(t)}{dt} &= K\Gamma_s\{u(t) - c(t,s)\} \end{aligned} \tag{12.2}$$

Here, K is the mass exchange coefficient between the dead (transient storage) zone and the main flow; Γ_c is ratio of interfacial area (between the main flow and the dead zone) to the main flow volume; and Γ_s is is the ratio of the interfacial area to dead zone volume. In equations (12.1) and (12.2), it is straightforward to add terms to allow for non-conservative or reactive solutes.

Beer and Young (1983) suggested, and provided evidence to show, that the effect of the dead zone may be dominant in many rivers and so the transient storage equation can be replaced by the single ADZ equation:

$$\frac{\partial c(t,s)}{\partial t} + V\frac{\partial c(t,s)}{\partial s} = \frac{1}{T}\{u(t) - c(t,s)\} \tag{12.3}$$

where the two terms on the left hand side represent 'plug' flow, which leads to a pure 'advective' time delay τ between the 'upstream' input $u(t)$ and the 'downstream' output $c(t)$; while the right hand term incorporates the dispersive effect of the, now 'aggregated', dead zone. For our purposes, equation (12.3) is better written in the ODE form relating directly only the temporal changes in concentration, $u(t)$ and $c(t)$, at the two locations:

$$\frac{dc(t)}{dt} = \frac{1}{T}\{u(t-\tau) - c(t)\} \tag{12.4}$$

where τ is the pure advective time delay and T is the time constant. This can then be written in the transfer function form:

$$c(t) = \frac{1}{1+Ts}u(t-\tau) = \frac{b_0}{s+a_1}u(t-\tau) \tag{12.5}$$

where, in this conservative situation, $b_0 = a_1 = 1/T$. In other words, in formulating this ODE/transfer function model, it is assumed that the length of the river reach is such that the distributed dead zones can be represented by a single 'aggregated' dead zone, where T is the 'effective' time constant or 'residence time'[2] that characterizes the ADZ dynamics.

Equation (12.4) will be recognized as simple mass balance equation, where it is assumed that the mass of solute being lost from the reach, because of the predominantly downstream flow, is proportional to the solute concentration in the reach (see e.g Young and Lees, 1993). Consequently, in a more general, possibly non-conservative situation, the TF equation (12.5) becomes:

$$c(t) = \frac{G}{1+Ts}u(t-\tau) \tag{12.6}$$

where the 'steady state gain' $G = 1.0$ in the conservative case; $G < 1.0$ if mass is being lost; and $G > 1.0$ if there is an accretion of mass for some reason. In addition, the ADZ model can be modified to allow for 'back-flow' caused, for example, by the physical nature of the river channel or, in the extreme case, by tidal effects. In this situation, the ADZ equation takes the following more general form:

$$T\frac{dc(t)}{dt} = -c(t) + Gu(t-\tau) + G_d c_d(t) \tag{12.7}$$

where $c_d(t)$ are the changes in the downstream solute concentration and G_d is the steady state gain for this downstream input that defines its effect on the concentration $c(t)$ of the solute in the reach. In this case, the model is conservative when $G + G_d = 1.0$.

Of course, the above ADZ equations only relate the temporal changes in solute concentration between two spatial locations on a river. However, a *Semi-Distributed* ADZ (SDADZ) model can be constructed rather simply by a chain of suitably small ADZ elements such as (12.7) connected in series, parallel or even feedback (should this relate to a physically meaningful situation). For example, in this SDADZ model, the $c_d(t)$ in equation (12.7) would be interpreted as the solute concentration in the immediate downstream reach, which would also be modelled as an ADZ element. In this manner, the equation for transport and dispersion of a conservative solute in the ith ADZ reach, of a uniform river system of n such identical reaches, with all the intermediate pure time delays set to zero, would then take the form:

---

[2] Note that although, in hydrology, the residence time is sometimes considered as $T + \tau$, most hydrologists believe this sum is better termed the total 'travel time'.

$$T\frac{dc_i(t)}{dt} = -c_i(t) + Gc_{i-1}(t) + G_d c_{i+1}(t) \quad i = 1,2,\ldots n \tag{12.8}$$

Alternatively, this model can be represented in TF form:

$$c_i(t) = \frac{G}{1+Ts}c_{i-1}(t) + \frac{G_d}{1+Ts}c_{i+1}(t) \quad i = 1,2,\ldots n \tag{12.9}$$

or as the following element in a n dimensional state space model:

$$\begin{bmatrix} \frac{dc_{i-1}(t)}{dt} \\ \frac{dc_i(t)}{dt} \\ \frac{dc_{i+1}(t)}{dt} \end{bmatrix} = \begin{bmatrix} -\frac{1}{T} & \frac{G_d}{T} & 0 \\ \frac{G}{T} & -\frac{1}{T} & \frac{G_d}{T} \\ 0 & \frac{G}{T} & -\frac{1}{T} \end{bmatrix} \begin{bmatrix} c_{i-1}(t) \\ c_i(t) \\ c_{i+1}(t) \end{bmatrix} \tag{12.10}$$

where the output is defined by the state variable $c_i(t)$. The input to this whole system is then the input to the farthest upstream reach, denoted by $u(t) = c_0(t)$, and the output is that of the farthest downstream reach, denoted by $x(t) = c_n(t)$. If it is assumed that any pure advective time delays are lumped into a single time delay τ at the input, the complete deterministic model can be represented in the following general TF form:

$$x(t) = \frac{B(s)}{A(s)}u(t-\tau) = \frac{b_0 s^{n-1} + b_1 s^{n-2} + \ldots + b_m}{s^n + a_1 s^{n-1} + \ldots + a_n}u(t-\tau) \tag{12.11}$$

If G and G_d are selected so that the system is conservative, then $b_m = a_n$ and the overall steady state gain is unity.

The original idea was to base this example around either the ADE or the transient storage ADE. Although analytical solutions of these models can be obtained for specified inputs (see e.g. De Smedt, 2006), the models are normally implemented on a computer using some form of numerical approximation that can, more usefully, apply for any specified input forcing functions. A popular approach is the Crank-Nicolson finite-difference solution developed by Runkle and Chapra (1993) for the solution of the transient storage model (12.2), which is the basis for the well known OTIS simulation model: see `http://csdms.colorado.edu/wiki/Model:OTIS`. However, at the time the analysis described below was carried out, this model was not available in Matlab and so, as a an interesting alternative, I used the SDADZ state space model, with elements defined by (12.10), since it was straightforward for me to program it in Matlab.

It is interesting to note that, more recently, Tych and Young (2011) have been able to run the OTIS model from Matlab and this new reference describes a similar emulation exercise to that described below applied to the OTIS model. Tych and Young also develop an efficient software framework in Matlab that makes the whole process of high order model emulation more systematic and efficient. However, the emulation results obtained for the OTIS model are qualitatively very similar to those obtained in the next sub-section for the SDADZ model.

Returning to the SDADZ model, the entire system needs to be conservative and uniform, so that all of the reach elements are of the same form (12.10), with $G + G_d = 1.0$, and the complete state space model matrix is 3-band diagonal, similar (12.10) but of dimension n. The model is simulated with $n = 40$, $\tau = 0$ and the $\{G_i \; T_i\}$ parameters defined as appropriate functions of D, V and the reach length dz: e.g. I used $1/T = -(2D/dz^2 + V/dz)$ but this not need to trouble the reader in the present, largely illustrative context. Note that, since this SDADZ model can be solved explicitly, it can be considered as a possible alternative to the transient storage/OTIS model (12.2). However, further research is continuing on the model and its relationship with both the ADE and the transient storage model.

12.5.2 Emulation modelling

Given the artificial assumption that, at this stage, we have no measured data from the wetland area, the simulation modelling is based on its physical characteristics and the measurement locations in the anticipated tracer experiment. This suggests model parameter values in the range of $D : \{0.030 - 1.5\} \; m^2/sec$ and $V : \{0.00045 - 0.0014\} \; m/sec$ (although note that these are parameters in the SSADZ model, not the ADE). For the specific measurement location considered later in section 12.5.3, reach 5 appears to be most suitable but, in order to ensure that a single model structure is possible for emulation at all reaches, if this is required, the initial nominal DMA considers both reach 5 and reach 40, with $D = 0.318$ and $V = 0.0013$ in both cases.

This nominal emulation analysis is carried out using an input signal in the form of two repeated pulses of period 1000 hours and amplitude 180 entering at reach 1. Although this is certainly not an optimal input from a statistical identification and estimation standpoint, it is sufficient to produce good TF emulation models. For instance, the continuous-time SRIVC identification results for reach 5, as produced by the CAPTAIN rivcbjid routine, are as follows (shown verbatim) for reach 5:

```
den    num    del   AR   MA      YIC         Rt2          BIC
 5      6      1     0    0   -25.7517   1.000000    -200025
 5      6      0     0    0   -18.6377   0.999998    -180477
 5      5      0     0    0   -22.4703   0.999998    -180288
 4      5      1     0    0   -19.1426   0.999976    -147983
 4      5      0     0    0   -11.6862   0.999976    -147983
 4      4      0     0    0   -18.1096   0.999976    -147990
```

Although this suggests that the [5 6 1] model is marginally better, the [5 5 0] model is not only almost as good but it has one less parameter and, most importantly, it proves superior in the case of reach 40, where the best identified model is [5 5 79].

To summarize, therefore, this initial, nominal DBM model emulation analysis, which examines the extremes of the large simulation model response characteristics, shows that a [5 5 τ] continuous-time TF model of the form;

$$\hat{x}(t) = \frac{\hat{b}_0 s^4 + \hat{b}_1 s^3 + \hat{b}_2 s^2 + \hat{b}_3 s + \hat{b}_4}{s^5 + \hat{a}_1 s^4 + \hat{a}_2 s^3 + \hat{a}_3 s^2 + \hat{a}_4 s + \hat{a}_5} u(t-\tau) \tag{12.12}$$

provides very good emulation at both extremes. For reach 5, $\tau = 0$ and the parameter estimates, obtained using the SRIVC option in the CAPTAIN rivcbj routine, are (note that $a_5 = b_4$ and mass is conserved):

$$\begin{aligned}
&\hat{a}_1 = 0.5003;\ \hat{a}_2 = 0.0772;\ \hat{a}_3 = 0.00335;\ \hat{a}_4 = 4.6033 \times 10^{-5};\\
&\hat{a}_5 = 1.7265 \times 10^{-7};\ \hat{b}_0 = 7.0278 \times 10^{-4};\ \hat{b}_1 = -4.5845 \times 10^{-4};\\
&\hat{b}_2 = 1.0403 \times 10^{-3};\ \hat{b}_3 = 3.1639 \times 10^{-5};\ \hat{b}_4 = 1.7265 \times 10^{-7}
\end{aligned} \tag{12.13}$$

while for reach 40, $\tau = 79$ and the estimates are:

$$\begin{aligned}
&\hat{a}_1 = 0.06116;\ \hat{a}_2 = 0.001475;\ \hat{a}_3 = 1.6720 \times 10^{-5};\ \hat{a}_4 = 8.7469 \times 10^{-8};\\
&\hat{a}_5 = 1.6736 \times 10^{-10};\ \hat{b}_0 = 1.2088 \times 10^{-5};\ \hat{b}_1 = 1.0735 \times 10^{-6};\\
&\hat{b}_2 = 9.9738 \times 10^{-8};\ \hat{b}_3 = 2.8188 \times 10^{-9};\ \hat{b}_4 = 1.6736 \times 10^{-10}
\end{aligned} \tag{12.14}$$

In both cases, the explanation of the large model response, when validated with a single impulse input of 70 for two hours, is almost perfect, as shown in Figure 12.5, with $R_T^2 = 0.999999$ in both cases. Similarly good validation results are obtained for other inputs but final emulation model validation is considered in more detail later.

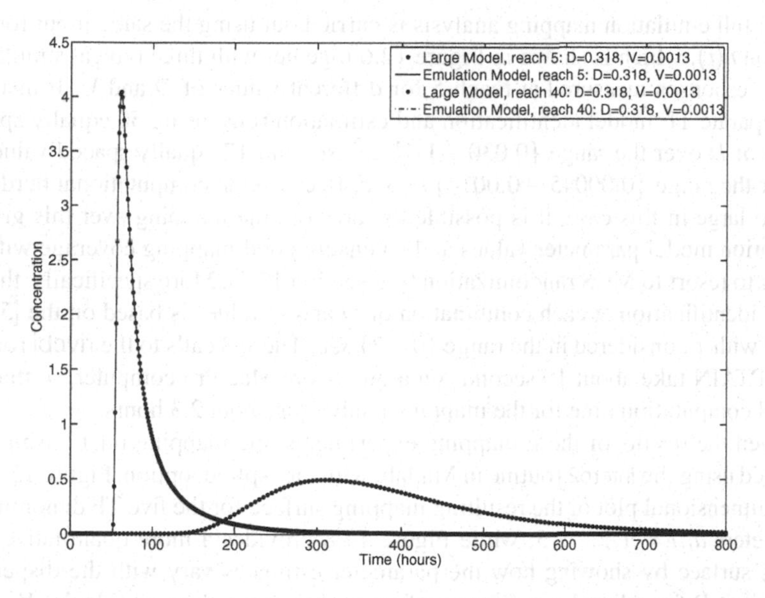

Fig. 12.5 Nominal emulation of the large simulation model at reaches 5 and 40 by continuous-time RIVC estimated 5th order TF models.

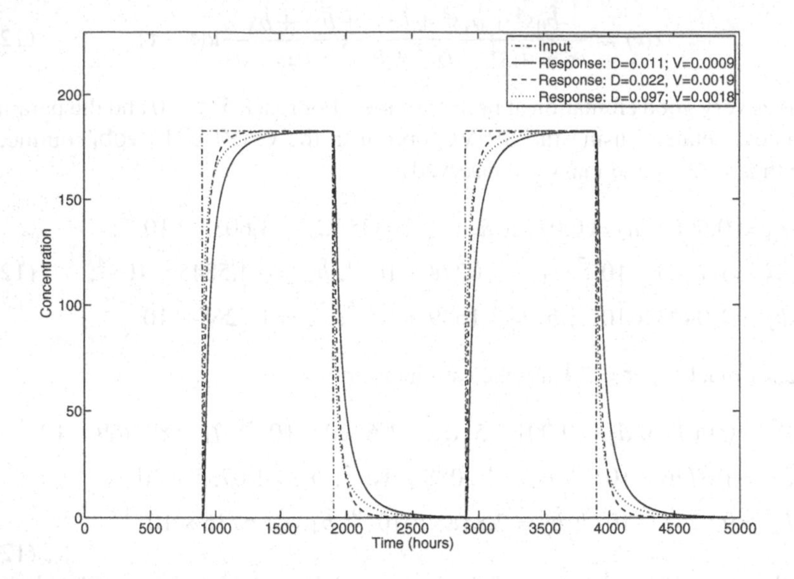

Fig. 12.6 Emulation experimental data at reach 5 for 3 combinations of the dispersion coefficient, D, and velocity, V.

The full emulation mapping analysis is carried out using the same input forcing function $u(t)$, which is shown in Figure 12.6 together with three typical simulation model responses measured at reach 5 for different values of D and V. It involves 850 separate TF model identification and estimation runs, using 50 equally spaced values of D over the range $\{0.030-1.5\}$ m^2/sec and 17 equally spaced values of V over the range $\{0.00045-0.0014\}$ m/sec. Because the computational burden is not too large in this case, it is possible to carry out the mapping over this grid of simulation model parameter values and so ensure good mapping coverage without having to resort to MCS randomization (see section 12.2). More specifically, the TF model identification at each combination of D and V values is based on the $[5\ 5\ \tau]$ model with τ considered in the range $\{0-2\}$ sec. These 3 calls to the rivcbj routine in CAPTAIN take about 10 seconds on a quad-core Mac Pro computer, so that the overall computation time for the mapping analysis is about 2.3 hours.

Given the results of these mapping experiments, the mapping relationships are obtained using the interp2 routine in Matlab, with the 'spline' option. Figure 12.7 is a three dimensional plot of the resulting mapping surface for the five TF denominator parameters $a_i, i = 1,2,\ldots 5$; while Figure 12.8 provides a more quantitative idea of this surface by showing how the parameter estimates vary with the dispersion coefficient D for 'slices' across the surface at different values of velocity V. Note that the mapping surface in Figure 12.7 is quite smooth for $D > 0.1m^2/sec$ but there is a quite sharp change at smaller values than this, suggesting that a finer grid might be necessary in this region. However this region is not important in the present case and this has not been investigated further.

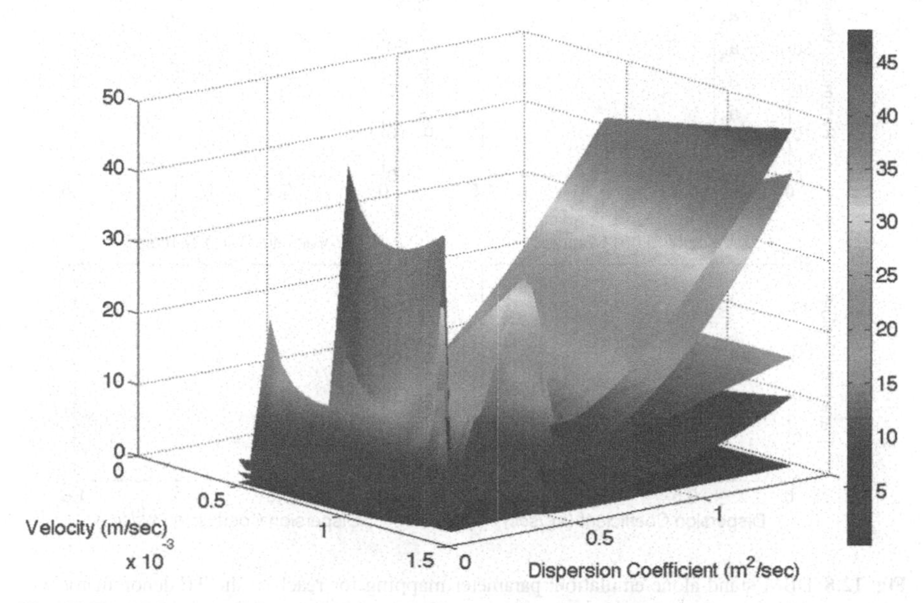

Fig. 12.7 DBM stand-alone emulation: parameter mapping for reach 5: 3 dimensional plot of the five TF denominator parameters $a_i, i = 1, 2, \ldots 5$, as functions of the dispersion coefficient, D, and the velocity, V.

The final stage of the full emulation analysis is validation on interpolated values of the parameters and two examples of such validation analysis are shown in Figure 12.9, for an impulse forcing function; and Figure 12.10 for a forcing function of the more general type that might be expected if the model was being employed for the evaluation of pollutant transport and dispersion. In both cases, the emulation is exceptional, with R_T^2 values greater than 0.999.

12.5.3 Modelling from real data

The standard DBM analysis of data such as those shown in Figure 12.4 is to use them to identify and estimate a TF model in either discrete or continuous-time form. Discrete-time DBM modelling of these data is described in Young (2001c), where a [4 2 22] model is identified and estimated in a constrained form to ensure physically interpretable real poles in the estimated TF model. In the present context, it is clear that a continuous time model DBM model makes more sense and so we employ hybrid continuous-time modelling, using the rivcbjid/rivcbj routines in CAPTAIN,

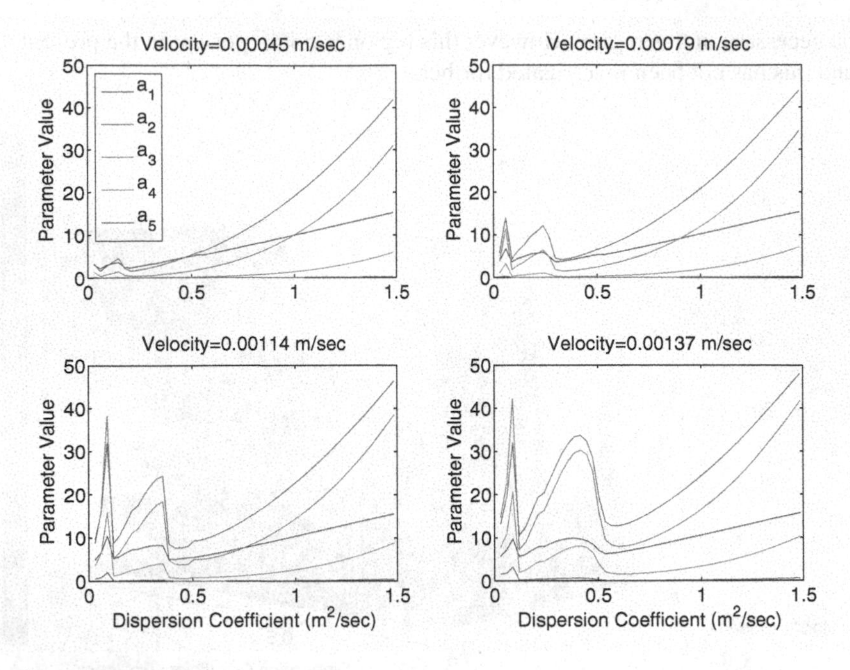

Fig. 12.8 DBM stand-alone emulation: parameter mapping for reach 5: the TF denominator parameters as changing functions of the dispersion coefficient, D, for four different velocity, V, values ('slices' from Figure 12.7).

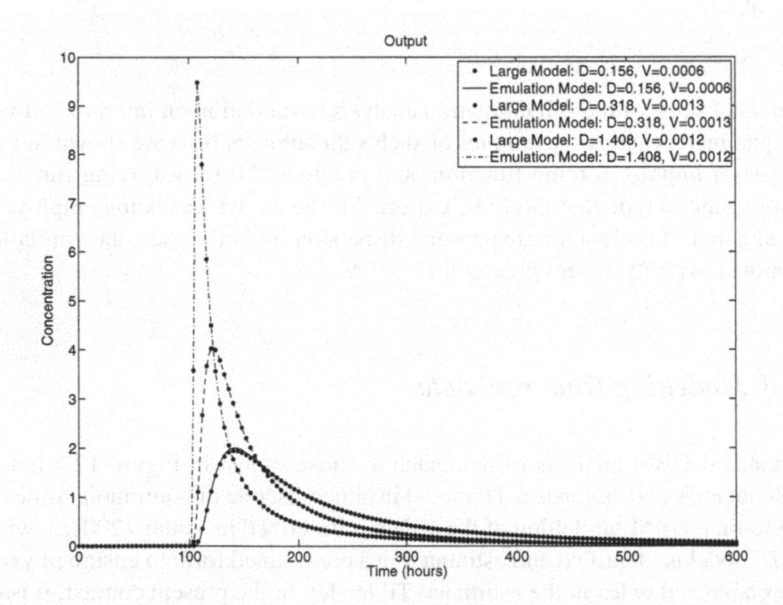

Fig. 12.9 Validation of DBM stand-alone emulation model with an impulsive-type input.

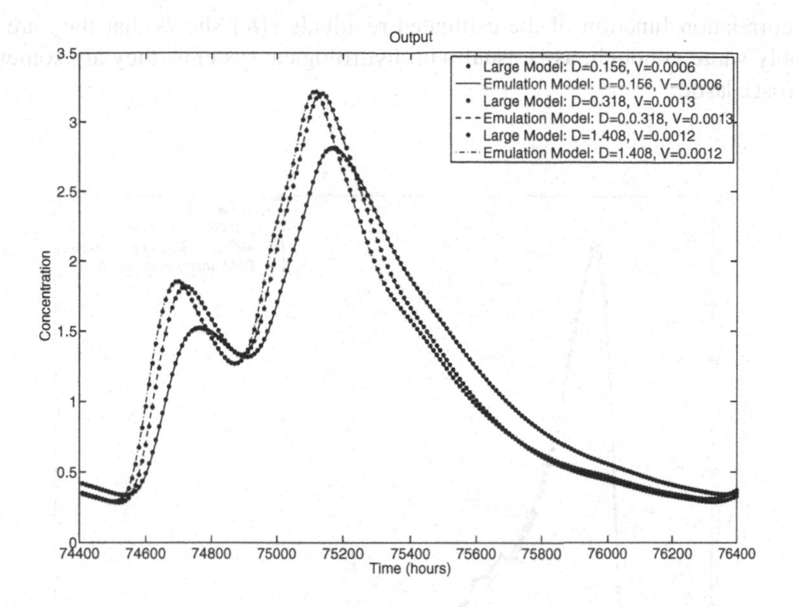

Fig. 12.10 Validation of the DBM stand-alone emulation model with a typical input.

as used in the above emulation modelling stage of the analysis, but this time using full RIVC estimation, rather than SRIVC, because these real data are noisy.

As in the discrete-time model case, the poles of this estimated continuous-time TF are complex so, using the same arguments as those used in the above reference, it is constrained to have real poles and, this time, it is identified by rivcbjid to have a [4 2 20] structure with additive coloured noise identified as an AR(20) process: i.e.,

$$\begin{aligned} y(t_k) &= \frac{b_0 s + b_1}{(s+\alpha_1)^3(s+\alpha_2)} u(t_k - 20) + \xi(t_k) \\ \xi(t_k) &= \frac{1}{1 + c_1 z^{-1} + \ldots + c_{20} z^{-20}} e(t_k) \qquad e(t_k) = N(0, \sigma^2) \end{aligned} \tag{12.15}$$

and the parameter estimates are as follows, with the standard errors shown in parentheses:

$$\begin{aligned} &\hat{\alpha}_1 = 0.07734(0.0005);\ \hat{\alpha}_2 = 0.0099(0.00015);\ \hat{\sigma}^2 = 0.0003 \\ &\hat{b}_0 = 1.9157 \times 10^{-4}(1.908 \times 10^{-6});\ \hat{b}_1 = 4.5415 \times 10^{-6}(3.157 \times 10^{-8}) \end{aligned} \tag{12.16}$$

Note that this DBM model is characterized by three real, equal modes with short time constants of 12.9 hours and one mode with a long time constant of 101 hours. As such, it can be considered as a special example of an ADZ model (see the previous examples in chapters 6 to 8). The tracer data are explained well, as shown in Figure 12.11, with $R_T^2 = 0.997$ and a standard coefficient of determination for the residual white noise (one-step-ahead prediction errors) of $R^2 = 0.9996$. The autocorrelation function of the estimated residuals $e(t_k)$ shows that they are reasonably white although, as is usual with hydrological systems, they are somewhat heteroscedastic.

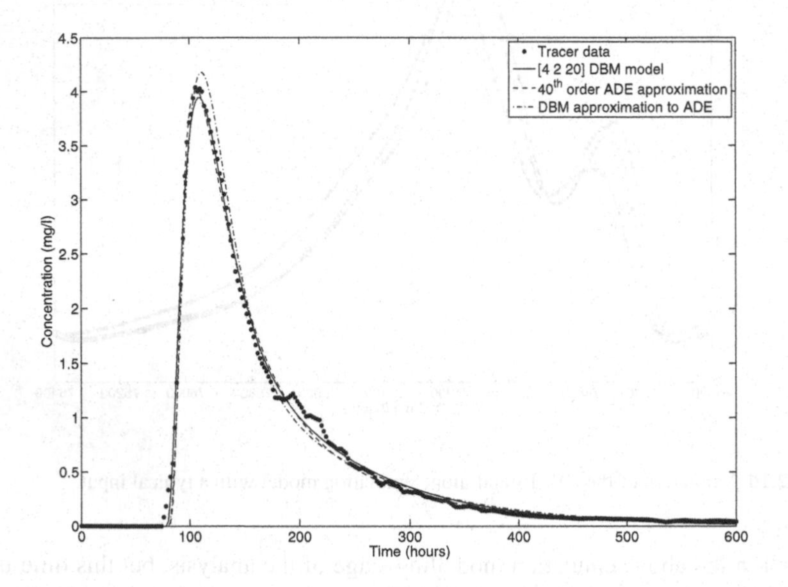

Fig. 12.11 The deterministic outputs of the three models compared with the tracer data.

But can the model (12.15) be reconciled with the large simulation model? The answer to this question is aided by the nature of this large model which has been chosen specially, for the purposes of this illustrative example, as a particular ODE approximation of the ADE that it is identifiable from the the tracer data in Figure 12.4. It should be noted that this is unusual for large simulation models because they are often over-parameterized. For instance, Wagener et al. (2002) have investigated the identifiability of a particular version of the transient storage equation (12.2) and they conclude that

> "It can be seen that very different combinations of the parameters yield identical performances in terms of the selected objective function (the weighted sum of squared differences between simulated and observed concentrations). From this analysis, only one parameter seems identifiable".

This despite the fact that their version of the model has only four parameters.

If, in the present case, the parameters D and V in our 40th order SDADZ model are optimized by simple nonlinear least squares, based on the tracer data in Figure 12.4 and using the optimization routines in Matlab, then the estimates are clearly defined at $\hat{D} = 0.3126\ m^2/sec$ and $\hat{V} = 0.00126\ m/sec$. However, as pointed out previously in section 12.5.1, these numerical values of D and V are not particularly important in the present context and, in any case, should not be interpreted directly in ADE terms: for our purposes, it is only required that the output of this high order SDADZ model is able to explain the tracer data well. This is certainly the case here, as shown in Figure 12.11, where the coefficient of determination is $R_T^2 = 0.995$.

Now, prompted by the above results and, in particular, the identified ADZ model (12.15), if a constrained [4 2 0] model, of similar form to (12.15), is estimated by the rivc routine *on the basis of deterministic data generated by this SDADZ model*, then the estimated parameters are as follows:

$$\begin{aligned} \hat{\alpha}_1 = 0.07366;\ \hat{\alpha}_2 = 0.00888;\ \hat{\sigma}_\xi^2 = 0.0021 \\ \hat{b}_0 = 1.9671 \times 10^{-4};\ \hat{b}_1 = 3.5308 \times 10^{-6} \end{aligned} \tag{12.17}$$

Note that, here, $\tau = 0$ since this is based on the SDADZ simulated data (see section 12.5.1); also no AR noise model is estimated and standard errors are omitted, because they are inappropriate in this simple least squares estimation situation, where the SDADZ simulated model output is noise free: i.e. here, $\hat{\sigma}_\xi^2$ is simply the variance of the fitted residuals and the coefficient of determination based on this is 0.990. The important point, however, is that the time constants of the estimated model, namely 13.6 and 113 hours, are quite similar to those of the DBM model based entirely on the tracer data.

But the reader will surely point out that this constrained [4 2 0] model is not fully reconciled with the emulation model because it is of lower order than the unconstrained [5 5 0] model identified for the SADZ in the emulation modelling. Indeed, this is why it does not explain the data as well as the emulation model. So we need to evaluate what happens if we select model orders greater than those in the identified [4 2 0] model. In fact, virtually the same results in terms of R_T^2 are obtained with a constrained [5 3 0] model, *but this is because the estimated TF numerator has a zero that cancels with one of the four identical poles*, suggesting that any constrained TF model over 4th order is likely to be over-parameterized. This is confirmed if constrained or unconstrained [5 5 0] models are estimated, because there are, again, clear signs of pole-zero cancellation. On the basis of these results, therefore, it is clear that the limited data used in this example will only properly support the [4 2 0] model and that high order models are not identifiable from these limited data. Of course, this does not mean that higher order models could not be justified if a larger data base was utilized.

With the above results in mind, we must conclude that, while the large simulation model can be reconciled with the DBM model based on real data, since they can both be fitted to the tracer experiment data and yield similar response characteristics, this reconciliation is only partial. In particular, although the large model has only two parameters, it is actually a 40th order dynamic system, with 40 dynamic modes, and so it has an enormous surplus capacity when we realize that the data can be explained marginally better ($R_T^2 = 0.997$ vs $R_T^2 = 0.995$) by the 4th order DBM model. There is clearly a case for considering how the large SDADZ model that we have emulated here could be simplified in cases where its enormous explanatory potential is not required: for instance, when the detail provided by the 40 short reaches is not essential to the solution of the problem at hand.

The results also raise the question of how this SDADZ model relates to its 'pure' ADE progenitor or the related storage zone model. The model is able to explain the rather elevated 'tail' of the tracer response shown in Figures 12.4 and 12.11 simply because of the 'numerical dispersion' that arises from its lumped parameter ODE approximation. But the pure, partial differential ADE model would not be able to explain this 'tail' at all because the numerical dispersion is not present. Indeed, this was the original reason why the ADZ model was developed. Finally, one might question also whether the additional complication of the transient storage model is required, particularly when the Wagener et al results mentioned previously suggest that it has severe identifiability problems.

Finally, it must be emphasized that this example is intended only to illustrate various aspects of the DBM approach and is in no sense a complete modelling study. For instance, the limited data availability has prevented any predictive validation of the models considered above. It could, however, constitute the first step in a much more comprehensive research project, such as that carried out into the ADZ model by Wallis et al. (1989). This established that, for many natural and man-made channels, the 'dispersive fraction', defined by the ratio $T/(T+\tau)$, is relatively invariant over most of the flow régime so that, once this is established, the ADZ model is able to describe pollution transport and dispersion for any defined flow conditions (which is not possible in the case of the OTIS model: see Martinez and Wise, page 217). It would be very interesting, therefore, to see whether the limited results presented in this chapter are confirmed in this wider context and with a much larger data base.

12.6 Exercises

1. The model reduction (nominal emulation) analysis for the ALSTOM Gasifier Example in section 12.3.1 is available in the gasifiermodeldemo.m demonstration example in CAPTAIN. Run this and then, by modifying the code, generate additional results with different reduced order models in order to see how different reduced orders affect the nominal emulation of the high order model.
2. Run the CAPTAIN demonstration example emuldemo.m which carries out emulation analysis for a distributed parameter dispersion model (The same example, called MOC-demo, is available in the 'Supplemental Material' that can be downloaded from Young and Ratto (2011)).

12.7 Conclusions

This chapter provides a brief outline of the procedures involved in DBM modelling and shows how the identification and estimation methods described in previous chapters of this book, which were all developed with DBM modelling in mind, provide an excellent basis for its application to real-world modelling problems. The main aim of the chapter, however, is to put the DBM approach to modelling in a philosophical context and demonstrate how this is reflected in an illustrative example, where DBM modelling is applied to the investigation of solute transport and dispersion in water bodies. From a philosophical standpoint, DBM modelling stresses the need to rely, whenever possible, on inductive inference from time series data, without over-reliance on pre-conceived notions about the structure of the model that can often lead to over-large simulation models with severe identifiability problems. But, by providing an emulation modelling bridge between such large simulation models, produced in a hypothetico–deductive manner, and parsimonious DBM models that are normally identifiable from the available data, it emphasizes the need to utilize both approaches, in an integrated manner, in order to meet multiple modelling objectives.

Epilogue
Good, Bad or Optimal?

The title of this epilogue comes from a short paper by Rosenbrock and McMorran (1971) which argues that the automatic control of an industrial plant in the vicinity of a steady state, using a control system design method based on optimal control with a quadratic performance index, is usually not feasible and, if feasible, can have serious defects. This paper had a profound effect on me when it was published forty years ago and has led me to always question, in a constructive manner, any methodological approach that is based on purely theoretical concepts and that has not been rigorously evaluated in practical terms. In other words, I believe theoretical analysis is essential, but it cannot be allowed to dominate the world of the scientist and engineer working on problems of the real world.

Of course, this does not mean that I am antipathetic to theoretical concepts and analysis. Indeed, this book has developed numerous recursive algorithms, most often on the basis of some criterion of optimality, such as *Linear Least Squares*, *Nonlinear Least Squares*, *Maximum Likelihood* and *Prediction Error Minimization* (PEM). In the case of transfer function models, the prevailing wisdom in the systems and control community is that the latter PEM approach to the nonlinear identification and estimation of linear transfer function models (see e.g. Ljung, 1987, 1999) should be considered as the norm when modelling linear or near linear systems from sampled time-series data. And this is reinforced in the present book, where the *Refined Instrumental Variable* (RIV) approach is based on the same PEM concepts.

But is PEM always the best approach to use in practical applications? In the stochastic context it, and the related ML approach, are certainly optimal *provided the time series data and the model satisfy the statistical assumptions on which they are based.* In which case, for instance, the parameter estimates will have the required consistency and minimum variance properties; and the minimization of the one-sample-ahead prediction error variance will lead also to the minimization of the multiple-step-ahead prediction error variance. *Such desirable properties are not guaranteed, however, if these statistical assumptions are not satisfied.*

In addition, optimality in the above, somewhat restricted, statistical sense is not necessarily optimal in a wider, less formal sense. For example, *Data-Based Mechanistic* (DBM) modelling, as described in chapter 12, tries to balance the requirements of strict statistical optimality with the need for a model that is 'credible' to the scientist and engineer; one that can be interpreted in physically meaningful terms. For example, the discrete-time, 'black-box' model, which has dominated research on PEM and ML methods, will not normally satisfy the requirements of the natural systems scientist, where a differential equation model of the kind used, for example, in the formulation of natural conservation laws (mass, energy, momentum etc.) has a much greater appeal. In particular, such a model, as produced by the RIVC estimation approach discussed in chapter 8, is immediately familiar and characterized by parameters that are not only independent of the sampling interval (provided this is not too large) but normally have a direct physical meaning and obvious physical units.

In this broader scientific context of the practicing scientist and engineer, the theoretical statistical assumptions required to establish PEM optimality can be violated and the ideal of white noise residuals is sometimes hard to achieve. The best data-based modelling methods are, therefore, those that are robust to the violation of the theoretical assumptions associated with the optimality criterion. Moreover, the real requirement is not that the one-step-ahead predictions should be minimized but that the 'noise-free' output of the model, in response to input perturbations, should be well defined, meaningful and explain as much of the output measurement as possible. Of course, if minimizing one-step-ahead prediction errors also achieves this objective, so much the better. But if not, then a model that can be interpreted in physical terms and whose output 'makes sense' to the scientist is often preferable.

Ep.1 An Illustrative Practical Example

This example shows how a nominally sub-optimal approach to model estimation, the SRIV-ARMA algorithm introduced briefly in section 7.3 of chapter 7, yields results that are, in most respects, superior in practical and interpretative terms to those obtained using nominally optimal methods (the RIV and PEM algorithms). It is concerned with the DBM modelling of the rainfall-level dynamics at Meifod on the River Severn in the UK, as considered previously in section 6.10.2 of Chapter 6.

Since the hourly sampling interval is well suited to discrete-time model estimation (see section 8.7 of chapter 8), this is used here since it allows for easier multi-hour-ahead forecasting analysis. However, continuous-time estimation, employing the SRIVC-ARMA algorithm, could have been used and it would yield similar results in this case.

Discrete-time SRIV-ARMA identification and estimation involves a combination of the SRIV algorithm, to estimate the system model, and the IVARMA algorithm, to separately estimate the ARMA noise model, based on the system model residuals. In the present example, the SRIV option of the CAPTAIN rivbjid routine, identifies a [2 2 4] TF model structure and the resulting rivbj estimated model performs well: 95% of the measured river level variance is explained by the simulated deterministic model output $\hat{x}(k)$. It is also capable of clear physical interpretation: if the TF is decomposed by partial fraction expansion, it results in the parallel flow model of the kind considered in chapters 7 and 10. This suggests that the rainfall can be construed as causing level changes in the river via two parallel pathways: a 'quick flow' pathway with residence time (time constant) of 13.1 hours and partition percentage 61.5%, that can be associated with the surface and near-surface processes; and a 'slow flow' pathway with residence time of 98 hours, and partition percentage 38.5%, arising from the displacement of groundwater into the river bed. Such a decomposition is well known and accepted in hydrological science.

The identified full BJ model structure is identified by rivbjid as [2 2 4 6 0] and the estimated model appears well defined, with low standard errors and clear convergence of the recursive estimates (see plots in Young, 2010d), so it appears perfectly acceptable at first sight. And this is confirmed by PEM estimation using the pem algorithm in the Matlab SID Toolbox, which yields virtually the same model parameter estimates. However, it is clear that these parameter estimates are considerably changed from those of the initial SRIV-ARMA estimated model; and the noise variance $\sigma_\xi^2 = 0.063$ is much higher, so that the coefficient of determination based on $\hat{x}(k)$ is considerably reduced from $R_T^2 = 0.95$ to $R_T^2 = 0.78$. On the other hand, the final residual variance $\sigma^2 = 0.0008$ is much less than the $\sigma^2 = 0.002$ obtained for the SRIV-ARMA estimated model. In other words, the full, nominally optimal, rivbj/pem estimated model has explained the data better in total stochastic terms, with significantly lower final residual error variance, *but only at the very detrimental cost of reducing considerably that part of the model output that can be attributed to the input signal* $u(k)$.

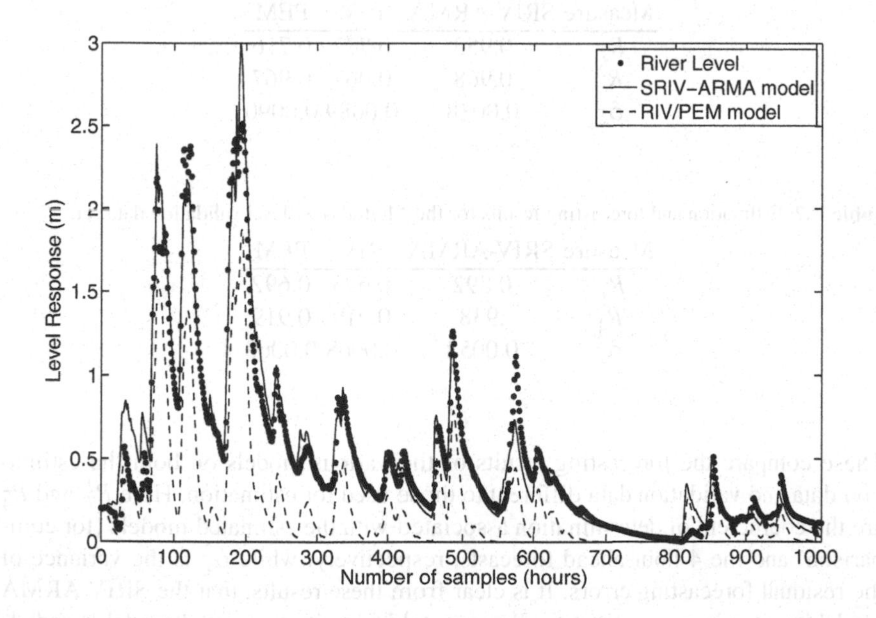

Fig. Ep.1 Comparison of the estimated Meifod model responses with the measured river level.

This inability to attach sufficient importance to the most important system part of the model, as illustrated by the comparison of the model outputs shown in Figure Ep.1, is not acceptable from a DBM modelling standpoint because the purely stochastic, noise part of the model then explains too much of the measured output river level. Moreover, the model decomposition makes little sense in physical terms: it has two real roots associated with two first order systems with residence times of 6.0 h and 3.5 h, so it is not explaining the long term dynamic mode that is so clear in the data and makes obvious hydrological sense. Indeed, this long term behaviour is now being explained by the noise model which, in DBM modelling terms, should explain *only* that part of the measured data that cannot be explained by the causative effect of the input. Moreover, the partition percentages of 233.6% and -133.6%, respectively, have different signs, which are physically meaningless in this context. By contrast, we have seen above that the nominally sub-optimal, constrained SRIV-ARMA estimated model can be interpreted in a physically meaningful manner and so makes a lot of sense for use in environmental management applications and 'what-if' investigations.

But environmental management is not the only reason for modelling rainfall-flow processes: another important use is in the design of flood forecasting systems, where we might expect that the inherent optimal predictive aspects of PEM and RIV estimation should lead to superior 4-hour-ahead forecasting performance. However, this is not the case in this example, as we see in Tables E.1 and E.2.

Table E.1 Estimation and forecasting results for the Meifod river level estimation data set.

Measure	SRIV-ARMA	RIV	PEM
R_T^2	0.953	0.777	0.776
R_4^2	0.968	0.967	0.967
$\hat{\sigma}_4^2$	0.0088	0.0089	0.0090

Table E.2 Estimation and forecasting results for the Meifod river level validation data set.

Measure	SRIV-ARMA	RIV	PEM
R_T^2	0.892	0.693	0.692
R_4^2	.938	0.919	0.919
$\hat{\sigma}_4^2$	0.0052	0.0068	0.0068

These compare the forecasting results of the various models on both the estimation data and validation data different to those used for estimation. Here R_T^2 and R_4^2 are the coefficient of determination associated with the estimated models (for comparison) and the 4-hour-ahead forecasts, respectively; while $\hat{\sigma}_4^2$ is the variance of the residual forecasting errors. It is clear from these results, that the SRIV-ARMA model is not only more scientifically acceptable, but its associated model-based, 4-hour-ahead forecasting performance is at least as good as and, in the more important case of the validation data, rather better than that of the RIV and PEM estimated models.

There are good reasons for the practical advantages of the SRIV and SRIVC algorithms. First, they are the pseudo-linear regression equivalent of the well known response or *Output Error* (OE) minimization method, which itself has proven practically useful and robust over many years. Moreover, as we see below, simulation results suggest that, in the case of the stiff dynamic system considered here, these algorithms can perform considerably better than the OE algorithm when applied to the same data, probably because they exploit instrumental variable estimation. This not only makes the algorithms more robust, because the estimation results are less sensitive to the violation of the prediction error minimization assumptions, it also concentrates the attention of the estimation algorithm on the causal relationship between the input and output measurements by its inclusion of the instrumental variables, which are based on $\hat{x}(k)$ (see chapters 6, 7 and 8).

Of course, it might be thought that the results obtained above from the analysis of the real rainfall-flow data are caused by the idiosyncrasies in the real measured data. However, this is not the case because similar results are obtained using a Monte Carlo simulation study based on the the SRIV-ARMA estimated model (Young, 2010d). These were generated using 500 stochastic realizations, with the random white noise input $e(k)$ to the AR(6) noise model selected randomly for each realization.

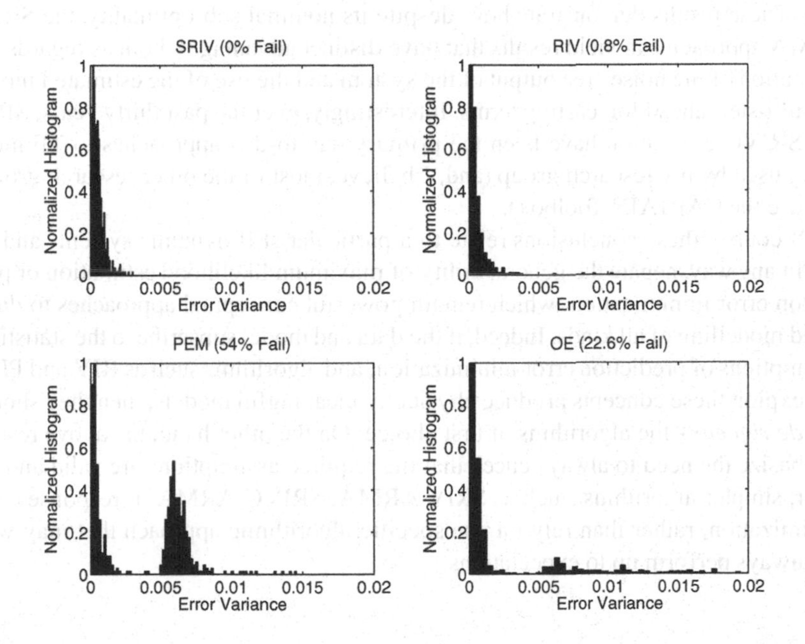

Fig. Ep.2 Histograms of error variance results for the models estimated by the SRIV, RIV, PEM and OE algorithms (obtained by MCS analysis using 500 stochastic realizations).

Figure Ep.2 shows the resulting normalized histograms of the variance σ_ε^2 of the error $\varepsilon(k)$ between the simulated deterministic output $x(k)$ and the model estimate of this, $\hat{x}(k)$, as obtained using the SRIV, RIV, PEM and OE estimation methods (as implemented in the CAPTAIN and SID Toolboxes) over the MCS sample set. These histograms reveal that the PEM and OE algorithm perform quite badly, with 54% and 22.6% failures, respectively; while the SRIV algorithm experiences no failures at all. The RIV algorithm, on the other hand, performs better than with the real data, experiencing only a small 0.8% of failures. The problem with the PEM algorithm and, to a lesser extent, the OE algorithm is that two types of estimation results are encountered: those that yield the correct model; and those that do not. It is believed that the latter problems are probably caused by initialisation problems in these gradient-based optimization algorithms (Ljung, 2003; Young, 2008b).

Ep.2 Conclusions

The primary aim of this epilogue is to emphasize a message that is inherent in most of the discussion and examples presented in the book but that has not been stated explicitly. In particular, it presents practical results, supported by related Monte Carlo simulation, that question an over-reliance on nominally optimal approaches to the identification and estimation of transfer function models from time series data. These results demonstrate how, despite its nominal sub-optimality, the SRIV-ARMA approach can yield results that have distinct advantages, both as regards the estimation of the noise-free output of the system and the use of the estimated model in multi-step-ahead forecasting terms. Interestingly, over the past thirty years, SRIV and SRIVC estimation have been the primary day-to-day approaches to TF modelling used by my research group (and, I believe, most of the other research groups that use the CAPTAIN Toolbox).

Of course, these conclusions relate to a particular stiff dynamic systems and do not, in any way, negate the general utility of maximum likelihood estimation or prediction error minimization, which remain powerful conceptual approaches to data-based modelling of all kinds. Indeed, if the data and model subscribe to the statistical assumptions of prediction error minimization, and algorithms such as RIV and PEM that exploit these concepts produce physically meaningful models, then they should be - *de rigueur* - the algorithms of first choice. On the other hand, the above results emphasize the need to always check that the required assumptions are valid and try other, simpler algorithms, such as SRIV-ARMA, SRIVC-ARMA or response error minimization, rather than rely on one specific algorithmic approach that may well not always perform up to expectations.

Appendix A
The K. F. Gauss Derivation of Recursive Least Squares

This Appendix is based on pages 53 to 55 of the book *Méthode des Moindres Carrés: Memoires sur la Combinaison des Observations*, which is the French translation by J. Bertrand of Gauss's collected works on least squares (1803-1826) and was published in 1855, with the authorization of Gauss. In Section 35, page 53, Bertrand's translation states:

> Nous traiterons particulièrement le problème suivant, tant à cause de son utilité pratique, que de la simplicité de la solution:
>
> *Trouver les changements que les valeurs les plus plausibles des inconnues subissent par l'adjonction d'une nouvelle équation, et assigner les poids de ces nouvelles determinations.*

In this Appendix, we will reproduce the following two and a half pages of Bertrand's book in greyboxes. These constitute the main part of Gauss's derivation but, in order to aid the reader both to appreciate the analysis and to draw comparisons with the equivalent vector-matrix analysis in the main text, comments are added at various stages in the analysis. These appear as ordinary text and, in all cases, the vector matrix nomenclature is similar to that used in chapters 3 and 4 of the present book. An English translation of Bertrand's book has been produced by Trotter (1957).

Gauss's analysis begins as follows:

Conservons les notations précédentes. Les équations primitives, réduites à avoir pour poids l'unité, seront

$$v = 0,\ v' = 0,\ v'' = 0,\ \ldots;$$

on aura

$$\Omega = v = 0,\ v'^2 = 0,\ v''^2 = 0, \tag{A.1}$$

ξ, η, ζ ,etc., seront les dérivées partielles

$$\frac{d\Omega}{2dx},\ \frac{d\Omega}{2dy},\ \frac{d\Omega}{2dz},\ldots, \tag{A.2}$$

et enfin on aura, par l'élimination

$$\begin{aligned}x &= A + (\alpha\alpha)\xi + (\alpha\beta)\eta + (\alpha\gamma)\zeta + \cdots \\ y &= B + (\alpha\beta)\xi + (\beta\beta)\eta + (\beta\gamma)\zeta + \cdots \\ z &= C + (\alpha\gamma)\xi + (\beta\gamma)\eta + (\gamma\gamma)\zeta + \cdots\end{aligned} \tag{A.3}$$

Comment:

Here Ω is the 'sum of squares cost function after a given number of observations, with the unknown parameters set to their true values. Gauss refers to these observations as 'equations' since he associates each new set of observations with the latest equation of the system under study. Using the nomenclature of the present book, each equation is of the form,

$$e(k) = y(k) - \mathbf{x}^T(k)\mathbf{a} \tag{A.4}$$

This is obtained from equation (3.3) of Chapter 3. Using the nomenclature of (4.37) in Chapter 4, the whole set of equations so constituted after $k-1$ observations can be written

$$\mathbf{e}(k-1) = \mathbf{y}(k-1) - \mathbf{X}^T(k-1)\mathbf{a} \tag{A.5}$$

where we have assumed initially $k-1$ observations in order to facilitate comparison of later equations with equivalent ones in the main text.

The variables ξ, η, ζ etc., are the partial derivatives (or gradients) of Ω with respect to the unknown parameters x, y, z etc., as defined in equation (A.2). Equation (A.3) (which, in the original, is Gauss's equation (1)) is simply a statement of the solution to the normal equations of least squares. This becomes clear if we consider equation (3.25) of Chapter 3 and obtain the following:

$$\begin{aligned}\mathbf{a} &= \hat{\mathbf{a}} + [\mathbf{X}^T(k-1)\mathbf{X}(k-1)]^{-1}\mathbf{X}^T(k-1)\mathbf{e}(k-1) \\ &= \hat{\mathbf{a}} + [\sum_{i=1}^{k-1} \mathbf{x}(i)\mathbf{x}(i)^T]^{-1} \sum_{i=1}^{k-1} \mathbf{x}(i)\mathbf{e}(i)\end{aligned} \tag{A.6}$$

In equation (A.3), x, y, z etc. are the elements of the true parameter vector $\mathbf{a}$; A, B, C etc., are the elements of the estimate vector $\hat{\mathbf{a}}$; $\alpha\alpha$, $\alpha\beta$, $\alpha\gamma$ etc., are the elements of the inverse matrix $[\mathbf{X}^T(k-1)\mathbf{X}(k-1)]^{-1} = \mathbf{P}(k-1)$; and ξ, η, ζ etc., are associated with the gradient vector $\mathbf{g}(k-1) = \mathbf{X}^T(k-1)\mathbf{e}(k-1)$. We can see that this is the gradient vector by referring to the least squares cost function in this case, i.e.

$$\Omega = \sum_{i=1}^{k-1} e^2(i) = [\mathbf{e}^T(k-1)][\mathbf{e}(k-1)] = [\mathbf{y}(k-1) - \mathbf{X}(k-1)\mathbf{a}]^T[\mathbf{y}(k-1) - \mathbf{X}(k-1)\mathbf{a}]$$

the gradient of which is (see Appendix B),

$$\mathbf{g}(k-1) = \mathbf{X}^T(k-1)[\mathbf{y}(k-1) - \mathbf{X}(k-1)\mathbf{a}]$$
$$= \mathbf{X}^T(k-1)[\mathbf{e}(k-1)] = \sum_{i=1}^{k-1} \mathbf{x}(i)[e(i)] \quad \text{(A.7)}$$

From this comparison, and noting that $\mathbf{X}^T(k-1)\mathbf{X}(k-1)]^{-1}$ is the $\mathbf{P}(k-1)$ matrix of the main text, we see that equation (A.3) can be written, in the nomenclature of the present book, as,

$$a_1 = \hat{a}_1 + p_{11}g_1 + p_{12}g_2 + p_{13}g_3 + \cdots$$
$$a_2 = \hat{a}_2 + p_{12}g_1 + p_{22}g_2 + p_{23}g_3 + \cdots$$
$$a_3 = \hat{a}_3 + p_{13}g_1 + p_{23}g_2 + p_{33}g_3 + \cdots$$

where a_1, a_2, a_3, etc., are the elements of the unknown parameter vector $\mathbf{a}$; $\hat{a}_1$, $\hat{a}_2$, $\hat{a}_3$ etc., are the elements of the estimate vector $\hat{\mathbf{a}}(k-1)$ after k-l samples; $p(i,j)$ are the elements of $\mathbf{P}(k-1)$ and g_1, g_2, g_3, etc. are the elements of $\mathbf{g}(k-1)$.

Supposons maintenant que l'on ait une nouvelle équation approximative,

$$v^* = 0 \quad \text{(A.8)}$$

dont nous supposerons le poids égal a l'unité. Cherchons les changements que subiront les valeurs les plus plausibles A, B, C, etc. et celles des coefficients $\alpha\alpha$, $\beta\beta$ etc.

Posons

$$\Omega + v^2 = \Omega^*,$$
$$\frac{1}{2}\frac{d\Omega^*}{dx} = \xi^*, \; \frac{1}{2}\frac{d\Omega^*}{dy} = \eta^*, \; \frac{1}{2}\frac{d\Omega^*}{dz} = \zeta^*, \cdots, \quad \text{(A.9)}$$

et soit

$$x - A^* + (\alpha\alpha)^*\xi^* + (\alpha\beta)^*\eta^* \mid (\alpha\gamma)^*\zeta^* \mid \quad \text{(A.10)}$$

le résultat de l'élimination

Comment:

Ω^* in equation (A.9) is the updated sum of squares with v^* denoting the latest error squared term. The associated new gradients are given as ξ^*, η^*, ζ^*, etc; and equation (A.10) is the new equation for x (with equations for y, z etc. not shown). This follows directly from equation (A.3) but with the updated values for the variables denoted by the star superscripts. Gauss is simply pointing out that all the variables need to be updated on receipt of new information in order to obtain new estimates A^*, B^*, C^* etc. This is the prelude to his development of the recursive equations, which now follows.

Soit enfin

$$v^* = fx + gy + hz + \cdots + k, \quad \text{(A.11)}$$

qui deviendra, en ayant égard aux équations (1)

$$v^* = f\xi + g\eta + h\zeta + \cdots + K, \quad \text{(A.12)}$$

et posons

$$Ff + Gg + Hh + \cdots = \omega \quad \text{(A.13)}$$

K sera évidemment la valeur la plus plausible de la fonction v^*, telle qu'elle résulte des équations primitives, sans avoir égard à la valeur fournie par la nouvelle observation, et $1/\omega$ sera le poids de cette détermination.

Comment:

Equation (A.11) is Gauss 's version of equation (A.4) with v^* denoting the latest error $e(k)$; f, g, h etc., the regressors [the elements of $\mathbf{x}(k)$]; and k, the new observation of the dependent variable $y(k)$. He obtains (A.12) by substituting for x, y, z etc. from equation (A.3), i.e.

$$\begin{aligned} v^* &= f(A + (\alpha\alpha)\xi + (\alpha\beta)\eta + \cdots) + g(B + (\alpha\beta)\,\xi + (\beta\beta)\eta + \cdots) \\ &\quad + h(C + (\alpha\gamma)\xi + (\beta\gamma)\eta + \cdots) + \cdots + k \\ &= [f(\alpha\alpha) + g(\alpha\beta) + h(\alpha\gamma) + \cdots]\xi + [f(\alpha\beta) + g(\beta\beta) + h(\beta\gamma) + \cdots]\eta \\ &\quad + fA + gB + hC + K = F\xi + G\eta + \cdots + k \end{aligned}$$

where K will be recognized as the latest recursive residual (innovations process). Using the nomenclature of the present paper, the reader can verify that the equivalent vector matrix expression is

$$e(k) = \mathbf{x}^T(k)\mathbf{P}(k-1)\mathbf{g}(k-1) + \{y(k) - \mathbf{x}^T(k)\hat{\mathbf{a}}(k-1)\} \quad \text{(A.14)}$$

where Gauss's F, G, H etc., are the elements of the vector $\mathbf{x}^T(k)P(k-1)$ andK is the latest recursive residual shown in curly brackets. Note also that ω defined by equation (A.13) is equivalent to $\mathbf{x}^T(k)\mathbf{P}(k-1)\mathbf{x}(k)$.

Or nous avons

$$\xi^* = \xi + fv^*, \ \eta^* = \eta + gv^*, \ \zeta^* = \zeta + hv^*, \cdots, \quad \text{(A.15)}$$

et, par suite,

$$F\xi^* + G\eta^* + H\zeta^* + K = v^*(1 + Ff + Gg + Hh + \cdots);$$

d'où l'on déduit:

$$\nu^* = \frac{F\xi^* + G\eta^* + H\zeta^* + K}{1+\omega} \tag{A.16}$$

Comment:

Here ξ^*, η^*, ζ^* etc., in (A.15), represent the updated gradient measures. In the vector terms used above, these equations are equivalent to

$$\mathbf{g}(k) = \mathbf{g}(k-1) + \mathbf{x}(k)e(k) \tag{A.17}$$

which follows from equation (A.7). Using equations (A.15), Gauss now defines $\nu*$ in terms of the updated gradients, i.e.

$$\nu^* = F\xi^* + G\eta^* + H\zeta^* + K = F(\xi^* + f\nu^*) + G(\eta^* + g\nu^*) + \cdots + K$$

so that,

$$\nu^*[1 + Ff + Gg + Hh + \cdots] = F\xi^* + G\eta^* + \cdots + K$$

and equation (A.16) for ν^* follows because of its definition in (A.12). The following vector-matrix equivalent of equation (A.16) is obtained straightforwardly by reference to equation (A.14) and the vector matrix definition of $\omega = \mathbf{x}^T(k)\mathbf{P}(k-1)\mathbf{x}(k)$

$$e(k) = \frac{\mathbf{x}^T(k)\mathbf{P}(k-1)\mathbf{g}(k) + \{y(k) - \mathbf{x}^T(k)\hat{\mathbf{a}}(k-1)\}}{\mathbf{x}^T(k)\mathbf{P}(k-1)\mathbf{x}(k)}$$

where $\mathbf{g}(k)$ is defined in (A.17).

On a, en outre,

$$\begin{aligned} x &= A + (\alpha\alpha)\xi^* + (\alpha\beta)\eta^* + (\alpha\gamma)\zeta^* + \cdots - \nu^*[f(\alpha\alpha) + g(\alpha\beta) + h(\alpha\gamma) + \cdots] \\ &= A + (\alpha\alpha)\xi^* + (\alpha\beta)\eta^* + \cdots - \nu^* \\ &= A + (\alpha\alpha)\xi^* + (\alpha\beta)\eta^* + \cdots - \frac{F}{1+\omega}(F\xi^* + G\eta^* + H\zeta^* + \cdots + K) \end{aligned} \tag{A.18}$$

Nous déduirons de là,

$$A^* + A - \frac{FK}{1+\omega} \tag{A.19}$$

qui sera la valeur la plus plausibles de x, déduite de toutes les observations. On aura aussi

$$(\alpha\alpha^*) = (\alpha\alpha) - \frac{F^2}{1+\omega} \tag{A.20}$$

par conséquent,

$$\frac{1}{(\alpha\alpha) - \frac{F^2}{1+\omega}}$$

sera le poids de cette determination.

On trouvera de la mème manière, pour valeur la plus plausible de y, déduite de toutes observations,

$$B^* = B - \frac{GK}{1+\varpi}$$

le poids de cette détermination sera

$$\frac{1}{(\beta\beta) - \frac{G^2}{1+\omega}}$$

et ainsi de suite.

Le problème est donc résolu.

Comment:

The above equations (A.19) and (A.20) for A^* and $\alpha\alpha^*$, respectively, constitute the recursive least squares update equations for the first unknown parameter and the associated diagonal element of the inverse matrix $\mathbf{P}(k)$. The associated equations for all other parameter estimates and the elements of the $\mathbf{P}(k)$ matrix follow in a similar manner to provide, finally, the complete recursive algorithm. Gauss does not continue further, however, since the subsequent derivation is obvious.

Equation (A.18) follows by substituting from equations (A.15) into equation (A.3) in the following manner,

$$\begin{aligned} x &= A + (\alpha\alpha)[\xi^* - f\nu^*] + (\alpha\beta)[\eta^* + g\nu^*] + \cdots \\ &= A + (\alpha\alpha)\xi^* + (\alpha\beta)\eta^* + \cdots \nu^*[f(\alpha\alpha) + g(\alpha\beta) + \cdots]; \end{aligned}$$

and then noting that $F = [f(\alpha\beta^*) + g(\alpha\beta^*) + \cdots]$, while ν^* is defined in terms of ξ^*, η^*, ζ^*,, etc. by equation (A.16).

Equation (A.20) which, taken together with similarly derived equations for $\alpha\beta^*$, $\alpha\gamma^*$,, etc. constitutes the equivalent of the matrix inversion lemma (see equation (3.12), section 3.1 of chapter 3), can be obtained quite straightforwardly but with rather lengthy algebraic manipulation. It is not clear from Gauss's reported analysis, however, exactly how he obtained these relationships since he does not include the details: in the classic phrase 'one has also' he parallels the over-used present day phrase 'it can be shown' and leaves the reader to his own devices. Alas poor reader, we will do the same'!

It remains to note that the vector-matrix equivalent of the above equations are, of course, the following recursive least squares equations of algorithm RLS in Chapter:

$$\hat{\mathbf{a}}(k) = \hat{\mathbf{a}}(k-1) + \frac{\mathbf{P}(k-1)\mathbf{x}(k)}{1+\mathbf{x}^T(k)\mathbf{P}(k-1)\mathbf{x}(k)}\{y(k) - \mathbf{x}^T(k)\hat{\mathbf{a}}(k-1)\}$$

$$\mathbf{P}(k) = \mathbf{P}(k-1) - \frac{\mathbf{P}(k-1)\mathbf{x}(k)\mathbf{x}^T(k)\mathbf{P}(k-1)}{1+\mathbf{x}^T(k)\mathbf{P}(k-1)\mathbf{x}(k)}$$

The above analysis, carried out at the beginning of the nineteenth century, serves to illustrate yet again the enormous contributions Gauss made to science and mathematics. While it may be arguable that Gauss and Lagrange evolved the method of least squares independently and at about the same time, it is clear that only Gauss was responsible for the development of the theory in its most elegant, recursive form. And while the development of the recursive form is fairly straightforward in these days of the digital computer and matrix analysis, we can only marvel at Gauss's derivation relying, as it had to, on the use of scalar algebra. Finally, it is nice to note that Gauss did not develop the method for its own sake (although he too was surely impressed by the elegance of the algorithmic form), but because it solved a very real practical problem. Gauss's practicality is demonstrated later in the analysis when he concludes (page 58 of Bertrand's book):

> *If, after the calculation is finished, several new equations should be adjoined to the original, or if the weights attributed to several of them were in error, the calculations of the corrections would become very complicated and it would be better to begin all over again.*

Of course, if he had access to the modern digital computer, he would not have needed to worry.

Appendix B
Basic Mathematical and Statistical Background

B.1 Matrix Algebra

This section presents some of the most useful results in matrix algebra that are required in the main text.

B.1.1 Matrices

A matrix is defined as a rectangular array of elements arranged in rows and columns; in this book it is denoted by a bold-face capital letter, e.g.

$$\mathbf{A} = \begin{bmatrix} a_{11} & a_{12} & \cdots & a_{1n} \\ a_{21} & a_{22} & \cdots & a_{2n} \\ \cdots & \cdots & \cdots & \cdots \\ a_{m1} & a_{m2} & \cdots & a_{mn} \end{bmatrix} \tag{B.1}$$

Often $\mathbf{A}$ is alternatively denoted by $[a_{i,j}]$ to indicate that it is characterized by elements $a_{i,j}, i = 1,2,\ldots,m; j = 1,2,\ldots,n$. If it has elements arranged in m rows and n columns like this, then it is said to be of order m by n, usually written $m \times n$. The following should be noted in relation to matrices:

1. A null matrix has all of its elements set to zero, i.e. $a_{i,j} = 0 \; \forall \; i, j$.
2. A symmetric matrix is a square matrix in which $a_{i,j} = a_{j,i}$; i.e. it is symmetric about the diagonal elements.
3. The trace of a square $n \times n$ matrix, denoted by *Tr.* or *tr.*, is the sum of its diagonal elements i.e. $Tr.\mathbf{A} = a_{11} + a_{22} + \ldots + a_{nn}$.
4. A diagonal matrix is a square matrix with all its elements *except those on the diagonal* set to zero i.e.,

$$\mathbf{A} = \begin{bmatrix} a_{11} & 0 & \cdots & 0 \\ 0 & a_{22} & \cdots & 0 \\ \dots & \dots & \dots & \dots \\ 0 & 0 & \cdots & a_{nn} \end{bmatrix}$$

5. An $n \times n$ diagonal matrix with elements set to unity is denoted by $\mathbf{I}_n$ and termed the identity (or unit) matrix of order n, e.g. for a 3×3 identity matrix

$$\mathbf{I}_3 = \begin{bmatrix} 1\,0\,0 \\ 0\,1\,0 \\ 0\,0\,1 \end{bmatrix}$$

sometimes the subscript is omitted if the order is obvious.
6. An idempotent matrix is a square matrix such that $\mathbf{A}^2 = \mathbf{AA} = \mathbf{A}$, i.e. it remains unchanged when multiplied by itself.

B.1.2 Vectors

A matrix of order $n \times 1$ contains a single column of n elements and is termed a column vector (or sometimes just a vector): in this book, it is denoted by a bold lower case letter, i.e. for a vector **b**

$$\mathbf{b} = \begin{bmatrix} b_1 \\ b_2 \\ \dots \\ b_n \end{bmatrix} \tag{B.2}$$

B.1.3 Matrix Addition (or Subtraction)

If two matrices **A** and **B** are of the same order, then we define $\mathbf{A}+\mathbf{B}$ to be a new matrix **C** where, $c_{ij} = a_{ij} + b_{ij}$. In other words, the addition of the matrices is accomplished by adding corresponding elements. $\mathbf{A} - \mathbf{B}$ is defined in an analogous manner.

B.1.4 Matrix or Vector Transpose

The transpose of a matrix **A** is obtained from **A** by interchanging the rows and columns: in this book, it is denoted by a superscript capital T; e.g. for **A** defined in (B.1),

$$\mathbf{A}^T = \begin{bmatrix} a_{11} & a_{21} & \cdots & a_{m1} \\ a_{12} & a_{22} & \cdots & a_{m2} \\ \cdots & \cdots & \cdots & \cdots \\ a_{1n} & a_{2n} & \cdots & a_{mn} \end{bmatrix}$$

The transpose of a column vector $\mathbf{b}$, denoted by $\mathbf{b}^T$, is termed a row vector, e.g. for $\mathbf{b}$ in (B.2), $\mathbf{b}^T = [b_1\ b_2\ \ldots\ b_n]$.

Note that

1. in the case of a symmetric matrix $\mathbf{A}^T = \mathbf{A}$.
2. $[\mathbf{A}^T]^T = \mathbf{A}$.
3. $[\mathbf{A}+\mathbf{B}]^T = \mathbf{A}^T + \mathbf{B}^T$.

B.1.5 Matrix Multiplication

If $\mathbf{A}$ is of order $m \times n$ and $\mathbf{B}$ is of order $n \times p$ then the product $\mathbf{AB}$ is defined to be a matrix of order $m \times p$ whose $(ij)^{th}$ element c_{ij} is given by

$$c_{ij} = \sum_{k=1}^{n} a_{ik} b_{kj}$$

i.e. the $(ij)^{th}$ element is obtained by, in turn, multiplying the elements of the ith row of the matrix $\mathbf{A}$ by the jth column of the matrix $\mathbf{B}$ and summing over all terms (therefore, the number of elements (n) in each row of $\mathbf{A}$ must be equal to the number of elements in each column of $\mathbf{B}$). Note that, in general, the commutative law of multiplication that applies for scalars does not apply for matrices, i.e.

$$\mathbf{AB} \neq \mathbf{BA}$$

so that pre-multiplication of $\mathbf{B}$ by $\mathbf{A}$ does not, in general, yield the same as post-multiplication of $\mathbf{B}$ by $\mathbf{A}$. However, pre-multiplying or post-multiplying by the identity matrix $\mathbf{I}_n$ leaves the matrix unchanged i.e.

$$\mathbf{AI}_n = \mathbf{I}_n\mathbf{A} = \mathbf{A}$$

Note also that for $\mathbf{A}$ of order $m \times n$, $\mathbf{B}$ of order $n \times p$ and $\mathbf{C}$ of order $p \times q$ the following results apply:

1. **(AB)C= A(BC)**.
2. With orders m, n, p and q chosen appropriately, **A(B+C) = AB+AC** and **(B+C)A= BA+CA**.
3. For **A**, **B** and **C**, the multiplication by a scalar λ yields a corresponding matrix with all its elements multiplied by λ , i.e. $\lambda\mathbf{A} = [\lambda a_{ij}]$.
4. $[\mathbf{AB}]^T = \mathbf{B}^T\mathbf{A}^T$.

5. $[\mathbf{ABC}]^T = \mathbf{C}^T\mathbf{B}^T\mathbf{A}^T$ (since $[\mathbf{ABC}]^T = [(\mathbf{AB})\mathbf{C}]^T = \mathbf{C}^T[\mathbf{AB}]^T = \mathbf{C}^T\mathbf{B}^T\mathbf{A}^T$ from 4.

Finally, it should be observed that, for a vector $\mathbf{x} = [x_1\, x_2\, \cdots\, x_n]^T$, the inner product $\mathbf{x}^T\mathbf{x}$ yields a scalar quantity that is the sum of the squares of the elements of $\mathbf{x}$, i.e.

$$[x_1\, x_2\, \cdots\, x_n] \begin{bmatrix} x_1 \\ x_2 \\ \cdots \\ x_n \end{bmatrix} = x_1^2 + x_2^2 + \cdots + x_n^2$$

The outer product $\mathbf{xx}^T$, on the other hand yields a symmetric square matrix of order $n \times n$, whose elements are the squares (on the diagonal) and cross products (elsewhere) of the $\mathbf{x}$ elements, i.e.

$$\begin{bmatrix} x_1 \\ x_2 \\ \cdots \\ x_n \end{bmatrix} [x_1\, x_2\, \cdots\, x_n] = \begin{bmatrix} x_1^2 & x_1x_2 & \cdots & x_1x_n \\ x_2x_1 & x_2^2 & \cdots & x_2x_n \\ \cdots & \cdots & \cdots & \cdots \\ x_nx_1 & x_nx_2 & \cdots & x_n^2 \end{bmatrix}$$

B.1.6 Determinant of a Matrix

The determinant of a square $n \times n$ matrix $\mathbf{A}$ is a scalar quantity, denoted by $|\mathbf{A}|$ or $\det\mathbf{A}$, obtained by performing certain systematic operations on the matrix elements. In particular, if the cofactors c_{ij} of $\mathbf{A}$ are defined as follows:

$$c_{ij} = (-1)^{i+j}|A_{ij}| \tag{B.3}$$

where $|A_{ij}|$ is the determinant of the sub-matrix obtained when the ith row and j^{th} column are deleted from $\mathbf{A}$, then $\det\mathbf{A}$ can be defined as follows in terms of the elements of the ith row or their cofactors.

$$|A| = a_{i1}c_{i1} + a_{i2}c_{i2} + \cdots + a_{in}c_{in} \tag{B.4}$$

$|\mathbf{A}|$ may be similarly expanded in terms of the elements of any row or column.

Note that, for a matrix of order greater than 2, it is necessary to nest the operations (B.3) and (B.4) and apply them repeatedly until $\mathbf{A}_{ij}$ is reduced to a scalar, in which case the determinant is equal to the scalar. The following example demonstrates this process:

$$A = \begin{bmatrix} a_{11} & a_{12} & a_{13} \\ a_{21} & a_{22} & a_{23} \\ a_{31} & a_{32} & a_{33} \end{bmatrix}$$

then,

$$|A| = a_{11}\begin{vmatrix} a_{22} & a_{23} \\ a_{32} & a_{33} \end{vmatrix} - a_{12}\begin{vmatrix} a_{21} & a_{23} \\ a_{31} & a_{33} \end{vmatrix} + a_{13}\begin{vmatrix} a_{21} & a_{22} \\ a_{31} & a_{32} \end{vmatrix}$$

so that, applying (B.3) and (B.4) again to the sub-determinants, we obtain,

$$|A| = a_{11}(a_{22}a_{33} - a_{32}a_{23}) - a_{12}(a_{21}a_{33} - a_{31}a_{23}) + a_{13}(a_{21}a_{32} - a_{31}a_{22})$$

B.1.7 Partitioned Matrices

Since a matrix is a rectangular array of elements, we may divide it up by means of horizontal and vertical dotted lines into smaller rectangular arrays of sub-matrices: e.g.

$$\mathbf{A} = \left[\begin{array}{ccc:c} a_{11} & a_{12} & a_{13} & a_{14} \\ a_{21} & a_{22} & a_{23} & a_{24} \\ \hdashline a_{31} & a_{32} & a_{33} & a_{34} \end{array}\right]$$

has been divided in this manner into 4 sub-matrices

$$\mathbf{A}_{11} = \begin{bmatrix} a_{11} & a_{12} & a_{13} \\ a_{21} & a_{22} & a_{23} \end{bmatrix} : \mathbf{A}_{12} = \begin{bmatrix} a_{14} \\ a_{24} \end{bmatrix} : \mathbf{A}_{21} = \begin{bmatrix} a_{31} & a_{32} & a_{33} \end{bmatrix}$$

$$\mathbf{A}_{22} = a_{34}$$

So that $\mathbf{A}_{11}$ is a 2×3 submatrix, $\mathbf{A}_{12}$ is a 2×1 column vector, $\mathbf{A}_{21}$ is a 1×3 row vector, and $\mathbf{A}_{22}$ is a scalar. As a result $\mathbf{A}$ can be denoted by

$$\mathbf{A} = \left[\begin{array}{c:c} \mathbf{A}_{11} & \mathbf{A}_{12} \\ \hdashline \mathbf{A}_{21} & \mathbf{A}_{22} \end{array}\right] \tag{B.5}$$

The basic operations for addition, multiplication and transposition apply for partitioned matrices but the matrices must be partitioned comformably to allow for such operations. A multiplicative example is:

$$\mathbf{AB} = \left[\begin{array}{c:c} \mathbf{A}_{11} & \mathbf{A}_{12} \\ \hdashline \mathbf{A}_{21} & \mathbf{A}_{22} \end{array}\right] \left[\begin{array}{c} \mathbf{B}_{11} \\ \hdashline \mathbf{B}_{21} \end{array}\right] = \left[\begin{array}{c} \mathbf{A}_{11}\mathbf{B}_{11} + \mathbf{A}_{12}\mathbf{B}_{21} \\ \hdashline \mathbf{A}_{21}\mathbf{B}_{11} + \mathbf{A}_{22}\mathbf{B}_{21} \end{array}\right]$$

The results of such operations will be the same as would be obtained by multiplying the unpartitioned matrices element by element but the partitioning approach may be extremely useful in simplifying the analysis.

One theorem for partitioned matrices that is useful concerns the determinant of a partitioned matrix $\mathbf{A}$ such as (B.5). It can be shown (e.g. Gantmacher (1960); Dhrymes (1970)) that

$$|\mathbf{A}| = |\mathbf{A}_{22}|.|\mathbf{A}_{11} - \mathbf{A}_{12}\mathbf{A}_{22}^{-1}\mathbf{A}_{21}|$$

or alternatively,

$$|\mathbf{A}| = |\mathbf{A}_{11}|.|\mathbf{A}_{22} - \mathbf{A}_{21}\mathbf{A}_{11}^{-1}\mathbf{A}_{12}|$$

where $\mathbf{A}_{11}^{-1}$ and $\mathbf{A}_{22}^{-1}$ are, respectively, the inverses of the matrices $\mathbf{A}_{11}$ and $\mathbf{A}_{22}$, respectively, as defined in the next sub-section.

B.1.8 Inverse of a Matrix

If a matrix $\mathbf{A}^{-1}$ exists such that $\mathbf{A}\mathbf{A}^{-1} = \mathbf{A}^{-1}\mathbf{A} = \mathbf{I}$, where $\mathbf{I}$ is an appropriately ordered identity matrix, then $\mathbf{A}^{-1}$ is termed the inverse (or reciprocal) of $\mathbf{A}$ by analogy with the scalar situation.

The inverse of a square matrix $\mathbf{A}$ of order $n \times n$ is obtained from $\mathbf{A}$ by means of the formula,

$$\mathbf{A}^{-1} = \frac{1}{|\mathbf{A}|}[\text{Adj.}\mathbf{A}] = \begin{bmatrix} \frac{c_{11}}{|\mathbf{A}|} & \frac{c_{21}}{|\mathbf{A}|} & \cdots & \frac{c_{n1}}{|\mathbf{A}|} \\ \cdots & \cdots & \cdots & \cdots \\ \frac{c_{1n}}{|\mathbf{A}|} & \frac{c_{2n}}{|\mathbf{A}|} & \cdots & \frac{c_{nn}}{|\mathbf{A}|} \end{bmatrix}$$

where Adj.$\mathbf{A}$ denotes the adjoint of the matrix $\mathbf{A}$ and is obtained as the transpose of an $n \times n$ matrix $\mathbf{C}$ with elements c_{ij} that are the co-factors of $\mathbf{A}$ as defined by (B.3) in section B.1.6., i.e.,

$$c_{11} = |\mathbf{A}_{11}|, \; c_{12} = -|\mathbf{A}_{12}|, \; c_{22} = |\mathbf{A}_{22}| \quad \text{etc.}$$

Note that, by definition, the inverse will only exist if $|\mathbf{A}| \neq 0$; otherwise the matrix is non-invertible or singular. A non-singular matrix is, therefore, invertible.

Several theorems on inverse matrices are useful, e.g.

1. $[\mathbf{AB}]^{-1} = \mathbf{B}^{-1}\mathbf{A}^{-1}$
2. $[\mathbf{AB}][\mathbf{B}^{-1}\mathbf{A}^{-1}] = \mathbf{A}[\mathbf{B}\mathbf{B}^{-1}]\mathbf{A}^{-1} = \mathbf{A}\mathbf{I}\mathbf{A}^{-1} = \mathbf{A}\mathbf{A}^{-1} = \mathbf{I}$
3. $[\mathbf{ABC}]^{-1} = \mathbf{C}^{-1}\mathbf{B}^{-1}\mathbf{A}^{-1}$
4. $[\mathbf{A}^T]^{-1} = [\mathbf{A}^{-1}]^T$
5. $|\mathbf{A}^{-1}| = 1/|\mathbf{A}|$

One of the most common uses of the inverse matrix is in solving a set of algebraic, simultaneous equations such as,

$$\mathbf{X}\mathbf{a} = \mathbf{b} \tag{B.6}$$

where $\mathbf{X}$ is a known $n \times n$ matrix, $\mathbf{a}$ is an $n \times 1$ vector of unknowns, and $\mathbf{b}$ is a known $n \times 1$ vector. The reader can easily verify that this represents a set of simultaneous equations in the elements of $\mathbf{a}$, where $\mathbf{a} = [a_1\, a_2\, \dots\, a_n]^T$, by defining $\mathbf{X} = [x_{ij}]$ and $\mathbf{b} = [b_1\, b_2\, \dots\, b_n]^T$. Premultiplying both sides of (B.6) by $\mathbf{X}^{-1}$, we obtain

$$\mathbf{X}^{-1}\mathbf{X}\mathbf{a} = \mathbf{X}^{-1}\mathbf{b} \text{ or } \mathbf{I}\mathbf{a} = \mathbf{X}^{-1}\mathbf{b}$$

so that

$$\mathbf{a} = \mathbf{X}^{-1}\mathbf{b} = \frac{1}{|X|}[\text{Adj.}\mathbf{X}]\mathbf{b}$$

This is the required solution for $\mathbf{a}$ and is an alternative to other methods of solution such as pivotal elimination.

B.1.9 Quadratic Forms

A *quadratic form* in a vector $\mathbf{e} = [e_1\ e_2\ \ldots\ e_n]^T$ is defined as $\mathbf{e}^T\mathbf{Q}\mathbf{e}$ where $\mathbf{Q}$ is a symmetric matrix of order $n \times n$. The reader can verify that, for a $\mathbf{Q} = [q_{ij}]$ with off-diagonal elements $q_{ij} = q_{ji}$, $\mathbf{e}^T\mathbf{Q}\mathbf{e}$ is a scalar given by

$$\begin{aligned}\mathbf{e}^T\mathbf{Q}\mathbf{e} &= q_{11}e_1^2 + 2q_{12}e_1e_2 + \ldots + 2q_{1n}e_1e_n + \\ &\qquad q_{22}e_2^2 + \ldots + 2q_{2n}e_2e_n + \ldots + q_{nn}e_n^2\end{aligned} \tag{B.7}$$

Note that if $\mathbf{Q}$ is *diagonal*, then this reduces to (cf. inner product)

$$\mathbf{e}^T\mathbf{Q}\mathbf{e} = q_{11}e_1^2 + q_{22}e_2^2 + \ldots + q_{nn}e_n^2$$

A quadratic form such as (B.7) is sometimes termed the *Weighted Euclidian Squared Norm* of the vector $\mathbf{e}$ and is denoted by $||\mathbf{e}||_Q^2$ As we see, it represents a very general or weighted (by the elements of $\mathbf{Q}$) 'sum of squares' type operation on the elements of $\mathbf{e}$. It proves particularly useful as a cost (or criterion function) if $\mathbf{e}$ represents a vector of errors (or lack of fit) associated with a model.

B.1.10 Positive Definite or Semi-Definite Matrices

A symmetric matrix $\mathbf{A}$ is said to be *positive definite* (p.d.) if $\mathbf{x}^T\mathbf{A}\mathbf{x} > 0$, where $\mathbf{x}$ is any non-null vector. It is termed *positive semi-definite* (p.s.d.) if $\mathbf{x}^T\mathbf{A}\mathbf{x} \geq 0$. For an $n \times n$ p.d. matrix $\mathbf{A}$, $a_ii > 0, i = 1,2,\ldots,n$; for a p.s.d. matrix $\mathbf{A}$, $a_{ii} \geq 0, i = 1,2,\ldots,n$. Note that if $\mathbf{A}$ is p.d. then $\mathbf{A}$ is non-singular and can be inverted; if $\mathbf{A}$ is p.s.d. (but not p.d.) then $\mathbf{A}$ is singular (Dhrymes, 1970).

B.1.11 The Rank of a Matrix

The *rank* of a matrix is the order of its largest sub-matrix that is non-singular and so has a non-zero determinant. Thus for a square $n \times n$ matrix, the rank must be n (i.e. the matrix must be *full rank*) for the matrix to be non-singular and invertible.

B.1.12 Differentiation of Vectors and Matrices

The differentiation of vectors and matrices is most important in optimization and statistical analysis. The main result concerns the differentiation of an inner product of two vectors with respect to the elements of one of the vectors.

Consider the inner product of two $(n \times 1)$ vectors $\mathbf{x}$ and $\mathbf{a}$, i.e.

$$\mathbf{x}^T\mathbf{a} = [x_1 x_2 \ldots x_n]\begin{bmatrix} a_1 \\ a_2 \\ \cdots \\ a_n \end{bmatrix} = x_1a_1 + x_2a_2 + \ldots + x_na_n$$

It is clear that for all i, $i = l, 2, \ldots, n$, the partial differentials with respect to a_i are given by

$$\frac{\partial\,(\mathbf{x}^T\mathbf{a})}{\partial\,a_i} = x(i)$$

As a result, if the partial differentials are arranged in order of their subscripts as a vector, then this vector is simply $\mathbf{x}$. Thus it is convenient to refer to the process of vector differentiation in shorthand as

$$\frac{\partial\,(\mathbf{x}^T\mathbf{a})}{\partial\,\mathbf{a}} = \mathbf{x} \quad \text{or} \quad \frac{\partial\,(\mathbf{x}^T\mathbf{a})}{\partial\,\mathbf{a}^T} = \mathbf{x}^T$$

The analogy with scalar differentiation is apparent from the above result. A particularly important example of vector differentiation that occurs in this book is concerned with the least squares optimization of the form

$$\hat{\mathbf{a}} = \arg\min_{\mathbf{a}} \sum_{i=1}^{k} \varepsilon^2(i)$$

associated with a *linear regression model* of the form

$$y(i) = \mathbf{x}(i)^T\mathbf{a} + e(i) \tag{B.8}$$

where the error $\varepsilon(i)$ is defined as $\varepsilon(i) = \mathbf{x}(i)^T\mathbf{a} - y(i)$, $i = 1, 2, \ldots, k$. In order to obtain the least squares estimate $\hat{\mathbf{a}}$ of $\mathbf{a}$, it is necessary to differentiate

$$\mathfrak{J}_2 = \sum_{i=1}^{k} \varepsilon^2(i)$$

with respect to all of the elements $a(i)$, $i = l, 2, \ldots n$ of $\mathbf{a}$. Using the above results, we see that since

$$\mathfrak{J}_2 = \sum_{i=1}^{k} [(\mathbf{x}^T(i)\mathbf{a})^2 - 2\mathbf{x}(i)^T\mathbf{a}y(i) + y^2(i)]$$

then

$$\begin{aligned} \frac{\partial \mathfrak{J}_2}{\partial \mathbf{a}} &= \sum_{i=1}^{k} [2\mathbf{x}(i)\mathbf{x}(i)^T\mathbf{a} - 2\mathbf{x}(i)y(i)] \\ &= 2\sum_{i=1}^{k} [\mathbf{x}(i)\mathbf{x}(i)^T\mathbf{a} - \mathbf{x}(i)y(i)] \end{aligned} \tag{B.9}$$

When set to zero in the usual manner, this constitutes a set of n simultaneous equations in the n unknowns $\hat{a}(i)$, $i = l, 2, \ldots n$: the so-called *normal equations* which must be satisfied by the optimal estimate $\hat{\mathbf{a}}$.

$$\sum_{i=1}^{k} \mathbf{x}(i)\mathbf{x}(i)^T \,\hat{\mathbf{a}} = \sum_{i=1}^{k} \mathbf{x}(i)y(i) \tag{B.10}$$

Alternatively, we can proceed by forming the $k \times n$ matrix $\mathbf{X}$ with rows defined by $\mathbf{x}^T(i),\ i = 1,2,\ldots,k$. The reader can then verify that the vector $\mathbf{e} = [e_1\, e_2\, \ldots\, e_n]^T$ is defined by

$$\mathbf{e} = \mathbf{Xa} - \mathbf{y}$$

where $\mathbf{y} = [y_1\, y_2\, \ldots\, y(k)]^T$ and so,

$$\begin{aligned} \mathfrak{J}_2 &= [\mathbf{Xa} - \mathbf{y}]^T[\mathbf{Xa} - \mathbf{y}] \\ &= \mathbf{a}^T\mathbf{X}^T\mathbf{Xa} - 2\mathbf{a}^T\mathbf{X}^T\mathbf{y} + \mathbf{y}^T\mathbf{y} \end{aligned}$$

since $\mathbf{y}^T\mathbf{Xa}$ is a scalar and so equal to its transpose $\mathbf{a}^T\mathbf{X}^T\mathbf{y}$. It now follows straightforwardly that

$$\frac{\partial \mathfrak{J}_2}{\partial \mathbf{a}} = 2\mathbf{X}^T\mathbf{Xa} - 2\mathbf{X}^T\mathbf{y}$$

which will be seen as identical to (B.9) by substituting for $\mathbf{X}$ in terms of $\mathbf{x}(i)$. The least squares estimate is then obtained by the solution of the normal equations, as above, which now take the alternative form

$$\mathbf{X}^T\mathbf{X}\hat{\mathbf{a}} = 2\mathbf{X}^T\mathbf{y} \tag{B.11}$$

If $\mathfrak{J}_2$ is replaced by the more general weighted least squares cost function (see section B.1.9), i.e.

$$\mathfrak{J}_2 = [\mathbf{Xa} - \mathbf{y}]^T\mathbf{Q}[\mathbf{Xa} - \mathbf{y}] = ||\mathbf{Xa} - \mathbf{y}||^2_{\mathbf{Q}}$$

where $\mathbf{Q}$ is a symmetric positive, definite weighting matrix, then it is straightforward to show that

$$\frac{\partial \mathfrak{J}_2}{\partial \mathbf{a}} = 2\mathbf{X}^T\mathbf{QXa} - 2\mathbf{X}^T\mathbf{Qy}$$

and the normal equations follow in an analogous manner to the above,

$$\mathbf{X}^T\mathbf{QX}\hat{\mathbf{a}} = 2\mathbf{X}^T\mathbf{Qy}$$

B.1.13 Cholesky Decomposition

The Cholesky decomposition expresses an $n \times n$, symmetric, positive definite matrix $\mathbf{P}$ as the product of a triangular matrix $\mathbf{T}$ and its transpose, i.e., $\mathbf{P} = \mathbf{TT}^T$, where $\mathbf{T}$ is a $n \times n$ upper triangular matrix. The matrix $\mathbf{P}$ must be positive definite, implying that all its diagonal elements are positive and that the off-diagonal elements are relatively small. This is a particularly useful decomposition in statistics: for example, if $\mathbf{P}$ is a $n \times n$ covariance matrix associated with a multivariate Gaussian probability distribution function (see next Section B.2), then the Cholesky decomposition factorizes this covariance matrix and allows for the simulation of correlated random sequences from this distribution on the basis of samples (draws) from n independent, purely random Gaussian sequences (i.e. from sequences generated by the randn routine in Matlab), as discussed later in Section B.3.1.

B.1.14 Singular Value Decomposition (SVD)

SVD is concerned with the factorization of an $N \times n$ matrix $\mathbf{X}$ into the following form: $\mathbf{X} = \mathbf{U}\boldsymbol{\sigma}\mathbf{V}$, where $\mathbf{U}$ and $\mathbf{V}$ are orthogonal matrices of dimension $N \times N$ and $n \times n$, respectively, while $\boldsymbol{\sigma}$ is $n \times N$ dimensional and has the form:

$$\boldsymbol{\sigma} = \begin{bmatrix} \mathbf{D} & \mathbf{0} \\ \mathbf{0} & \mathbf{0} \end{bmatrix}; \mathbf{D} = \begin{bmatrix} \sigma_1 & & & & \\ & \sigma_2 & & & \\ & & \sigma_3 & & \\ & & & \ddots & \\ & & & & \sigma_r \end{bmatrix}$$

where $\sigma_1 \geq \sigma_2 \geq \cdots \geq \sigma_r > 0$ and $r \leq \min(N,n)$. This can be used in the context of linear regression analysis when N (the data sample size) is normally much greater than n (the number of terms in the regression). In this situation, the elements $\sigma_1, \sigma_2, \ldots, \sigma_r$ and possibly the additional $N - n$ zero diagonal elements of $\mathbf{D}$ are called *singular values* of $\mathbf{X}$. The columns $\mathbf{u}(i)$ of $\mathbf{U}$ and $\mathbf{v}(i)$ of $\mathbf{V}$ are called the *singular vectors* of $\mathbf{X}$. Note that these singular vectors are also the eigenvectors of $\mathbf{X}\mathbf{X}^T$ and $\mathbf{X}^T\mathbf{X}$, respectively.

SVD decomposition can be useful in many different applications. For instance, in linear least squares (regression) analysis, it can indicate whether the condition number of the matrix $\mathbf{X}^T\mathbf{X}$, as defined by the ratio of the largest and the smallest singular value, is very large, so that the matrix is poorly conditioned and near to singularity. In this situation, inversion of the matrix, as required in the solution of the normal equations (B.9) will be difficult. Other applications are for *Principal Component Analysis* (see e.g. Young, 1993) and model structure identification (see e.g. Wang, and Young, 2008, where it is utilized within an instrumental variable estimation context that is relevant to chapters 6 to 10 of the present book).

B.2 Statistics and Probability

B.2.1 Discrete Random Variables

A *discrete-valued random variable* x is defined as a discrete valued function $x(j)$ with a probability of occurrence of the jth value given by $p(j)$: $p(j)$ is the *probability mass function* of the random variable $x(j)$. For simplicity $x(j)$ and $p(j)$ are denoted by x and $p(x)$. The random variable x can be characterized approximately in probabilistic terms by specifying a finite number of moments of $p(j)$. The best known are the first two moments, defined as follows:

(i) the *mean value* or *first moment* of $p(x)$ which is defined as the *expected value* (or expectation) of x, denoted by μ, i.e.

$$E\{x\} = \mu = \sum_j x(j)p(j)$$

(ii) the *variance* or *second central moment* of $p(x)$ which is defined as the expected value of the square of the difference between $x(j)$ and its mean value μ , i.e.

$$E\{(x-\mu)^2\} = \sigma^2 = \sum_j \{x(j) - \mu\}^2 p(j)$$

If $x_1, x_2, \ldots, x_n$ is a sample sequence of a random variables, then:

(iv) the variously named *arithmetic mean, sample mean* or *average value*, $\bar{x}$, is given by

$$\bar{x} = \frac{1}{n}\sum_{i=1}^{i=n} x_i \tag{B.12}$$

(v) the *sample variance*, s^2, is given by

$$s^2 = \frac{1}{n}\sum_{i=1}^{i=n} (x_i - \bar{x})^2 \tag{B.13}$$

(vi) and the *root mean square deviation* or *sample standard deviation* is given by

$$s = \left(\frac{1}{n}\sum_{i=1}^{i=n} (x_i - \bar{x})^2\right)^{\frac{1}{2}} \tag{B.14}$$

However, this nomenclature is not universal and, as in the main text of the present book, the sample variance and sample standard deviation are often denoted by σ^2 and σ, respectively, unless the context of the computations makes it important to differentiate between the sample variance and the variance of the probability distribution.

B.2.2 Law of large Numbers

If $X_1,\ X_2,\ \ldots,\ X_n$ is a sample sequence of independent identically distributed (iid) random variables each having a mean μ and variance σ^2, then let

$$\bar{X}_n = \frac{1}{n}\sum_{i=1}^{i=n} X_i; \quad n = 1,\ 2,\ \ldots \tag{B.15}$$

Since X_i are independent random variables,

$$\mathrm{var}\left[\sum_{i=1}^{i=n} X_i\right] = \sum_{i=1}^{i=n} \mathrm{var}(X_i) = n\sigma^2$$

so that,

$$\mathrm{var}(\bar{X}_n) = \sigma^2/n \tag{B.16}$$

Then it can be shown (e.g. Priestley, 1981), that

$$\lim_{n\to\infty} \Pr(|\bar{X}_n - \mu| > \varepsilon) = 0 \text{ for any } \varepsilon > 0 \tag{B.17}$$

or the *probability in the limit* of the sequence $\bar{X}_n$ is μ, i.e.,

$$\underset{n\to\infty}{\mathrm{p.lim}}\ \bar{X}_n = \mu \tag{B.18}$$

This is called the *Weak Law of Large Numbers* which, in words, states that *as n increases, the mean $\bar{X}_n$ of a random sample of n observations gets ever closer to the mean of the associated probability distribution μ.*

B.2.3 Discrete Random Vectors

A *random vector* is a column vector **x** whose elements are discrete random variables, e.g. $x(i), i = 1,2,\ldots,n$. If each component $x(i)$ can take on a discrete set of values $x(i)(j_i)$ where $j_i = 1,2,\ldots,m_i$, then there are $m_1, m_2, \ldots, m_n$ possible vectors. The *joint probability mass function* $p(j_1,\ j_2, \ldots, j_n)$ is the probability that x_1 has its j_1^{th} value, x_2, has its j_2^{th} value, etc. For simplicity, the joint probability mass function is usually written $p(\mathbf{x}) = p(x_1 . x_2, \ldots, x_n)$. The *marginal probability mass function* $p(j_1)$ is the probability that x_1 takes on its j_1^{th} value while $x_2, \ldots, x_n$ take on any possible values, i.e. in general

$$P(j_i) = \sum_{j_2=1}^{m2} \sum_{j_3=1}^{m3} \cdots \sum_{j_n=1}^{mn} p(j_1,\ j_2, \ldots,\ j_n)$$

As in the scalar case, it is possible to characterize $\mathbf{x}$ approximately by specifying moments of $p(x)$, i.e.

1. the mean of $\mathbf{x}$:

$$E(\mathbf{x}) = \boldsymbol{\mu} = \sum_{j_1=1}^{m_1} \cdots \sum_{j_n=1}^{m_n} \begin{bmatrix} x_1(j_1) \\ \cdots \\ x_n(j_n) \end{bmatrix} p(j_1, j_2, \ldots, j_n)$$

2. the covariance of $\mathbf{x}$: since $\mathbf{x}$ is a vector, it has n variances and $n(n-1)/2$ covariances associated with it, where the covariances are defined as the expected value of the cross products of the elements with the means removed. Thus the covariance is specified by an $n \times n$ symmetric covariance matrix $\mathbf{P}$ defined by

$$\begin{aligned}
\mathbf{P} &= E\{[\mathbf{x}-\boldsymbol{\mu}][\mathbf{x}-\boldsymbol{\mu}]^T\} \\
&= E\{\begin{bmatrix} (x_1-\mu_1)^2 & \cdots & \cdots & (x_1-\mu_1)(x_n-\mu_n) \\ \cdots & (x_2-\mu_2)^2 & \cdots & \cdots \\ \cdots & \cdots & \cdots & \cdots \\ (x_n-\mu_n)(x_1-\mu_1) & \cdots & \cdots & (x_n-\mu_n)^2 \end{bmatrix}\} \\
&= \sum_{j_1=1}^{m_1} \cdots \sum_{j_n=1}^{m_n} [\mathbf{x}-\boldsymbol{\mu}][\mathbf{x}-\boldsymbol{\mu}]^T p(j_1, j_2, \ldots, j_n)
\end{aligned}$$

Any such covariance matrix is at least positive semi-definite. The sample mean and covariance matrix are defined by analogy with the scalar case.

B.2.4 Conditional Probabilities

If a random vector $\mathbf{x}$ is characterized by a covariance matrix $\mathbf{P} = [p_{ij}]$ with $p_{ij} \neq 0$ for $i \neq j$, then the elements of $\mathbf{x}$ are correlated. The elements of $\mathbf{x}$ are said to be *dependent* if knowledge of $p(x_1), p(x_2), \ldots, p(x_n)$ does not determine $p(x_1, \ldots, x_n)$ completely. If, on the other hand,

$$p(x_1, \ldots, x_n) = p(x_1).p(x_2).\ldots.p(x_n)$$

for all possible values of $x_1, \ldots, x_n$, then the elements are said to be *independent*.

If two random vectors $\mathbf{x}$ and $\mathbf{y}$ are dependent and $\mathbf{x}$ takes on a particular value, it should be possible to predict $\mathbf{y}$ better than if this additional information was not available. This leads to the concept of the *conditional probability mass function* $p(\mathbf{y}|\mathbf{x})$, where

$$p(\mathbf{y}|\mathbf{x}) = p(\mathbf{y},\mathbf{x})/p(\mathbf{x}), \text{ for } p(\mathbf{x}) \neq 0$$

is the probability of $\mathbf{y}$ conditioned on a given value of $\mathbf{x}$. The conditional mean and covariance are defined in a similar manner to that shown above, with the joint probability mass function replaced by the conditional probability mass function. The conditional mean and covariance are random variables because they are a function of the conditioning random variable $\mathbf{x}$. Since

$$p(\mathbf{x}|\mathbf{y}) = p(\mathbf{y},\mathbf{x})/p(\mathbf{y})$$

then

$$p(\mathbf{y}|\mathbf{x}) = p(\mathbf{x},\mathbf{y}).p(\mathbf{x})$$

and so

$$p(\mathbf{y}|\mathbf{x}) = \frac{p(\mathbf{x}|\mathbf{y}).p(\mathbf{y})}{p(\mathbf{x})}$$

This is known as the Bayes Rule for conditional probabilities and is a most important concept in recursive estimation theory when considered from the standpoint of Bayesian estimation: see e.g. Bryson and Ho (1969). If we consider that $p(\mathbf{y})$ is the *a priori* probability of $\mathbf{y}$ without any knowledge of $\mathbf{x}$ then $p(\mathbf{y}|\mathbf{x})$ can be considered the *a posteriori* probability of $\mathbf{y}$ given that $\mathbf{x}$ has taken on a certain value. Of course if $\mathbf{x}$ and $\mathbf{y}$ are independent then $p(\mathbf{y}|\mathbf{x}) = p(\mathbf{y})$, so that knowledge of $\mathbf{x}$ is not useful in predicting the value of $\mathbf{y}$.

B.2.5 Continuous Random Variables and Vectors

The concepts described in the above sections can be extended to continuous random variables and vectors. So, for example,

$$\boldsymbol{\mu} = \int_{-\infty}^{\infty} \cdots \int_{-\infty}^{\infty} \begin{bmatrix} x_1 \\ \cdots \\ x_n \end{bmatrix} p(x_1,\ldots,x_n)\, dx_1 \ldots dx_n$$

$$\mathbf{P} = \int_{-\infty}^{\infty} \cdots \int_{-\infty}^{\infty} [\mathbf{x}-\boldsymbol{\mu}][\mathbf{x}-\boldsymbol{\mu}]^{\mathrm{T}}\, p(x_1,\ldots,x_n)\, dx_1 \ldots dx_n$$

where the function p is the *probability density function* $p(x_1,\ldots,x_n)\, dx_1 \ldots dx_n$ and is defined as the probability that the random vector $\mathbf{x}$ will lie in the *differential volume* $(dx_1 \ldots dx_n)$ with centre at $(x_1,\ldots,x_n)$.

B.2.6 The Normal or Gaussian Density Function

A *normally distributed random variable (scalar)* has an amplitude density function $p(x)$ defined by

$$p(x) = \frac{1}{(2\pi)^{\frac{1}{2}}\sigma} \exp.\frac{(x-\mu)^2}{2\sigma^2}$$

so that

$$\int_{-\infty}^{\infty} p(x)\,dx = 1.0$$

$$E\{x\} = \mu$$

$$E\{(x-\mu)^2\} = \sigma^2$$

The distribution is, therefore, completely specified by its mean and variance; and so it is usual to summarize the distribution as $\mathcal{N}(\mu, \sigma^2)$.

A *normally distributed random vector* $\mathbf{x} = [x_1 x_2 \ldots x_n]^T$ has a multivariate normal probability density function pdf defined by

$$p(x) = \frac{1}{(2\pi)^{\frac{n}{2}}|p|^{\frac{1}{2}}} \exp.\{-\frac{1}{2}[\mathbf{x}-\boldsymbol{\mu}]p^{-1}[\mathbf{x}-\boldsymbol{\mu}]^T\}$$

so that

$$\int_{-\infty}^{\infty} \cdots \int_{-\infty}^{\infty} p(\mathbf{x})\,dx_1 \ldots dx_n = 1$$

$$E\{\mathbf{x}\} = \boldsymbol{\mu}$$

$$E\{(\mathbf{x}-\boldsymbol{\mu})(\mathbf{x}-\boldsymbol{\mu})^T\} = \mathbf{P}$$

and $p(\mathbf{x})$ is completely characterized by its mean $\boldsymbol{\mu}$ and covariance matrix $\mathbf{P}$. Once again, it is usual to summarize the distribution as $\mathcal{N}(\boldsymbol{\mu}, \mathbf{P})$.

The concept of the normally distributed random variable or vector is most useful in analytical terms: often the density functions, for random variables or vectors, can be considered as approximately normally distributed and so characterized almost completely by their mean and covariance matrix properties. As a result, mathematical analysis can be made much more straightforward by making the normal distribution assumption.

B.2.7 Properties of Estimators

Suppose a model is characterized by a single unknown parameter θ and we need to find an *estimate* $\hat{\theta}$ of θ based on N observations $y_1, \ldots, y_N$. The rule, or algorithm, for processing the observations is termed the *estimator* of θ and the estimate $\hat{\theta}$ is a function of the observations. A reasonable estimator should produce estimates for different sample sizes that are reasonably close, in some sense, to the true value θ. Such an estimator is said to be *unbiased* if

$$E\{\hat{\theta}\} = \theta \text{ for all } \theta$$

The estimate is sometimes said to be *asymptotically unbiased* if the estimate $\hat{\theta}(k)$ based on k samples is unbiased for $k \rightarrow \infty$. A more rigorous definition is that $\hat{\theta}(k)$ is an asymptotically unbiased estimate of θ if the mean of the limiting distribution of $\sqrt{k}[\hat{\theta}(k) - \theta]$ is zero. Clearly unbiasedness is a more desirable property than asymptotic unbiasedness but the latter may often be acceptable if sample sizes are reasonably high.

The *mean square error* of $\hat{\theta}(k)$ is defined simply as

$$MSE(\hat{\theta}) = E\{(\hat{\theta} - \theta)^2\}$$

and, as we shall see later, it is possible to design estimates that produce unbiased estimates which attain the lowest possible or minimum value of the *MSE*. Such estimators are termed *minimum variance unbiased estimators*.

A *consistent estimator* is one which produces estimates $\hat{\theta}(k)$ that become more accurate, in the sense that the probability of its being close to the true value $\theta(k)$ increases as the sample size k increases. Or, in mathematical terms,

$$\underset{k \to N}{\text{p.lim}}\ \hat{\theta}(k) = \theta$$

where the *probability in the limit*, p.lim, has been defined earlier in (B.18).

B.2.8 The Likelihood Function and Maximum Likelihood Estimation

Suppose that we have a set of observations on N random variables that compose a vector $\mathbf{y} = [y_1, y_2, \ldots, y_N]^T$. It is possible to consider a *joint density function* $L(\boldsymbol{\theta}, y_1, y_2, \ldots, y_N)$ that depends upon the n unknown parameters in the vector $\boldsymbol{\theta} = [\boldsymbol{\theta}_1, \boldsymbol{\theta}_2, \ldots, \boldsymbol{\theta}_n]^T$ and which can be interpreted as the probability of obtaining particular values of $y_1, y_2, \ldots, y_N$. Once a sample has been taken, then $y_1, y_2, \ldots, y_N$ becomes a set of fixed numbers and the expression for L can be reinterpreted as a function of $\hat{\boldsymbol{\theta}}(k)$, where $\hat{\boldsymbol{\theta}}(k)$ is any admissable value of the parameter vector rather than the true value. In this sense L can be considered as a means of assessing the relative merits of different values of $\hat{\boldsymbol{\theta}}(k)$, given the sample. L is, therefore, termed the *Likelihood Function* and denoted by $L(\boldsymbol{\theta})$ *or* $L(\boldsymbol{\theta}, \mathbf{y})$.

The maximum likelihood (ML) approach to the problem of estimating $\boldsymbol{\theta}$ is one of investigating which value of $\hat{\boldsymbol{\theta}}$ is most likely given the observations: the ML estimate is then given by $\hat{\boldsymbol{\theta}}(0)$, where

$$L[\hat{\boldsymbol{\theta}}(0)] \geq L(\hat{\boldsymbol{\theta}})$$

in which $\hat{\boldsymbol{\theta}}$ is any other admissable estimate of $\boldsymbol{\theta}$.

The classical theory of maximum likelihood (see e.g. Kendal and Stuart, 1961, Cox and Hinkley, 1974; Priestley, 1981) is based on the situation in which the N observations are drawn independently of each other from the same distribution so that

$$L[\boldsymbol{\theta}; y_1, \ldots, y_N] = \prod_{k=1}^{N} p[y(k); \boldsymbol{\theta}]$$

where $\prod$ denotes the multiplication operator. If, for example, we consider the likelihood function for a sample of N independent observations from a normal distribution with mean μ and variance σ^2 then

$$\log_e L[\mu, \sigma^2; y_1, \ldots, y_N] = -\frac{N}{2}\log_e 2\pi - \frac{N}{2}\log_e \sigma^2 - \frac{1}{2\sigma^2}\sum_{k=1}^{N}[y(k) - \mu]^2$$

where, in this case, $\boldsymbol{\theta} = [\mu, \sigma^2]^T$. Note that, as is usual in this kind of analysis, the natural logarithm of L is considered here so that the analysis is easier: this is quite allowable since if $\hat{\boldsymbol{\theta}}(0)$ satisfies $L\{\boldsymbol{\theta}(0)\} \geq L(\hat{\boldsymbol{\theta}})$, it also satisfies $\log_e L\{\theta(0)\} \geq \log_e L(\hat{\theta})$. The maximum likelihood estimates are, therefore, obtained by finding those estimates of μ and σ^2 which simultaneously maximize $log_e L$. These can be obtained in the usual manner by differentiating $log_e L$ with respect to μ and σ^2 in turn and setting the resultant expression to zero, i.e.

$$\frac{\partial \log_e L}{\partial \boldsymbol{\theta}} = \nabla_{\boldsymbol{\theta}} \log_e L = 0$$

where $\nabla_{\boldsymbol{\theta}}$ denotes the partial differential with respect to each element of $\boldsymbol{\theta}$ in turn. In the present example, this yields the following where, for convenience, L is used to represent the log likelihood:

$$\frac{\partial L}{\partial \mu} = \frac{1}{\sigma^2}\sum (y(k) - \mu)^2 = 0$$

$$\frac{\partial L}{\partial \sigma^2} = \frac{N}{2\sigma^2} + \frac{1}{2\sigma^4} + \sum (y(k) - \mu)^2 = 0$$

As a result,

$$\bar{x}(0) = \frac{1}{N}\sum_{k=1}^{N} y(k) \tag{B.19}$$

and

$$\sigma^2(0) = \frac{1}{N}\sum_{k=1}^{N}[y(k) - \bar{x}]^2$$

which are the ML expressions for the estimated (sample) mean and variance of a random variable. We see that, in maximising the likelihood function, these estimates also minimize the sum of the squares function. Also, the reader can verify easily that the estimates do indeed maximize L, since the matrix of second partial derivatives $H(\theta)$ (the 'Hessian' of the log-likelihood function $\log_e L$) given by

$$
\begin{aligned}
H(\theta) &= \nabla^2_{\boldsymbol{\theta}} \log_e L \\
&= \begin{bmatrix} \frac{\partial^2 \log_e L}{\partial \mu^2} & \frac{\partial^2 \log_e L}{\partial \mu \partial \sigma^2} \\ \frac{\partial^2 \log_e L}{\partial \sigma^2 \partial \mu} & \frac{\partial^2 \log_e L}{\partial \sigma^2} \end{bmatrix} \\
&= \begin{bmatrix} -\frac{N}{\sigma^2} & -\frac{1}{\sigma^4}\Sigma[y(k)-\mu] \\ \frac{1}{\sigma^4}\Sigma[y(k)-\mu] & \frac{N}{2\sigma^4} - \frac{1}{\sigma^6}\Sigma[y(k)-\mu]^2 \end{bmatrix}
\end{aligned}
$$

Consequently, substituting from above

$$
H(\theta) = \begin{bmatrix} -\frac{N}{\sigma^2(0)} & 0 \\ 0 & \frac{N}{2\sigma^4(0)} \end{bmatrix}
$$

which is negative definite because $\sigma^2(0) > 0$ for a non-zero random variable.

B.2.9 The Cramer-Rao Lower Bound

If the vector $\boldsymbol{\theta}$ is of order one, so that the unknown parameter is a scalar θ, then the amount of information in the sample is defined by

$$
I(\theta) = -E[H(\theta)] = -E[\nabla^2 \log_e L]
$$

The value $1/I(\theta)$ is termed the *minimum variance bound* (MVB) for any unbiased estimate $\hat{\theta}$ of θ ; in other words, the variance of $\hat{\theta}$ must be greater than or equal to $1/I(\theta)$ or

$$
E\{(\hat{\theta} - \theta)\} \geq 1/I(\theta)
$$

which is known as the *Cramer-Rao inequality*. Because maximum likelihood estimators attain the MVB for large samples, the estimate $\hat{\theta}$, obtained in this manner, is referred to as an *asymptotically efficient* estimate of θ.

B.2.10 Maximum Likelihood Estimators: the Vector Case

The major results in this vector case are as follows:

(i) Consistency: $\text{p.lim}(\hat{\boldsymbol{\theta}}) = \boldsymbol{\theta}$
(ii) Asymptotic Normality: $\hat{\boldsymbol{\theta}} \to \mathcal{N}(\boldsymbol{\theta}, \mathbf{I}(\boldsymbol{\theta})^{-1})$ where $\mathbf{I}(\boldsymbol{\theta})$ is the vector equivalent of $I(\theta)$ and is called the *information matrix*. This matrix can be defined in two equivalent ways:

$$
\mathbf{I}(\boldsymbol{\theta}) = E\left[\left(\frac{\partial L}{\partial \boldsymbol{\theta}}\right)\left(\frac{\partial L}{\partial \boldsymbol{\theta}}\right)^T\right] = -E\left[\frac{\partial^2 L}{\partial \boldsymbol{\theta}\, \partial \boldsymbol{\theta}^T}\right] \tag{B.20}
$$

where the Hessian matrix (see above) is given by:

$$\frac{\partial^2 L}{\partial \boldsymbol{\theta}\, \partial \boldsymbol{\theta}^T} = \begin{bmatrix} \frac{\partial^2 L}{\partial \boldsymbol{\theta}_1^2} & \frac{\partial^2 L}{\partial \boldsymbol{\theta}_1 \partial \boldsymbol{\theta}_2} & \cdots & \frac{\partial^2 L}{\partial \boldsymbol{\theta}_1 \partial \boldsymbol{\theta}_n} \\ \frac{\partial^2 L}{\partial \boldsymbol{\theta}_2 \partial \boldsymbol{\theta}_1} & \frac{\partial^2 L}{\partial \boldsymbol{\theta}_2^2} & \cdots & \frac{\partial^2 L}{\partial \boldsymbol{\theta}_2 \partial \boldsymbol{\theta}_n} \\ \vdots & \vdots & \ddots & \vdots \\ \frac{\partial^2 L}{\partial \boldsymbol{\theta}_n \partial \boldsymbol{\theta}_1} & \frac{\partial^2 L}{\partial \boldsymbol{\theta}_n \partial \boldsymbol{\theta}_2} & \cdots & \frac{\partial^2 L}{\partial \boldsymbol{\theta}_n^2} \end{bmatrix}$$

It is usually easier, in practice, to evaluate the second expression.

For the maximum likelihood estimator,

$$\sqrt{n}\,(\hat{\boldsymbol{\theta}} - \boldsymbol{\theta}) \rightarrow \mathcal{N}(\mathbf{0}, \mathbf{P})$$

for some positive definite matrix $\mathbf{P}$. If $\tilde{\mathbf{P}}$ denotes the covariance matrix associated with any other consistent, asymptotically normal estimator, then $\tilde{\mathbf{P}} - \mathbf{P}$ is a positive definite matrix. Under the above assumptions of normality, therefore, the maximum likelihood estimator obtained in this manner, is *asymptotically efficient* and it achieves the MVB. However, this is a theoretical result and it must be remembered that, in practical situations, the conditions required for its fulfilment are quite often not established. Nevertheless it provides a useful relative measure on which the asymptotic behaviour of estimators can be evaluated.

B.3 Time Series

If we consider a simple time series of a random variable

$$x(i),\ i = -\infty, \ldots, -1, 0, 1, \ldots, \infty$$

where the subscript i denotes the sampled value of the variable $x(i)$ at the ith instant of time, then the mean, variance and covariance (the first and second statistical moments) are defined as follows:

1. mean : $E\{x(i)\} = \mu$
2. variance : $E\{[x(i) - \mu]^2\} = \sigma^2$
3. covariance : $E\{[x(i) - \mu][x(j) - \mu]\} = \gamma(i-j) = \gamma_\tau,\ \tau = i - j$

The time series is said to be stationary in the wide sense or stationary to second order if the mean, variance and covariance are not functions of time. Note that the covariance is defined as the expected value of the random variable multiplied by itself lagged by a given number of time instants; this is sometimes termed the *autocovariance* and non-zero values for $\tau \neq 0$ indicates that the variable is *autocorrelated* in time. A *white noise variable* is defined here as one that is *serially uncorrelated* in time, i.e. $\gamma_\tau = 0\ \forall\ \tau \neq 0$. The autocorrelation ρ_τ of a time series variable $x(i)$ at lag τ is simply the normalized autocovariance of the variable, where normalization is based on the autocovariance at lag zero, γ_0 , i.e.

$$\rho_\tau = \frac{\gamma_\tau}{\gamma_0}; \; \tau = 0, 1, 2, \ldots$$

so that ρ_0, the instantaneous autocorrelation is normalized to unity.

For a vector of time series variables $\mathbf{x} = [x_1, \ldots, x_n]^T$, it is necessary to allow for the possibility of the *serial correlation* in time of the individual elements and *cross correlation* between elements at different lag values. A *white noise vector* is one whose elements are *serially uncorrelated* in time but may be correlated with other elements of the vector *at the same instant of time*. The covariance matrix of such a white noise vector is usually defined as

$$\mathbf{R}_{\mathbf{x},\mathbf{x}} = E\{[\mathbf{x}(i) - \boldsymbol{\mu}][\mathbf{x}(j) - \boldsymbol{\mu}]^T\} = \mathbf{Q}\delta_{ij}$$

where δ_{ij} is the *Kronecker delta function*, which is equal to unity if $i = j$ and zero if $i \neq j$:

$$\delta_{ij} = \begin{cases} 1.0 & \text{if } i = j; \\ 0 & \text{if } i \neq j; \end{cases}$$

Often, where the mean value $\boldsymbol{\mu} = 0$, it is omitted from the definition. If $\mathbf{Q}$ is a diagonal matrix, then the elements are mutually uncorrelated white noise variables. A vector of time series variables $\mathbf{e} = [e_1 \; e_2 \; \ldots \; e_n]^T$ with zero mean and covariance matrix $\mathbf{Q}$ i.e.

$$E\{\mathbf{e}(i)\} = 0$$

$$E\{\mathbf{e}(i)\mathbf{e}^T(j)\} = \mathbf{Q}\delta_{ij}$$

provides a useful source of random variables in the mathematical description of stochastic dynamic systems (see later): in effect the system is seen to 'process' the vector in some manner to yield other vectors composed of correlated random variables (of greater or less dimension than $\mathbf{e}(i)$) which will, in general, be composed of '*coloured noise*' components; i.e. each element will be serially correlated in time and crosscorrelated with all other elements of the vector at all instants of time and all lags.

The sample statistics for scalar and vector time series are defined in a similar manner to those in the previous sections on discrete discrete random variables. So, for example, the sample autocorrelation function (ACF) r_τ for lag τ and sample size N is given by:

$$r_\tau = \frac{c_\tau}{c_0}; \quad c_\tau = \frac{1}{N} \sum_{i=1}^{i=N-\tau} (x(i) - \bar{x})(x(i+\tau) - \bar{x}); \; \tau = 0, 1, 2, \ldots$$

where $\bar{x}$ is the sample mean. A plot of the sample autocorrelation function for increasing lag τ is a very useful statistical device for evaluating whether a time series has no serial correlation ('white' noise) or significant serial correlation ('coloured' noise). This requires a measure of statistical significance, as discussed by Box and

Jenkins (1970) (later edition: Box et al, 1994), who point out that Bartlett Bartlett (1946) has shown that an approximate expression for the variance of the autocorrelations ρ_τ at lags greater that some value q *beyond which the theoretical autocorrelation may be deemed to have 'died out'* is given by

$$\operatorname{var}(r_\tau) \simeq \frac{1}{N}\left\{1+2\sum_{v=1}^{q}\rho_v^2\right\}, \qquad k>q \tag{B.21}$$

where, in practice, the estimated auto correlations $r_\tau, \tau = 1,2, \ldots, q$, are substituted for the theoretical autocorrelations ρ_τ. Then, the square root of (B.21) is termed the *large lag standard error* on the ACF. This is used in the CAPTAIN Toolbox acf routine, which also returns the 'partial autocorrelation function' that helps to define the order of an autoregressive model for the time series. These and the associated sample cross correlation function (CCF: ccf routine in CAPTAIN), which measures the correlation between two time series at increasing lags, are employed to evaluate models in examples that are discussed in the main text.

Note that the Box-Jenkins book (see above) also discusses the PACF and CCF, which are key tools in their approach to time series estimation, forecasting and control: it is an excellent and readable resource on such correlation analysis, as well as *AutoRegressive* (AR), *Moving Average* (MA), and *AutoRegressive Moving Average* (ARMA) modelling. It is strongly recommended to all readers of the present book, which is not able to deal with these topics in full detail because of size limitations.

B.3.1 Gauss-Markov Random Processes

To describe a random time series vector (or scalar) sequence $\mathbf{x}(i), i = 1,2,\ldots,N$, completely, the joint probability density function

$$p[\mathbf{x}(N), \mathbf{x}(N-1), \ldots, \mathbf{x}(0)]$$

of all the elements in the sequence must be specified. Although this involves an enormous amount of information in general terms, it is possible to simplify the situation by assuming that the sequence is a *Markov sequence* where the *conditional (or transition) probability density function* has the special property

$$p[\mathbf{x}(i)\,|\,\mathbf{x}(i-1), \mathbf{x}(i-2), \ldots, \mathbf{x}(0)] \;=\; p[\mathbf{x}(i)|\mathbf{x}(i-1)]$$

for all i. In other words, the probability density function of $\mathbf{x}(i)$ depends only on knowledge of $\mathbf{x}(i-1)$ at the previous instant and not on any previous values $\mathbf{x}(i-\tau)$, $\tau = 2, 3, \ldots$. The knowledge of $\mathbf{x}(i-1)$ can be either deterministic, in which the exact value of $\mathbf{x}(i-1)$ is known, or probabilistic, where only $p[\mathbf{x}(i-1)]$ is known.

The *joint probability density function* of a Markov random sequence can be described completely by specifying its initial density function $p[\mathbf{x}_0]$ and the transition density functions $p[\mathbf{x}(i)|\mathbf{x}(i-1)]$ i.e.,

$$p[\mathbf{x}(N),\mathbf{x}(N-1),\ldots,\mathbf{x}(0)] = p[\mathbf{x}(N)|\mathbf{x}(N-1)].p[\mathbf{x}(N-1)|\mathbf{x}(N-2)]\cdots p[\mathbf{x}(1)|\mathbf{x}(0)].p[\mathbf{x}(0)]$$

A purely random (or white noise) sequence is defined by the property that

$$p[\mathbf{x}(i)|\mathbf{x}(i)] = p[\mathbf{x}(i)] \ \forall\, i$$

A Gauss-Markov random sequence is a Markov random sequence with the additional requirement that $p[\mathbf{x}(i)]$ and $p[\mathbf{x}(i)|\mathbf{x}(i-1)]$ are Gaussian probability density functions for all i. The density function for a Gauss-Markov random sequence is, in this manner, described completely by the mean value vector $\bar{\mathbf{x}}(i) = E\{\mathbf{x}(i)\}$ and covariance matrix

$$\mathbf{P}(i) = E\{[\mathbf{x}(i)-\bar{\mathbf{x}}(i)][\mathbf{x}(i)-\bar{\mathbf{x}}(i)]^T\}$$

At any sampling instant k, a Gauss-Markov random sequence of nth order vectors $\mathbf{x}(k)$ can always be represented by the following vector-matrix model:

$$\mathbf{x}(k) = \mathbf{A}\mathbf{x}(k-1) + \mathbf{D}\boldsymbol{\eta}(k-1) \tag{B.22}$$

where $\mathbf{A}$ is an $n \times n$ *transition matrix*, $\mathbf{D}$ is an $n \times q$ *input matrix* associated with the *system noise* vector $\boldsymbol{\eta}(k)$, which is an qth order, white noise vector with mean $\bar{\boldsymbol{\eta}}$ and covariance matrix $\mathbf{Q}$, i.e.,

$$E\{\boldsymbol{\eta}(k)\} = \bar{\boldsymbol{\eta}}(k); \quad E\{[\boldsymbol{\eta}_j - \bar{\boldsymbol{\eta}}(j)][\boldsymbol{\eta}(k)-\bar{\boldsymbol{\eta}}(k)]^T\} = \mathbf{Q}\delta_{jk}$$

The Gauss-Markov model (B.22) describes a purely stochastic, dynamic system, driven by the white noise vector $\boldsymbol{\eta}(k)$.

B.3.2 The State Space Model of a Linear, Discrete-time, Stochastic Dynamic System

In systems analysis, the stochastic model (B.22) is considered as a set of *stochastic state equations*. However, this description is of limited utility in systems terms for two main reasons. First, it is purely stochastic and does not include the effects of any deterministic or control inputs. Second, it assumes that whole of the vector $\mathbf{x}(k)$ can be observed. A more comprehensive representation, from the systems standpoint, is the *general, linear, stochastic state space model*:

$$\begin{aligned} \mathbf{x}(k) &= \mathbf{A}(k-1)\mathbf{x}(k-1) + \mathbf{B}(k-1)\mathbf{u}(k-1) + \mathbf{D}(k-1)\boldsymbol{\eta}(k-1) \quad &\text{(i)}\\ y(k) &= \mathbf{C}(k)\mathbf{x}(k) + \mathbf{B}_I(k)\mathbf{u}(k) + \mathbf{e}(k) \quad &\text{(ii)} \end{aligned} \tag{B.23}$$

Here, $\mathbf{B}(k-1)$ is a an $n \times m$ input matrix associated with a measured, deterministic $m \times 1$ input vector $\mathbf{u}(k)$, which could constitute a natural input vector in an environmental system or, at the other extreme, a control input vector in an automatic control context. In the output or observation equation (ii), $\mathbf{y}(k)$ is a p dimensional vector of output measurements that are assumed to be linearly related to the state vector $\mathbf{x}(k)$ by the $p \times n$ output matrix $\mathbf{C}(k)$ and may also be affected instantaneously by the input vector $\mathbf{u}(k)$ modulated by the $p \times m$ matrix $\mathbf{B}_I(k)$. And $\mathbf{y}(k)$ is further assumed to be contaminated by a p dimensional white noise vector $\mathbf{e}(k)$ that has mean value $\boldsymbol{\mu}$ and covariance matrix $\mathbf{R}$, i.e.,

$$E\{\mathbf{e}(k)\} = \boldsymbol{\mu}; \quad E\{[\mathbf{e}(j) - \boldsymbol{\mu}][\mathbf{e}(k) - \boldsymbol{\mu}]^T\} = \mathbf{R}\delta_{jk}$$

Often, $\mathbf{e}(k)$ is also assumed to be statistically independent of the system noise $\boldsymbol{\eta}(k)$ i.e.

$$E\{[\mathbf{e}(j) - \bar{\mathbf{e}}][\boldsymbol{\eta}(k) - \bar{\boldsymbol{\eta}}]^{\mathrm{T}}\} = \mathbf{0} \quad \forall j,k$$

Note that, in the above equations, the matrices $\mathbf{A}(k)$, $\mathbf{B}(k)$, $\mathbf{C}(k)$, $\mathbf{D}(k)$ and $\mathbf{B}_I(k)$ are all nominally time variable, making the description nonstationary if this is appropriate. In practice, however, they are often constant over time and so, for convenience, this will be assumed from hereon. The interested reader should consult a text such as Bryson and Ho (1969) for a more complete background on this subject.

B.3.3 The Discrete-Time Transfer Function Model

In order to obtain the *Transfer Function* (TF) representation of the state space model (B.23), we need to introduce the backward shift operator z^{-r}, i.e. $z^{-r}\mathbf{y}(k) = \mathbf{y}(k-r)$ into the equation (B.23)(i). This yields:

$$[\mathbf{I} - \mathbf{A}z^{-1}]\mathbf{x}(k) = \mathbf{B}z^{-1}\mathbf{u}(k) + \mathbf{D}z^{-1}\boldsymbol{\eta}(k)$$

so that,

$$\mathbf{x}(k) = [\mathbf{I} - \mathbf{A}z^{-1}]^{-1}\mathbf{B}z^{-1}\mathbf{u}(k) + [\mathbf{I} - \mathbf{A}z^{-1}]^{-1}\mathbf{D}z^{-1}\boldsymbol{\eta}(k)$$

As a result, from (B.23)(ii),

$$\mathbf{y}(k) = \mathbf{C}\,[\mathbf{I} - \mathbf{A}z^{-1}]^{-1}\mathbf{B}z^{-1}\mathbf{u}(k) + \mathbf{C}\,[\mathbf{I} - \mathbf{A}z^{-1}]^{-1}\mathbf{D}z^{-1}\boldsymbol{\eta}(k) + \mathbf{B}_I(k)\mathbf{u}(k) + \mathbf{e}(k)$$

This constitutes a *Multivariable Transfer Function* (MTF) representation of the model (A.11) and can be expanded to the form (see section B.1.8 for the inverse of a matrix):

$$
\begin{aligned}
\mathbf{y}(k) &= \left[\mathbf{C}\frac{1}{A(z^{-1})}\,\mathrm{Adj}[\mathbf{I}-\mathbf{A}z^{-1}]\,\mathbf{B}z^{-1}+\mathbf{B}_I\right]\mathbf{u}(k) \\
&\quad +\mathbf{C}\frac{1}{A(z^{-1})}\,\mathrm{Adj}[\mathbf{I}-\mathbf{A}z^{-1}]\mathbf{D}z^{-1}\boldsymbol{\eta}(k)+\mathbf{e}(k) \qquad \text{(B.24)} \\
&= \mathbf{L}(z^{-1})\mathbf{u}(k)+\mathbf{M}(z^{-1})\boldsymbol{\eta}(k)+\mathbf{e}(k)
\end{aligned}
$$

where $\mathbf{L}(z^{-1})$ and $\mathbf{M}(z^{-1})$ are, respectively, $p \times m$ and $p \times q$ *Transfer Function Matrices* (TFMs) in which each elemental TF in the matrices has a common denominator $A(z^{-1})$, where

$$
A(z^{-1}) = a_1 z^{-1} + a_2 z^{-2} \ldots + a_n z^{-n}
$$

The equation $A(z^{-1}) = 0$ constitutes the *Characteristic Equation* of the model; and the model is *stable* if the roots of this characteristic equation lie *outside* the unit circle of the *complex* z^{-1} *domain*; or, alternatively, they lie *within* the unit circle of the *complex z domain*. Note that the roots of the characteristic equation are also the *eigenvalues* of the matrix $\mathbf{A}$.

In the *Single-Input, Single Output* (SISO) case, where $p = 1$, $m = 1$ and $q = 1$, (B.24) is simplified to:

$$
\begin{aligned}
y(k) &= \left[\frac{1}{A(z^{-1})}\mathbf{c}^T\,\mathrm{Adj}[\mathbf{I}-\mathbf{A}z^{-1}]\,\mathbf{b}z^{-1}+b_I\right]u(k) \\
&\quad +\frac{1}{A(z^{-1})}\mathbf{c}^T\,\mathrm{Adj}[\mathbf{I}-\mathbf{A}z^{-1}]\mathbf{d}z^{-1}\eta(k)+e(k)
\end{aligned} \qquad \text{(B.25)}
$$

where $\mathbf{c}^T$ is now the $1 \times n$ output vector; $\mathbf{b}$ is the $n \times 1$ input vector associated with the scalar input $u(k)$; b_I is the scalar coefficient that specifies the instantaneous effect of $u(k)$ on $y(k)$; and $\mathbf{d}$ is the $n \times 1$ input vector associated with the scalar noise $\eta(k)$. If we choose a specific 'canonical' (minimally parameterized and uniquely identifiable) form for these matrices and then evaluate (B.25) directly using these definitions, then the TF form of the equation can be obtained easily.

Numerous such canonical forms are available but the most convenient in the present context is the following:

$$
\mathbf{A} = \begin{bmatrix} -a_1 & 1 & 0 & \ldots & 1 \\ -a_2 & 0 & 1 & \ldots & 0 \\ \ldots & \ldots & \ldots & \ldots & \ldots \\ \ldots & 0 & 0 & 0 & 1 \\ -a_n & 0 & 0 & 0 & 0 \end{bmatrix} \qquad \mathbf{b} = \begin{bmatrix} b_1 \\ b_2 \\ \ldots \\ b_n \end{bmatrix}
$$

$$
\mathbf{d} = \begin{bmatrix} d_1 \\ d_2 \\ \ldots \\ d_n \end{bmatrix} \qquad \mathbf{c}^T = [1\,0\,\ldots\,0] \qquad \text{(B.26)}
$$

The reader can verify by evaluation of (B.25) that, with these definitions and the instantaneous input gain term removed from the output equation,

$$y(k) = \frac{B(z^{-1})}{A(z^{-1})}u(k) + \frac{C(z^{-1})}{A(z^{-1})}\eta(k) + \xi(k) \tag{B.27}$$

where $B(z^{-1})$ and $C(z^{-1})$ are polynomials in the z^{-1} operator of the form

$$B(z^{-1}) = b_1 z^{-1} + b_2 z^{-2} \ldots + b_m z^{-n}$$
$$C(z^{-1}) = c_1 z^{-1} + c_2 z^{-2} \ldots + c_n z^{-n}$$

From (B.27), we see that

$$A(z^{-1})y(k) = B(z^{-1})u(k) + C(z^{-1})\eta(k) + A(z^{-1})\xi(k) \tag{B.28}$$

so that, by expanding this equation in discrete-time terms, we obtain:

$$y(k) + a_1 y_{k-1} + \ldots + a_n y_{k-n} = b_1 u_{k-1} + \ldots + b_n u_{k-n} + \mu(k) \tag{B.29}$$

where,

$$\mu(k) = c_1 \eta_{k-1} + \ldots + c_n \eta_{k-n} + a_1 \xi_{k-1} + \ldots + a_n \xi_{k-n} \tag{B.30}$$

Equation (B.29) represents a simple linear relationship between the measured variables $y(k-i)$ and $u(k-j)$, for $i = 0,1,2,\ldots,n$, and $j = 1,2,\ldots,n$, defined by the unknown system parameters $a_i, i = 1,2,\ldots,n$; and $b_j, j = 1,2,\ldots,n$. However, the noise term $\mu(k)$ is not simple in statistical terms, being a linear function of the two white noise source terms $\eta(i), i = k-1,\ldots,k-n$ and $\xi(j), j = k-1,\ldots,k-n$. Or, in vector terms,

$$\mu(k) = \mu_1(k) + \mu_2(k)$$

where,

$$\mu_1(k) = c_1\eta(k-1) + c_2\eta(k-2) + \ldots + c_n\eta(k-n)$$
$$\mu_2(k) = \xi(k) + a_1\xi(k-1) + \ldots + a_n\xi(k-n)$$

Since it is well known (e.g. Granger and Morris, 1976) that the sum of any number of independent moving average processes is itself a moving average process, then we can simplify the expression (B.29) somewhat by replacing $\mu(k)$ by a moving average in a single white noise source $e(k)$ which is different from both $\eta(k)$ and $\xi(k)$. This results in the TF model:

$$y(k) = \frac{B(z^{-1})}{A(z^{-1})}u(k) + \frac{C(z^{-1})}{A(z^{-1})}e(k) \tag{B.31}$$

or,

$$A(z^{-1})y(k) = B(z^{-1})u(k) + C(z^{-1})e(k) \tag{B.32}$$

where $e(k)$ is assumed to be a zero mean white noise sequence with variance σ^2 and so $C(z^{-1})$ is defined as a *monic polynomial* with a leading coefficient of unity, i.e. $C(z^{-1}) = 1 + c_1 z^{-1} + c_2 z^{-2} \ldots + c_p z^{-p}$. This TF model is normally referred to as the *AutoRegressive, Moving Average eXogenous variables* or ARMAX model.

An alternative TF form that has certain advantages in estimation terms is the *Box-Jenkins* (BJ) model (Box and Jenkins, 1970):

$$y(k) = \frac{B(z^{-1})}{A(z^{-1})} u(k) + \frac{D(z^{-1})}{C(z^{-1})} e(k)$$

We see that, in contrast to the ARMAX model (B.31), where the transfer functions of the 'system' and the 'noise' parts share a common denominator polynomial $A(z^{-1})$, this BJ model has different denominators defined by the polynomials $A(z^{-1})$ and $C(z^{-1})$. Both the ARMAX and BJ models can be generalized to allow for different order polynomials and the presence of a pure time delay δ defined in sampling intervals and allowing for a delay between the application of the input $u(k)$ and its first effect on the output $y(k)$. For instance, in the BJ model case, this general model takes the form,

$$y(k) = \frac{B(z^{-1})}{A(z^{-1})} u(k-\delta) + \frac{D(z^{-1})}{C(z^{-1})} e(k) \tag{B.33}$$

where the polynomials in z^{-1} are defiined as follows:

$$\begin{aligned} A(z^{-1}) &= 1 + a_1 z^{-1} + a_2 z^{-2} + \ldots + a_n z^{-n} \\ B(z^{-1}) &= b_0 + b_1 z^{-1} + b_2 z^{-2} + \ldots + b_m z^{-m} \\ C(z^{-1}) &= 1 + c_1 z^{-1} + c_2 z^{-2} + \ldots + c_p z^{-p} \\ D(z^{-1}) &= 1 + d_1 z^{-1} + d_2 z^{-2} + \ldots + d_q z^{-q} \end{aligned} \tag{B.34}$$

Note that, in control systems terminology, the roots of the characteristic equation $A(z^{-1}) = 0$ are called the *poles* of the polynomial $A(z^{-1})$, or the poles of the transfer function $B(z^{-1})/A(z^{-1})$; while the roots of the equation $B(z^{-1}) = 0$ are called the *zeros* of the polynomial $B(z^{-1})$, or the transfer function zeros. Note also that both $C(z^{-1})$ and $D(z^{-1})$ in the *AutoRegressive Moving Average* (ARMA) noise model are defined as a monic polynomials. This is made possible because the variance of $e(k)$ is specified separately as σ^2: in effect, the use of unity leading coefficients, together with the definition of this variance, completely specifies the statistical characteristics of the noise $\xi(k)$ as an ARMA process. This is the conventional method of defining models for noise with rational spectral density (see the Box-Jenkins book) and it can be contrasted with the deterministic part of the model, where only the $A(z^{-1})$ polynomial in the TF has a unity leading coefficient. $B(z^{-1})$, on the other hand, has a leading coefficient b_0 which has to be estimated and which controls the instantaneous effect that the input $u(k)$ has on the output $y(k)$.

Finally it is useful to consider the BJ model in the following decomposed form:

$$
\begin{aligned}
x(k) &= \frac{B(z^{-1})}{A(z^{-1})}u(k-\delta) \\
\xi(k) &= \frac{D(z^{-1})}{C(z^{-1})}e(k) \\
y(k) = x(k)+\xi(k);& \qquad v(k) = u(k)+\eta(k)
\end{aligned}
\tag{B.35}
$$

where $x(k)$ is the deterministic (noise-free) output of the system and $\xi(k)$ is the additive ARMA noise on the output variable. Note here that we have taken the opportunity to insert an observation equation for the input to the system $u(k)$, showing that this could also be subject to noise contamination $\eta(k)$ that is purely observational and does not pass through the system. When this input measurement noise is *not* present, the problem of estimating the model (B.35) from the observations $y(k)$ and $v(k)$ is relatively straightforward since the noise model estimation problem is converted to the simpler problem of estimating the discrete-time ARMA process for the output measurement noise. The details of the resulting *Refined Instrumental Variable* (RIVBJ) and *Prediction Error Minimization* (PEM) model estimation algorithms are discussed in chapters 7 and 10. When the additive input noise $\eta(k)$ is present, however, the model (B.35) is referred to as an *Errors-in-Variables* (EIV) model and the estimation of this EIV model is much more difficult, as discussed in section 7.4 of chapter 7.

Finally, note that in control systems analysis, a transfer function such as (B.35), where the $B(z^{-1})$ polynomial has a leading coefficient b_0, so that the input $u(k)$ can have an instantaneous effect on the output $y(k)$, is sometimes, inappropriately, called 'improper'. This derives from the analogous situation with continuous-time transfer functions (see next section B.3.4), where it *could* (but does not often), imply free differentiators that, for input signals of arbitrary high frequency, would deliver corresponding output signals of arbitrary high amplitude, which are physically not realisable. But, except in automatic control system terms, there is nothing improper in this discrete-time situation: it just means that there will be input effects that occur within one sampling interval of Δt time units that will affect the output instantaneously, something which is often encountered in practical situations. For instance, in rainfall-flow modelling (see main text), the rainfall can have such an immediate effect on flow. The reason it is 'improper' in automatic control terms is because the control input, which is normally calculated by a computational device of some kind, cannot have an immediate effect on the output.

B.3.4 Continuous-time, Stochastic Dynamic Models

Much of the algebra connected with discrete-time systems presented in the previous section carries over to the alternative continuous-time models. The *Continuous-Time* (CT) model has several advantages in the modelling of physical systems (see main text), not the least being its independence of the sampling interval: unlike the discrete-time state space model (B.23), the parametric matrices $\mathbf{F}(k)$, $\mathbf{G}(k)$, $\mathbf{H}(k)$ and $\mathbf{J}(k)$ in the following continuous-time model are not functions of the sampling interval Δt:

$$\begin{aligned} \frac{d\mathbf{x}(t)}{dt} &= \mathbf{F}(k)\mathbf{x}(t) + \mathbf{G}(k)\mathbf{u}(t) + \mathbf{J}(k)\boldsymbol{\eta}(t) \qquad (i) \\ y(t) &= \mathbf{H}(t)\mathbf{x}(t) + \mathbf{G}_I\mathbf{u}(t) + \mathbf{e(t)} \qquad (ii) \end{aligned} \tag{B.36}$$

Using similar algebraic manipulation to that followed above in the discrete-time case, the equivalent CT form of the *Box-Jenkins* (BJ) SISO transfer function model (F.1) is obtained:

$$y(t) = \frac{G(s)}{F(s)}u(t-\tau) + \xi(t) \tag{B.37}$$

or, alternatively, in the useful decomposed form:

$$\begin{aligned} x(t) &= \frac{G(s)}{F(s)}u(t-\tau) \\ y(t) &= x(t) + \xi(t) \end{aligned} \tag{B.38}$$

where $x(t)$ is the deterministic (noise-free) output of the system, $\xi(t)$ is the additive noise, which is not modelled specifically at this time (see next section B.3.5). The TF polynomials $F(s)$ and $G(s)$ are defined as follows in terms of the derivative operator $s^r = \frac{d^r}{dt^r}$:

$$\begin{aligned} F(s) &= s^n + f_1 s^{n-1} + f_2 s^{n-2} \ \ldots + f_n \\ G(s) &= g_0 s^m + g_1 s^{m-1} + g_2 s^{m-2} \ \ldots + g_m \end{aligned} \tag{B.39}$$

where $m \leq n$. The underlying continuous-time TF model can be written straightforwardly in the equivalent differential equation form:

$$\begin{aligned} &\frac{d^n x(t)}{dt^n} + f_1 \frac{d^{n-1}x(t)}{dt^{n-2}} + f_2 \frac{d^{n-2}x(t)}{dt^{n-2}} + \cdots + f_n x(t) = \\ &g_0 \frac{d^m u(t-\tau)}{dt^m} + g_1 \frac{d^{m-1}u(t-\tau)}{dt^{m-2}} + g_2 \frac{d^{m-2}u(t-\tau)}{dt^{m-2}} + \cdots + g_m u(t-\tau) \end{aligned} \tag{B.40}$$

B.3.5 Hybrid Stochastic Dynamic Models

The relationships between the continuous and discrete-time TF models is discussed in detail by, for example, Stoten and Harrison (1994). The Matlab functions `c2d` and `d2c` implement the conversions in either direction for both transfer function and state space models; and they are, of course, dependent on the sampling interval Δt associated with the time series data. This emphasizes the fact that, for any specified continuous-time model, there are an infinite number of discrete-time models depending on Δt. The main problem of the such conversion is the technique used to interpolate the input signal over the sampling interval.

The most common interpolation is the *Zero Order Hold* (ZOH): i.e. $u(k)$ is maintained constant between samples. However, the c2d and d2c routines allow for the specification of other techniques: the bilinear or 'Tustin' approximation to the derivative; the Tustin approximation with 'frequency prewarping'; and the 'matched pole-zero' method. Interpolation is also required when simulating a continuous-time system that has a sampled input: e.g. the Matlab lsim routine allows for the specification of ZOH or *First Order Hold* (FOH) (linear interpolation). And more sophisticated interpolation procedures, such as fixed interval smoothing (see chapter 4 of the main text) or the equivalent use of 'spline smoothing' (see e.g. Young and Pegregal, 1999b) could be used. Of course, from a statistical standpoint, the ideal approach would be some form of optimal interpolation based on the properties of the input time series

From an identification and estimation standpoint, the superficial algebraic similarity between the discrete and continuous-time TF models tends to conceal the increased complexity of the differential equations and, in particular, two quite fundamental statistical problems.

1. When considered in linear regression terms similar to that outlined for the discrete-time model case, we see that the estimation of the CT differential equation model (B.40) requires the measurement of not only the input and output signals but also the time derivatives of these signals up to order n. Direct differentiation of these signals is impossible in practice because the unlimited gain of the differentiation operation at high frequencies amplifies the high frequencies components in the noise signal to an unacceptable degree. Fortunately, this problem can be obviated by the introduction of 'prefilters' that not only attenuate the high frequency noise but also enhance the statistical efficiency of the parameter estimates. This is discussed fully in chapter 8 of the main text.
2. The white noise input $e(t)$ is normally defined as a continuous-time Gaussian process with a constant spectrum. However, this presents a major difficulty since the noise does not exist in a rigorous sense because it implies infinite variance (see Priestley (1981), page 156 *et seq*), i.e.,

$$E(e(t)=0; \quad E(e(t)\,e(t-\tau))=\sigma^2\delta(\tau) \tag{B.41}$$

where $\delta(\tau)$ is the Dirac delta function, i.e.,

$$\delta(\tau)=\begin{cases}\infty & \text{if } \tau=0;\\ 0 & \text{if } \tau\neq 0;\end{cases}$$

The practical consequence of this theoretical singularity is that it makes the direct estimation of the parameters in the ARMA noise model very difficult (see e.g. Garnier and Wang (2008)).

One way of avoiding this second problem, for the purposes of identification and estimation, is to consider a 'hybrid' BJ model with a continuous-time system model and a discrete-time noise model. If we consider a 'snapshot' of this model at a time instant t_k, then it can be specified in the following rather informal (mixing s and z^{-1} operators) but clear manner:

$$y(t_k) = \frac{B(s)}{A(s)}u(t_k - \tau) + \frac{D(z^{-1})}{C(z^{-1})}e(t_k) \tag{B.42}$$

or, alternatively, in the decomposed form:

$$\begin{aligned} x(t_k) &= \frac{B(s)}{A(s)}u(t_k - \tau) \\ \xi(t_k) &= \frac{D(z^{-1})}{C(z^{-1})}e(t_k) \\ y(t_k) = x(t_k)+\xi(t_k); &\qquad v(t_k) = u(t_k) + \nu(t_k) \end{aligned} \tag{B.43}$$

where $x(t_k)$ is the uniformly sampled value of the deterministic (noise-free) output of the CT system measured at the t_k^{th} sampling instant; $\xi(t_k)$ is the additive noise associated with this measurement, now assumed to be generated by a discrete-time ARMA model with a zero mean, white noise input $e(t_k)$; and, once again for completeness, we have introduced the measurement equation, where the input variable $u(t_k)$ is contaminated by noise $\nu(t_k)$, which could be coloured.

As in the purely discrete-time case, when this input measurement noise $\nu(t_k)$ is not present, the noise estimation problem is converted to the simpler problem of estimating the discrete-time ARMA process for the output measurement noise. The details of the *Refined Instrumental Variable estimation algorithm* (RIVCBJ) for estimating the Continuous-Time Hybrid model (B.42) in this situation are discussed in chapter 8. However, when $\nu(t_k)$ is present, then the model is in the EIV form and needs to handled accordingly (see section 7.4 of chapter 7).

B.3.6 Multivariable (Multi-Input, Multi-Output) TF Models

Previous subsections have concentrated on SISO transfer function models. A similar analysis can be carried out in discrete, continuous or hybrid terms, to produce the more general *Multi-Input, Multi-Output* (MIMO) transfer function model. This results in two related model forms: the multivariable TFM given in equation (B.24) or the following *Polynomial Matrix Description* (PMD):

$$\begin{aligned} \mathbf{A}(z^{-1})\mathbf{x}(k) &= \mathbf{B}(z^{-1})\mathbf{u}(k) \\ \mathbf{C}(z^{-1})\xi(k) &= \mathbf{D}(z^{-1})\mathbf{e}(k) \\ \mathbf{y}(k) &= \mathbf{x}(k) + \xi(k) \end{aligned}$$

where, $\mathbf{A}(z^{-1})$, $\mathbf{B}(z^{-1})$, $\mathbf{C}(z^{-1})$ and $\mathbf{D}(z^{-1})$ are polynomials with appropriately dimensioned matrix coefficients. This PMD, which is referred to as a *Multivariate*

Time Series or *Vector Time Series* model in the time series literature, is useful in deriving full multivariable versions of the estimation algorithms considered in the main text: see e.g. Jakeman and Young (1979b). However, this analysis is beyond the scope of the present book, requiring the use of Kronecker products. The interested reader who wishes to pursue this topic further is referred to the above paper, as well as: Priestley (1981), page 685 *et seq*); Ljung (1987), page 115 *et seq*; and chapter 8 of Harvey (1989).

B.3.7 Physical Interpretation of TF Models

As shown in the chapter 12 of the main text, an important aspect of *Data-Based Mechanistic* (DBM) modelling is the interpretation of the estimated dynamic model in physically meaningful terms. This sub-section considers certain aspects of TF models that facilitate such interpretation: the time and frequency domain properties of first and second order TFs; how TF models can be decomposed into series, parallel and feedback connections; how they can be reconstructed from such decompositions (block diagram algebra); and the structural implications of a decomposed TF, i.e. how the decomposed TF elements can be interconnected to yield physically interpretable structures. These topics are considered mainly in terms of continuous-time TF models in the derivative operator $s^r = \frac{d^r}{dt^r}$, since continuous-time models have a much more transparent interpretation in physical terms than discrete-time models. However, since TF model manipulation is algebraic, all of the topics, except the time and frequency properties considered first, apply if the s operator is replaced by the backward shift operator z^{-1}; so the results are equally meaningful and useful in discrete-time modelling terms.

1. **Time and Frequency Properties** The two fundamental building blocks of transfer function emerge when the denominator polynomial is factorized. These building blocks are the following first and second order transfer functions:

$$\begin{aligned} G_{fo}(s) &= \frac{b_0}{s+a_1} = \frac{G}{1+sT} && (i) \\ G_{so}(s) &= \frac{b_0 s + b_1}{s^2+a_1 s+a_2} = \frac{b_0 s + G\omega_n^2}{s^2+2\zeta\omega_n s+\omega_n^2} && (ii) \end{aligned} \qquad \text{(B.44)}$$

where in (i) $G = b_0/a_1$ is the *Steady State Gain*; $T = 1/a_1$ is the *Time Constant*; while in (ii), $\omega_n = \sqrt{a_2}$ is the *Natural Frequency* of oscillation of the second order system. If the *Damping Factor* $\zeta > 0$, where $\zeta = a_1/2\omega_n$, the dynamic system defined by the TF model is stable; if $\zeta = 1.0$, then the TF denominator has two equal, real roots at $-\omega_n$; if $\zeta > 1.0$, these negative roots become unequal; and in the situation where $0 < \zeta < 1.0$, it will have a complex conjugate pair of imaginary roots, with negative real parts.

In the situation when there are two negative real roots, it is clearly composed of two first order systems and each of these can be considered as in case (i). In the case of imaginary roots, the transfer function exhibits oscillatory response with the natural frequency of oscillation determined by ω_n and the damping of the oscillations determined by ζ. If $0 < \zeta < 1.0$ these oscillations are damped; if $\zeta = 0$, the roots are purely imaginary and the osillations are sustained (a 'limit cycle'); while if $\zeta < 0$, the roots have positive real parts, the system is unstable and the oscillations diverge.

The step and impulse responses of (*i*) and (*ii*) can be obtained manually using Laplace transform methods (see any book on control systems, such as Kuo and Golnaraghi (2002); or, computationally, by using the Matlab routines `step` and `impulse`, which also apply to the discrete-time equivalent models. The response to more general inputs can be generated using the `lsim` routine, or the `filter` routine for equivalent discrete-time systems. In fact, the continuous-time models can be converted to discrete-time models using the `c2d` routine and `filter` can then be used (or vice-versa using the `d2c` routine and `lsim`): see section B.3.7.1). Finally, the frequency response properties are discussed in section B.3.7.2.

2. **Analysis of TF Models using Block Diagram Algebra** First and second order TF models such as those discussed above in item 1., can be interconnected to produce multi-order TF models of wide variety. The most important, fundamental connections are the serial, parallel and feedback connections shown in Figure B.1, where $H_1(s)$ and $H_2(s)$ denote transfer functions of any order (e.g. each could represent other combinations of first and second order TFs). Transfer function sub-systems connected in parallel are additive; while those connected in series are multiplicative.

Referring to Figure B.1, therefore, the connections (a) and (b) can be represented by the following composite TFs:

$$H_s(s) = H_1(s).H_2(s) \tag{B.45a}$$

$$H_p(s) = H_1(s) + H_2(s) \tag{B.45b}$$

so, if $H_1(s)$ and $H_2(s)$ are both first order TFs of the form G_{fo} given in equation (B.44), i.e.

$$H_1(s) = \frac{b_{10}}{s+a_{11}}; H_2(s) = \frac{b_{20}}{s+a_{21}} \tag{B.46}$$

then,

$$H_s(s) = \frac{b_{10}b_{20}}{(s+a_{11})(s+a_{21})} = \frac{b_{10}b_{20}}{s^2+\{a_{11}+a_{21}\}s+a_{11}a_{21}}$$

$$H_p(s) = \frac{b_{10}}{s+a_{11}} + \frac{b_{20}}{s+a_{21}} = \frac{b_{10}(s+a_{21})+b_{20}(s+a_{11})}{(s+a_{11})(s+a_{21})}$$

$$= \frac{\{b_{10}+b_{20}\}s+b_{10}a_{21}+b_{20}a_{11}}{s^2+\{a_{11}+a_{21}\}s+a_{11}a_{21}}$$

The feedback connection is not quite so simple but it is obtained by noting that

$$e(t) = u(t) - H_2(s)x(t) \text{ and } x(t) = H_1(s)e(t)$$

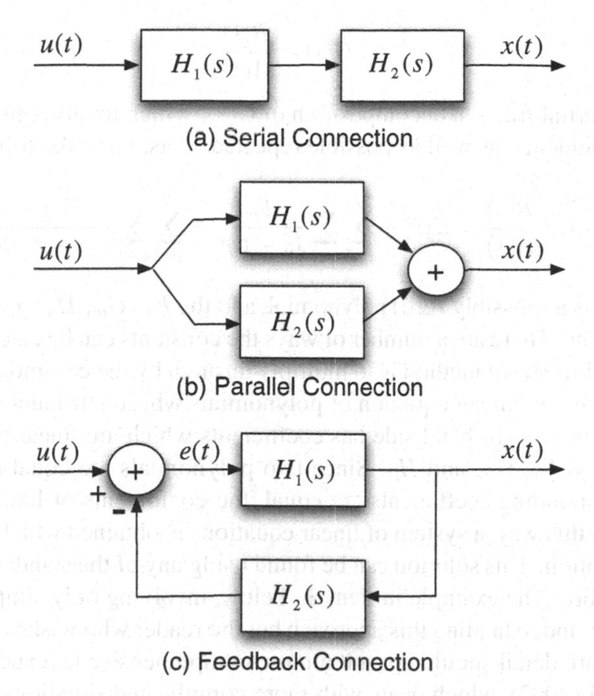

Fig. B.1 Series, parallel and feedback connection of TF sub-models.

so that,

$$x(t) = H_1(s)\{u(t) - H_2(s)\}x(t)$$

and so,

$$x(t) = \frac{H_1(s)}{1 + H_1(s)H_2(s)}u(t) \tag{B.47}$$

Applying this to the feedback system in Figure B.1(c), with the $H_1(s)$ and $H_2(s)$ defined in (B.46), yields the following closed loop TF:

$$H_{fb}(s) = \frac{b_{10}s + b_{10}a_{21}}{s^2 + \{a_{11} + a_{21}\}s + a_{11}a_{21} + b_{10}b_{20}}$$

3. **General TF Decomposition by Partial Fraction Expansion** Not surprisingly, given item 2., above, high order TFs can be decomposed into their first and second order elements. TF decomposition is well known in control and systems analysis and it is aided considerably by the use of the 'partial fraction expansion'. Fortunately Matlab has a routine residue that automates the procedure of partial fraction expansion, so facilitating the analysis. However, let us consider the nature of the expansion in general terms. Let $G(s)$ be any rational transfer function of the form:

$$G(s) = \frac{B(s)}{A(s)}$$

Then the partial fraction decomposition of $G(s)$, which involves first and second order TF elements, as well as possible repeated roots, takes the following form:

$$G(s) = \frac{B(s)}{A(s)} = E(s) + \sum_{i=1}^{j} \sum_{r=1}^{l} \frac{F_{ir}}{(s-f_i)^r} + \sum_{i=1}^{k} \sum_{r=1}^{m} \frac{G_{ir}s + H_{ir}}{(s^2 + g_i s + h_i)^r}$$

Here, $E(s)$ is a (possibly zero) polynomial, and the F_{ir}, G_{ir}, H_{ir}, f_i, g_i, and h_i are real constants. There are a number of ways the constants can be calculated but the most straightforward method is to multiply through by the common denominator $A(s)$. We then obtain an equation of polynomials whose left-hand side is simply $B(s)$ and whose right-hand side has coefficients which are linear expressions of the constants F_{ir}, G_{ir}, and H_{ir}. Since two polynomials are equal if and only if their corresponding coefficients are equal, the coefficients of like terms can be equated. In this way, a system of linear equations is obtained which always has a unique solution. This solution can be found using any of the standard methods of linear algebra. The example in item 3., below, involving only simple, real roots will help in understanding this approach but the reader who wishes to pursue this topic in more detail should consult a more comprehensive text such as Kuo and Golnaraghi (2002), which deals with more complicated situations involving TF denominators with multiple order roots and second order factors characterized by complex conjugate roots.

4. **Example 1: Parallel Pathway Decomposition of 2nd Order TF models** Consider the following TF model of the relationship between the concentration of a dye tracer $u(t)$, measured every 15 minutes (0.25 h), at one location in the River Conder, North West England, and the resulting output concentration $x(t)$ at a downstream location (see also the example concerning the modelling of tracer experiment data considered in chapters 6, 7 and 8):

$$x(t) = \frac{1.1939s + 0.64281}{s^2 + 2.0513s + 0.60323} u(t) = \frac{1.1939s + 0.64281}{(s + 1.6955)(s + 0.35578)} u(t) \quad \text{(B.48)}$$

where this continuous-time model has been estimated with time units of hours. We wish to convert this TF to the partial fraction form:

$$G(s) = \frac{1.1939s + 0.64281}{(s + 1.6955)(s + 0.35578)} = \frac{R(1)}{s + 1.6955} + \frac{R(2)}{s + 0.35578}$$

where $R(1)$ and $R(2)$ are the elements of the *residue* vector $\mathbf{R}$. Now, multiplying throughout by the common denominator and comparing this with the numerator in (B.48), we obtain

$$R(1)s + 0.35578R(1) + R(2)s + 1.6955R(2) = 1.1939s + 0.64281$$

Comparing like terms on each side of this equation produces the following linear simultaneous equations in the unknown *residues* $R(1)$ and R(2), as written in vector form:

$$\begin{bmatrix} 0.35578 & 1.6955 \\ 1 & 1 \end{bmatrix} \begin{bmatrix} R(1) \\ R(2) \end{bmatrix} = \begin{bmatrix} 0.64281 \\ 1.1939 \end{bmatrix}$$

which leads to the solution

$$R(1) = 1.0311; \quad R(2) = 0.16275$$

and the finally decomposed TF model:

$$x(t) = \left\{ \frac{1.0311}{s+1.6955} + \frac{0.16275}{s+0.35578} \right\} u(t) \tag{B.49}$$

Not surprisingly, this is the same result generated by the application of the `residue` routine in Matlab. The decomposition in (B.49) shows that the output $x(t)$ of the second order TF (B.48) can be considered as the summed output of two first order TFs *connected in parallel*. The *Time Constants* $\{T_1, T_2\}$, *Steady State Gains* $\{G_1, G_2\}$ and *Partition Percentages* $\{P_1, P_2\}$ associated with the first order elements are immediately interpretable in physical terms:

$$T_1 = \frac{1}{1.6955} = 0.59 \text{ hours}; \quad T_2 = \frac{1}{0.35578} = 2.81 \text{ hours}$$

$$G_1 = \frac{1.0311}{1.6955} = 0.608; \quad G_2 = \frac{0.16275}{0.35578} = 0.457$$

$$P_1 = \frac{G_1 \times 100}{G_1 + G_2} = 0.57.1\%; \; P_2 = 100\frac{G_2 \times 100}{G_1 + G_2} = 0.429\%$$

And further hydrological interpretation is possible, as discussed in the main text examples mentioned above. As pointed out, although we have considered continuous-time TFs in the above analysis, it applies equally well to discrete-time TFs, although the physical interpretation is not so obvious since it requires the introduction on conversions depending on the sampling interval Δt.

5. **Example 2: Serial and Feedback Decomposition of 2nd Order TF Models**
When a TF is decomposed into separate, additive elements by partial fraction expansion, it reveals possible parallel connection of the TF elements. However serial and feedback decompositions are possible, in addition to parallel. These are easy to compute in the case of 2nd order systems, as we see from the simple example below.

Consider again the TF model in (B.48). While parallel decomposition involves the addition of the TF elements, serial decomposition involves their *multiplication* . Such a decomposition is not unique and not particularly interesting in this case and it is unlikely to have any physical interpretation. For example, the decomposition

$$x(t) = \left\{ \frac{1.1939s + 0.64281}{s + 1.6955} \cdot \frac{1}{s + 0.35578} \right\} u(t)$$

is a serial connection of two first order systems: one with a steady state gain of 0.379, a time constant of 0.59 hours and, when responding to a unit step input. it has a considerable initial overshoot to 1.19, before settling to the steady state level of 0.379; the other has a standard, first order response with no overshoot, characterized by a steady state gain of 2.811 and time constant of 2.8 hours.

The feedback decomposition is also not unique unless the feedback structure is defined. For example, consider the case where it is assumed that the feedback structure involves first order TF in both the forward and feedback paths (as in Figure B.1(c)). This is more capable of a physical interpretation, since the 'closed loop' transfer function for the system in this case is given by (see item 2., above):

$$\frac{\beta_1 s + \beta_1 \alpha_2}{s^2 + (\alpha_1 + \alpha_2) + (\alpha_1 \alpha_2 + \beta_1 \beta_2)} \tag{B.50}$$

Comparing coefficients between (B.50) and (B.48), it is straightforward to solve for α_1, β_1. α_2 and β_2 in terms of the known TF coefficients $a_1 = 2.0513$, $a_2 = 0.60323$ $b_0 = 1.1939$ and $b_1 = 0.64281$. This yields:

$$\alpha_1 = 1.1529 \; \beta_1 = 1.1939 \; \alpha_2 = 0.53841 \; \beta_2 = 0.177$$

which corresponds to a first order TF with steady state gain $G_1 = 0.789$ and time constant $T_1 = 0.661$ hours in the forward path; and a first order TF with steady state gain $G_1 = 0.33$ and time constant $T_1 = 1.86$ hours in the negative feedback path. Note how the dynamics of these first order transfer functions are considerably different from the first order transfer functions in the parallel decomposition of (B.49), demonstrating how the feedback structure, unlike the parallel structure, has little obvious resemblance to the original transfer function (indeed, this is the reason why feedback is important to control system design; the natural dynamics of the system can be changed in this manner to make them more desirable in some sense: for instance, an naturally unstable system can be stabilized; or natural oscillations can be damped). In this particular case, however, the feedback decomposition is not much use: it suggests that the output tracer is being fed back negatively to the input and being subtracted from the input, something that makes little physical sense. So, in this example, serial and feedback decomposition would be rejected on physical grounds, leaving the parallel decomposition, which makes a lot of physical sense, as shown in chapters 6, 7 and 8 of the main text.

6. **Decomposition of Higher Order TF Models** While serial, parallel and feedback decompositions are easy to compute in the case of second order systems, they become increasingly difficult as the order of the input-output transfer function increases, with many permutations and combinations of interconnections being possible. More research is required on this topic but the Matlab routine shown in Figure B.2 on the next page, which implements a specific 4th order example, will

illustrate the complexity of the computation for just one of the possible feedback decompositions.

7. **The General, Multi-Order TF Model** It is clear from the above items that, in general, serial, parallel and feedback connections of elemental first or second order TF models, such as those in equation (B.44), lead to a multi-order TF model that takes the general TF form:

$$x(t) = \frac{G(s)}{F(s)} u(t-\tau)$$

$$F(s) = s^n + f_1 s^{n-1} + \ldots + f_n; \quad G(s) = g_0 s^m + g_1 s^{m-1} + \ldots + g_m$$

in which n and m can take on any positive integer values.

```
Figure B.2 Matlab routine for decomposition of a
fourth order TF model

function [r,p,k] = TFdecomp4(b,a);
% function [r,p,k] = 4thorderTF_decomp(b,a);
% [4 5 0] CT model polynomials denoted by [a b]
% foward path gain denoted by k
% Each feedback TF denoted by ri/(s+pi), i=1,2,3,4
% Form of numerator in TF with these definitions:
% k(s+p1)(s+p2)(s+p3)(s+p4)
% This can be obtained by factorization of the
% b polynomial in the [4 5 0] model
% with b(1) in the b polynomial providing k
p=-roots(b)
p1=p(1);p2=p(2);p3=p(3);p4=p(4);
k=b(1);
a1=a(2);a2=a(3);a3=a(4);a4=a(5);
A11=k;A12=k;A13=k;A14=k;
A21=k*p4+k*p3+k*p2;
A22=k*p1+k*p3+k*p4;
A23=k*p1+k*p2+k*p4;
A24=k*p1+k*p2+k*p3;
A31=k*p3*p4+k*p4*p2+k*p2*p3;
A32=k*p1*p3+k*p1*p4+k*p3*p4;
A33=k*p1*p2+k*p1*p4+k*p2*p4;
A34=k*p2*p3+k*p1*p2+k*p1*p3;
A41=k*p4*p3*p2;
A42=k*p4*p3*p1;
A43=k*p4*p1*p2;
A44=k*p1*p2*p3;
AA=[A11 A12 A13 A14;A21 A22 A23 A24; ...
A31 A32 A33 A44;A41 A42 A43 A44];
B1=a1-p4-p1-p2-p3;%-k*r4;
B2=a2-p2*p1-p3*p1-p3*p2-p4*p1-p4*p2-p4*p3;
B3=a3-p3*p2*p1-p4*p2*p1-p4*p3*p1-p4*p3*p2;
B4=a4-p4*p3*p2*p1;
BB=[B1 B2 B3 B4]';
r=inv(AA)*BB;r=r';p=p';
```

8. **Discrete-Time Models** All of the operations discussed above in items 2. to 7. are algebraic so that they can be applied equally well to discrete-time transfer function models. However, the physical interpretation of the discrete-time TFs is not so instantly recognizable as in the continuous-time case because the discrete-time TF is a function of the sampling interval Δt. Given a discrete-time TF, therefore, there are two ways of considering TF decomposition: first, transforming the TF to continuous-time, either manually (a daunting task in all but the simplest cases) or, more sensibly, by employing the Matlab routine d2c, and then decomposing this transformed TF; or second, by decomposing the discrete-time TF directly and then converting the decomposed TF elements to continuous-time. In the second case, the discrete-time TF elements will be either first or second order and these are easily converted manually to continuous time.

B.3.7.1 TF Conversion Between Continuous and Discrete-Time

This book deals with both continuous-time and discrete-time systems and it is necessary to consider the relationship between such models, as well as how one converts from continuous to discrete-time models and vice versa. This is best considered in general terms from the standpoint of the deterministic state space models, such as the deterministic version of the continuous-time state space model (B.36):

$$\begin{aligned} \frac{d\mathbf{x}(t)}{dt} &= \mathbf{F}(k)\mathbf{x}(t) + \mathbf{G}(k)\mathbf{u}(t) \qquad (i) \\ y(t) &= \mathbf{H}(t)\mathbf{x}(t) + \mathbf{G}_I\mathbf{u}(t) \qquad (ii) \end{aligned} \tag{B.51}$$

The evolution of the state vector $\mathbf{x}(t)$ from a sampling instant t_{k-1} to a future time t is obtained by solving (B.51) in the following manner:

$$\mathbf{x}(t) = e^{\mathbf{F}(t-t_{k-1})}\mathbf{x}(t_{k-1}) + \int_{t_{k-1}}^{t} e^{\mathbf{F}(t-\tau)}\mathbf{G}\mathbf{u}(\tau)d\tau \tag{B.52}$$

where $e^{\mathbf{F}(t)}$ is the 'matrix exponential' function which can be evaluated in various ways (see Moler and Loan, 2003, Higham, 2005 and `http://www.mathworks.com/help/techdoc/ref/expm.html`): for example by taking sufficient terms in the infinite expansion

$$e^{\mathbf{F}} = \mathbf{I} + \mathbf{F} + \frac{\mathbf{F}^2}{2!} + \cdots + \frac{\mathbf{F}^n}{(n)!} + \cdots, \tag{B.53}$$

although this not necessarily the best method (see the cited references). As a result, the state at the next sampling instant t_k is given by:

$$\begin{aligned}\mathbf{x}(t_k) &= e^{\mathbf{F}(t_k-t_{k-1})}\mathbf{x}(t_{k-1}) + \int_{t_{k-1}}^{t_k} e^{\mathbf{F}(t_k-\tau)}\mathbf{G}\mathbf{u}(\tau)d\tau \\ &= e^{\mathbf{F}(t_k-t_{k-1})}\mathbf{x}(t_{k-1}) + \int_{t_{k-1}}^{t_k} e^{\mathbf{F}(t_k-\tau)}d\tau\, \mathbf{G}\mathbf{u}(t_{k-1}) \\ &= \mathbf{A}(t_k,t_{k-1})\mathbf{x}(t_{k-1}) + \mathbf{B}_I(t_k,t_{k-1})\mathbf{u}(t_{k-1})\end{aligned}$$

where the final equation here is obtained under the assumption that the input vector $\mathbf{u}(t_k)$ is constant over the sampling interval: the zero-order hold (ZOH) assumption.

Converting the above result to our simpler discrete-time nomenclature (i.e. $\mathbf{x}(k) = \mathbf{x}(t_k)$ etc.), in order to emphasize that we are now completely within a discrete-time context, the state equations for the discrete-time state space equivalent of (B.51) take the form:

$$\begin{aligned}\mathbf{x}(k) &= \mathbf{A}(k)\mathbf{x}(k-1) + \mathbf{B}(k)\mathbf{u}(k-1) && \text{(i)} \\ y(k) &= \mathbf{C}(k)\mathbf{x}(k) + \mathbf{B}_I(k)\mathbf{u}(k) && \text{(ii)}\end{aligned} \tag{B.54}$$

where, since $(t_k - t_{k-1})$ is the sampling interval Δt,

$$\begin{aligned}\mathbf{A}(k) &= e^{\mathbf{F}\Delta t} \\ \mathbf{B}(k) &= \int_0^{\Delta t} e^{\mathbf{F}\tau}d\tau\, \mathbf{G}\end{aligned} \tag{B.55}$$

which is the deterministic version of the discrete-time state equations (B.23) and is the *exact* equivalent of the continuous-time state space model (B.51) *provided the* ZOH *assumption is valid*. For this reason, it is often called the ZOH equivalent of (B.51). The reason why this assumption (or the alternative first order hold, FOH) is usually made relates to the normal situation, when confronted with sampled data, that there is no information about how the input variables are changing over the sampling interval. As a result, some form of interpolation over the sampling interval is necessary and the ZOH interpolation (the signal remains constant) is the simplest. Given the stochastic nature of the model estimation problem, it would be possible to use some form optimal interpolation (e.g. based upon the fixed interval smoothing algorithms discussed in chapters 4 and 5 of the main text) but this would add considerable complexity and, as suggested in chapter 8, it does not seem to be justified.

As a simple example, if we consider the stable, first order differential equation,

$$\frac{dx(t)}{dt} = -fx(t) + gu(t), \tag{B.56}$$

then, under the ZOH assumption, the equivalent discrete-time model takes the form

$$x(k) = ax(k-1) + bu(k-1) \tag{B.57}$$

where,

$$a = e^{-f\Delta t} = 1 - f\Delta t + \frac{f^2}{2!}\Delta t^2 - \frac{f^3}{3!}\Delta t^3 + \cdots;$$
$$b = g\Delta t$$

so that, considering only the first order terms in the expansion of $e^{-f\Delta t}$, $a = 1 - f\Delta t$. Note that this is the same as that derived by simple discretization of equation (B.56):

$$\frac{x(k) - x(k-1)}{\Delta t} = -fx(k-1) + gu(k-1)$$
$$\therefore\ x(k) = \{1 - f\Delta t\}x(k-1) + \{g\Delta t\}u(k-1) \tag{B.58}$$

so that, comparing (B.58) with (B.57), $a = 1 - f\Delta t$, or $f = (1-a)/\Delta t$. Consequently, the time constant $T = 1/f$ is given by $\Delta t/(1-a)$. Finally, note that since $a = e^{-f\Delta t}$, f and the associated time constant $T = 1/f$ are given exactly by

$$f = -\frac{log_e(a)}{\Delta t}; \quad T = -\frac{\Delta t}{log_e(a)} \tag{B.59}$$

which, following discrete-time model parameter estimation (yielding estimates of a and b), provides a convenient quick conversion to these physically more meaningful continuous-time parameters (see the main text).

B.3.7.2 Frequency Domain Representations of TF Models

Operators such as z^{-1} and s are strange mathematical objects. At the superficial level considered above, they are simply a shorthand form that allows us to form transfer functions in a convenient manner. However, they can be interpreted as much more than this if the system defined by the transfer function is perturbed by a purely sinusoidal-type of input interpreted in terms of complex numbers. This interpretation is quite easy to develop by using again the series expansion of the exponential function, which now includes the imaginary number j and the frequency ω: i.e.,

$$e^{j\omega} = 1 + j\omega - \frac{\omega^2}{2!} - \frac{j\omega^3}{3!} + \frac{\omega^4}{4!} + \cdots$$

whose real ($\Re$) and imaginary ($\Im$) parts are simply:

$$\Re\{e^{j\omega}\} = 1 - \frac{\omega^2}{2!} + \frac{\omega^4}{4!} + \cdots$$
$$\Im\{e^{j\omega}\} = \omega - \frac{\omega^3}{3!} + \frac{\omega^5}{5!} + \cdots$$

But these are the series expansions of the *cos* and *sine* functions, respectively, so that,

$$e^{j\omega} = cos(\omega) + j\,sin(\omega) \tag{B.60}$$

Now let us consider the linear differential equation (B.40) with $\tau = 0$, for convenience, i.e.,

$$\begin{aligned} &\frac{d^n x(t)}{dt^n} + f_1 \frac{d^{n-1} x(t)}{dt^{n-2}} + f_2 \frac{d^{n-2} x(t)}{dt^{n-2}} + \cdots + f_n x(t) = \\ &g_0 \frac{d^m u(t)}{dt^m} + g_1 \frac{d^{m-1} u(t)}{dt^{m-2}} + g_2 \frac{d^{m-2} u(t)}{dt^{m-2}} + \cdots + g_m u(t) \end{aligned} \tag{B.61}$$

or, in TF form,

$$x(t) = \frac{g_0 s^m + g_1 s^{m-1} + \cdots + g_m}{s^n + f_1 s^{n-1} + f_2 s^{n-2} + \cdots + f_n} u(t) \tag{B.62}$$

with the input $u(t)$ defined as $u(t) = e^{j\omega t}$, so that the system is perturbed by a complex sinusoidal signal defined at all frequencies ω. If the input is complex like this, then the output will also be complex, with the the real part of the output being the response of the linear model to the real part of the input; and similarly for the complex parts of the input and output. As a result, the output will take the form $G(j\omega)e^{j\omega t}$, where $G(j\omega)$ is the transfer function of the model defined in the frequency domain and is often called the *frequency response function* (FRF) of the model. Introducing these into (B.61) and carrying out the differentiations, we obtain

$$\begin{aligned} \left[(j\omega)^n + f_1(j\omega)^{n-1} + f_2(j\omega)^{n-2} + \ldots + f_n\right] G(j\omega)e^{j\omega t} &= \\ \left[g_0(j\omega)^m + g_1(j\omega)^{m-1} + \ldots + g_m\right] e^{j\omega t} \end{aligned}$$

so that

$$G(j\omega) = \frac{g_0(j\omega)^m + g_1(j\omega)^{m-1} + \cdots + g_m}{(j\omega)^n + f_1(j\omega)^{n-1} + +f_2(j\omega)^{n-2} \cdots + f_n} \tag{B.63}$$

which is simply the TF in the model (B.62) with $j\omega$ substituted for the differential operator s. $G(j\omega)$ is a function of the complex variable and so its evaluation for any value of the frequency ω results in a complex number $\alpha + j\beta$, where α and β will depend upon the value of ω. And since the complex number defines an amplitude and phase, defined by in the normal manner, the amplitude A and phase shift φ for the model as a function of ω are given by:

$$\begin{aligned} A(\omega) &= |G(j\omega)| = [G(j\omega)G^*(j\omega)]^{1/2} = \sqrt{(}[\alpha(\omega)^2 + \beta(\omega)^2]) \\ \varphi(\omega) &= \tan^{-1} \frac{1}{j}\left[\frac{G(j\omega) - G^*(j\omega)}{G(j\omega) + G^*(j\omega)}\right] = \tan^{-1} \frac{\beta(\omega)}{\alpha(\omega)} \end{aligned} \tag{B.64}$$

where $G^*(j\omega)$ denotes the complex conjugate of $G(j\omega)$. For example, in the case of a first order system:

$$G(s) = \frac{1}{sT + 1} \tag{B.65}$$

the reader can show that the amplitude and phase are as follows:

$$A(\omega) = (1 - \omega^2 T^2)^{-1/2}; \quad \varphi(\omega) = tan^{-1}(-\omega T)$$

In the control and systems literature, it is normal to refer to the Gain $G(\omega) = 20log_{10}A(\omega)$, in decibels (db). The plot of $G(\omega)$ against $log_{10}(\omega)$, together with the associated plot of φ, in degrees, against $log_{10}(\omega)$, is called the Bode Plot. For example, the Bode plot in the case of the first order system (B.65), with $T = 10$ time units, is shown below in Figure B.2 (full lines).

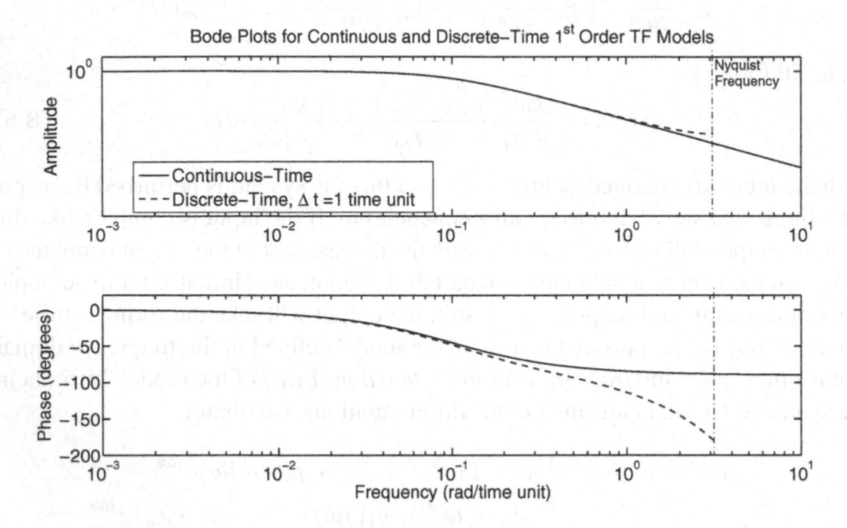

Fig. B.2 Bode frequency response plots for continuous and discrete-time, first order TF models.

The frequency domain representation of a discrete-time TF can be handled in a similar manner to the above by noting that, using a Taylor expansion,

$$\begin{aligned} z^{-1}x(t) = x(t-\Delta t) &= x(t) - \frac{\Delta t}{1!}\frac{dx(t)}{dt} + \frac{\Delta t^2}{2!}\frac{d^2x(t)}{dt^2} + \cdots \\ &= \left[1 - \frac{\Delta t\, s}{1!} + \frac{\Delta t^2\, s^2}{2!} + \cdots\right] x(t) \\ &= e^{-s\Delta t}x(t) \end{aligned}$$

where Δt is the sampling interval. As a result, the Bode plot for discrete-time TF models can be obtained by substituting

$$e^{-j\omega\Delta t} = cos(\omega\Delta t) + jsin(\omega\Delta t)$$

for z^{-1} in the discrete-time TF and then following a similar procedure to that used in the case of the continuous-time TF. The resulting Bode diagram is normally plotted for $0 \leq \omega \leq \omega_N$, where $\omega_N = \frac{\pi}{\Delta T}$ is called the *Nyquist frequency*: see section 8.7 of chapter 8. The equivalent continuous-time frequency ω is then used as the x-axis variable and the plot can be compared directly with the equivalent continuous time plot. However, the response is plotted only up to the ω_N since then aliasing can occur. The Bode plot for the first order discrete-time equivalent of (B.65), at a sampling interval of $\Delta t = 1$ time unit is shown in Figure B.2 (dashed lines), with the Nyquist frequency limit, $\omega_N = \pi/\Delta t = \pi$ rad/sec., shown as a dash-dot vertical line. Note the distortion appearing in the discrete-time diagram as the Nyquist Frequency is approached, which raises the question of how the sampling interval affects the estimation of TF models. This is addressed in section 8.7 of chapter 8 in the main text.

B.3.8 Differentiation of a TF with Respect to a Given Parameter

When considering Maximum Likelihood estimation of the parameters of a SISO TF model, such as the BJ or ARMAX models, it is necessary to differentiate expressions written in TF terms with respect to each coefficient in turn. This is accomplished quite easily by reference to the rule for differentiating a product or quotient. For example for $i = 1, 2, \ldots, n$

$$\frac{\partial}{\partial b_i}\{\frac{B(z^{-1})}{A(z^{-1})}u(k)\} = \frac{\partial}{\partial b_i}\{\frac{b_1 z^{-1} + \ldots + b_n z^{-n}}{1 + a_1 z^{-1} + \ldots + a_n z^{-n}}u(k)\} = \frac{z^{-i}}{A(z^{-1})}u(k)$$

while

$$\begin{aligned}\frac{\partial}{\partial a_i}\{\frac{B(z^{-1})}{A(z^{-1})}u(k)\} &= \frac{A(z^{-1})\frac{\partial}{\partial a_i}\{B(z^{-1})\} - B(z^{-1})z^{-i}}{[A(z^{-1})]^2}u(k) \\ &= \frac{B(z^{-1})}{[A(z^{-1})]^2}z^{-i}u(k)\end{aligned}$$

since $\frac{\partial}{\partial a_i}\{B(z^{-1})\} = 0$; as $B(z^{-1})$ has no terms in a_i.

B.3.9 A Simple Introduction to Monte Carlo Simulation

The advent of fast digital computers has enabled engineers, scientists and social scientists to carry out analysis that would not have been possible a few decades ago. One of the most useful procedures of this type is Monte Carlo Simulation (MCS) analysis and an early example of its use in practice is described in Young and Whitehead (1975). There are many publications on MCS methods but an easily accessible source is the Wikipedia entry for 'Monte Carlo method' at `http://en.wikipedia.org/wiki/Monte_Carlo_method`.

MCS methods use random sampling, as in the toss of a dice, in order to compute their results and it is from this that the name derives. In particular, they utilize the repeated computation of random or pseudo-random numbers that are able to simulate the effects of stochastic phenomena associated with the model, which are normally defined by the results obtained from estimation procedures: e.g. covariance matrices associated with parameter estimates from the RIVBJ/RIVCBJ algorithms described in chapters 7 and 8 and implemented as the rivbj and rivcbj routines in the CAPTAIN Toolbox. It is this need for repeated sampling and simulation, often many thousands (or even hundreds of thousands) of time, that means MCS methods can only be implemented realistically within a digital computer.

B.3.9.1 The Generation of Random Realizations

MCS analysis can be applied with any defined probability distribution function (pdf) but the most commonly used and useful pdf is the multivariate Gaussian normal distribution, which is defined completely by its first two moments: the mean vector and the covariance matrix (see section B.2.6). In order to explain the rudiments of MCS analysis, therefore, let us consider its use in relation to stochastic phenomena that can be described by the normal pdf. For instance, in the case of the parameter estimates produced by the RIVBJ and RIVCBJ algorithms, the pdf of the parameter estimates is normal, where the mean is the estimated parameter vector $\hat{\boldsymbol{\theta}}$ and the associated estimate of the covariance matrix is $\mathbf{P}$. So, given these, how do we generate random realizations of $\hat{\boldsymbol{\theta}}$ and how could we use these to our advantage?

First, it is straightforward to generate Gaussian random variables using readily available routines, such as randn in Matlab. When using MCS in the context of RIVBJ estimation, therefore, a sequence of N random vectors $\mathbf{e}(k)$, each of the same order $p = m+n+1$ as the estimated parameter vector $\hat{\boldsymbol{\theta}}$, are generated. From the properties of $\mathbf{e}(k)$ generated in this manner, we know that,

$$E\{\mathbf{e}(k)\} = \mathbf{0}; E\{\mathbf{e}(k)\mathbf{e}^T(j)\} = \delta_{kj}\mathbf{I}_p \tag{B.66}$$

where δ_{kj} is the Kronecker delta function (see earlier). From this sequence of purely random vectors, we wish to generate a sequence that has the same covariance properties as $\hat{\boldsymbol{\theta}}$. This can be accomplished by exploiting the Cholesky decomposition (see section B.1.13), which is able to factorize the covariance matrix $\mathbf{P}$ into the form $\mathbf{P} = \mathbf{T}\mathbf{T}^T$, where $\mathbf{T}$ is an upper triangular matrix. If we now form a sequence of vectors $\mathbf{n}(k) = \mathbf{T}^T\mathbf{e}(k)$, then the mean and covariance of $\mathbf{n}(k)$ are given by:

$$\begin{aligned}
E\{\mathbf{n}(k)\} &= E\{\mathbf{T}^T\mathbf{e}(k)\} = \mathbf{0} \\
\mathrm{cov}\{\mathbf{n}(k), \mathbf{n}(j)\} &= E\{\mathbf{n}(k)\mathbf{n}^T(j)\} = E\{\mathbf{T}^T\mathbf{e}(k)\mathbf{e}^T(j)\mathbf{T}\} \\
&= \mathbf{T}^T\mathbf{T} = \mathbf{P}
\end{aligned}$$

because of the covariance properties of $\mathbf{e}(k)$ in (B.66). As required, therefore, the $\mathbf{n}(k)$ sequence generated in this manner will be characterized by the same covariance matrix $\mathbf{P}$ as the RIVBJ vector parameter estimate $\hat{\boldsymbol{\theta}}$.

On the basis of the above result, it is straightforward to generate the requisite number of samples $\mathbf{n}(k)$, $k = 1, 2, \ldots., N$, where N if the number of Monte Carlo draws that we require to conduct MCS analysis. Suppose, for example, we have estimated the parameter vector $\hat{\boldsymbol{\theta}}$ associated with a TF and, on the basis of the estimated uncertainty in $\hat{\boldsymbol{\theta}}$, as defined by the estimated covariance matrix $\mathbf{P}$, it is required to investigate the uncertainty in the poles of the denominator polynomial $A(s)$ or $A(z^{-1})$, depending on the type of TF, as well as the uncertainty in the output response to an input $u(k)$. This can be accomplished very simply by the simple Matlab m-file function script in Figure B.3 on the following page, which returns the MCS results for the poles and output response, as well as the denominator and numerator polynomials.

If the model has real poles, then the histograms for the associated time constants can then be drawn by reference to the realizations collected in the *amcs* matrix returned from this routine. Of course, the uncertainty on *any* derived parameter or property of the estimated TF can be investigated in this manner. For instance, the steady state gain, the Bode plot or, in the case of a TF with complex poles, the natural frequency and damping. Moreover, the effects of the uncertainty in $\hat{\boldsymbol{\theta}}$ on a simulated system that uses the model can also be investigated: e.g. the effects of uncertainty in a water pollution simulation model designed on the basis of the estimated TF model, as in Young and Whitehead (1975). The potential applications of this MCS methodology are enormous and its general utility as an investigative tool in identification, estimation, forecasting and control system design is obvious.

There has been a great expansion of research on MCS-based methods over the past few years as digital computers have continued their exponential growth in computing power. Statisticians, in particular, have seen such approaches as liberation from the need for the limiting statistical assumptions that are normally required to generate theoretical analytical results. Some of these methods are outlined in the next sub-section and are discussed.

B.3.9.2 Advanced Monte Carlo Methods

Amongst the Monte Carlo-based methods, great interest has been aroused by *Markov Chain Monte Carlo* (MCMC) methods. These are algorithms for sampling from probability distributions based on constructing a Markov chain that has the desired distribution as its equilibrium distribution. The state of the chain after a large number of steps is then used as a sample from the desired distribution. The quality of the sample improves as a function of the number of steps.

Random Walk Monte Carlo (RWMC) methods include the Metropolis-Hastings algorithm and Gibbs sampling. The former generates a random walk using a 'proposal' density and is based on a method for rejecting proposed moves. The latter requires that all the conditional distributions of the target distribution can be sampled exactly: it is practically useful because it requires less expertise and 'tuning'. Other methods of this general type are discussed and described in `http://en.wikipedia.org/wiki/Monte_Carlo_method`.

Fig B.3 MCS analysis of discrete or continuous-time SISO TF model

```
% function [amcs,bmcs,poles,Ymcs]=mcs(th,N,u,type,SI)
%
% INPUTS
% th: results matrix from RIVBJ
% N: number of MCS realization
% u: system input
% type: Discrete TF (1); Continuous TF (0)
% SI: sampling interval
%
% OUTPUTS
% amcs: MCS denom. poly
% bmcs: MCS num. poly
% poles: MCS poles of den. poly
% Ymcs: MCS model outputs
%
%
%recovers results from RIVBJ th function
[a,b,c,d,P]=getparbj(th);
b=b(1,:);
n=length(a)-1;
m=length(b);
%define thetahat from RIVBJ results
thetahat=[a(2:end) b];
%select system part of cov. matrix
P=P(1:length(thetahat),1:length(thetahat));
T= chol(P);T=T'; %Cholesky decomposition
for i = 1:N  % MCS analysis
    e= randn(n+m,1);
    nm= (T*e)';
    if i==1;% 1st draw is nominal estimate from RIVBJ
        theta=thetahat;
    else
        theta=thetahat+nm;
    end
    amcs(i,:) =  [1 theta(1:n)];% save TF denom. poly
    bmcs(i,:)=theta(n+1:end);% save TF num. poly
    % calculate and save roots of den. poly
    poles(i,:)=roots(amcs(i,:));
    if type
        Ymcs(:,i)=filter(bmcs(i,:),amcs(i,:),u);
    else
        M=tf(bmcs(i,:),amcs(i,:));
        Ymcs(:,i)=lsim(M,u,(0:SI:SI*(length(u)-1)'));
    end
end
```

Finally, note that, within the Control and Systems Community, there has been growing interest in *Sequential Monte Carlo* (SMC) methods because they have been developed as a computationally intensive alternative to the Kalman Filter in complex nonlinear and non-Gaussian situations, where its purely analytical solution is no longer appropriate. SMC methods can be considered as sequential (recursive or 'on-line') alternatives to the MCMC batch or 'en bloc' methods and they are designed specially for application to stochastic, dynamic systems. There is more discussion on this topic in section 10.4 of chapter 10.

Appendix C
Stochastic Approximation

Stochastic Approximation (SA) is a technique for general optimization in the presence of noise and is not restricted to the kind of recursive data processing discussed in the main parts of this book. It is specifically this kind of application with which this book is primarily concerned, however, and so this Appendix concentrates on the interpretation of SA in these terms.

Much has been written of stochastic approximation in the technical literature and it would serve little useful purpose to review the topic in detail since several useful essays on the subject are available elsewhere for the interested reader: see e.g. Wilde (1964); Sakrison (1966); Mendel and Fu (1970); Tsypkin (1971, 1973); Young (1976a). It will suffice here merely to point out the major characteristics of stochastic approximation, which are relevant to the present book, in sufficient detail for the reader to appreciate its significance as a general tool for recursive analysis.

Search techniques that are able to reach a pre-assigned goal despite the effects of noise or uncertainty were first termed stochastic approximation procedures in a seminal paper by Robbins and Monro (1951). However, they were primarily concerned with the problem of finding the root of a noisy function, and it was left to Kiefer and Wolfowitz (1952) to point out that a similar approach could be applied to general problems of estimating the maximum of a unimodal function obscured by noise. Blum (1954) extended the approach to the multidimensional case and Dvoretsky (1956) considered in detail the conditions for the convergence of a stochastic approximation scheme; conditions which, as Wilde (1964) has pointed out, are of special significance in helping the practitioner to design his own SA procedure.

In order to generalize the treatment of stochastic approximation a little, let us use the approach of Tsypkin (1971, 1973) who considers the problem of finding the value of an estimate $\hat{\mathbf{a}}$ of a parameter vector $\mathbf{a} = [a_1, a_2, \ldots, a_n]^T$ which is optimal in the sense that it defines the extremum (maximum or minimum) of a performance index $\mathcal{J}(\mathbf{a})$, which is given in the form of an expectation[1]

[1] Tsypkin refers to these general procedures of stochastic approximation as 'probabilistic iterative methods'.

$$\mathcal{J}(\mathbf{a}) = \int_t Q[\mathbf{z}(i), \mathbf{a}]p(z)dz \tag{C.1}$$

or, in shorthand form

$$\mathcal{J}(\mathbf{a}) = E\{Q[\mathbf{z}(i), \mathbf{a}]\} \tag{C.2}$$

where E is the expectation operation and the realization $Q[\mathbf{z}(i), \mathbf{a}]$ is a function of the vector $\mathbf{a}$ which also depends upon a vector of *random* sequences or processes $\mathbf{z}(i) = [z_1(i), z_2(i), \ldots, z_n(i)]^T$ with (usually unknown) probability density function $p(\mathbf{z})$.

A simple form of such a cost function (performance index or criterion of optimality), is one based on the averaging of the realization $Q[\mathbf{z}(i), \mathbf{a}]$ with respect to time: if $\mathbf{z}(i)$ is a random sequence $\mathbf{z}(i)$, $i = 1, 2, \ldots, \infty$: i.e.,

$$\mathcal{J}(\mathbf{a}) = \lim_{k \to \infty} \frac{1}{k} \sum_{i=1}^{k} Q[\mathbf{z}(i), \mathbf{a}]$$

while, if $\mathbf{z}(t)$ is a continuous-time random process $\mathbf{z}(t): 0 \leqslant \mathbf{z}(t) < \infty$, it becomes

$$\mathcal{J}(\mathbf{a}) = \lim_{k \to \infty} \frac{1}{T} \int_t Q[\mathbf{z}(i), \mathbf{a}]dt$$

A typical example of the discrete-time criterion function (C.3) is the least squares cost function (3.5) in chapter 3, where $Q[\mathbf{z}(i), \mathbf{a}]$ is simply a quadratic function and the infinite summation is replaced by the finite summation over k samples and $\mathbf{z}(i)$, in this context, represents the available data, in the form of the dependent and regression variables. In other words, the true probability density function is replaced or approximated by the empirical one based on the finite data set $\mathbf{z}(i) = \{\mathbf{x}(i); y(i)\}$, $i = 1, 2, \ldots, k$.

Posed in the above terms the solution to the problem of optimality is basically one of finding, in some manner, an estimate $\hat{\mathbf{a}}$ of $\mathbf{a}$ that defines the extremum (normally minimum) of the function $\mathcal{J}(\mathbf{a})$ in

$$\hat{\mathbf{a}} = \arg\min_{\mathbf{a}} \mathcal{J}(\mathbf{a}) \qquad \mathcal{J}(\mathbf{a}) = E\{Q[\mathbf{z}(i), \mathbf{a}]\} \tag{C.3}$$

In the usual manner, the condition of optimality for $\mathcal{J}(\mathbf{a})$ can be written as

$$\nabla_{\mathbf{a}} \mathcal{J}(\mathbf{a}) = E\{\nabla_{\mathbf{a}} Q[\mathbf{z}(i), \mathbf{a}]\} = 0 \tag{C.4}$$

where $\nabla_{\mathbf{a}} \mathcal{J}(\mathbf{a})$ represents the gradient of the cost function and where

$$\nabla_{\mathbf{a}} Q(\mathbf{z}(i), \mathbf{a}) = \left[\frac{\partial Q(\mathbf{z}(i), \mathbf{a})}{\partial a_1}, \frac{\partial Q(\mathbf{z}(i), \mathbf{a})}{\partial a_2}, \ldots, \frac{\partial Q(\mathbf{z}(i), \mathbf{a})}{\partial a_n} \right]^T$$

is the gradient of $Q[\mathbf{z}(i), \mathbf{a}]$ with respect to $\mathbf{a}$.

The basic idea of solving equation (C.4) using stochastic approximation is to define a sequence of estimated parameter vectors $\hat{\mathbf{a}}(k)$ by the recursive algorithm

$$\hat{\mathbf{a}}(k) = \hat{\mathbf{a}}(k-1) - \gamma(k)\{\nabla_{\hat{\mathbf{a}}}Q[\mathbf{z}(k), \hat{\mathbf{a}}(k-1)]\} \quad \text{(C.5)}$$

where the variable scalar $\gamma(k)$ is called the SA gain sequence. This can be compared directly with the simple recursive algorithm for estimating the mean value of a random variable given by algorithm I in chapter 2 and the RLS algorithm in chapter 3. It will also be noted that the algorithm is similar in form to the equivalent gradient algorithms used in deterministic optimization (Wilde, 1964).

Unlike the deterministic algorithm, however, we know that even if $\hat{\mathbf{a}}(k) = \mathbf{a}\ \forall k$, $\nabla_{\mathbf{a}}Q[\mathbf{z}(k), \hat{\mathbf{a}}] \neq 0$ because of the random nature of $\mathbf{z}(k)$ that introduces 'noise' effects into the sample measure of the gradient at each recursive step and which makes the vector sequence $\hat{\mathbf{a}}(k)$, $k = 0, 1, 2, ...,$ itself a random sequence. As a result, it is no longer possible to consider the convergence of equation (C.5) in the ordinary (deterministic) sense; rather it is necessary to introduce new concepts of *stochastic* convergence.

This can be accomplished by introducing the concept of *almost sure convergence* in the sense that the sequence of random vectors $\hat{\mathbf{a}}(k)$, $k = 0, 1, 2, ...$ converges to $\mathbf{a}$ as $k \to \infty$ *almost surely*, or *with probability one*, if the probability of ordinary convergence of $\hat{\mathbf{a}}(k)$ to $\mathbf{a}$ is equal to unity, i.e. if

$$p\{\lim \|\hat{\mathbf{a}}(k) - \hat{\mathbf{a}}^*\| = 0\} = 1$$

where $\|.\|$ denotes the norm of the error vector (see Appendix B and Fröberg, 1970). Normally, of course, the cost function is chosen so that asymptotically as $k \to \infty$ so $\hat{\mathbf{a}}(k)$ tends to the $\mathbf{a}$, with the estimation error vector denoted by $\tilde{\mathbf{a}}(k) = \hat{\mathbf{a}}(k) - \mathbf{a}$. Clearly the *rate* of convergence depends on the particular sample sequence generated with specific $\mathbf{z}(k)$ sequences and, therefore, it also has a random character.

As Tsypkin points out, in order to design SA algorithms of the form (C.5), it is necessary to conform with the necessary and sufficient conditions for convergence in the above sense. If we consider (C.5) in the following form

$$\Delta\hat{\mathbf{a}}(k) = -\gamma(k)\{\nabla_{\hat{\mathbf{a}}}Q[\mathbf{z}(k), \hat{\mathbf{a}}(k-1)]\} \quad \text{(C.6)}$$

where Δ is the difference operator, $\Delta\hat{\mathbf{a}}(k) = \hat{\mathbf{a}}(k) - \hat{\mathbf{a}}(k-1)$, then it is clearly necessary that, for $\hat{\mathbf{a}}(k)$ to converge almost surely to $\mathbf{a}$, then the right hand side of (C.6) should converge to zero as $k \to \infty$ for practically any $\mathbf{z}(k)$. And since the gradient is generally different from zero because of the presence of $\mathbf{z}(k)$, it is necessary that *the SA gain sequence $\gamma(k)$ should tend towards zero as k increases.*

This is consistent with our experience of the recursive least squares estimation algorithms considered in the main text and is evidenced by the strictly decreasing nature of the gain factors $g(k)$ and $\mathbf{g}(k)$ (scalar and vector) which occur naturally in these algorithms. Indeed, on the basis of this experience, particularly with the scalar gain $1/k$ in equation (2.14) and the equivalent $g(k)$ in equation (I-1) of chapter 2, we might guess that the gain sequence should be a strictly decreasing function of k which reduces in some specific manner, perhaps as a harmonic sequence $1/k$.

In fact, such intuition is confirmed by theoretical analysis: as Tsypkin shows, the algorithm (C.5) will converge almost surely if the following conditions (which are basically derived from the conditions of Dvoretzky (1956)) are satisfied (the introductory reader can ignore conditions (b) and (c)):

$$\text{(a)} \quad \text{(i) } \gamma(\text{k}) > 0; \quad \text{(ii) } \sum_{\text{i}=1}^{\infty} \gamma(\text{k}) = \infty; \quad \text{(iii) } \sum_{k=1}^{\infty} \gamma^2(k) < \infty$$

$$\text{(b)} \quad \frac{\inf}{\varepsilon < \|\hat{\mathbf{a}}-\mathbf{a}\| < \frac{1}{\varepsilon}} E\{[\hat{\mathbf{a}}-\mathbf{a}]^T \nabla_{\hat{\mathbf{a}}} Q(\mathbf{z}(k),\hat{\mathbf{a}}\} > 0\,(\varepsilon > 0)$$

$$\text{(c)} \quad \text{E}\{\nabla_{\hat{\mathbf{a}}}^T Q(\mathbf{z}(k),\hat{\mathbf{a}})\},\, \nabla_{\hat{\mathbf{a}}} Q(\mathbf{z}(k),\hat{\mathbf{a}})\}\} \leq \alpha(1+\hat{\mathbf{a}}\hat{\mathbf{a}}^T) \quad (\alpha > 0)$$

In physical terms, condition (a) requires that $\gamma(k)$ tends towards zero and the rate of decrease is such that the variance of the estimate is also reduced to zero. In particular, although we want the corrective effect of $\gamma(k)$ to be small when $\hat{\mathbf{a}}(k)$ is close to $\mathbf{a}$, we do not want it to vanish for $k < \infty$, for otherwise the procedure could stop before it had reached its goal. Thus condition (a)(ii) is introduced, i.e., that the sum of the $\gamma(k)$ should be infinite so that if $\hat{\mathbf{a}}(k)$ tried to converge to other than the optimum, an infinite amount of corrective effort would be available. Of course, this is not to say that convergence might not take an inordinately long time.

For a good heuristic discussion on the need for condition (a)(iii) namely that the infinite sum of the variances is finite, see Wilde (1964); Chapter 6. Basically, however, this property is required because, together with the requirement that the noise variance is finite, it ensures that the residual fluctuations caused by noise will die out in the long term, with the individual random errors caused by the noise at each step in the recursive algorithm, tending to cancel each other out.

Condition (b) is required to define the behaviour of the surface $E\{\nabla_{\mathbf{a}} Q(\mathbf{z}(k),\mathbf{a})\}$ near zero, and thus the sign of the increments in $\hat{\mathbf{a}}(k)$; while condition (c) ensures that the rate of increase of the expectation of the quadratic form

$$\nabla_{\mathbf{a}} Q(\mathbf{z}(k),\mathbf{a})^T \nabla_{\hat{\mathbf{a}}} Q(\mathbf{z}(k),\mathbf{a})$$

with $\hat{\mathbf{a}}(k)$ is smaller than in a parabola of second degree. For further consideration of these conditions, however, see Tsypkin (1971, 1973) or Mendel and Fu(1970).

From the design standpoint, perhaps the most important of the convergence conditions discussed above is (a) which defines the characteristics of the SA gain $\gamma(k)$. The harmonic sequence l/k : $1, 1/2, 1/3, \ldots$. etc., is, as we conjectured, the best known example of a sequence which satisfies these properties. Other examples exist, of course, but it is necessary to be careful in choosing sequences to ensure that they do satisfy conditions (a). Some sequences, e.g., $1/k^2$: $1, 1/4, 1/9, \ldots$, etc., do not satisfy condition (a) (ii); indeed for all $p > 1$

$$\sum_{k=1}^{\infty} \frac{1}{k^p} < \infty$$

and an SA procedure based on such a gain sequence might well not converge under all circumstances, since the total correction effort is limited. On the other hand, the sequence $1/\sqrt{k}$: 1, $1/\sqrt{2}, 1/\sqrt{3}, \ldots$ etc., would converge too slowly to dissipate the cumulative error, although a faster sequence with the power of $k > 0.5$ would be suitable.

One final comment on the basic principles of SA concerns the nature of the realization $Q(\mathbf{z}(i), \mathbf{a})$. Although Tsypkin, for example, gives only brief information on how the realization should be chosen, it is clearly extremely important in relation to the performance of the SA algorithm. In most of the algorithms discussed in this book $Q(\mathbf{z}(i), \mathbf{a})$ is the form of a squared error function, as in the least squares regression example examined in the previous section. In this situation, we have a general cost function of the form

$$\mathcal{J}(\mathbf{a}) = \frac{1}{k} \sum_{i=1}^{k} [e(\mathbf{z}(i), \mathbf{a})]^2 \tag{C.7}$$

where $e(\mathbf{z}(i), \mathbf{a})$, $i = 1, 2, \ldots, k$ is the error function and the SA algorithm (C.5) takes the specific form

$$\hat{\mathbf{a}}(k) = \hat{\mathbf{a}}(k-1) - \gamma(k) \{\nabla_{\hat{\mathbf{a}}} \hat{e}(\mathbf{z}(k), \hat{\mathbf{a}}(k-1)) [\hat{e}(\mathbf{z}(k), \hat{\mathbf{a}}(k-1))]\} \tag{C.8}$$

where the contents of the curly brackets represent the gradient of $Q(\mathbf{z}, \hat{\mathbf{a}})$ with respect to $\hat{\mathbf{a}}$ and with $\hat{e} = \hat{e}(k|k-1)$ denoting the *a priori* error at the kth instant, i.e.

$$\nabla_{\hat{\mathbf{a}}} Q[\mathbf{z}(k), \hat{\mathbf{a}}(k-1)] = 2\nabla_{\hat{\mathbf{a}}} [\hat{e}(\mathbf{z}(k), \hat{\mathbf{a}}(k-1))] \hat{e}(\mathbf{z}(k), \hat{\mathbf{a}}(k-1))$$

We see therefore, that the gradient term in the SA algorithm is simply the gradient of the error function multiplied by the error function itself, with both evaluated by substitution of the latest value of $\mathbf{z}(k)$ and estimate $\hat{\mathbf{a}}(k-1)$ obtained at the previous, $(k-1)^{th}$ recursive update (cf. equation (3.17) in chapter 3). So, as mentioned above, special scalar example of this is the recursive mean estimator (2.14) in chapter 2, where

$$\nabla_{\hat{\mathbf{a}}} \hat{e}[\mathbf{z}(k), \hat{\mathbf{a}}(k-1)] = 1.0 \quad \hat{e}[\mathbf{z}(k), \hat{\mathbf{a}}(k-1)] = [y(k) - \hat{\mathbf{a}}(k-1)]$$

and $\gamma(k)$ defined as the harmonic sequence $1/k$.

The important error measure $\hat{e}[\mathbf{z}(k), \hat{\mathbf{a}}(k-1)]$ is the recursive residual or 'innovations' process (see below), since it represents the latest residual error (or *a priori* prediction error) computed in the recursive algorithm: its appearance in the algorithm makes obvious sense since the update of parameter estimates should clearly be a function of this latest residual error, which will be due to a combination of the real estimation error and noise on the data. The function of the SA gain is then to 'filter' the random noise progressively from the recursive residual so that estimation error tends to zero as $k \rightarrow \infty$.

From the previous discussion it seems reasonable that the recursive residual should possess certain desirable statistical properties: for example, it should be zero mean and have finite variance so that the smoothing effect of the SA gain can be fully effective. In addition, and by consideration of other methods of statistical inference, including the recursive algorithms considered in the main text, we might also assume that the recursive residuals should form a serially uncorrelated sequence of random variables since the problem becomes much simpler in statistical terms. Also, as we shall see, it is much easier to make the solution statistically efficient, in the sense that the estimation error variance is minimized.

With these intuitive notions in mind, it is not surprising to find that all of these properties are desirable in theory and that one of the primary functions of the SA algorithm designer is to ensure that the realization is such that, if possible, the recursive residuals constitute a zero mean, serially uncorrelated sequence of random variables; in other words, they represent a discrete 'white noise' sequence.

In the control and systems literature, this function of a recursive estimator in yielding a white noise sequence of recursive residuals is termed the 'innovations property' and the sequence itself is termed the 'innovations sequence' (see the main text). As pointed out in chapter 3, the term 'innovations' has arisen because, at each recursive step, the term represents the new information on which statistical inference is to be based. The usual innovations approach is, therefore, to operate first on the observed process by means of a linear transformation (although it is possible to conceive of analogous operations which might be nonlinear) such that the resultant transformed or 'filtered' noise sequence has discrete white noise properties. This 'pre-whitened' noise sequence then provides the basis for the definition of the error in the cost function and determines the nature of the innovations sequence $\hat{e}(k) = \hat{e}(k|k-1)$ in the SA algorithm.

In fact, theoretical analysis (Kumar and Moore, 1980) shows that convergence can be assured if the noise process satisfies rather more general requirements than those of zero mean, white noise: in particular, for standard convergence theory, the sequence is taken to be a sample sequence from a noise process with

$$E[\hat{e}(k)|A(k-1) = 0] \quad \text{and} \quad E[\hat{e}^2(k)|A(k-1)] = \sigma^2$$

for all k, where $A(k-1)$ represents past information and is in the form of a σ-algebra (Goodwin and Payne, 1977; Loeve, 1963) generated by $\hat{e}(1), \hat{e}(2), \ldots, \hat{e}(k-1)$. The less mathematical reader need not be confused by this, however, since zero mean, white noise satisfies these conditions and serves to exemplify the requirements in more digestible and well known terms.

The problem of statistical inference with white noise innovations is clearly much easier to solve and, if it is required, the solution to the simpler problem can then be expressed in terms of the original observations by means of the inverse of the original 'whitening' filter. Similar arguments apply in the continuous-time case considered briefly in a subsequent section.

As a specific example of this innovations approach, consider again the mean value estimator: here the innovations sequence is simply $y(k) - \hat{\mathbf{a}}(k-1)$ and it is clear that asymptotically as $\hat{a}(k) \rightarrow a$ the sequence $\hat{e}(k|k-1)$ is, by definition, serially uncorrelated if $e(k)$ is itself serially uncorrelated. What we are saying here is that, in a good estimator, if $e(k)$ is serially uncorrelated, then $y(k) - \hat{\mathbf{a}}(k-1)$ will also be serially uncorrelated for all k, although it will not necessarily have constant variance because of the effects of convergence and the varying gain (Brown et al., 1975).

C.1 Some Extensions to Stochastic Approximation

So far, this Appendix has discussed the basic elements of 'scalar gain' SA applied to optimization problems and tried to provide a physical interpretation of the SA procedure. There are, however, a number of generalizations and extensions to the basic procedure which make it extremely flexible in application.

C.1.1 Matrix gain SA and optimum algorithms

The first point to note is that the algorithm (C.5), although multidimensional in the sense that it updates the n vector $\hat{\mathbf{a}}(k)$, is characterized by a scalar SA gain $\gamma(k)$. By reference to the RLS algorithm in chapter 2, it seems reasonable that we might, in general, replace this by a matrix gain $\Gamma(k)$, i.e.,

$$\hat{\mathbf{a}}(k) = \hat{\mathbf{a}}(k-1) - \Gamma(k)\nabla_{\hat{\mathbf{a}}}Q\{\mathbf{z}(k), \hat{\mathbf{a}}(k-1)\} \qquad \text{(C.9)}$$

The problem is, of course, how to determine the form of $\Gamma(k)$, since it could have as many as n^2 elements. Heuristically, we might simplify the problem by constraining $\Gamma(k)$ to the diagonal form,

$$\Gamma(k) = \begin{bmatrix} \gamma_1(k) & 0 & \dots & 0 \\ 0 & \gamma_2(k) & \dots & 0 \\ 0 & 0 & \dots & 0 \\ 0 & 0 & \dots & \gamma_n(k) \end{bmatrix}$$

and then make each of the elements $\gamma_i(k), i = 1,2,....,n$ satisfy the conditions (a). But even so, we do not have any rules for choosing the differences between the elements and it might be a hazardous trial and error business.

Tsypkin considers this problem in theoretical terms and define an SA gain matrix sequence that is optimal in a statistical sense. However, he restricts attention to deterministic algorithms[2] in which *a priori* assumptions and information about the noise are not used to infer anything about the statistical nature of the estimates other than that $\hat{\mathbf{a}}(k)$ converges almost surely to $\mathbf{a}$ as $k \to \infty$.

---

[2] In deterministic algorithms, no estimates of the second statistical moment (or variance) of the estimation errors are generated. And more generally, in multi-parameter algorithms such as RLS, no estimates of second moment or covariance matrix of the estimation errors are obtained.

There has been a great deal of research on the design of deterministic, matrix gain, SA algorithms of this general kind (see Saridis (1974) for an early review) but the paper by Kumar and Moore (1979) considers a specific class of problems that are related to the RLS algorithm. With insight gained both from the nature of the RLS solution and a useful geometric interpretation of the recursive procedure (see also Young (1965a)), they suggest SA algorithms in which the scalar SA gain is retained with its usual characteristics, but is modulated by a time variable matrix, so that the product can be considered as a crude approximation to $\mathbf{P}(k)$, and is computed by reference to the $\mathbf{z}(k)$ vector.

Kumar and Moore's work serves to emphasize that the RLS algorithm can be considered as a naturally occurring SA algorithm with optimum properties. There is, perhaps, a moral in this observation: if we are dealing with a stochastic optimization problem in which a non-recursive analytic solution already exists, we may do better to look for a recursive version of this solution, in which the stochastic approximation properties are implicit, rather than concocting an SA algorithm and then attempting to make it optimal in some manner. This is particularly important in the matrix gain situation. This approach is exploited in the present book where recursive algorithms are most often synthesized on the basis of analogies with RLS and similar procedures. This does not, of course, negate the utility of SA both as a conceptual basis for recursive analysis and a useful starting point for stochastic convergence analysis: see e.g. Solo (1978, 1980). Moreover, the simpler SA versions of algorithms may provide better practical solutions when limited complexity is demanded: simplicity of implementation may be worth the price of slower convergence.

C.1.2 Continuous-time algorithms

Although in this age of the digital computer we are often concerned with sampled data and discrete-time formulations, it may sometimes be advantageous to use a continuous-time approach. To obtain a continuous-time version of the SA algorithm (C.9) we first note that it can be written in the form of equation (C.6). It is then easy to see, by a limiting argument, that a continuous version of the algorithm (C.9) will be of the form

$$\frac{d\mathbf{a}(t)}{dt} = -\Gamma(t)\,\{\nabla_{\hat{\mathbf{a}}}Q(\mathbf{z},\hat{\mathbf{a}}),\,\hat{\mathbf{a}}(t)\} \tag{C.10}$$

where $\Gamma(t)$ may be scalar. Algorithms such as this are analogous to the constant gain, continuous-time steepest (or steep) descent algorithms that have been in use in the control and computer field for some considerable time: see e.g. Young (1965a). For example, such equations with constant gains were often used on analog computers to solve sets of algebraic equations. The importance of the link between these well known deterministic algorithms and the lesser known continuous-time SA algorithms is that it may often be advisable to incorporate such a variable gain device to counter the inevitable noise that affects any electronic mechanization.

Indeed, it is quite likely that variable gain modifications to continuous steepest descent solutions are used in practice, although these modifications, if they exist, will probably be based on some heuristic reasoning rather than a knowledge of stochastic approximation. Although the theorist may not wish to admit it, often techniques developed with great mathematical rigour and effort have been used by a practitioner who has found by trial and error and a detailed knowledge of his problem that they 'work'. His reasoning may have been completely heuristic and based totally on intuition, but the solution is not necessarily any the worse for that.

C.1.3 Search algorithms

If, for some reason, it is not possible to obtain the gradient of the realization $\nabla_{\hat{\mathbf{a}}} Q(\mathbf{z}(k), \hat{\mathbf{a}})$ but sample values of $Q(\mathbf{z}(k), \hat{\mathbf{a}})$ are available, then it is still possible to use SA by replacing the gradient in the algorithm by an estimate of the gradient obtained from the sample information; in fact this is the basis of the approach used by Kiefer and Wolfowitz (1952) mentioned earlier. How this estimate should be obtained is to a large degree open to the experimenter to choose: for instance, one of the many methods that have been developed for deterministic hill climbing could be chosen. To quote but one possible example, Tsypkin (1971) suggests perturbing $\hat{\mathbf{a}}$ in a certain manner and then computing the average gradient over the perturbation from the sample values of the realizations $Q(\mathbf{z}(k), \hat{\mathbf{a}})$ obtained during the perturbation experiments.

C.1.4 Acceleration of convergence

Little has been said up to this point on the rate of convergence of SA algorithms. If designed properly they are guaranteed to converge 'almost surely' as $k \rightarrow \infty$. But such asymptotic convergence may be intolerably slow. Here, as is often the case in the design of hill climbing algorithms, there are two approaches open to the designer: one heuristic, the other analytic. There are many examples of the first type but only one will be mentioned here, namely the method of Kesten and Lapidus et al., (see Wilde, 1964). They note that the ideal SA scheme will take large steps when far from the optimum but should shorten its steps rapidly as it approaches the optimum. With this in mind, they monitor the gradient (actually the average gradient for they use a Kiefer-Wolfowitz scheme) and argue that, when far from the optimum, there will be a few reversals of the sign of the gradient but as the optimum is approached, the random effects of the noise will cause repeated reversals of sign. They then apply the simple rule that the SA gain will only be reduced when the direction of search changes, as indicated by changes of sign in the recursive residual: in this way there is a built-in 'adaption' of the SA gain sequence and the scheme seems to be very effective in promoting more rapid convergence compared with that obtained using the standard SA gain sequence.

As far as analytic approaches are concerned, it would appear that, if an optimal non-recursive analytic solution exists and is converted into an equivalent recursive solution, then the resulting implicitly generated SA gain sequence yields rapid convergence and it is not usually necessary to attempt any acceleration. Thus, for example, the RLS solution is nearly always superior in convergence to a scalar gain SA algorithm applied to the same problem (Ho and Blaydon, 1966), simply because the implicity generated matrix gain sequence $\mathbf{P}(k)$ contains within it information gleaned from the sampled data which allows it to adjust the estimates in a mutually dependent fashion and so achieve faster convergence. In fact, as shown in chapter 4 of the main text , if the noise sequence in a regression model has the statistical properties of white noise, then $\mathbf{P}(k)$ is directly proportional to the covariance matrix of the errors on the parameter estimates. The advantages of using such information in guiding the search are obvious. An attractive general approach to the analytic evaluation of the convergence properties of recursive algorithms has been suggested by Ljung (see Ljung (1987) and the prior references therein), largely within a time-series context. This is discussed further in chapter 10.

Some scalar gain SA algorithms can be exceptionally slow to converge and the potential user should realize this. The point is, however, that heuristically designed SA algorithms are, just because of their simplicity, extremely flexible and widely applicable. And therein lies their attraction: better an algorithm which takes a long time to obtain an answer than no algorithm at all. After all, there is at least the possibility with an SA algorithm that, having designed it and found it too slow to converge, the designer can modify it in some manner in order to improve its convergence properties.

C.2 Summary

Stochastic approximation provides additional insight into the behaviour of recursive algorithms and provides a theoretical basis both for convergence analysis and the design of SA-like algorithms. But stochastic approximation theory is not a panacea: while it can be used to design algorithms with 'almost sure' convergence, these may have rates of convergence which are intolerably slow. In this sense, naturally occurring recursive SA algorithms derived from optimal *en bloc* estimators or some approximation of these, such as the recursive estimation algorithms in chapters 3 and 4 are usually superior in performance.

Appendix D
Deterministic Regularization and Stochastic Fixed Interval Smoothing

Chapter 5 in the main text considers *Unobserved Component* (UC) models, as well as their application to both forecasting and signal extraction, in great detail. This Appendix presents some additional, interesting aspects of UC modelling and associated signal extraction.

Consider the simplest UC model where the time series $y(k)$ can be represented by the sum of an IRW trend $T(k)$ and an irregular component $e(k)$:

$$y(k) = T(k) + e(k) \tag{D.1}$$

The non-recursive approach, known in numerical analysis as *Deterministic Regularization* (DR) and in the statistical literature as *spline smoothing*, is based on the computation of an estimate $\hat{T}(k|N)$ which minimizes a deterministic cost function of the general form,

$$J = \sum_{k=1}^{N} \{(y(k) - \hat{T}(k|N)^2 + \lambda(\nabla^2\hat{T}(k|N)^2\} \tag{D.2}$$

In other words, the trend component is computed so that it minimizes the sum of the squares of the difference between the trend and the series, with the constraint that it should have a given level of smoothness, as measured by the square of the 2nd difference of the estimated trend. This constraint is introduced by the *Lagrange Multiplier*, or smoothing parameter λ, which is selected in some manner, such as cross validation, that is defined by the user.

The solution to the optimization problem (D.2) is obtained in the standard manner as,

$$\hat{T}(k|N) = \left[\mathbf{I} + \lambda\mathbf{U}^T\mathbf{U}\right]^{-1}\mathbf{y} = \mathbf{W}\mathbf{y} \tag{D.3}$$

where $\mathbf{I}$ is the $N \times N$ identity matrix; $\mathbf{y} = [y(1)\ y(2)\ \ldots\ ,y(N)]^T$ is the $N \times 1$ vector of data; and $\mathbf{U}$ is the following $N \times N$ matrix:

$$\mathbf{U} = \begin{bmatrix} 0 & 0 & 0 & 0 & \cdots & 0 & 0 & 0 \\ 0 & 0 & 0 & 0 & \cdots & 0 & 0 & 0 \\ 1 & -2 & 1 & 0 & \cdots & 0 & 0 & 0 \\ 0 & 1 & -2 & 1 & \cdots & 0 & 0 & 0 \\ \vdots & \vdots & \vdots & \vdots & \ddots & \vdots & \vdots & \vdots \\ 0 & 0 & 0 & 0 & \cdots & 1 & -2 & 1 \end{bmatrix}$$

The $N \times N$ matrix $\mathbf{W}$ has rows which define the nature of the moving average weighting applied to the data to yield the estimate of the trend. Figure D.1 shows a plot of every fourth row in the matrix for a data set with $N = 110$ and $\lambda = 90$. The weighting is symmetric in the centre of the data but becomes asymmetric at the ends to handle the end effects in this 2-point boundary value problem. An enlarged version of the plot at the left hand side shows this asymmetry in more detail (note that the area under each curve is always maintained at unity).

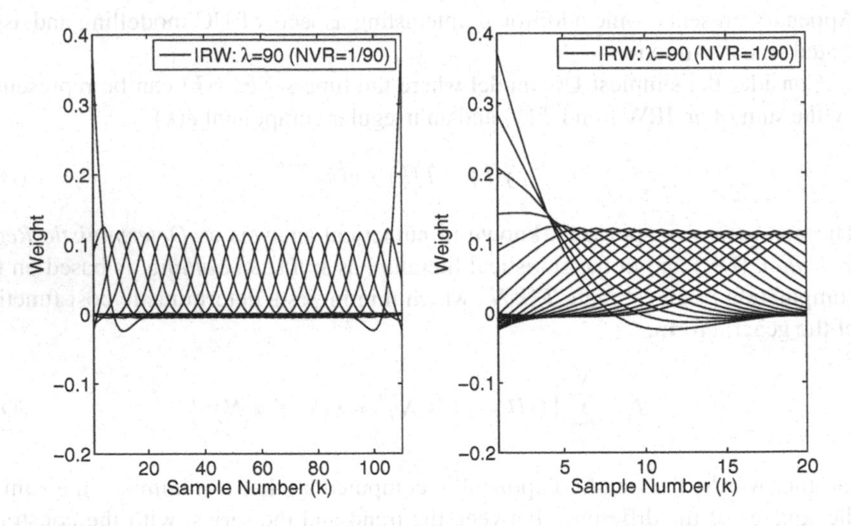

Fig. D.1 Plot of selected rows of the **W** smoother matrix (left) and enlarged version of the left hand side (right) showing how the rows change at the beginning (and similarly at the end) of the series.

Remarks

1. It is clear from equation (D.3) that DR requires the inversion of a matrix with the same dimension as the data, so that it can be very computationally intensive and requires a large memory size for long series unless some device for handling large matrices is introduced at the computational stage (in the above Matlab code the `spdiags` sparse matrix routine). This is in contrast to the recursive FIS algorithm which, in the case of the present trend estimation problem, requires no matrix inversion at all.

2. As in the case of FIS, it is straightforward to generalize the above DR formulation. For example, the Lagrange multiplier term could be defined in terms of the jth difference of the trend estimate, where $j = 1, 2, \ldots$, etc. And other more complex generalizations are possible: see next section D.1.
3. The so-called 'Hodrick-Prescott' (HP) detrending method (Hodrick and Prescott, 1997) used in economics is a special case of this DR approach obtained when $\lambda = 1600$, which is well suited to the estimation of trends in quarterly macroeconomic time series. However, as Pedregal and Young (2001) point out, the equivalent UC trend estimation procedure (the irwsm routine in the CAPTAIN Toolbox, with the smoothing parameter NVR (the inverse of λ: see chapter 5) optimized by maximum likelihood using the associated irwsmopt routine) seems a more sensible and flexible approach.

D.1 Wiener-Kolmogorov-Whittle Optimal Signal Extraction

The original derivations of optimal filter theory appeared in Kolmogorov (1941) and Wiener (1941, 1949). More recently, *Optimal Signal Extraction* (OSE) and filter theory for stationary time series has been presented by Whittle (1983); and for non-stationary time series by Bell (1984). In contrast to FIS estimation, optimal signal extraction is normally formulated in the frequency domain, based on a transfer function (TF) model of the stochastic process (see also section 5.2.1 of chapter 5).

For example, the TF form of the IRW trend mode (D.1) is given by,

$$T(k) = \frac{1}{\nabla^2}\eta(k) \tag{D.4}$$

The optimal smoothed estimate $\hat{T}(k|N)$ of the trend is then given by the expression,

$$\begin{aligned}\hat{T}(k|N) &= \frac{g\{T(k)\}}{g\{T(k)+e(k)\}}y(k)\\ &= \frac{\frac{\sigma_\eta^2}{(1-z^{-1})^2(1-z)^2}}{\frac{\sigma_\eta^2}{(1-z^{-1})^2(1-z)^2}+\sigma^2}y(k) = \frac{NVR}{NVR+(1-z^{-1})^2(1-z)^2}y(k)\end{aligned} \tag{D.5}$$

where $g\{T(k)\}$ and $g\{T(k)+e(k)\}$ are, respectively, the spectral generating functions of the signal $T(k)$ and the signal+noise $T(k)+e(k)$; $NVR = \sigma_\eta^2/\sigma^2$ is the Noise Variance Ratio: see main text), and z is the lead (or forward shift) operator, i.e. $z^r y(k) = y(k+r)$.

Remarks

1. The IRW smoothing filter defined by equation (D.5) is a low-pass filter, in which the spectral characteristics are defined as follows by the NVR parameter: i.e.,

$$f_T(\omega) = \frac{NVR}{NVR + \{2(1-\cos\omega)\}^2}$$

2. More general formulations than that given above for the IRW trend model are clearly possible. For example, the OSE smoother for the whole family of multiple IRW smoothing filters (i.e. IRW, Double Integrated Random Walk (DIRW), Triple Integrated Random Walk (TIRW), etc.) is obtained simply by replacing the 2nd power in the IRW trend definition by a general power $i = 1, 2, 3 \ldots$ etc.. The spectral characteristics of the OSE smoother defined in this more general manner are given by,

$$f_T(\omega) = \frac{NVR}{NVR + \{2(1-\cos\omega)\}^i}$$

From this formula, the cut-off frequency that corresponds to a given power of the spectrum is given by Pedregal (1995):

$$\omega = \arccos\left[1 - \left\{\frac{NVR(1-\alpha)}{4\alpha}\right\}^{\frac{1}{i}}\right]$$

in which α is the specified percentage cut-off (usually 50% in signal processing literature). The spectral characteristics of these filters for different levels of integration i are plotted in Figure D.2. Note that the special case of this filter when $i = 4$ (i.e. a TIRW modelled trend) is interesting since it can be compared directly with the well known Henderson family of CMA filters used in the X-11 seasonal adjustment program: see Young and Pedregal (1999b).

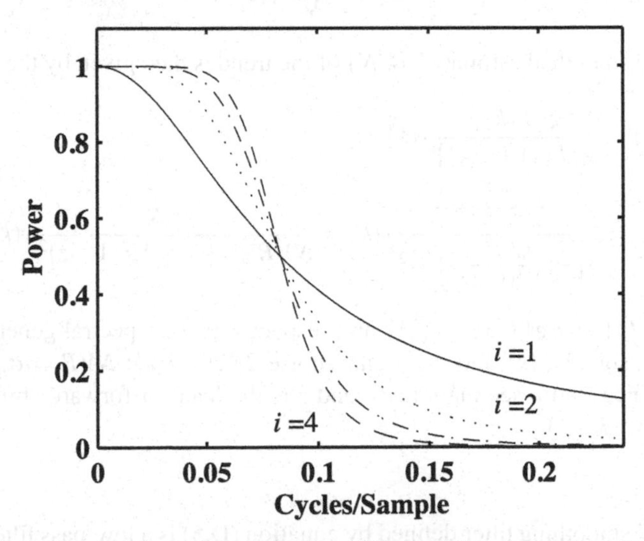

Fig. D.2 Spectral characteristics of filters based on the integrated random walk family of stochastic models for integration levels $i = 1 \rightarrow 4$.

3. Clearly, as in the case of FIS and DR, it is straightforward to develop still more general formulations of OSE: see Young and Pedregal (1999b).
4. It must be stressed that, unlike the FIS and DR approaches, the primary function of the above OSE analysis is to provide a theoretical basis for smoothing, rather than a computational methodology. Although it is possible to use the theory to develop a smoothing algorithm from the theoretical results, this is not entirely satisfactory, since the theory applies only to infinitely long series. Consequently, it is necessary to introduce approximations to handle end effects: see Young and Pedregal (1999b). Whittle (1983) concludes that, for most applications, the state space smoothing algorithms, as discussed in the main text, have greater generality and are computationally much more attractive.
5. Finally, it is interesting to note that Jakeman and Young (1984) demonstrated the equivalence between the FIS and DR approaches by considering FIS within the context of Maximum Likelihood estimation: see also Young (1991) and Young and Pedregal (1999b). Their analysis shows that maximization of the log-likelihood function is the same than minimising the following expression (see also Appendix E):

$$J = \sum_{k=1}^{N} \|y(k) - \mathbf{C}(k)\mathbf{x}(k)\|^2_{1/\sigma^2} + \sum_{k=1}^{N} \|\mathbf{x}(k) - \mathbf{A}\mathbf{x}(k-1)\|^2_{\mathbf{Q}^{-1}}$$

Comparing this cost function with the DR cost function in (D.2), it is clear that, if $\mathbf{A}$ and $\mathbf{C}(k)$ are defined as for the IRW model in equation (D.1), they are identical if σ^2 is normalized to unity and

$$Q^{-1} = \begin{pmatrix} 0 & 0 \\ 0 & \lambda \end{pmatrix}$$

In other words, although very different in computational form, the algorithms are entirely equivalent in their smoothing effect on the time series. In particular, the CMA operation that is explicit in the non-recursive DR algorithm is implicit in the recursive FIS algorithm. Moreover, this equivalence is quite general: for any FIS algorithm defined by a selected state space model, it is possible to formulate an equivalent DR cost function that will yield identical results. An example of this is general equivalence is revealed in the work of Akaike (1980) who uses a DR formulation for monthly data, which includes an annual seasonal component, in addition to the trend (equivalent to the DHR model: see main text). In this case the cost function takes the following form:

$$\begin{aligned} J &= \sum_{k=1}^{N} \left\{ \left(y(k) - \hat{T}(k|N) - \hat{S}(k|N)\right)^2 + [\lambda_1 f_1 + \lambda_2 f_2 + \lambda_3 f_3] \right\} \\ f_1 &= \left(\nabla \hat{T}(k|N)\right)^2; \quad f_2 = \left(\nabla_s \hat{S}(k|N)\right)^2 \\ f_3 &= \left(\hat{S}(k|N) + \hat{S}(k-1|N) + \cdots + \hat{S}(k-s+1|N)\right)^2 \end{aligned} \qquad \text{(D.6)}$$

Here $\hat{S}(k|N)$ is the estimate of the seasonal component; s is the seasonal period; $\nabla_s = (1 - z^{-s})$ is the seasonal difference operator; and λ_1, λ_2, λ_3 are Lagrange multipliers which Akaike optimizes within a Bayesian framework. If the term $\lambda_2 \left(\nabla_s \hat{S}(k|N)\right)^2$ is dropped from the previous equation, the problem formulated in these terms yields a similar result to that obtained using the Basic Structural Model with seasonal dummies as defined by Harvey (1984).

Appendix E
The Instantaneous Cost Function Associated with the Recursive Least Squares Algorithm

The derivation of the recursive least squares regression algorithm used in the main text does not directly address the situation at the start of the algorithm, when it is necessary to choose the initial estimate of the parameter vector $\hat{\mathbf{a}}(k)$ and the associated matrix $\mathbf{P}^*(k)$. In this connection, it is interesting to consider the following instantaneous cost function at the kth sampling instant (Young, 1965b; Rauch et al., 1965):

$$\hat{\mathbf{a}} = \arg\min_{\mathbf{a}} \mathcal{J}(\mathbf{a}) \tag{E.1}$$

where:

$$\begin{aligned}\mathcal{J}(\mathbf{a}) &= \left\|y(k) - \mathbf{x}^T(k)\hat{\mathbf{a}}(k)\right\|^2_{1/\sigma^2} + \|\hat{\mathbf{a}}(k) - \hat{\mathbf{a}}(k-1])\|^2_{\mathbf{P}^*(k-1)^{-1}} \\ &= \frac{1}{\sigma^2}[y(k) - \mathbf{x}^T(k)\hat{\mathbf{a}}(k)]^2 + [\hat{\mathbf{a}}(k) - \hat{\mathbf{a}}(k-1)]^T[\mathbf{P}^*(k-1)]^{-1}[\hat{\mathbf{a}}(k) - \hat{\mathbf{a}}(k-1)]\end{aligned}$$

As usual, the conditions for a minimum of $\mathcal{J}$ with respect to the unknown vector $\mathbf{a}(k)$, are

$$\frac{\partial \mathcal{J}(\mathbf{a})}{\partial \mathbf{a}} = \nabla_{\mathbf{a}} \mathcal{J} = 0$$

Consequently, we can obtain from the solution of the following equation,

$$\frac{1}{\sigma^2}\mathbf{x}(k)[y(k) - \mathbf{x}^T(k)\hat{\mathbf{a}}(k)] + [\mathbf{P}^*(k-1)]^{-1}\hat{\mathbf{a}}(k) - [\mathbf{P}^*(k-1)]^{-1}\hat{\mathbf{a}}(k-1) = 0$$

or,

$$\left[[\mathbf{P}^*(k-1)]^{-1} + \frac{1}{\sigma^2}\mathbf{x}(k)\mathbf{x}^T(k)\right]\hat{\mathbf{a}}(k) = \frac{1}{\sigma^2}\mathbf{x}(k)y(k) + [\mathbf{P}^*(k-1)]^{-1}\mathbf{a}(k-1)$$

so that,

$$\hat{\mathbf{a}}(k) = \left[[\mathbf{P}^*(k-1)]^{-1} + \frac{1}{\sigma^2}\mathbf{x}(k)\mathbf{x}^T(k)\right]^{-1}\left\{\frac{1}{\sigma^2}\mathbf{x}(k)y(k) + [\mathbf{P}^*(k-1)]^{-1}\hat{\mathbf{a}}(k-1)\right\}$$

Referring now to the matrix inversion lemma of equation (3.12) in section 3.1 of the main text, this latter equation can be written in the form (cf. the expression above in the first square brackets with that in equation (3.10)(a) and note the subsequent analysis for the SRLS algorithm),

$$\begin{aligned}\hat{\mathbf{a}}(k) = &\{\mathbf{P}^*(k-1) - \mathbf{P}^*(k-1)\mathbf{x}(k)[\sigma^2 + \mathbf{x}^T(k)\mathbf{P}^*(k-1)\mathbf{x}(k)]^{-1}\mathbf{x}^T(k)\mathbf{P}^*(k-1)\}\\ &\times\left\{\frac{1}{\sigma^2}\mathbf{x}(k)y(k) + [\mathbf{P}^*(k-1)]^{-1}\hat{\mathbf{a}}(k-1)\right\}\end{aligned}$$

As a result, we can obtain the recursive algorithm for $\hat{\mathbf{a}}(k)$ in terms of $\hat{\mathbf{a}}(k-1)$ by multiplying out the expression on the right hand side of this equation and re-arranging the terms, i.e.

$$\begin{aligned}\hat{\mathbf{a}}(k) = &\,\hat{\mathbf{a}}(k-1) + \mathbf{P}^*(k-1)\mathbf{x}(k)\frac{y(k)}{\sigma^2}\\ &- \mathbf{P}^*(k-1)\mathbf{x}(k)[\sigma^2 + \mathbf{x}^T(k)\mathbf{P}^*(k-1)\mathbf{x}(k)]^{-1}\mathbf{x}^T(k)\mathbf{P}^*(k-1)\mathbf{x}(k)\frac{y(k)}{\sigma^2}\\ &- \mathbf{P}^*(k-1)\mathbf{x}(k)[\sigma^2 + \mathbf{x}^T(k)\mathbf{P}^*(k-1)\mathbf{x}(k)]^{-1}\mathbf{x}^T(k)\hat{\mathbf{a}}(k-1)\\ = &\,\hat{\mathbf{a}}(k-1) - \mathbf{P}^*(k-1)\mathbf{x}(k)[\sigma^2 + \mathbf{x}^T(k)\mathbf{P}^*(k-1)\mathbf{x}(k)]^{-1}\varepsilon(k)\end{aligned}$$

where,

$$\begin{aligned}\varepsilon(k) = &-[\sigma^2 + \mathbf{x}^T(k)\mathbf{P}^*(k-1)\mathbf{x}(k)]\frac{y(k)}{\sigma^2}\\ &+ \mathbf{x}^T(k)\mathbf{P}^*(k-1)\mathbf{x}(k)\frac{y(k)}{\sigma^2} + \mathbf{x}^T(k)\hat{\mathbf{a}}(k-1)\end{aligned}$$

Therefore, collecting terms and simplifying,

$$\begin{aligned}\hat{\mathbf{a}}(k) &= \hat{\mathbf{a}}(k-1) + \mathbf{g}(k)[y(k) - \mathbf{x}^T(k)\hat{\mathbf{a}}(k-1)]\\ \mathbf{g}(k) &= \mathbf{P}^*(k-1)\,\mathbf{x}(k)[\sigma^2 + \mathbf{x}^T(k)\mathbf{P}^*(k-1)\mathbf{x}(k)]^{-1}\end{aligned} \tag{E.2}$$

which is the parameter estimation update of the SRLS equation in section 3.2 of chapter 3.

This analysis reveals that each step in the SRLS algorithm can be considered as minimising the instantaneous cost function $\mathcal{J}(\mathbf{a})$ defined in (E.1), with $\mathbf{P}^*(k-1)$ and σ^2 defined as in chapter 3. Considering the situation at the beginning of the algorithm, therefore, we see that the initial estimate $\hat{\mathbf{a}}(0)$ and its associated covariance matrix $\mathbf{P}^*(0)$ appear in the cost function via the additive quadratic form (see Appendix 1 and Section 5.3, Chapter 5),

$$[\mathbf{a}(1) - \hat{\mathbf{a}}(0)]^T(\mathbf{P}^*(0))^{-1}[\mathbf{a}(1) - \hat{\mathbf{a}}(0)]$$

Thus, at its initiation, the algorithm is selecting $\hat{\mathbf{a}}(0)$ in order to minimize not only the normal least square cost term, i.e.

$$[y(k) - \mathbf{x}^T(k)\hat{\mathbf{a}}(k-1)]^2; \quad k = 1$$

but also a quadratic form in the difference between $\mathbf{a}(1)$ and the initial *a priori* estimate $\hat{\mathbf{a}}(0)$, weighted by the inverse of the associated *a priori* covariance matrix $\mathbf{P}^*(0)$.

From this simple analysis, we see that, if the analyst has little confidence in the *a priori* estimate $\hat{\mathbf{a}}(0)$ and so chooses $\mathbf{P}^*(0)$ to be large (e.g. 10^6 diagonal), then little notice will be taken of $\hat{\mathbf{a}}(0)$ in determining $\hat{\mathbf{a}}(1)$. On the other hand, if there is good prior knowledge of the parameter values, then the algorithm can be informed of this by the analyst choosing a suitably smaller $\mathbf{P}^*(0)$ covariance matrix which reflects the increased confidence associated with the assumed knowledge of $\hat{\mathbf{a}}(0)$. In this manner, the second term in the cost function will be given more weight and the estimate $\hat{\mathbf{a}}(1)$ will be much more dependent upon $\hat{\mathbf{a}}(0)$.

The Bayesian statistical interpretation of the above procedure is obvious, with the *a priori* information ($\hat{\mathbf{a}}(0)$; $\mathbf{P}^*(0)$) playing an important role in the computation of the *a posteriori* estimate $\hat{\mathbf{a}}(1)$. But the algorithm remains essentially the same without these statistical interpretations: in the deterministic, recursive least squares algorithm RLS of section 3.1, chapter 3, for example, we see that $\sigma^2 = 1.0$ and $\mathbf{P}^*(0) = \mathbf{P}(0)$), but the above algebraic results still apply. Consequently, on purely deterministic, numerical grounds, the $\mathbf{P}(0)$ matrix should be chosen by the analyst so that $(\mathbf{P}(0))^{-1}$ suitably reflects the 'weight' (to use the term favoured by Gauss) to be associated with the *a priori* choice of the initial estimate vector $\hat{\mathbf{a}}(0)$. And the choice of $\mathbf{P}(0)$ as a diagonal matrix with large elements is clearly consistent with the usual situation of low confidence in $\hat{\mathbf{a}}(0)$, since then the diagonal elements of $(\mathbf{P}(0))^{-1}$ will be very small and the quadratic form in the cost function (E.1) will play little part in the recursive update of $\hat{\mathbf{a}}(0)$ to yield $\hat{\mathbf{a}}(1)$.

Finally, it should be noted that the evaluation of the recursive least squares algorithm in the manner shown in this Appendix can be made much more general, since it applies to all recursive algorithms of the 'least squares-like' form. For example, as in Chapter 5, we could consider the case of a regression function with vector measurements $\mathbf{y}(k)$ and replace the cost function in (E.1) by the following cost function:

$$\begin{aligned}\mathcal{J}(\mathbf{a}) = {} & [\mathbf{y}(k) - \mathbf{H}(k)\mathbf{a}(k)]^T\mathbf{R}^{-1}[\mathbf{y}(k) - \mathbf{H}(k)\mathbf{a}(k)] \\ & + [\mathbf{a}(k) - \hat{\mathbf{a}}(k|k-1)](\mathbf{P}^*(k|k-1))^{-1}[\mathbf{a}(k) - \hat{\mathbf{a}}(k|k-1)]\end{aligned}$$

minimization of which gives rise to algorithm VTVP section 4.3 of chapter 4 if $\hat{\mathbf{a}}(k|k-1)$ and $\mathbf{P}^*(k|k-1)$ are suitably defined. Or again, we could consider the Kalman filter algorithm by repeating the analysis for a cost function:

$$\begin{aligned}\mathcal{J}(\mathbf{x}) = {} & [\mathbf{y}(k) - \mathbf{C}(k)\mathbf{x}(k)]\mathbf{R}^{-1}(k)[\mathbf{y}(k) - \mathbf{C}(k)\mathbf{x}(k)] \\ & + [\mathbf{x}(k) - \hat{\mathbf{x}}(k|k-1)](\mathbf{P}^*(k|k-1))^{-1}[\mathbf{x}(k) - \hat{\mathbf{x}}(k|k-1)]\end{aligned}$$

This yields the KF algorithm in section 4.4 of chapter 4 if $\hat{\mathbf{x}}(k|k-1)$ and $\mathbf{P}^*(k|k-1)$ are suitably defined in relation to the state equations (4.44).

Appendix F
Maximum Likelihood Derivation of the Refined Instrumental Variable Algorithm

This Appendix shows how the standard IV procedure described in chapter 6 can be considered within the context of *Maximum Likelihood* (ML) estimation and so can be refined to yield optimal statistical performance. The analysis starts with the BJ time series model (7.1) from chapter 7. This is re-written here in order to define the vector form of the white noise input:

$$\begin{aligned} y(k) &= \frac{B(z^{-1})}{A(z^{-1})}u(k) + \frac{D(z^{-1})}{C(z^{-1})}e(k) \\ \mathbf{e}(k) &= \mathcal{N}(\mathbf{0}, \sigma^2\mathbf{I}); \quad \mathbf{e}(k) = [e_1(k)\, e_2(k)\, \ldots,\, e_N(k)]^T \end{aligned} \tag{F.1}$$

For simplicity of exposition, the time delay δ will be set to zero, the z^{-1} argument will be dropped from the polynomials and it is assumed initially that the C and D polynomials are known *a priori*. Since the white noise input $\mathbf{e}(k)$ is assumed to be independent and identically distributed with zero mean value, the ML approach to estimation is straightforward. It requires the formulation of a log-likelihood function for the N observations given in (F.1) of the form (see section B.2.8, Appendix B),

$$\begin{aligned} \mathcal{L}(\mathbf{a}, \mathbf{b}, \sigma^2, \mathbf{y}, \mathbf{u}) = & -\frac{N}{2}\log_e(2\pi) - \frac{N}{2}\log_e\sigma^2 - \\ & \frac{1}{2\sigma^2}[\frac{C}{D}\mathbf{y} - \frac{BC}{AD}\mathbf{u}]^T[\frac{C}{D}\mathbf{y} - \frac{BC}{AD}\mathbf{u}] \end{aligned} \tag{F.2}$$

where,

$$\begin{aligned} \mathbf{a} &= [a_1\, a_2\, \ldots,\, a_n]^T; \mathbf{b} = [b_1\, b_2\, \ldots,\, b_n]^T \\ \mathbf{y} &= [y_1\, y_2\, \ldots,\, y_N]^T; \mathbf{u} = [u_1\, u_2\, \ldots,\, u_N]^T \end{aligned}$$

Note that, for convenience, the system model parameter vectors **a** and **b** are assumed to be of the same order n.

Maximization of this log likelihood function clearly requires the minimization of the final term on the RHS of (F.2), which the reader will recognize as simply the sum of the squares of the residuals $e(k)$. Thus it is necessary to find those estimates of the **a** and **b** parameters in the polynomials A and B, together with the variance σ^2, which minimize the sum of the squares of the residuals - i.e. a classical but clearly nonlinear least squares problem. More formally this requires:

$$\hat{\boldsymbol{\theta}} = \arg\min_{\boldsymbol{\theta}} \mathcal{L}(\boldsymbol{\theta}, \mathbf{y}, \mathbf{u}) \tag{F.3}$$

where $\boldsymbol{\theta}$ is the vector of unknown parameters (**a**, **b** and σ^2). The conditions for this are obtained in the usual manner by partially differentiating with respect to a_i, $i = 1, 2, \ldots, n$; b_i, $i = 1, 2, \ldots, n$ and σ^2, respectively, and setting these derivatives to zero. This yields the following three equations:

$$\frac{\partial \mathcal{L}}{\partial a_i} = \frac{1}{\sigma^2} \sum_{k=2n+1}^{N} [\frac{C}{D} y(k) - \frac{BC}{AD} u(k)] \frac{BC}{A^2 D} z^{-i} u(k) = 0 \; ; \quad i = 1, 2, \ldots, n \tag{F.4a}$$

$$\frac{\partial \mathcal{L}}{\partial b_i} = \frac{1}{\sigma^2} \sum_{k=2n+1}^{N} [\frac{C}{D} y(k) - \frac{BC}{AD} u(k)] \frac{C}{AD} z^{-i} u(k) = 0 \; ; \quad i = 1, 2, \ldots, n \tag{F.4b}$$

$$\frac{\partial L}{\partial \sigma^2} = -\frac{T}{\sigma^2} + \frac{1}{\sigma^4} \sum_{k=2n+1}^{N} [\frac{C}{D} y(k) - \frac{BC}{AD} u(k)]^2 = 0 \tag{F.4c}$$

These equations, for given C and D (assumed known) polynomials, are highly nonlinear in the a_i and b_i parameters but are linear in the b_i for a given a_i; and they are also linear in the σ^2. And, moreover, if estimates of the a_i and b_i are available, an estimate of σ^2 can be obtained easily from condition (F.4)c. One method of obtaining estimates of the a_i and b_i that satisfy conditions (F.4) is to search the parameter space in some manner; for example, using a hill-climbing scheme such as the Newton-Raphson procedure employed by Åström and Bohlin (1966) or a 'relaxation' technique such as that described by Goodwin and Payne (1977).

An alternative procedure becomes apparent if we define the following discrete 'prefiltered' variables as shown in Figure 7.1 of chapter 7.

$$y_{f_1}(k) = \frac{C}{DA} y(k); \quad u_{f_1}(k) = \frac{C}{DA} u(k); \quad \hat{x}_{f_1}(k) = \frac{B}{A} u_{f_1}(k) \tag{F.5}$$

and note then that equations (F.4)a and (F.4)b can be rewritten in terms of these prefiltered variables:

$$\sum [A y_{f_1}(k) - B u_{f_1}(k)] \hat{x}_{f_1}(k-i) = 0 \; ; \quad i = 1, 2, \ldots, n \tag{F.6a}$$

$$\sum [A y_{f_1}(k) - B u_{f_1}(k)] u_{f_1}(k-i) = 0 \; ; \quad i = 1, 2, \ldots, n \tag{F.6b}$$

which are linear in the a_i and b_i parameters, provided we assume knowledge of the prefiltered variables $y_{f_1}(k)$, $u_{f_1}(k)$ and $\hat{x}_{f_1}(k)$.

F.1 RIV System Model Estimation Within the Context of Maximum Likelihood

Perhaps the most interesting aspect of the equations (F.6) is their remarkable similarity to the equivalent IV equations for the same problem. This becomes clear if we examine the IV normal equations (c.f equation (3.38) in chapter 3) i.e.,

$$[\sum_{k=1}^{N} \hat{\mathbf{x}}(k)\hat{\mathbf{z}}^T(k)]\hat{\mathbf{a}} - \sum_{k=1}^{N} \hat{\mathbf{x}}(k)y(k) = 0 \tag{F.7}$$

where $\hat{\mathbf{x}}(k)$ and $\mathbf{z}(k)$ are defined as follows:

$$\begin{aligned} \hat{\mathbf{x}}(k) &= [-\hat{x}(k-1)\ -\hat{x}(k-2)\ ...\ -\hat{x}(k-n)\ u(k)\ u(k-1)\ ...\ u(k-n)]^T \\ \mathbf{z}(k) &= [-y(k-1)\ -y(k-2)\ ...\ -y(k-n)\ u(k)\ u(k-1)\ ...\ u(k-n)]^T \end{aligned} \tag{F.8}$$

This vector-matrix equation represents a set of $2n+1$ linear simultaneous equations. Upon expansion, this yields the following

$$\begin{aligned} \sum y(k) + \hat{a}_1 y(k-1) + \ldots + \hat{a}_n y(k-n) - \hat{b}_0 u(k) - \ldots - \hat{b}_n u(k-n)]\,\hat{x}(k-i) &= 0 \\ \sum y(k) + \hat{a}_1 y(k-1) + \ldots + \hat{a}_n y(k-n) - \hat{b}_0 u(k) - \ldots - \hat{b}_n u(k-n)]\,u(k-i) &= 0 \end{aligned}$$

for $i = 1, 2, \ldots, n$. On the introduction of the operator notation and the A and B polynomials, these become

$$\begin{aligned} \sum [Ay(k) - Bu(k)]\,\hat{x}(k-i) &= 0\,; \quad i = 1, 2, \ldots, n \\ \sum [Ay(k) - Bu(k)]\,u(k-i) &= 0\,; \quad i = 1, 2, \ldots, n \end{aligned} \tag{F.9}$$

Comparison of equations (F.6) and (F.9) shows immediately that equation (F.6) can be interpreted as the IV normal equations for the system, with the input $u(k)$, the output $y(k)$ and the auxiliary model output $\hat{x}(k)$ replaced by their prefiltered equivalents $u_{f_1}(k)$, $y_{f_1}(k)$ and $\hat{x}_{f_1}(k)$, respectively. And not surprisingly, if these prefilters are introduced, then the resulting normal equations are identical to the expanded form of the RIV normal equations (7.16) for the system model parameters, as developed in section 7.1.2 of chapter 7 (changing the nomenclature accordingly, compare (F.7) with (7.16) and carry out analysis similar to that shown above).

Finally note that, in the original version of this book, a similar approach to the above was used to develop an estimation algorithm for the ARMA noise model parameters, based on the analysis in Young (1976b). While this *IV Approximate Maximum Likelihood* (IVAML) algorithm had reasonable performance it has now been replaced by the superior IVARMA algorithm (see section 7.1.3 of chapter 7).

Appendix G
The CAPTAIN Toolbox for Matlab: an Overview

The **C***omputer-***A***ided* **P***rogram for* **T***ime Series* **A***nalysis and* **I***dentification of* **N***oisy Systems* (CAPTAIN) Toolbox provides access to novel algorithms for various important aspects of identification, estimation, nonstationary time series analysis and signal processing, adaptive forecasting and automatic control system design. These have been developed between 1981 and the present at Lancaster University, UK, based on research carried out by the author at the University of Cambridge, UK (1965-1975) and the Australian National University, Canberra, Australia (1975-1981). Although it has its origins in the CAPTAIN (Young and Shellswell, 1973; Young and Jakeman, 1979a) and micro-CAPTAIN (Young and Benner, 1991) packages, the CAPTAIN Toolbox for Matlab is much more flexible and provides access to many more algorithms that have been added over the 1990s and continue to be developed up to the present day, most of them only available in this Toolbox. A new Toolbox Website, at `http://captaintoolbox.co.uk/`, provides news on recent developments, publication and report downloads, most of the data sets used for examples in the present book, advice on the use of CAPTAIN routines and short articles on items of interest. The fully functional CAPTAIN Toolbox can be downloaded for a three month long trial from a linked website at `http://www.cs.lancs.ac.uk/cres/captain/`. Further usage requires payment of a licence fee, which entitles the licensee to a very useful e-Handbook (Pedregal et al., 2005) and the provision of all future additions and enhancements. The Handbook can be considered as a companion volume to the present book: it contains a description of each routine available in the Toolbox, as well as numerous examples that include verbatim snapshots of command line operations and explanations of how the various program parameters are specified. The modest, non-profit license fee, available to legitimate students, is required to help cover the overhead costs required for continuing development of the Toolbox.

Toolbox Overview

Essentially, the CAPTAIN toolbox is a collection of routines (Matlab m-file scripts) for the identification and estimation of the various model types discussed in the main body of the present book; and their subsequent use with various application routines available in the Toolbox. These are organized around the core of the *Recursive Least Squares* (RLS), *Kalman Filter* (KF), *Fixed Interval Smoothing* (FIS) and *Refined Instrumental Variable* (RIV) recursive algorithms. However, in order to allow for straightforward user access, CAPTAIN consists of numerous 'shells' i.e. top level functions that automatically generate the necessary model structures, with default options based on the experience of the developers and the extensive practical experience with the algorithms (but user-adjustable to override the default settings). In this regard, the main areas of functionality are listed below.

Unobserved Component (UC) Models

- The *Dynamic Harmonic Regression* (DHR) model in the form of a harmonic regression with time variable parameters (Young et al, 1999) . This is estimated using the hyper-parameter optimization function dhropt followed by the DHR algorithm dhr primed with these optimized hyper-parameters. It is a particularly useful algorithm for signal extraction, forecasting and back-casting of periodic or quasi-periodic series. The same function allows for the estimation of other, related models, such the *Basic Structural Model* (BSM: Harvey et al, 1989).
- Other related TVP models are: *Dynamic Linear Regression* (DLR), *Dynamic Auto-Regression* (DAR), *Dynamic AR with eXogenous inputs* (DARX), and *Dynamic Transfer Function* (DTF). These are accessible via the dlr, dar, darx and dtfm routines; and the associated dlropt, daropt, darxopt optimization routines. dtfmopt.

As this book shows, these UC routines provide a general resource for nonstationary time series analysis, forecasting and signal extraction, as well as time-frequency analysis; and they have been applied to a wide range of engineering, environmental, biological and socio-economic systems, as discussed in numerous earlier publications.

Transfer Function Model Identification and Estimation

- The functional pairs rivid/riv (for discrete-time SISO and MISO TF models) and rivcid/rivc (for continuous-time SISO and MISO TF models estimated from discrete-time sampled data) are provided for order/structure identification (rivid, rivcid) and parameter estimation (riv, rivc) in the case of constant parameter, linear TF models. These routines include, as options, both recursive and *en-bloc* versions of the optimal RIV and SRIV algorithms, in addition to conventional least squares-based approaches. Recent enhancements are the introduction of the rivbj/rivcbj and rivbjid/rivcbjid routines for full BJ models with ARMA additive noise. These also produce the estimation results in an 'object' coded form that is compatible with routines in the Matlab SID Toolbox.

- In all of the order/structure identification routines (the above routines ending in 'id'), the model listing, for the selection of models chosen, can be reported in the order of various statistics, such as the coefficient of determination R_T^2 and various identification statistics, including the AIC, BIC and our own YIC (see main text).
- The ivarmaid and ivarma routines that implement the IVARMA algorithm for the identification and estimation of ARMA noise models. The ivarma routine is part of the rivbj/rivcbj routines but is made available separately for univariate noise model estimation.
- New rivbjdd and rivcbjdd for estimating parameters in multi-input models with different denominators are being developed and it is hoped that they will be available to all users during the next year.
- New clrivbj and clrivcbj routines for closed loop system identification and estimation (see chapter 9) are being developed and it is hoped that these, as well as rriv and rrivc routines for real-time recursive TVP estimation (see chapter 10), will be available sometime during the next two years (we are a very small team and the 'CAPTAINization' tasks required to prepare foolproof routines for CAPTAIN is a time-consumptive operation).

Nonlinear State Dependent Parameter Model Identification

- State-dependent parameter (SDP) nonlinear modelling is based around the sdp routine, which yields non-parametric (graphical) estimates of state-dependent parameters. If required, the user can parameterize these graphically defined nonlinearities using specified nonlinear functions (*e.g.* exponential, power law, radial basis functions etc.) that can then be optimized using standard Matlab functions (see chapter 11 in the main text).

Proportional-Integral-Plus (PIP) Control System Design Routines

- The PIP control system design routines include the pip algorithm for pole assignment design and the pipopt for PIP-LQ design, together with all other required support routines. These are not discussed in this book but details can be found in Taylor et al. (2000) and the prior references therein. Also a book, *True Digital Control* on this topic is currently in preparation. These very powerful, multivariable PIP control system design routines are based on TF models identified and estimated using the above routines in CAPTAIN, which are then converted to the *Non-Minimum State Space* (NMSS) form that is the basis for the design of the PIP non-minimal state variable feedback control laws. They are also integrated into Simulink objects that can be used in simulation studies and, potentially, for on-line use.

Other Routines

- Various conventional models, identification tools and auxiliary functions, too numerous to list individually here. Of these, the largest is the kalmanfis routine, which provides a shell to the KF/FIS algorithms for general state space filtering, smoothing and forecasting purposes. System identification is inherent to the modelling approach utilized by most of the functions already discussed. Other routines include: acf to determine the sample and partial autocorrelation function; ccf for the sample cross-correlation; period to estimate the periodogram; and statist for some sample descriptive statistics. Additional statistical diagnostics include: boxcox (optimal Box-Cox transformation for homoscedasticity); cusum (cusum recursive test for time varying mean and cusumsq recursive test for time varying variance); and histon (histogram over normal distribution and Bera-Jarque statistical test); while useful general routines are: del for generating a matrix of delayed variables; irwsm for smoothing, decimation or for fitting a simple trend to a time series; prepz to prepare data for TF modelling (e.g. baseline removal and input scaling); scaleb to rescale the estimated TF model numerator polynomial following initial prepz use; stand to standardise or destandardise a matrix by columns; and reconst to reconstruct a time series by removing any dramatic jumps in the trend.

Finally, note that almost all of the recursive estimation and smoothing procedures outlined above will automatically handle missing data in the time series, represented in Matlab by special Not-a-Number (NaN) variables. Indeed, by appending or prepending such variables to the data set using the fcast function, the UC routines will forecast, interpolate or back-cast as appropriate, without requiring further user intervention. Finally, the toolbox is supported by on-line help and demonstration examples (see below).

Demonstration examples

- The demonstration examples are invoked by typing the instruction captdemo in the Matlab command line, which displays a GUI Menu with various 'button' options. In addition to a Captain Overview, these include: the DHR show, DAR show and RIV show options for free running demonstration examples in a special graphics window, as well as other 'command line' options for TVP command line demos; TF and PIP command line demos; Upgraded (BJ) TF demos and Handbook demos. These various command line demos will be the most useful for readers of this book and some their options are referred to in various of the exercises at the end of the book chapters. Each of the command line demo buttons accesses another GUI sub-menu that has self-explanatory options.

Glossary

The listing below is in alphabetical order but similar topics are grouped out of alphabetical order when this is advantageous.

ACF AutoCorrelation Function: also **PACF** Partial AutoCorrelation Function; **CCF** Cross-Correlation Function.

ADE Advection Dispersion Equation.

ADZ Aggregated Dead Zone: also **AMZ** Aggregated Mixing Zone ; **AMV** Aggregated Mixing Volume.

AIC Akaike Information Criterion: also, **BIC** Bayesian Information Criterion (sometimes **SIC**, the related Schwarz Information Criterion); **YIC** Young heuristic Information Criterion.

ARMAX Auto-Regressive, Moving Average with eXogenous variables model: also **ARMA** Auto-Regressive, Moving Average model; **AR** Auto-Regressive model; **MA** Moving Average model; **ARIMA** Auto-Regressive, Integrated Moving Average model; **DA** Dynamic Adjustment model with **AR** residuals.

CAPTAIN Computer-Aided Program for Time-series Analysis and Identification of Noisy systems: also CAPTAIN Toolbox (a toolbox for the Matlab™software environment).

CT Continuous-Time: also **DT** Discrete-Time; **CC** Continuous-time prediction, Continuous-time update in the **KF**; **CD** Continuous-time prediction, Discrete-time-time update in the **KF**.

BSM Basic Structural Model.

DAR Dynamic Auto-Regressive model: also **DARX** Dynamic Auto-Regression with eXogenous variables model and **RARX** Recursive **ARX** algorithm; **DHR** Dynamic Harmonic Regression model; **LDHR** Linear **DHR** algorithm; **HR** Harmonic Regression; **DLR** Dynamic Linear Regression model; **DTF** Dynamic Transfer Function model.

DBM Data-Based Mechanistic modelling philosophy and methodology.

DC 'Direct Current' but used in the context of a zero frequency component in spectral analysis.

DFT Discrete Fourier Transform.

DMA Dominant Mode Analysis.

DR Deterministic Regularization.

EWP Exponentially-Weighted-Past algorithm: also **RWP** Rectangularly-Weighted-Past algorithm; **DF** Directional Forgetting algorithm.

KF Kalman Filter algorithm : also other, related algorithms **KFX** Kalman Filter with exogenous inputs; **EKF** Extended **KF**; **EnKF** Ensemble **KF**; **UKF** Unscented **KF**; **PF** Particle Filter.

FIR Finite Impulse Response model.

FIS Fixed Interval Smoothing algorithm: also **FISIV** Fixed Interval Smoothing IV algorithm.

GAM Generalized Additive Model.

GCM Global Circulation Model.

GLR General Linear Regression model.

GM Gauss-Markov model.

GRW Generalized Random Walk model: also specific examples of this **RW** Random Walk model; **IRW** Integrated random Walk model; **SRW** Smoothed Random Walk model.

HP Hodrick-Prescott filter.

KALMSMO Kalman Filter and Fixed Interval Smoothing algorithm.

MDV Modified Dependent Variable used in **SDP** model estimation.

NVR Noise Variance Ratio.

PLR Pseudo-Linear Regression model.

PEM Prediction Error Minimization algorithm (*en bloc*): also **RPEM** Recursive Prediction Error Minimization algorithm.

ML Maximum Likelihood estimation: also **RML** Recursive Maximum Likelihood algorithm; and various forms of this **RML1** and **RML2**.

MCS Monte Carlo Simulation: also **RWMC** Random Walk Monte Carlo.

ODE Ordinary Differential Equation; also **PDE** Partial Differential Equation.

OSE Optimal Signal Extraction.

PCA Principal Component Analysis: also **POD** Proper Orthogonal Decomposition (**PCA** applied to state space model reduction).

PIP Proportional-Integral-Plus control system design: also **NMSS** Non-Minimal State Space form used in PIP design. PMD Polynomial Matrix Description.

IV Instrumental Variable algorithm: also **SIV** Standard Instrumental Variable algorithm; **IVARMA** Instrumental Variable **ARMA** algorithm with recursive form denoted **RIVARMA**; **IVAML** Instrumental Variable, Approximate Maximum Likelihood algorithm; **IV4** 4-step **IV** algorithm.

RIV Refined Instrumental Variable algorithm: also **RIVC** Refined Instrumental Variable for continuous-time models; **RIVBJ** is the **RIV** algorithm for **BJ** models; **RIV-AR** is the RIVBJ algorithm with the noise modelled as an AR rather than a full ARMA process; **RIVCBJ** is the **RIVC** algorithm for hybrid (continuous-discrete time) **BJ** models; **SRIV**, **SRIVC** Simplified **RIV** and **RIVC** algorithms; **CLRIV** Closed Loop **RIV** algorithm; **CLRIVC** Closed Loop **RIVC** algorithm; **RIVDD**, **RIVCDD** Different Denominator forms of **RIV** and **RIVC** algorithms; **RRIV** Recursive **RIV** algorithm for on-line, real-time use.

SISO Single Input, Single Output model: also . **MISO** Multiple Input, Single Output model; **MIMO** Multiple Input, Multiple Output model; **PMD** Polynomial Matrix Description.

RBF Radial Basis Functions.

RLS Recursive Least Squares algorithm: also **SRLS** Stochastic form of **RLS** algorithm; **VRLS** vector form of **RLS** algorithm.

SIR Sequential Importance Re-sampling: also **RR** Residual Re-sampling Filter.

SID System Identification Toolbox in Matlab^{TM} software environment.

SDP State-Dependent Parameter, nonlinear, transfer function model estimation algorithm.

SVD Singular Value Decomposition.

TF Transfer Function: also **MTF** Multivariable **TF**.

STF Stochastic Transfer Function.

SA Stochastic Approximation.

TVP Time Variable Parameter model or estimation: also **STVP** Standard recursive **TVP** algorithm; **VTVP** vector form of **TVP** algorithm.

UC Unobserved Component model.

VI Variance Intervention.

ZOH Zero Order Hold: also **FOH** First Order Hold.

References

Agénor, P. R., J. McDermott, and E. S. Prasad (2000). Macroeconomic fluctuations in developing countries: Some stylized facts. *World Bank Econ Rev 14*, 251–285.

Agüero, J. C. and G. C. Goodwin (2008). Identifiability of errors in variables dynamic systems. *Automatica 44*, 371–382.

Akaike, H. (1974). A new look at statistical model identification. *IEEE Transactions on Automatic Control 19*, 716–723.

Akaike, H. (1979). A Bayesian extension of the minimum AIC procedure of autoregressive model fitting. *Biometrika 66*, 237–242.

Akaike, H. (1980). Seasonal adjustment by a Bayesian modelling. *Journal of Time Series Analysis 1*, 1–13.

Anderson, B. D. O. and M. Deistler (1984). Identifiability in dynamic errors-in-variables model. *Journal of Time Series Analysis 5*, 1–13.

Åström, K. J. (1969). On the choice of sampling rates in parametric identification of time series. *Information Sci. 1*(1), 273–278.

Åström, K. J. (1970). *Introduction to Stochastic Control Theory*. New York: Academic Press.

Åström, K. J. and R. D. Bell (2000). Drum-boiler dynamics. *Automatica 36*, 363–378.

Åström, K. J. and T. Bohlin (1966). Numerical identification of linear dynamic systems from normal operating records. In P. H. Hammond (Ed.), *Theory of Self Adaptive Control Systems*, New York, pp. 96–111. Plenum Press.

Åström, K. J. and P. Eykhoff (1971). System identification - a survey. *Automatica 7*, 123–162.

Åström, K. J. and B. Wittenmark (1973). On self tuning regulators. *Automatica 9*, 185–199.

Bartlett, M. S. (1946). On the theoretical specification of sampling properties of autocorrelated time series. *Journal of Royal Statistical Society, Series B 8*, 27.

Bastogne, T., H. Noura, P. Sibille, and A. Richard (1998). Multivariable identification of a winding process by subspace methods for a tension control. *Control Engineering Practice 6*(9), 1077–1088.

Beer, T. and P. C. Young (1983). Longitudinal dispersion in natural streams. *Am. Soc. of Civil Eng., Jnl. Env. Eng. 109*, 1049–1067.

Bell (1984). Signal extraction for non-stationary time series. *Annals of Statistics 12*, 646–664.

Bertrand, J. (1855). *Méthode des Moindres Carrés: translation into French of 'Memoirs on the Combination of Observations' by K.F. Gauss, published with authorization of Gauss*. Paris: Mallet-Bachelier.

Beven, K. J., D. T. Leedal, P. J. Smith, and P. C. Young (2011). Identification and representation of state dependent non-linearities in flood forecasting using the DBM methodology. In L. Wang, H. Garnier, and A. J. Jakeman (Eds.), *System Identification, Environmetric Modelling and Control*, Berlin. Springer-Verlag (in press).

Beven, K. J. and P. C. Young (1988). An aggregated mixing zone model of solute transport through porous media. *J. Contam. Hydrol. 3*, 129–143.

Beven, K. J., P. C. Young, and D. Leedal (2008). Computationally efficient flood water level prediction (with uncertainty). In *Proceedings European Conference on Flood Risk Management*, Oxford.

Bloomfield, P. (1976). *Fourier Analysis of Time Series: An Introduction*. New York: John Wiley & Sons.

Blum, J. A. (1954). Multidimensional stochastic approximation methods. *Ann. Math Stat. 25*, 737–744.

Bodewig, E. (1956). *Matrix Calculus*. Amsterdam: North Holland.

Bohlin, T. (1970). On the maximum likelihood method of identification. *IBM Jnl. Res. and Dev. 14*, 41–51.

Box, G. E. P. and G. M. Jenkins (1970). *Time Series Analysis Forecasting and Control*. Holden-Day: San Francisco.

Box, G. E. P., G. M. Jenkins, and G. C. Reinsel (1994). *Time Series Analysis: Forecasting and Control* (3rd ed.). Upper Saddle River, NJ, USA: Prentice Hall PTR.

Bray, J. W., R. J. High, A. D. McCann, and H. Jemmeson (1965). On-line model making for chemical plant. *Trans. Soc. Inst. Tech. 17*.

Brown, R. L., J. Durbin, and J. M. Evans (1975). Techniques for testing the constancy of regression relationships over time. *Journal of Royal Statistical Society 37*(Series B), 149–192.

Brownlee, K. A. (1965). *Statistical Theory and Methodology in Science and Engineering*. New York: John Wiley.

Bryson, A. E. and Y. C. Ho (1969). *Applied Optimal Control*. Waltham, Mass.: Blaisdell.

Buhmann, M. D. (2003). *Radial Basis Functions*. Cambridge: Cambridge University Press.

Bujosa, M., A. García-Ferrer, and P. C. Young (2007). Linear dynamic harmonic regression. *Computational Statistics & Data Analysis 52*, 999–1024.

Burman, J. P. (1980). Seasonal adjustment by signal extraction. *Journal of the Royal Statistical Society, Series A 143*, 321–337.

Chatfield, C. (1975). *The Analysis of Time Series: Theory and Practice.* London: Chapman and Hall.

Chen, S. and S. A. Billings (1989). Representation of non-linear systems: the NARMAX model. *International Journal of Control 49*, 1012–1032.

Chotai, A., P. C. Young, P. G. McKenna, and W. Tych (1998). Proportional-Integral-Plus (PIP) design for delta operator systems: Part 2, MIMO systems. *International Journal of Control 70*, 149–168.

Chow, G. C. (1960). A test for equality between sets of observations in two linear regressions. *Econometrica 28*, 591–605.

Clark, M. P., D. E. Rupp, R. A. Woods, X. Zheng, R. P. Ibbitt, A. G. Slater, J. Schmidt, and M. J. Uddstrom (2008). Hydrological data assimilation with the ensemble Kalman filter: Use of streamflow observations to update states in a distributed hydrological model. *Advances in Water Resources 31*(10), 1309 – 1324.

Clarke, D. W. (1967). Generalised least squares estimation of the parameters of a dynamic model. In *Proc. First IFAC Symposium on Identification in Automatic Control Systems, Prague, paper 3.17.*

Co, T. B. and S. Ungarala (1997). Batch scheme recursive parameter estimation of continuous-time systems using the modulating functions method. *Automatica 33*(6), 1185–1191.

Cox, D. R. and D. V. Hinkley (1974). *Theoretical Statistics.* London: Chapman and Hall.

De Smedt, F. (2006). Analytical solutions for transport of decaying solutes in rivers with transient storage. *Journal of Hydrology 330*, 672– 680.

Deistler, M. (1986). Linear dynamic errors-in-variables models. In J. Gani and M. Priestley (Eds.), *Essays in Time Series and Allied Processes. Papers in honour of E.J. Hannan*, pp. 23–39. Applied Probability Trust.

Dempster, A. P., N. M. Laird, and D. B. Rubin (1977). Maximum likelihood from incomplete data via the EM algorithm. *Jnl. Royal Stat. Soc., Series B 39*, 1–38.

Deutsch, R. (1965). *Estimation Theory.* New Jersey: Prentice Hall.

Dhrymes, P. J. (1970). *Econometrics: Statistical Foundations and Applications.* New York: Harper and Row.

Doucet, A., N. de Freitas, and N. Gordon (2001). *Sequential Monte Carlo Methods in Practice.* Springer-Verlag, New York.

Duncan, D. B. and S. D. Horn (1972). Linear dynamic recursive estimation from the viewpoint of regression analysis. *Jnl. Am. Statist. Assoc. 67*, 815–821.

Durbin, J. (1954). Errors in variables. *Review Int. Stat. Inst 22*, 23–32.

Durbin, J. (1960). Estimation of parameters in time-series regression models. *Jnl. Roy. Stat. Soc. Series B 22*, 139–153.

Durbin, J. and S. J. Koopmans (2001). *Time Series Analysis by State Space Methods.* Oxford University Press: Oxford.

Dvoretsky, A. (1956). On stochastic approximation. In J. Neyman (Ed.), *Proc. 3rd Berkeley Symp. Math. Statist. Prob.*, Berkeley. Univ. Calif. Press.

Elgerd, O. I. (1967). *Control Systems Theory.* New York: McGraw Hill.

Evensen, G. (2007). *Data Assimilation: The Ensemble Kalman Filter*. Springer-Verlag: New York.

Eykhoff, P. (1974). *System Identification - Parameter and State Estimation*. John Wiley and Sons.

Findley, D. F., B. C. Monsell, W. R. Bell, M. C. Otto, and B. C. Chen (1996). New capabilities and methods of the X-12 ARIMA seasonal adjustment program. Technical report, U.S. Bureau of the Census.

Findley, D. F., B. C. Monsell, M. C. Otto, W. R. Bell, and M. Pugh (1992). Towards X-12 ARIMA. Technical report, Bureau of the Census.

Fitzgerald, W. J., A. Walden, R. Smith, and P. C. Young (Eds.) (2000). *Nonlinear and Nonstationary Signal Processing*. Cambridge: Cambridge University Press.

Fröberg, C. E. (1970). *Introduction to Numerical Analysis*. Reading, Mass.: Addison-Wesley.

Gammerman, D. (1997). *Markov Chain Monte Carlo: Stochastic Simulation for Bayesian Inference*. Chapman and Hall: London.

Gantmacher, F. R. (1960). *Matrix Theory*, Volume 1. New York: Chelsea:.

Garnier, H., M. Gilson, P. C. Young, and E. Huselstein (2007). An optimal IV technique for identifying continuous-time transfer function model of multiple input systems. *Control Engineering Practice 15*, 471–486.

Garnier, H., M. Mensler, and A. Richard (2003). Continuous-time model identification from sampled data. Implementation issues and performance evaluation. *International Journal of Control 76*(13), 1337–1357.

Garnier, H. and L. Wang (Eds.) (2008). *Identification of Continuous-Time models from Sampled data*. Springer-Verlag: London.

Gauss, K. (1826). Theoria combinationis observationum erroribus minimis obnoxiae. In *Werke 4, Parts 1, 2 and supplement*, pp. 1–108.

Gelb, A. (Ed.) (1974). *Applied Optimal Estimation*. MIT Press.

Gevers, M. and X. Bombois (2006). Input design : from open-loop to control-oriented design. In *Proc. of 14th IFAC Symposium on System Identification (SYSID, 2006)*, pp. 1329–1334.

Giarre, L., D. Bauso, P. Falugi, and B. Bamieh (2006). LPV model identification for gain scheduling control: An application to rotating stall and surge control problem. *Control Engineering Practice 14*, 351–361.

Gillberg, J. and L. Ljung (2009). Frequency-domain identification of continuous-time ARMA models from sampled data. *Automatica 45*(6), 1371 – 1378.

Gilson, M., H. Garnier, P. C. Young, and P. V. den Hof (2008). Instrumental variable methods for closed-loop continuous-time model identification. In H. Garnier and L. Wang (Eds.), *Identification of Continuous-Time Models from Sampled Data*, pp. 133–160. Springer-Verlag: London.

Gilson, M., H. Garnier, P. C. Young, and P. V. den Hof (2009). Refined instrumental variable methods for closed-loop system identification. In *Proceedings 15th IFAC Symposium on System Identification SYSID09, St Malo, France*.

Gilson, M. and P. Van den Hof (2005). Instrumental variable methods for closed-loop system identification. *Automatica 41*, 241–249.

Gómez, V. and A. Maravall (1996). Programs TRAMO and SEATS, instructions for the user. Technical Report No. 9628, Bank of Spain.

Goodwin, G. C. and R. L. Payne (1977). *Dynamic System Identification: Experiment Design and Data Analysis*. Academic Press: New York.

Gordon, N., A. Marrs, and D. Salmond (2000). Sequential analysis of nonlinear dynamic systems using particles and mixtures. In W. J. Fitzgerald, A. Walden, R. Smith, and P. C. Young (Eds.), *Nonlinear and Nonstationary Signal Processing*, pp. 42–73. Cambridge University Press: Cambridge.

Gordon, N., D. Salmond, and A. F. M. Smith (1993). Novel approach to nonlinear and non-Gaussian Bayesian state estimation. *Proc. Inst. Elect. Eng., F 140*, 107–113.

Granger, C. W. J. and T. Teräsvirta (1993). *Modelling Nonlinear Economic Relationships*. Oxford: Oxford University Press.

Graybill, F. A. (1961). *An Introduction to Linear Statistical Models*. New York: McGraw Hill.

Green, H. M., K. J. Beven, K. Buckley, and P. C. Young (1994). Pollution incident prediction with uncertainty. In K. J. Beven, P. Chatwin, and J. Millbank (Eds.), *Mixing and Transport in the Environment*, Chichester, pp. 113–137. John Wiley and Sons.

Gu, C. (2002). *Smoothing Spline ANOVA Models*. Berlin: Springer-Verlag.

Gurney, W. S. C., S. P. Blythe, and R. M. Nisbet (1980). Nicholson's blowflies revisited. *Nature 287*, 17–21.

Hannan, E. J. (1976). The convergence of some recursions. *Ann. Statist. 4*, 1258–1270.

Hannan, E. J. and B. G. Quinn (1979). The determination of the order of an autoregression. *Journal of Royal Statistical Society, Series B 41*, 190–195.

Hannan, E. J. and J. Rissanen (1982). Recursive estimation of mixed autoregressive moving average order. *Biometrika 69*, 81–94.

Harrison, P. J. and C. F. Stevens (1976). Bayesian forecasting. *Journal of the Royal Statistical Society, Series B 38*, 207–247.

Harvey, A. C. (1976). An alternative proof and generalisation of a test for structural change. *American Statistician 30*, 122–23.

Harvey, A. C. (1981). *Time Series Models*. Oxford: Phillip Allen.

Harvey, A. C. (1984). A unified view of statistical forecasting procedures. *Journal of Forecasting 3*, 245–275.

Harvey, A. C. (1989). *Forecasting Structural Time Series Models and the Kalman Filter*. Cambridge University Press: Cambridge.

Harvey, A. C. and J. Durbin (1986). The effect of seat belt legislation on British road casualties: A case study in structural time series modeling. *Journal of Royal Statistical Society, Series B B 149*, 187–227.

Hastie, T. J. and R. J. Tibshirani (1996). *Generalized Additive Models*. Chapman and Hall: London.

Hastings-James, R. and M. W. Sage (1969). Recursive generalised least squares procedure for on-line identification of process parameters. *Proc. Inst. of Electrical Engineers 116*, 2057–2062.

Haupt, G. T., N. J. Kasdin, G. M. Keiser, and B. W. Parkinson (1996.). An optimal recursive iterative algorithm for discrete nonlinear least-squares estimation. *J. Guid. Control Dyn 19*, 643–649.

Higham, N. J. (2005). The scaling and squaring method for the matrix exponential revisited. *SIAM J. Matrix Anal. Appl. 26*, 1179–1193.

Ho, Y. C. (1962). On the stochastic approximation method and optimal filtering theory. *Jnl. Math. Anal. App. 6*, 152.

Ho, Y. C. and C. Blaydon (1966). On the abstraction problem in pattern classification. In *Proc. Nat. Elect. Conf.*

Hoberock, L. L. and R. H. Kohr (1966). An experimental determination of differential equations to describe simple nonlinear systems. In *Proceedings, Joint Automatic Control Conference, Washington, Seattle*, pp. 616–623.

Hodrick, R. J. and E. C. Prescott (1980). Postwar U.S. business cycles: An empirical investigation. Technical report, Carnegie Mellon University discussion paper no. 451.

Hodrick, R. J. and E. C. Prescott (1997). Postwar U.S. business cycles: an empirical investigation. *Journal of Money, Credit and Banking 29*, 1–16.

Hof, P. M. J. V. and R. J. P. Schrama (1993). An indirect method for transfer function estimation from closed loop data. *Automatica 29*, 1523–1527.

Holst, U., O. Hössjer, C. Björklund, P. Ragnarsson, and H. Edner (1996). Locally weighted least squares kernel regression and statistical evaluation of LIDAR measurements. *Environmetrics, 7*, 401–406.

Hong, M. and T. Söderström (2009). Relations between bias-eliminating least squares, the Frisch scheme and extended compensated least squares methods for identifying errors-in-variables systems. *Automatica 45*, 277–282.

Hu, J., K. Kumamaru, and K. Hirasawa (2001). A quasi-ARMAX approach to modelling nonlinear systems. *International Journal of Control 74*, 1754–1766.

Imbrie, J., E. A. Boyle, S. C. Clemens, A. Duffy, W. R. Howard, G. Kukla, J. Kutzback, D. G. Martinson, A. McIntyre, A. C. Mix, B. Molfino, J. J. Morley, L. C. Peterson, N. G. Pisias, W. L. Prell, M. E. Raymo, N. J. Shackleton, and J. R. Toggweiler (1992). On the structure and origin of major glaciation cycles: 1. linear responses to Milankovitch forcing. *Paleoceanography 7*, 701–738.

Isermann, R., U. Baur, W. Bamberger, and P. Kneppo (1974). Comparison of six on-line identification and parameter estimation methods. *Automatica 10*, 81–103.

Jakeman, A. J., L. P. Steele, and P. C. Young (1980). Instrumental variable algorithms for multiple input systems described by multiple transfer functions. *IEEE Transactions on Systems, Man, and Cybernetics SMC-10*, 593–602.

Jakeman, A. J. and P. C. Young (1979a). Joint parameter/state estimation. *Electronics Letters 15*(19), 582–583.

Jakeman, A. J. and P. C. Young (1979b). Refined instrumental variable methods of time-series analysis: Part II, multivariable systems. *International Journal of Control 29*, 621–644.

Jakeman, A. J. and P. C. Young (1981). On the decoupling of system and noise model parameter estimation in time-series analysis. *International Journal of Control 34*, 423–431.

Jakeman, A. J. and P. C. Young (1983). Advanced methods of recursive time-series analysis. *International Journal of Control 37*, 1291–1310.

Jakeman, A. J. and P. C. Young (1984). Recursive filtering and the inversion of ill-posed causal problems. *Utilitas Mathematica 35*, 351–376.

Jang, J.-S. R., C.-T. Sun, and E. Mizutani (1997). *Neuro-Fuzzy and Soft Computing*. Upper Daddle River, NJ: Prentice Hall.

Jarque, C. M. and A. K. Bera. (1987). A test for normality of observations and regression residuals. *International Statistical Review 55*, 163–172.

Jazwinski, A. H. (1970). *Stochastic Processes and Filtering Theory*. San Diego: Academic Press.

Johnston, J. (1963). *Econometric Methods*. New York: McGraw Hill.

Johnston, J. and J. DiNardo (1997). *Econometric Methods*. New York: McGraw Hill.

Joseph, P., J. Lewis, and J. Tou (1961). Plant identification in the presence of disturbances and application to digital adaptive systems. *AIEE Trans. (Applications and Industry) 80*, 18–24.

Julier, S., J. Uhlmann, and H. F. Durrant-Whyte (2000). A new method for the nonlinear transformation of means and covariances in filters and estimators. *IEEE Transactions on Automatic Control 45*, 477 – 482.

Kailath, T. (1968, December). An innovations approach to least-squares estimation–Part 1: Linear filtering in additive white noise. *Automatic Control, IEEE Transactions on 13*(6), 646 – 655.

Kalman, R. E. (1960). A new approach to linear filtering and prediction problems. *Transactions American Society of Mechanical Engineers, Journal Basic Eng. 82-D*, 35–45.

Kendall, M. G. and A. Stuart (1961). *The Advanced Theory of Statistics*, Volume 2. London: Griffin.

Kiefer, J. and J. Wolfowitz (1952). Stochastic estimation of the maximum of a regression function. *Ann. Math. Stat. 23*, 462–466.

Kitagawa, G. (1981). A non-stationary time series model and its fitting by a recursive filter. *Jnl. of Time Series Anal. 2*, 103–116.

Kitagawa, G. and W. Gersch (1996). *Smoothness Priors Analysis of Time Series*. New York: Springer-Verlag.

Kitagawa, G. and T. Takanami (1985). Extraction of signal by a time series model and screening out micro earthquakes. *Signal Processing 8*(3), 303 – 314.

Kitagawa, G. and T. Takanami (2003). Extraction of small seismic signal by state space modeling. In T. Takanami and G. Kitagawa (Eds.), *Methods and Applications of Signal Processing in Seismic Network Operations*. Springer-Verlag: Berlin.

Kolmogorov, A. N. (1941). Interpolation and extrapolation. *Bulletin de l'Academie des Sciences de U.S.S.R., Ser. Math 5*, 3–14.

Konikow, L. and J. Bredehoeft (1992). Ground water models cannot be validated. *Advances in Water Resources 15*, 75–83.

Koopmans, S. J., A. C. Harvey, J. A. Doornik, and N. Shephard (1995). *STAMP 5.0: Structural Time Series Analyser, Modeller and Predictor*. London: Chapman and Hall.

Kopp, R. E. and R. J. Orford (1963). Linear regression applied to system identification for adaptive control systems. *American Institution of Aeronautics and Astronautics Journal 1*, 2300–2306.

Kuersteiner, G. M. (1999). Optimal instrumental variables estimation for ARMA models. Technical Report Working papers 99-07, Department of Economics, Massachusetts Institute of Technology.

Kuersteiner, G. M. (2001). Optimal instrumental variables estimation for ARMA models. *Journal of Econometrics 104*(2), 359 – 405.

Kuhn, T. (1962). *The Structure of Scientific Revolutions*. Chicago: University of Chicago.

Kulhavy, R. (1987). Restricted exponential forgetting in real time identification. *Automatica 23*, 589–600.

Kumar, R. and J. Moore (1980). State inverse and decorrelated state stochastic approximation. *Automatica 16*, 295–311.

Kuo, B. C. and F. Golnaraghi (2002). *Automatic Control Systems*. New York, NY, USA: John Wiley & Sons, Inc.

Kushner, H. J. (1967). Dynamical equations for optimum non-linear filtering. *J. Differential Equations 3*, 179–190.

Landau, I. D. (1976). Unbiased recursive estimation using model reference adaptive techniques. *IEEE Transactions on Automatic Control AC-21*, 194–202.

Larsson, E. K., M. Mossberg, and T. Söderström (2008). Estimation of continuous-time stochastic system parameters. In H. Garnier and L. Wang (Eds.), *Identification of Continuous-time Models from Sampled Data*, pp. 31–66. Springer-Verlag.

Laurain, V., M. Gilson, H. Garnier, and P. C. Young (2008). Refined instrumental variable methods for identification of Hammerstein continuous-time Box–Jenkins models. In *47th Conference on Decision and Control, Cancun, Mexico*. http://hal.archives-ouvertes.fr/ docs/00/42/41/35/PDF/ LaurainetAlCLIVfinalsubmission.pdf.

Laurain, V., R. Toth, M. Gilson, and H. Garnier (2010). Refined instrumental variable methods for identification of LPV Box–Jenkins models. *Automatica 46*(6), 959–967.

Lawson, C. L. and R. J. Hanson (1974). *Solving Least Squares Problems*. Englewood Cliffs, NJ.: Prentice-Hall.

Lee, R. C. K. (1964). *Optimal Identification, Estimation and Control*. MIT Press: Cambridge, Mass.

Leedal, D., K. Beven, A. Weerts, and P. Young (2010). An operational application of DBM catchment modelling for real-time flood forecasting. In L.Wang, H. Garnier, and A. J. Jakeman (Eds.), *System Identification, Environmetric Modelling and Control System Design*, U.K. Springer (in press).

Lees, M., P. C. Young, and S. Ferguson (1993). Adaptive flood warning. In P. C. Young (Ed.), *Concise Encycolpedia of Environmental Systems*, Oxford, pp. 234–236. Pergamon.

Lees, M. J. (2000). Data-based mechanistic modelling and forecasting of hydrological systems. *Journal of Hydroinformatics 2*, 15–34.

Lees, M. J., P. C. Young, S. Ferguson, K. J. Beven, and J. Burns (1994). An adaptive flood warning scheme for the River Nith at Dumfries. In W. R. White and J. Watts (Eds.), *International Conference on River Flood Hydraulics*, pp. 65–77. J. Wiley: Chichester.

Levadi, V. S. (1964). Parameter estimation of linear systems in the presence of noise. In *Proceedings, International Conference on Microwaves, Circuit theory and Information Theory, Tokyo*.

Levin, M. J. (1960). Optimal estimation of impulse response in the presence of noise. *I R E Trans. Circuit Theory CT-7*, 50–56.

Levin, M. J. (1964). Estimation of system pulse transfer function in the presence of noise. *IEEE Trans.Auto. Control AC-9*, 229–235.

Liaw, C. M. (1986). Model reduction of discrete systems using the power decomposition method. *Proc. Inst. of Electrical Engineers 133, Pt.D*, 30–34.

Lin, Z. and M. B. Beck (2007). On the identification of model structure in hydrological and environmental systems. *Water Resources Research 43*, doi:10.1029/2005WR004796.

Linden, J. G., P. C. Young, T. Larkowski, and K. J. Burnham (2009). Identification of a catchment model via errors-in-variables approaches - a preliminary study. In *Proc. 20th Int. Conf. on Systems Engineering, Coventry, UK*, pp. 312–320.

Littlewood, I. G., P. C. Young, and B. F. W. Croke (2010). Preliminary comparison of two methods for identifying rainfall–streamflow model parameters insensitive to data time-step: the Wye at Cefn Brwyn, Plynlimon, Wales. In C. Walsh (Ed.), *Role of Hydrology in Managing Consequences of a Changing Global Environment*, University of Newcastle, U.K., pp. 539–543. British Hydrological Society.

Liu, X., J. Wang, and W. X. Zheng (2011). Convergence analysis of refined instrumental variable method for continuous-time system identification. *IET Control Theory and Applications, in press*.

Ljung, L. (1976). On the consistency of prediction error methods. In R. Mehra and D. Lainiotis (Eds.), *System Identification: Advances and Case Studies*, New York, pp. 121–164. Academic Press.

Ljung, L. (1977a). Analysis of recursive stochastic algorithms. *IEEE Trans. Auto Control AC-22*, 551–575.

Ljung, L. (1977b). On positive-real functions and convergence of some recursive schemes. *IEEE Trans. Auto. Control AC-22*, 539–551.

Ljung, L. (1978). Convergence analysis of parametric identification methods. *IEEE Trans. Auto. Control AC-23*, 770–783.

Ljung, L. (1979a). Asymptotic behaviour of the extended Kalman filter as a parameter estimator for linear systems. *IEEE Trans. Auto. Control AC-24*, 36–50.

Ljung, L. (1979b). Convergence of recursive estimators. In R. Isermann (Ed.), *Identification and System Parameter Estimation*, Oxford, pp. 131–144. Pergamon.

Ljung, L. (1987). *System Identification. Theory for the User*. Englewood Cliffs, NJ.: Prentice Hall.

Ljung, L. (1999). *System identification. Theory for the user* (2nd ed.). Upper Saddle River, N.J.: Prentice Hall.

Ljung, L. (2003, September). Initialisation aspects for subspace and output-error identification methods. In J. Maciejowski (Ed.), *Proc. European Control Conference, ECC 03*, Cambridge, UK.

Ljung, L. and T. Söderström (1983). *Theory and Practice of Recursive Estimation*. MIT Press: Cambridge, Mass.

Loeve, M. M. (1963). *Probability Theory*. New York: Von Nostrand.

Lu, C. X., N. W. Rees, and P. C. Young (2011). Simulation model emulation in control system design. In L. Wang, H. Garnier, and A. J. Jakeman (Eds.), *System Identification, Environmetric Modelling and Control*, Berlin. Springer-Verlag (in press).

Mann, H. B. and A. Wald (1943). On the statistical treatment of linear stochastic difference equations. *Econometrica 11*, 173–220.

Maravall, A. (1993). Stochastic linear trends. models and estimators. *Journal of Econometrics 56*, 5–37.

Markovsky, I., J. C. Willems, S. V. Huffel, B. D. Moor, and R. Pintelon (2005). Application of structured total least squares for system identification and model reduction. *IEEE Transactions on Automatic Control 50*, 1490–1500.

Martinez, C. J. and W. R. Wise (2003). Analysis of constructed treatment wetland hydraulics with the transient storage model OTIS. *Ecological Engineering 20*(3), 211 – 222.

Mayne, D. Q. (1967). A method for estimating discrete-time transfer functions. In *Advances in Computer Control, Second UKAC Control Convention*, University of Bristol.

McIntyre, N., P. C. Young, B. Orellana, M. Marshall, B. Reynolds, and H. Wheater (2011). Identification of nonlinearity in rainfall-flow response using data-based mechanistic modeling. *Water Resour. Res. 47*, W03515, doi:10.1029/2010WR009851.

Mees, A. I. (1991). Dynamical systems and tesselations: detecting determinism in data. *Int. Jnl. of Bifurcation and Chaos 1*, 777–794.

Mees, A. I. (1993). Parsimonious dynamical reconstruction. *Int. Jnl. of Bifurcation and Chaos 3*, 669–675.

Mees, A. I., K. Aihara, M. Adachi, K. Judd, T. T. Ikeguchi, and G. Massumoto (1992). Deterministic prediction and chaos in squid axon response. *Physics Letters A 169*, 41–45.

Mehra, R. K. (1974). Optimal input signals for parameter estimation in dynamic systems-survey and new results. *IEEE Trans Auto. Control AC-19*, 753–768.

Mendel, J. (1970). Gradient identification for linear systems. In J. M. Mendel and K. S. Fu (Eds.), *Adaptive Learning and Pattern Recognition Systems*, New York, pp. 209–242. Academic Press.

Mendel, J. M. (1969). *A priori* and *a posteriori* identification of time varying parameters. In *Proceedings, 2nd IEEE Conference on System Sciences*, Hawaii (USA).

Mendel, J. M. and K. S. Fu (1970). *Adaptive Learning and Pattern Recognition Systems*. New York: Academic Press.

Middleton, R. H. and G. C. Goodwin (1990). *Digital Control and Estimation - A Unified Approach*. Prentice Hall.

Milankovitch, M. (1920). *Théorie Mathématique des Phénomènes Thermiques produits par la Radiation Solaire*. Paris: Gauthier-Villars.

Moler, C. B. and C. F. V. Loan (2003). Nineteen dubious ways to compute the exponential of a matrix. *SIAM Review 45*, 3–49.

Moore, B. (1981). Principal component analysis in linear systems: Controllability, observability, and model reduction,. *IEEE Transactions on Automatic Control AC-26*(1), 17–31.

Moore, J. B. (1973). Discrete-time fixed-lag smoothing algorithms. *Automatica 9*, 175–184.

Moradkhani, H., K.-L. Hsu, H. Gupta, and S. Sorooshian (2005). Uncertainty assessment of hydrologic model states and parameters: sequential data assimilation using the particle filter. *Water Resources Research 41*(W05012, doi:10.1029/2004WR003604.).

Moradkhani, H., S. Sorooshian, H. V. Gupta, and P. R. Houser (2005). Dual state-parameter estimation of hydrological models using ensemble Kalman filter. *Advances in Water Resources 28*, 135–147.

Ng, C. N. and P. C. Young (1990). Recursive estimation and forecasting of non-stationary time series. *Journal of Forecasting: Special Issue on State-Space Forecasting and Seasonal Adjustment 9*, 173–204.

Nicholson, A. J. (1954). An outline of the dynamics of animal populations. *Austral. Zool. Jnl. 2*, 9–65.

Norton, J. P. (1975). Optimal smoothing in the identification of linear time-varying systems. *Proceedings Institute Electrical Engineers 122*, 663–668.

Norton, J. P. (1986 (reprinted by Dover Publications, Inc., 2009)). *An Introduction to Identification*. New York: Academic Press.

Ochieng, G. and F. Otieno (2009). Data-based mechanistic modelling of stochastic rainfall-flow processes by state dependent parameter estimation. *Environmental Modelling & Software 24*, 279–284.

Oreskes, N., K. Shrader-Frechette, and K. Belitz (1994). Verification, validation, and confirmation of numerical models in the earth sciences. *Science 263*, 641–646.

Panuska, V. (1969). An adaptive recursive least squares identification algorithm. In *Proceedings 8th IEEE Symposium on Adaptive processes*, pp. Paper 6e.

Parkinson, S. and P. C. Young (1998). Uncertainty and sensitivity in global carbon cycle modelling. *Climate Research 9*, 157–174.

Pedregal, D. J. (1995). *Comparación Teórica, Estructural y Predictiva de Modelos de Componentes no Observables y Extesiones del Modelo de Young*. Ph. D. thesis, Universidad Autónoma de Madrid.

Pedregal, D. J., C. J. Taylor, and P. C. Young (2005). *The CAPTAIN Handbook*. Lancaster University: Centre for Research on Environmental Systems and Statistics.

Pedregal, D. J. and P. C. Young (2001). Some comments on the use and abuse of the Hodrick-Prescott filter. *Review on Economic Cycles 2*(1).

Pedregal, D. J. and P. C. Young (2006). Modulated cycles, an approach to modelling periodic components from rapidly sampled data. *International Journal of Forecasting 22*, 181–194.

Penrose, R. (1955). A generalized inverse for matrices. *Proc. Cambridge Phil. Soc. 51*, 406–413.

Percival, D. B. and A. T. Walden (1993). *Spectral Analysis for Physical Applications : Multitaper and Conventional Univariate Techniques*. Cambridge: Cambridge University Press.

Pierce, D. A. (1972). Least squares estimation in dynamic disturbance time-series models. *Biometrika 5*, 73–78.

Plackett, R. L. (1950). Some theorems in least squares. *Biometrika 37*, 149–157.

Popper, K. (1959). *The Logic of Scientific Discovery*. London: Hutchinson.

Previdi, F. and M. Lovera (2004). Identification of non-linear parametrically varying models using separable least squares. *International Journal of Control 77*, 1382–1392.

Price, L., P. C. Young, D. Berckmans, K. Janssens, and J. Taylor (1999). Data-based mechanistic modelling and control of mass and energy transfer in agricultural buildings. *Annual Reviews in Control 23*, 71–82.

Priestley, M. B. (1981). *Spectral Analysis and Time Series*. London: Academic Press.

Priestley, M. B. (1988). *Nonlinear and Nonstationary Time Series Analysis*. London: Academic Press.

Quandt, R. E. (1960). Tests of the hypothesis that a linear regression system obeys two separate regimes. *Jnl. Am. Statist. Assoc. 55*, 324–330.

Rauch, H. E., F. Tung, and C. T. Streibel (1965). Maximum likelihood estimates of linear dynamic systems. *American Institute of Aeronautics and Astronautics (AIAA) Journal 3*, 1445–1450.

Robbins, H. and S. Monro (1951). A stochastic approximation method. *Ann. Math. Statist. 22*, 400–407.

Romanowicz, R. J., P. C. Young, and K. J. Beven (2006). Data assimilation and adaptive forecasting of water levels in the River Severn catchment. *Water Resources Research 42*, W06407, doi:10.1029/2005WR004373.

Rosenbrock, H. H. and P. D. McMorran (1971). Good, bad or optimal? *IEEE Trans. Auto. Control AC-16*, 552–554.

Ruanaidh, J. J. and W. Fitzgerald (1996). *Numerical Bayesian Methods Applied to Signal Processing*. New York: Springer Verlag.

Runkel, R. L. and S. C. Chapra (1993, doi:10.1029/92WR02217). An efficient numerical solution of the transient storage equations for solute transport in small streams. *Water Resour. Res. 29*(1), 211–215.

Sadeghi, J., W. Tych, A. Chotai, and P. C. Young (2010). Multi-state dependent parameter model identification and estimation for control system design. *Electronics Letters 46*(18), 1265–1266.

Sakrison, D. (1966). Stochastic approximation: a recursive method for solving regression problems. In A. V. Balakrishnan (Ed.), *Advances in Communication Theory, 2*, New York. Academic Press.

Saridis, G. N. (1974). Comparison of six on-line identification algorithms. *Automatica 10*, 69–79.

Schwarz, G. (1978). Estimating the dimension of a model. *Annal. Statist. 6*, 461–464.

Schweppe, F. (1965). Evaluation of likelihood functions for gaussian signals. *IEEE Trans. on Information Theory 11*, 61–70.

Shamma, J. and M. Athans (1991). Guaranteed properties of gain scheduled control for linear parameter varying plants. *Automatica 27*, 559–564.

Shibata, R. (1985). Various model selection techniques in time series analysis. In E. J. Hannan, P. R. Krishnaiah, and M. Rao (Eds.), *Handbook of Statistics 5: Time Series in the Time Domain*, pp. 179–187.

Sinha, N. and G. Rao (Eds.) (1991). *Identification of Continuous-Time Systems: Methodology and Computer Implementation*. Dordrecht: Kluwer.

Sjöberg, J., Q. Zhang, L. Ljung, A. Benveniste, B. Deylon, P. Y. glorennec, H. Hjalmarsson, and A. Juditsky (1995). Nonlinear black-box models in system identification : a unified overview. *Automatica 31*(12), 1691–1724.

Smith, P. J., K. J. Beven, and J. A. Tawn (2006). Detection of structural inadequacy in process-based hydrological models: A particle-filtering approach. *Water Resour. Res. 44*, W01410, doi:10.1029/2006WR005205.

Söderström, T. (2007). Errors-in-variables methods in system identification. *Automatica 43*, 939–958.

Söderström, T., L. Ljung, and I. Gustavsson (1974). A comparative study of recursive identification methods. Technical Report TR 7308, Lund Institute of Technology, Division of Automatic Control.

Söderström, T. and P. Stoica (1983). *Instrumental variable methods for system identification*. New York: Springer Verlag.

Söderström, T. and P. Stoica (1989). *System Identification*. Series in Systems and Control Engineering. New York: Prentice Hall.

Solo, V. (1978). *Time Series Recursions and Stochastic Approximation*. Ph. D. thesis, Australian National University, Canberra, Australia.

Solo, V. (1980). Some aspects of recursive parameter estimation. *Int. Jnl. Control 32*, 395–410.

Solo, V. (1986). Identifiabilityof time series models with errors in variables. In J. Gani and M. Priestley (Eds.), *Essays in Time Series and Allied Processes. Papers in honour of E.J. Hannan*, pp. 63–71. Applied Probability Trust.

Staley, R. M. and P. C. Yue (1970). On system parameter identifiability. *Information Sciences 2*, 127–138.

Steiglitz, K. and L. E. McBride (1965, October). A technique for the identification of linear systems. *IEEE Transactions on Automatic Control 10*, 461–464.

Stoica, P. and T. Söderström (1981). The Steiglitz-McBride identification algorithms revisited. Convergence analysis and accuracy aspects. *IEEE Transactions on Automatic Control AC-26*, 712–717.

Stoica, P., T. Söderström, and B. Friedlander (1985). Optimal instrumental variable estimates of the AR parameters of an ARMA process. *IEEE Trans on Automatic Control AC-30*, 1066–1074.

Stoten, D. P. and A. J. L. Harrison (1994). Generation of discrete and continuous time transfer function coefficients. ,*International Journal of Control 59*, 1159 — 1172.

Strang, G. and T. Nguyen (1996). *Wavelets and Filter Bank*. Wellesley-Cambridge Press: Wellesley, MA, USA.

Talmon, J. L. and A. J. W. van den Boom (1973). On the estimation of the transfer function parameters of process and noise dynamics using a single stage estimator. In P. Eykhoff (Ed.), *Identification and System Parameter Estimation*, Amsterdam/New York. North Holland/American Elsevier.

Taylor, C. J., A. Chotai, and P. C. Young (2000). State space control system design based on non-minimal state variable feedback: further generalization and unification results. *International Journal of Control 73*, 1329–1345.

Taylor, C. J., A. Chotai, and P. C. Young (2008). Non-linear control by input–output state variable feedback pole assignment. *International Journal of Control 10.1080/00207170802400970*, 1–16.

Taylor, C. J., P. C. Young, and A. Chotai (1994). On the relationship between GPC and PIP control. In D. W. Clarke (Ed.), *Advances in Model-Based Predictive Control*, Oxford, pp. 53–68. Oxford University Press.

Taylor, G. I. (1954). The dispersion of matter in turbulent flow through a pipe. *Proc. Roy. Soc. A 223*, 446–468.

Tian, X., Z. Xie, and A. Dai (2008). A land surface soil moisture data assimilation system based on the dual-UKF method and the community land model. *J. Geophys. Res. 113*, D14127, doi:10.1029/2007JD009650.

Trotter, H. F. (1957). Gauss's work 1803-1826 on theory of least squares; an english translation. Technical report, Statist. Techniques Research Group, Dept. of Maths., Univ. of Princeton, N. J.

Truong, N.-V., L. Wang, and P. C. Young (2006). Nonlinear system modeling based on non-parametric identification and linear wavelet estimation of SDP models. In *45th IEEE Conference on Decision and Control*, San Diego, CA, USA.

Truxal, T. G. (1955). *Control System Synthesis*. New York: McGraw Hill.

Tsay, R. S. (1988). Outliers, level shifts, and variance changes in time series. *Journal of Forecasting 7*, 1–20.

Tsypkin, Y. Z. (1971). *Adaption an Learning in Automatic Systems*. New York: Academic Press.

Tsypkin, Y. Z. (1973). *Foundations of the Theory of Learning Systems*. New York and London: Academic Press.

Tych, W., D. J. Pedregal, P. C. Young, and J. Davies (2002). An unobserved component model for multi-rate forecasting of telephone call demand: the design of a forecasting support system. *International Journal of Forecasting 18*, 673–695.

Tych, W. and P. C. Young (2011). A Matlab software framework for dynamic model emulation. *Environmental Modelling & Software (submitted: available from authors)*.

Unbehauen, H. and G. P. Rao (1987). *Identification of Continuous Systems*. Systems and Control Series. Amsterdam: North-Holland.

Unbehauen, H. and G. P. Rao (1990). Continuous-time approaches to system identification - a survey. *Automatica 26*(1), 23–35.

Unbehauen, H. and G. P. Rao (1998). A review of identification in continuous-time systems. *Annual Reviews in Control 22*, 145–171.

Vajk, I. (2005). Identification methods in a unified framework. *Automatica 41*, 1385–1393.

Valentine, E. M. and I. R. Wood (1977). Longitudinal dispersion with dead zones. *Journal of the Hydraulics Division, ASCE 103*(9), 975–990.

Valstar, J. E. (1963). In flight dynamic checkout. *IEEE Transactions on Aerospace AS1*(2), 213–221.

Van den Hof, P. M. J. (1998). Closed-loop issues in system identification. *Annual Reviews in Control 22*, 173–186.

Van Huffel, S., P. Lemmerling, and S. V. Huffel (2002). *Total Least Squares And Errors-In-Variables Modeling*. Netherlands: Springer.

Verhaegen, M. (1993). Application of a subspace model identification technique to identify LTI systems operating in closed-loop. *Automatica 29*, 1027–1040.

Wagener, T., L. A. Camacho, and H. S. Wheater (2002). Dynamic identifiability analysis of the transient storage model for solute transport in rivers. *Journal of Hydroinformatics 4*, 199–211.

Wahba, G. (1990). *Spline Models for Observational Data*. CBMS-NSF Regional Conference Series in Applied Mathematics.

Wallis, S. G., P. C. Young, and K. J. Beven (1989). Experimental investigation of the aggregated dead zone model for longitudinal solute transport in stream channels. *Proc. Inst. of Civil Engrs, Part 2 87*, 1–22.

Wang, L. and P. C. Young (2008). Model order identification for continuous-time models. In H. Garnier and L. Wang (Eds.), *Identification of Continuous-Time models from Sampled data*, London, pp. 162–187. Springer-Verlag.

Weerts, A. H. and G. Y. H. E. Serafy (2006). Particle filtering and ensemble Kalman filtering for state updating with hydrological conceptual rainfall-runoff models. *Water Resources Research 42*(W09403, doi:10.1029/2005WR004093).

Wei, X., L. Del Re, and J. Tan (2005). Robust adaptive control of quasi-LPV systems. In *Advanced Intelligent Mechatronics. Proceedings, 2005 IEEE/ASME International Conference on*, pp. 1617 –1622.

Wellstead, P. E. (1978). An instrumental product moment test for model order estimation. *Automatica 14*, 89–91.

Wellstead, P. E. and M. B. Zarrop (1991). *Self-Tuning Systems: Control and Signal Processing*. New York: John Wiley.

West, M. and J. Harrison (1989). *Bayesian Forecasting and Dynamic Models*. Springer-Verlag: New York.

Weston, P. F. and J. P. Norton (1997). Detection and estimation of abrupt changes in input or state. *Int. Jnl. Control 67*, 699–711.

Whittle, P. (1963 (2nd Ed. Blackwell: Oxford, 1983)). *Prediction and Regulation*. English Universities Press.

Wiener, N. (1941). Extrapolation, interpolation and smoothing of stationary time series. Technical report, Services Reasearch Project DIC-6037.

Wiener, N. (1949). *Extrapolation, Interpolation and Smoothing of Stationary Time Series with Engineering Applications.* New York: MIT Press and J. Wiley and Sons.

Wilde, D. J. (1964). *Optimum Seeking Methods.* N.J.: Prentice-Hall.

Willcox, K. and J. Peraire (2002). Balanced model reduction via the proper orthogonal decomposition. *American Institute of Aeronautics and Astronautics (AIAA) Journal 40*(11), 2323–2330.

Wong, K. Y. and E. Polak (1967). Identification of linear discrete-time systems using the instrumental variable approach. *IEEE Transactions on Automatic Control AC-12*, 707–718.

Yaglom, A. M. (1955). The correlation theory of processes whose nth. difference constitute a stationary process. *Matem. Sb 37*, 141.

Young, P. C. (1964). In flight dynamic checkout - a discussion. *IEEE Transactions on Aerospace AS2*(3), 1106–1111.

Young, P. C. (1965a). The determination of the parameters of a dynamic process. *Radio and Electronic Eng. (Jnl. of IERE) 29*, 345–361.

Young, P. C. (1965b). On a weighted steepest descent method of process parameter estimation. Report, Department of Engineering, Cambridge University (available from author).

Young, P. C. (1966). Process parameter estimation and self adaptive control. In P. H. Hammond (Ed.), *Theory of Self Adaptive Control Systems*, pp. 118–140. Plenum Press: New York.

Young, P. C. (1968a). Process parameter estimation. *Control and Automation Progress 12*, 931–37.

Young, P. C. (1968b). Regression analysis and process parameter estimation: a cautionary message. *Simulation 10*, 125–128.

Young, P. C. (1968c). The use of linear regression and related procedures for the identification of dynamic processes. In *Proc. 7th IEEE Symp. on Adaptive Processes, San Antonia, Texas*, New York, pp. 501–505. IEEE.

Young, P. C. (1969a). Applying parameter estimation to dynamic systems: Part I - theory. *Control Engineering 16*(10), 119–125.

Young, P. C. (1969b). Applying parameter estimation to dynamic systems: Part II - applications. *Control Engineering 16*(11), 118–124.

Young, P. C. (1970a). *Differential Equation Error Method of Real-Time Process Identification.* Cambridge, U.K.: Ph.D Thesis, University of Cambridge.

Young, P. C. (1970b). An instrumental variable method for real-time identification of a noisy process. *Automatica 6*, 271–287.

Young, P. C. (1972). Comments on "On-line identification of linear dynamic systems with applications to Kalman Filtering". *IEEE Trans. Auto. Control AC-17*, 269–270.

Young, P. C. (1974). Recursive approaches to time-series analysis. *Bulletin Inst. Maths and its Applications 10*, 209–224.

Young, P. C. (1976a). Optimization in the presence of noise - a guided tour. In L. C. W. Dixon (Ed.), *Optimization in Action*, London, pp. 517–573. Academic Press.

Young, P. C. (1976b). Some observations on instrumental variable methods of time-series analysis. *International Journal of Control 23*, 593–612.

Young, P. C. (1978a). A general theory of modeling for badly defined dynamic systems. In G. C. Vansteenkiste (Ed.), *Modeling, identification and Control in Environmental Systems*, pp. 103–135. North Holland: Amsterdam.

Young, P. C. (1979). Self-adaptive Kalman filter. *Electronics Letters 15*(12), 358–360.

Young, P. C. (1981a). Parameter estimation for continuous-time models - a survey. *Automatica 17*(1), 23–39.

Young, P. C. (1981b). A second generation adaptive autostabilization system for airborne vehicles. *Automatica 17*, 459–470.

Young, P. C. (1984). *Recursive Estimation and Time-Series Analysis*. Springer-Verlag, Berlin. (earlier version of present book).

Young, P. C. (1985a). The instrumental variable method: a practical approach to identification and system parameter estimation. In H. A. Barker and P. C. Young (Eds.), *Identification and System Parameter Estimation: Vols. 1 & 2*, pp. 1–15. Pergamon Press: Oxford.

Young, P. C. (1985b). Recursive identification, estimation and control. In E. J. Hannan, P. R. Krishnaiah, and M. M. Rao (Eds.), *Handbook of Statistics: Time Series in the Time Domain*, Volume 5, pp. 213–255. North Holland: Amsterdam.

Young, P. C. (1989). Recursive estimation, forecasting and adaptive control. In C. T. Leondes (Ed.), *Control and Dynamic Systems*, pp. 119–166. Academic Press: San Diego.

Young, P. C. (1991). Comments on likelihood and cost as path integrals. *Journal Royal Statistical Society, Series B 53*, 529–531.

Young, P. C. (Ed.) (1993a). *Concise Encyclopaedia of Environmental Systems*. Pergamon Press: Oxford.

Young, P. C. (1993b). Time variable and state dependent modelling of nonstationary and nonlinear time series. In T. Subba Rao (Ed.), *Developments in Time Series Analysis*, pp. 374–413. Chapman and Hall: London.

Young, P. C. (1994). Time-variable parameter and trend estimation in non-stationary economic time series. *Journal of Forecasting 13*, 179–210.

Young, P. C. (1996). A general approach to identification, estimation and control for a class of nonlinear dynamic systems. In M. I. Friswell and J. E. Mottershead (Eds.), *Identification in Engineering Systems*. University of Wales: Swansea.

Young, P. C. (1998). Data-based mechanistic modeling of environmental, ecological, economic and engineering systems. *Environmental Modelling and Software 13*, 105–122.

Young, P. C. (1999a). Data-based mechanistic modelling, generalised sensitivity and dominant mode analysis. *Computer Physics Communications 117*, 113–129.

Young, P. C. (1999b). Nonstationary time series analysis and forecasting. *Progress in Environmental Science 1*, 3–48.

Young, P. C. (2000). Stochastic, dynamic modelling and signal processing: time variable and state dependent parameter estimation. In W. J. Fitzgerald, A. Walden,

R. Smith, and P. C. Young (Eds.), *Nonlinear and Nonstationary Signal Processing*, pp. 74–114. Cambridge University Press: Cambridge.

Young, P. C. (2001a). Comment on 'a quasi-ARMAX approach to modelling nonlinear systems' by J. Hu *et al. International Journal of Control 74*, 1767–1771.

Young, P. C. (2001b). Data-based mechanistic modelling and validation of rainfall-flow processes. In M. G. Anderson and P. D. Bates (Eds.), *Model Validation: Perspectives in Hydrological Science*, pp. 117–161. John Wiley: Chichester.

Young, P. C. (2001c). Data-based mechanistic modelling of environmental systems. In *Proceedings, IFAC Workshop on Environmental Systems, first plenary session keynote paper*, Yokohama (Japan).

Young, P. C. (2001d). The identification and estimation of nonlinear stochastic systems. In A. I. Mees (Ed.), *Nonlinear Dynamics and Statistics*, pp. 127–166. Birkhauser: Boston.

Young, P. C. (2002a). Advances in real-time flood forecasting. *Philosophical Trans. Royal Society, Physical and Engineering Sciences 360*(9), 1433–1450.

Young, P. C. (2002b). Identification of time varying systems. In H. Unbehauen (Ed.), *Encyclopedia of Life Support Systems (EOLSS)*, Volume 6.43: Control Systems, Robotics and Automation. EOLSS Publishers: Oxford.

Young, P. C. (2003a). Top-down and data-based mechanistic modelling of rainfall-flow dynamics at the catchment scale. *Hydrological Processes 17*, 2195–2217.

Young, P. C. (2003b). Transfer function identification and estimation in non-standard situations. Centre for Research on Environmental Systems and Statistics, Report TR/205, Lancaster University.

Young, P. C. (2005). Comments on 'identification on non-linear parametrically varying models using separable least squares' by F. Previdi and M. Lovera: black-box or open box? *International Journal of Control 78*, 122–127.

Young, P. C. (2006a). The data-based mechanistic approach to the modelling, forecasting and control of environmental systems. *Annual Reviews in Control 30*, 169–182.

Young, P. C. (2006b). An instrumental variable approach to ARMA model identification and estimation. In *Proceedings 14th IFAC Symposium on System Identification SYSID06, Newcastle, NSW, Australia*, pp. 410–415.

Young, P. C. (2006c). New approaches to volcanic time series analysis. In H. M. Mader, S. G. Coles, C. B. Connor, and L. J. Connor (Eds.), *Statistics in Volcanology*, Bath, UK, pp. 143–160. Geological Society of London.

Young, P. C. (2008a). Real-time flow forecasting. In H. Wheater, S. Sarooshian, and K. D. Sharma (Eds.), *Hydrological modelling in arid and semi-arid areas*, Cambridge, UK, pp. 113–138. Cambridge University Press.

Young, P. C. (2008b). The refined instrumental variable method: unified estimation of discrete and continuous-time transfer function models. *Journal Européen des Systèmes Automatisés 42*, 149–179.

Young, P. C. (2010a). The estimation of continuous-time rainfall-flow models for flood risk management. In C. Walshe (Ed.), *Role of Hydrology in Managing Consequences of a Changing Global Environment*, University of Newcastle, U.K., pp. 303–310. British Hydrological Society.

Young, P. C. (2010b). Gauss, Kalman and advances in recursive parameter estimation. *Journal of Forecasting (special issue celebrating 50 years of the Kalman Filter) 30*, 104–146.

Young, P. C. (2010c). Real-time updating in flood forecasting and warning. In G. J. Pender and H. Faulkner (Eds.), *Flood Risk Science and Management*, Oxford, UK, pp. 163–195. Wiley-Blackwell.

Young, P. C. (2010d). Some observations on prediction error minimization. In K. J. Burnham and V. E. Ersanilli (Eds.), *Proceedings UKACC International Conference on CONTROL 2010*, Coventry University, UK, pp. 1241–1246.

Young, P. C. (2011). Data-based mechanistic modelling: natural philosophy revisited? In L. Wang and H. Garnier (Eds.), *System Identification, Environmetric Modelling and Control*, Berlin. Springer-Verlag (in press).

Young, P. C. and M. B. Beck (1974). The modelling and control of water quality in a river system. *Automatica 10*, 455–468.

Young, P. C., M. A. Behzadi, C. L. Wang, and A. Chotai (1987). Direct digital and adaptive control by input-output, state variable feedback. *International Journal of Control 46*, 1861–1881.

Young, P. C. and S. Benner (1991). *microCAPTAIN 2 User Handbook*. Lancaster University: Centre for Research on Environmental Systems and Statistics.

Young, P. C. and K. J. Beven (1994). Data-based mechanistic (DBM) modelling and the rainfall-flow nonlinearity. *Environmetrics 5*, 335–363.

Young, P. C., A. Castelletti, and F. Pianosi (2007). The data-based mechanistic approach in hydrological modelling. In A. Castelletti and R. S. Sessa (Eds.), *Topics on System Analysis and Integrated Water Resource Management*, pp. 27–48. Elsevier: Amsterdam.

Young, P. C., A. Chotai, and K. J. Beven (2004). Data-based mechanistic modelling and the simplification of environmental systems. In J. Wainwright and M. Mullgan (Eds.), *Environmental Modelling: Finding Simplicity in Complexity*, pp. 371–388. Wiley: Chichester.

Young, P. C., A. Chotai, P. G. McKenna, and W. Tych (1998). Proportional-Integral-Plus (PIP) design for delta operator systems: Part 1, SISO systems. *International Journal of Control 70*, 123–147.

Young, P. C., A. Chotai, and W. Tych (1991). Identification, estimation and control of continuous-time systems described by delta operator models. In N. Sinha and G. Rao (Eds.), *Identification of Continuous-Time Systems*, pp. 363–418. Kluwer: Dordrecht.

Young, P. C., M. Foster, and M. Lees (1993). A direct approach to the identification and estimation of continuous-time systems from discrete-time data based on fixed interval smoothing. In *Proceedings, 12th Triennial IFAC World Congress on Automatic Control, Vol. 10*, Sydney (Australia), pp. 27–30.

Young, P. C., H. Garnier, and M. Gilson (2006). An optimal instrumental variable approach for identifying hybrid continuous-time Box-Jenkins models. In *Proceedings 14th IFAC Symposium on System Identification SYSID06, Newcastle, NSW, Australia*, pp. 225–230.

Young, P. C., H. Garnier, and M. Gilson (2008). Refined instrumental variable identification of continuous-time hybrid Box-Jenkins models. In H. Garnier and L. Wang (Eds.), *Identification of Continuous-Time Models from Sampled Data*, pp. 91–131. Springer-Verlag: London.

Young, P. C., H. Garnier, and M. Gilson (2009). A simple refined IV method of closed-loop system identification. In *Proceedings 15th IFAC Symposium on System Identification SYSID09, St. Malo, France*.

Young, P. C. and A. J. Jakeman (1979a). The development of CAPTAIN: A computer aided program for time-series analysis and identification of noisy systems. In *Proc IFAC Symposium on Computer Aided Design of Control Systems, Zurich, 1979*.

Young, P. C. and A. J. Jakeman (1979b). Refined instrumental variable methods of time-series analysis: Part I, SISO systems. *International Journal of Control 29*, 1–30.

Young, P. C. and A. J. Jakeman (1979, 1980b). Refined instrumental variable methods of time-series analysis: Parts I, II and III. *International Journal of Control 29, 29, 31*, 1–30; 621–644; 741–764.

Young, P. C. and A. J. Jakeman (1980a). Refined instrumental variable methods of time-series analysis: Part III, extensions. *International Journal of Control 31*, 741–764.

Young, P. C., A. J. Jakeman, and R. McMurtrie (1980). An instrumental variable method for model order identification. *Automatica 16*, 281–296.

Young, P. C., D. Leedal, and K. J. Beven (2009). Reduced order emulation of distributed hydraulic models. In *Proceedings 15th IFAC Symposium on System Identification SYSID09, St Malo, France*.

Young, P. C. and M. J. Lees (1993). The active mixing volume: a new concept in modelling environmental systems. In V. Barnett and K. Turkman (Eds.), *Statistics for the Environment*, pp. 3–43. J. Wiley: Chichester.

Young, P. C., P. McKenna, and J. Bruun (2001). Identification of nonlinear stochastic systems by state dependent parameter estimation. *International Journal of Control 74*, 1837–1857.

Young, P. C. and C. N. Ng (1989). Variance intervention. *Journal of Forecasting 8*, 399–416.

Young, P. C., C. N. Ng, and P. Armitage (1989). A systems approach to recursive economic forecasting and seasonal adjustment. *Computers Math. Applic. 18*, 481–501.

Young, P. C. and S. Parkinson (2002). Simplicity out of complexity. In M. B. Beck (Ed.), *Environmental Foresight and Models: A Manifesto*, pp. 251–294. Elsevier: Oxford.

Young, P. C., S. Parkinson, and M. J. Lees (1996). Simplicity out of complexity: Occam's razor revisited. *Journal of Applied Statistics 23*, 165–210.

Young, P. C. and D. J. Pedregal (1996). Recursive fixed interval smoothing and the evaluation of LIDAR measurements. *Environmetrics 7*, 417–427.

Young, P. C. and D. J. Pedregal (1999a). Macro-economic relativity: government spending, private capital investment and unemployment in the USA 1948-1998. *Structural Change and Economic Dynamics 10*, 359–380.

Young, P. C. and D. J. Pedregal (1999b). Recursive and en-bloc approaches to signal extraction. *Journal of Applied Statistics 26*, 103–128.

Young, P. C., D. J. Pedregal, and W. Tych (1999). Dynamic harmonic regression. *Jnl. of Forecasting 18*, 369–394.

Young, P. C., L. Price, D. Berckmans, and K. Janssens (2000). Recent developments in the modelling of imperfectly mixed airspaces. *Computers and Electronics in Agriculture 26*, 239–254.

Young, P. C. and M. Ratto (2009). A unified approach to environmental systems modeling. *Stochastic Environmental Research and Risk Assessment 23*, 1037–1057.

Young, P. C. and M. Ratto (2011). Statistical emulation of large linear dynamic models. *Technometrics 53*, 29–43.

Young, P. C., R. J. Romanowicz, and K. J. Beven (2006, December). Updating algorithms in flood forecasting. Technical Report UR5, Flood Risk Management Research Consortium, www.floodrisk.org.uk, U.K.

Young, P. C. and S. H. Shellswell (1973). CAPTAIN - a computer aided program for time series analysis and the identification of noisy systems. In *Proc. IEE Conference on Computer Aided Control System Design*, London. IEE.

Young, P. C. and S. G. Wallis (1993). Solute transport and dispersion in channels. In K. J. Beven and M. J. Kirkby (Eds.), *Channel Network Hydrology*, Chichester, pp. 129–173. J. Wiley.

Young, P. C. and P. G. Whitehead (1975). Water quality in river systems: Monte carlo analysis and planning. *Water Resources Research 15*, 451–459.

Young, P. C. and P. G. Whitehead (1977). A recursive approach to time-series analysis for multivariable systems. *Int. Jnl. Control 25*, 457–482.

Index